# INDUSTRIAL FURNACES

**CHRONOLOGY of Trinks and Mawhinney books on furnaces**

INDUSTRIAL FURNACES

Volume I First Edition, by W. Trinks, 1923
6 chapters, 319 pages, 255 figures

Volume I Second Edition, by W. Trinks, 1926

Volume I Third Edition, by W. Trinks, 1934
6 chapters, 456 pages, 359 figures, 22 tables

Volume I Fourth Edition, by W. Trinks, 1951
6 chapters, 526 pages, 414 figures, 26 tables

Volume I Fifth Edition, by W. Trinks and M. H. Mawhinney, 1961
8 chapters, 486 pages, 361 figures, 23 tables

Volume I Sixth Edition, by W. Trinks, M. H. Mawhinney,
R. A. Shannon, R. J. Reed, and J. R. V. Garvey, 2000
9 chapters, 490 pages, 199 figures,* 40 tables

Volume II First Edition, by W. Trinks, 1925

Volume II Second Edition, by W. Trinks, 1942
6 chapters, 351 pages, 337 figures, 12 tables

Volume II Third Edition, by W. Trinks, 1955
7 chapters, 358 pages, 303 figures, 4 tables

Volume II Fourth Edition, by W. Trinks and M. H. Mawhinney, 1967**
9 chapters, 358 pages, 273 figures, 13 tables

PRACTICAL INDUSTRIAL FURNACE DESIGN, by M. H. Mawhinney, 1928
9 chapters, 318 pages, 104 figures, 28 tables

*This 6th Edition also includes 3 equations, 20 examples, 54 review questions, 4 problems, and 5 suggested projects. The 199 figures consist of 43 graphs, 140 drawings and diagrams, and 16 photographs.

**No further editions of Volume II of *INDUSTRIAL FURNACES* are planned because similar, but up-to-date, material is covered in this 6th Edition of *INDUSTRIAL FURNACES* and in Volumes I and II of the North American *COMBUSTION HANDBOOK.*

# INDUSTRIAL FURNACES, SIXTH EDITION

W. Trinks

M. H. Mawhinney

R. A. Shannon

R. J. Reed

J. R. Garvey

WILEY

JOHN WILEY & SONS, INC.

This book is printed on acid-free paper. ♾



Published by John Wiley & Sons, Inc., Hoboken, New Jersey
Published simultaneously in Canada



***Library of Congress Cataloging-in-Publication Data:***

Industrial furnaces / Willibald Trinks . . . [et al.]. — 6th ed.
p. cm.
Previous ed. cataloged under: Trinks, W. (Willibald), b. 1874.
Includes bibliographical references and index.
ISBN 0-471-38706-1 (Cloth)
1. Furnaces—Design and construction. 2. Furnaces—Industrial applications. I. Trinks, W. (Willibald), b. 1874. II. Trinks, W. (Willibald), b. 1874. Industrial furnaces.
TH7140 .I48 2003
621.402'5—dc21

2003007736

10 9 8 7 6 5 4 3 2 1

This 6th Edition is dedicated to our wives:
Emily Jane Shannon and Catherine Riehl Reed
whom we thank for beloved encouragement and
for time away to work on this 6th Edition.

| ROBERT A. SHANNON | RICHARD J. REED |
|---|---|
| *Avon Lake, Ohio* | *Willoughby, Ohio* |

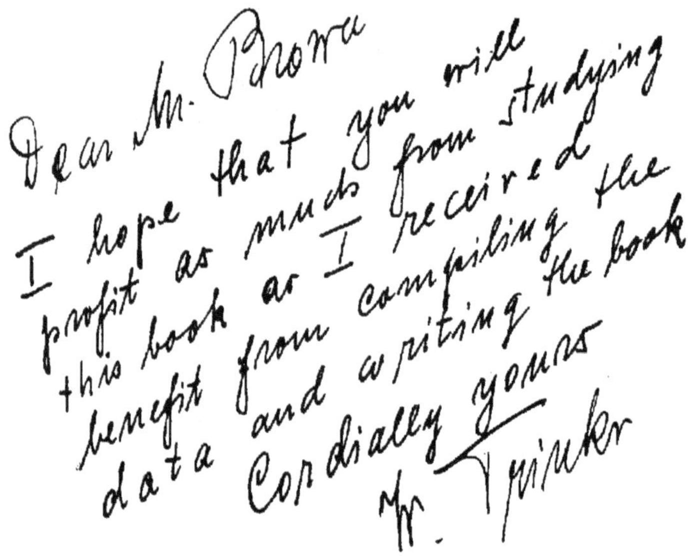

Dear Mr. Brown

I hope that you will profit as much from studying this book as I received benefit from compiling the data and writing the book

Cordially yours

W. Trinks

Photostat copy of a hand-written note from Prof. W. Trinks to Mr. Brown, founder of North American Mfg, Co. . . . about 1942.

# EXCERPTS FROM THE PREFACE TO THE 5TH EDITION

*Industrial Furnaces, Volume I,* has been on the market for 40 years. The book, which together with Volume II, is known as the "furnace-man's bible," was originally written to rationalize furnace design and to dispel the mysteries (almost superstitions) that once surrounded it. Both volumes have been translated into four foreign languages and are used on every continent of this globe.

The 5th Edition of Volume I is the result of the combined efforts of the original author, W. Trinks, and of M. H. Mawhinney, who has brought to the book a wealth of personal experience with furnaces of many different types. While retaining the fundamental features of the earlier editions, the authors made many changes and improvements.

We acknowledge with thanks the contributions of A. F. Robbins for many of the calculations and of A. S. Sobek for his assistance in the collection of operating data.

W. TRINKS
*Ohiopyle, Pennsylvania*

M. H. MAWHINNEY
*Salem, Ohio*

April 15, 1961

# PREFACE

There has not been a new text/reference book on industrial furnaces and industrial process heating in the past 30 years. Three retired engineers have given much time and effort to update a revered classic book, and to add many facets of their long experience with industrial heating processes—for the benefit of the industry's future and as a contribution to humanity.

The sizes, shapes, and properties of the variety of furnace loads in the world should encourage furnace engineers to apply their imagination and ingenuity to their own particular situations. Few industrial furnaces are duplicates. Most are custom-made, so their designs present many unique and enjoyable challenges to engineers.

***As Professors Borman and Ragland imply in Chapter 1 of their 1998 textbook, "Combustion Engineering," improving industrial furnaces requires understanding chemistry, mathematics, thermodynamics, heat transfer, and fluid dynamics. They cite, as an example, that a detailed understanding of even the simplest turbulent flame requires a knowledge of turbulence and chemical kinetics, which are at the frontiers of current science. They conclude that "the engineer cannot wait for such an understanding to evolve, but must use a combination of science, experiment, and experience to find practical solutions."***

This 6th Edition of Trinks' *Industrial Furnaces, Volume I* deals primarily with the practical aspects of furnaces as a whole. Such discussions must necessarily touch on combustion, loading practice, controls, sensors and their positioning, in-furnace flow patterns, electric heating, heat recovery, and use of oxygen. The content of Professor Trinks' Volume II is largely covered by Volumes I and II of the *North American Combustion Handbook.*

While Professor Trinks' stated objective of his book was to "rationalize furnace design," he also helped operators and managers to better **understand how best to load and operate furnaces.** Readers of this 6th Edition will realize that the current authors have greatly extended the coverage of **how to best use furnaces**, providing valuable insight in areas where experience counts as much as analytical skills.

Coauthors Shannon, Reed, and Garvey have lived through many tough years, dealing with furnace problems that may occur again and again. If others can find help with their furnace problems by reading this book, our goal will be reached.

The lifetime of most furnaces extends through a variety of sizes and types of loads, through a number of managers and operators, and through a number of reworks with newly developed burners and controls, and sometimes changed fuels; so it is essential that everyone involved with furnaces have the **know-how to adjust to changing modes of furnace operation**.

In this edition, particular emphasis has been given to a very thorough Glossary and an extensive Index. The Glossary is a schoolbook in itself. For the benefit of readers from many lands, a host of abbreviations are included. Thanks to John Wiley and Sons, Inc. for assistance in making the Index very complete so that this book can be an easily usable reference.

The authors thank Pauline Maurice, John Hes, Sandra Bilewski, and many others who helped make possible this modern continuation of a proud tradition dating from 1923 in Germany.

ROBERT A. SHANNON

RICHARD J. REED

J. R. VERNON GARVEY

# BRIEF BIOGRAPHIES OF THE AUTHORS

**Professor W. Trinks** was born Charles Leopold Willibald Trinks on December 10, 1874 in Berlin, Germany. He was educated in Germany, and graduated with honors from Charlottenburg Technical Institute in 1897. After two years as a Mechanical Engineer at Schuchstermann & Kremen, he emigrated to the United States of America, where he was an engineer at Cramps Shipyard, at Southwark Foundry and Machine Company, and then Chief Engineer at Westinghouse Machine Co.

One of the first appointments to the faculty of Carnegie Institute of Technology, Professor Trinks organized the Mechanical Engineering Department, and headed that department for 38 years, in what became Carnegie-Mellon University. During that time, he was in touch with most of his department's 1500 graduates. A witty philosopher, he kept his students thinking with admonitions such as: "A college degree seldom hurts a chap, if he is willing to learn something after graduation." "If a college student is right 85 percent of the time, he gets a B, may be on the honor roll. In industry, if a man is wrong 15 percent of the time, he gets fired."

During his long academic career, Professor Trinks was a Consulting Engineer for many companies and Associated Engineers, American Society of Mechanical Engineers, and the U.S. Government. An authority on steel mill roll pass design, governors, and industrial furnaces, he published three, two, and two books on each subject, respectively, some translated from English into German, French, Spanish, and Russian. Professor Trinks died in 1966 at the age of 92, an eminent engineer and the world authority on industrial furnaces.

**Matthew Holmes Mawhinney** was a graduate of Peabody High School near Pittsburgh. While attending Carnegie Tech (now Carnegie-Mellon University), he became a member of Sigma Nu, an invitational honorary scientific fraternity. He received B.S. and M.S. degrees in Mechanical Engineering, in 1921 and 1925, respectively, both from Carnegie Tech. Mr. Mawhinney became a Senior Design Engineer with Salem Furnace Company, Salem, Ohio (later Salem-Brosius). He authored *Practical Industrial Furnace Design* (316 pages) in 1928. He also wrote a famous technical paper on heating steel that he presented before the American Society of Mechanical Engineers and the Association of Iron and Steel Engineers.

Mr. Mawhinney formed and led his own consulting engineering company. He collaborated with Professor Trinks on his *Industrial Furnaces, Volume I, 5th Edition,* published in 1961, and on *Volume II,* and *4th Edition* published in 1967.

**Robert A. Shannon** has more than 50 years experience with engineering work. He has been North American Mfg. Co.'s authority on steel reheat furnaces, soaking pits, and forging furnaces. He continues private consulting relative to his extensive experience with steel reheat, pelletizing, forging, heat treating, catenary furnaces, and industrial boilers.

Mr. Shannon was previously a world-wide consultant for USSteel Engineers and Consultants. Before that, he was Superintendent of Utilities at USSteel's Lorain Works (now USS-Kobe).

Mr. Shannon has a B.S. degree in Chemical Engineering from Carnegie Institute of Technology (now Carnegie-Mellon University) in Pittsburgh and is a registered Professional Engineer. He has several patents relating to industrial heating processes. Mr. Shannon served in the U.S. Merchant Marines during World War II.

**Richard J. Reed** is a Consulting Engineer, recently retired after 47 years at North American Mfg. Co. as the Technical Information Director. Prior to that, he served on the Engineering faculties of Case-Western Reserve University and Cleveland State University teaching Fuels, Combustion, Heat Transfer, Thermodynamics, and Fluid Dynamics. He is a registered Professional Engineer in Ohio and was an officer in the U.S. Navy. He has an M.S. degree from Case-Western Reserve University and a B.S. degree in Mechanical Engineering from Purdue University.

Mr. Reed was the second of six persons "Leaders in Thermal Technology" listed by *Industrial Heating Journal* in February 1991. He is the author of both volumes of the North American *Combustion Handbook,* technical papers on heat transfer and combustion in industrial heating, four chapters for the *Mechanical Engineers' Handbook* (by John Wiley & Sons), and a chapter for McGraw-Hill's *Handbook of Applied Thermal Design.* At the Center for Professional Advancement, Mr. Reed was director of courses in "Applied Combustion Technology" and "Moving Air and Flue Gas" (United States and Europe). At the University of Wisconsin, Mr. Reed has been involved with three courses, and led "Optimizing Industrial Heating Processes."

**J. R. Vern Garvey** is a Consultant, retired from Director of Steelmaking Projects at H. K. Ferguson Company. His responsibilities included supervision, coordination, and technical quality of steel plant design and construction projects. Mr. Garvey's technical experience involved upgrading many facilities—basic oxygen processes, electric furnaces, continuous casting, waste disposal, reheat furnaces, bar mill, rolling practice, cooling beds, gauging, and material handling. He planned a Cascade Steel plant reported by the International Trade Commission to be the finest mini-mill in operation at that time.

Mr. Garvey served in the Air Force Corps of Engineers and is a registered Professional Engineer. He has degrees in Mechanical Engineering, Electrical Engineering, and Business Administration from the University of Wisconsin.

# NO-LIABILITY STATEMENT

This is a textbook and reference book of engineering practice and suggestions—all subject to local, state, and federal codes, to insurance requirements, and to *good common sense.*

No patent liability may be assumed with respect to the use of information herein. While every precaution has been taken in preparing this book, neither the publisher nor the authors assume responsibility for errors, omissions, or misjudgments. No liability can be assumed for damages incurred from use of this information.

**WARNING:** Situations dangerous to personnel and property can develop from incorrect operation of furnaces and combustion equipment. The publisher and the authors urge compliance with all safety standards and insurance underwriters' recommendations. With all industrial equipment, think twice, and consider every operation and situation.

# CONTENTS

# 1

# INDUSTRIAL HEATING PROCESSES

## 1.1. INDUSTRIAL PROCESS HEATING FURNACES

Industrial process heating furnaces are insulated enclosures designed to deliver heat to loads for many forms of heat processing. *Melting* ferrous metals and glasses requires *very high temperatures,*[*] and may involve erosive and corrosive conditions. *Shaping* operations use *high temperatures*[*] to soften many materials for processes such as forging, swedging, rolling, pressing, bending, and extruding. *Treating* may use *midrange temperatures*[*] to physically change crystalline structures or chemically (metallurgically) alter surface compounds, including hardening or relieving strains in metals, or modifying their ductility. These include aging, annealing, austenitizing, carburizing, hardening, malleablizing, martinizing, nitriding, sintering, spheroidizing, stress-relieving, and tempering. Industrial processes that use *low temperatures*[*] include drying, polymerizing, and other chemical changes.

Although Professor Trinks' early editions related mostly to metal heating, particularly steel heating, his later editions (and especially this sixth edition) broaden the scope to heating other materials. Though the text may not specifically mention other materials, readers will find much of the content of this edition applicable to a variety of industrial processes.

Industrial furnaces that do not "show color," that is, in which the temperature is below 1200 F (650 C), are commonly called "ovens" in North America. However, the dividing line between ovens and furnaces is not sharp, for example, coke ovens operate at temperatures above 2200 F (1478 C). In Europe, many "furnaces" are termed "ovens." In the ceramic industry, furnaces are called "kilns." In the petrochem and CPI (chemical process industries), furnaces may be termed "heaters," "kilns," "afterburners," "incinerators," or "destructors." The "furnace" of a boiler is its 'firebox' or 'combustion chamber,' or a fire-tube boiler's 'Morrison tube.'

[*]In this book, "very high temperatures" usually mean >2300 F (>1260 C), "high temperatures" = 1900–2300 F (1038–1260 C), "midrange temperatures" = 1100–1900 F (593–1038 C), and "low temperatures" = < 1100 F (<593 C).

**TABLE 1.1 Temperature ranges of industrial heating processes**

| Material | Operation | Temperature, F/K |
|---|---|---|
| Aluminum | Melting | 1200–1400/920–1030 |
| Aluminum alloy | Aging | 250–460/395–510 |
| Aluminum alloy | Annealing | 450–775/505–685 |
| Aluminum alloy | Forging | 650–970/616–794 |
| Aluminum alloy | Heating for rolling | 850/728 |
| Aluminum alloy | Homogenizing | 850–1175/720–900 |
| Aluminum alloy | Solution h.t. | 820–1080/708–800 |
| Aluminum alloy | Stress relieving | 650–1200/615–920 |
| Antimony | Melting point | 1166/903 |
| Asphalt | Melting | 350–450/450–505 |
| Babbitt | Melting[1] | 600–800/590–700 |
| Brass | Annealing | 600–1000/590–811 |
| Brass | Extruding | 1400–1450/1030–1060 |
| Brass | Forging | 1050–1400/840–1030 |
| Brass | Rolling | 1450/1011 |
| Brass | Sintering | 1550–1600/1116–1144 |
| Brass, red | Melting[1] | 1830/1270 |
| Brass, yellow | Melting | 1705/1200 |
| Bread | Baking | 300–500/420–530 |
| Brick | Burning | 1800–2600/1255–1700 |
| Brick, refractory | Burning | 2400–3000/1589–1920 |
| Bronze | Sintering | 1400–1600/1033–1144 |
| Bronze, 5% aluminum | Melting[1] | 1940/1330 |
| Bronze, manganese | Melting | 1645/1170 |
| Bronze, phosphor | Melting | 1920/1320 |
| Bronze, Tobin | Melting | 1625/1160 |
| Cadmium | Melting point | 610/595 |
| Cake (food) | Baking | 300–350/420–450 |
| Calcium | Melting point | 1562/1123 |
| Calender rolls | Heating | 300/420 |
| Candy | Cooking | 225–300/380–420 |
| Cement | Calcining kiln firing | 2600–3000/1700–1922 |
| China, porcelain | Bisque firing | 2250/1505 |
| China, porcelain | Decorating | 1400/1033 |
| China, porcelain | Glazing, glost firing | 1500–2050/1088–1394 |
| Clay, refractory | Burning | 2200–2600/1480–1700 |
| Cobalt | Melting point | 2714/1763 |
| Coffee | Roasting | 600–800/590–700 |
| Cookies | Baking | 375–450/460–505 |
| Copper | Annealing | 800–1200/700–920 |
| Copper | Forging | 1800/1255 |
| Copper | Melting[1] | 2100–2300/1420–1530 |
| Copper | Refining | 2100–2600/1420–1700 |
| Copper | Rolling | 1600/1144 |
| Copper | Sintering | 1550–1650/1116–1172 |
| Copper | Smelting | 2100–2600/1420–1700 |

**TABLE 1.1 (*Continued*)**

| Material | Operation | Temperature, F/K |
|---|---|---|
| Cores, sand | Baking | 250–550/395–560 |
| Cupronickel, 15% | Melting | 2150/1450 |
| Cupronickel, 30% | Melting | 2240/1500 |
| Electrotype | Melting | 740/665 |
| Enamel, organic | Baking | 250–450/395–505 |
| Enamel, vitreous | Enameling | 1200–1800/922–1255 |
| Everdur 1010 | Melting | 1865/1290 |
| Ferrites | | 2200–2700/1478–1755 |
| Frit | Smelting | 2000–2400/1365–1590 |
| German silver | Annealing | 1200/922 |
| Glass | Annealing | 800–1200/700–920 |
| Glass | Melting, pot furnace | 2300–2500/1530–1645 |
| Glass, bottle | Melting, tank furnace | 2500–2900/1645–1865 |
| Glass, flat | Melting, tank furnace | 2500–3000/1645–1920 |
| Gold | Melting | 1950–2150/1340–1450 |
| Iron | Melting, blast furnace tap | 2500–2800/1645–1810 |
| Iron | Melting, cupola[1] | 2600–2800/1700–1810 |
| Iron, cast[2] | Annealing | 1300–1750/978–1228 |
| Iron, cast | Austenitizing | 1450–1700/1060–1200 |
| Iron, cast | Malleablizing | 1650–1800/1170–1255 |
| Iron, cast | Melting, cupola[2] | 2600–2800/1700–1800 |
| Iron, cast | Normalizing | 1600–1725/1145–1210 |
| Iron, cast | Stress relieving | 800–1250/700–945 |
| Iron, cast | Tempering | 300–1300/420–975 |
| Iron, cast | Vitreous enameling | 1200–1300/920–975 |
| Iron, malleable | Melting[1] | 2400–3100/1590–1980 |
| Iron, malleable | Annealing, long cycle | 1500–1700/1090–1200 |
| Iron, malleable | Annealing, short cycle | 1800/1255 |
| Iron | Sintering | 1283–1422/1850–2100 |
| Japan | Baking | 180–450/355–505 |
| Lacquer | Drying | 150–300/340–422 |
| Lead | Melting[1] | 620–750/600–670 |
| Lead | Blast furnace | 1650–2200/1170–1480 |
| Lead | Refining | 1800–2000/1255–1365 |
| Lead | Smelting | 2200/1477 |
| Lime | Burning, roasting | 2100/1477 |
| Limestone | Calcining | 2500/1644 |
| Magnesium | Aging | 350–400/450–480 |
| Magnesium | Annealing | 550–850/156–728 |
| Magnesium | Homogenizing | 700–800/644–700 |
| Magnesium | Solution h.t | 665–1050/625–839 |
| Magnesium | Stress relieving | 300–1200/422–922 |
| Magnesium | Superheating | 1450–1650/1060–1170 |
| Meat | Smoking | 100–150/310–340 |
| Mercury | Melting point | 38/234 |
| Molybdenum | Melting point | 2898/47 |

(*continued*)

**TABLE 1.1 (*Continued*)**

| Material | Operation | Temperature, F/K |
|---|---|---|
| Monel metal | Annealing | 865–1075/1100–1480 |
| Monel metal | Melting[1] | 2800/1810 |
| Moulds, foundry | Drying | 400–750/475–670 |
| Muntz metal | Melting | 1660/1175 |
| Nickel | Annealing | 1100–1480/865–1075 |
| Nickel | Melting[1] | 2650/1725 |
| Nickel | Sintering | 1850–2100/1283–1422 |
| Palladium | Melting point | 2829/1827 |
| Petroleum | Cracking | 750/670 |
| Phosphorus, yellow | Melting point | 111/317 |
| Pie | Baking | 500/530 |
| Pigment | Calcining | 1600/1150 |
| Platinum | Melting | 3224/2046 |
| Porcelain | Burning | 2600/1700 |
| Potassium | Melting point | 145/336 |
| Potato chips | Frying | 350–400/450–480 |
| Primer | Baking | 300–400/420–480 |
| Sand, cove | Baking | 450/505 |
| Silicon | Melting point | 2606/1703 |
| Silver | Melting | 1750–1900/1225–1310 |
| Sodium | Melting point | 208/371 |
| Solder | Melting[1] | 400–600/480–590 |
| Steel | Annealing | 1250–1650/950–1172 |
| Steel | Austenitizing | 1400–1700/1033–1200 |
| Steel | Bessemer converter | 2800–3000/1810–1920 |
| Steel | Calorizing (baking in aluminum powder) | 1700/1200 |
| Steel | Carbonitriding | 1300–1650/778–1172 |
| Steel | Carburizing | 1500/1750 |
| Steel | Case hardening | 1600–1700/1140–1200 |
| Steel | Cyaniding | 1400–1800/1030–1250 |
| Steel | Drawing forgings | 850/725 |
| Steel | Drop-forging | 2200–2400/1475–1590 |
| Steel | Forging | 1700–2150/1200–1450 |
| Steel | Form-bending | 1600–1800/1140–1250 |
| Steel | Galvanizing | 800–900/700–760 |
| Steel | Heat treating | 700–1800/650–1250 |
| Steel | Lead hardening | 1400–1800/1030–1250 |
| Steel | Melting, open hearth[1] | 2800–3100/1810–1975 |
| Steel | Melting, electric furnace[1] | 2400–3200/1590–2030 |
| Steel | Nitriding | 950–1051/783–838 |
| Steel | Normalizing | 1650–1900/1170–1310 |
| Steel | Open hearth | 2800–2900/1810–1866 |
| Steel | Pressing, die | 2200–2370/1478–1572 |
| Steel | Rolling | 2200–2300/1478–1533 |
| Steel | Sintering | 2000–2350/1366–1561 |

**TABLE 1.1 (*Continued*)**

| Material | Operation | Temperature, F/K |
|---|---|---|
| Steel | Soaking pit, heating for rolling | 1900–2100/1310–1420 |
| Steel | Spheroidizing | 1250–1330/950–994 |
| Steel | Stress relieving | 450–1200/505–922 |
| Steel | Tempering (drawing) | 300–1400/422–1033 |
| Steel | Upsetting | 2000–2300/1365–1530 |
| Steel | Welding | 2400–2800/1590–1810 |
| Steel bars | Heating | 1900–2200/1310–1480 |
| Steel billets | Rolling | 1750–2275/1228–1519 |
| Steel blooms | Rolling | 1750–2275/1228–1519 |
| Steel bolts | Heading | 2200–2300/1480–1530 |
| Steel castings | Annealing | 1300–1650/978–1172 |
| Steel flanges | Heating | 1800–2100/1250–1420 |
| Steel ingots | Heating | 2000–2200/1365–1480 |
| Steel nails | Blueing | 650/615 |
| Steel pipes | Butt welding | 2400–2600/1590–1700 |
| Steel pipes | Normalizing | 1650/1172 |
| Steel rails | Hot bloom reheating | 1900–2050/1310–1400 |
| Steel rivets | Heating | 1750–2275/1228–1519 |
| Steel rods | Mill heating | 1900–2100/1310–1420 |
| Steel shapes | Heating | 1900–2200/1310–1480 |
| Steel, sheet | Blue annealing | 1400–1600/1030–1140 |
| Steel, sheet | Box annealing | 1500–1700/1090–1200 |
| Steel, sheet | Bright annealing | 1250–1350/950–1000 |
| Steel, sheet | Job mill heating | 2000–2100/1365–1420 |
| Steel, sheet | Mill heating | 1800–2100/1250–1420 |
| Steel, sheet | Normalizing | 1750/1228 |
| Steel, sheet | Open annealing | 1500–1700/1090–1200 |
| Steel, sheet | Pack heating | 1750/1228 |
| Steel, sheet | Pressing | 1920/1322 |
| Steel, sheet | Tin plating | 650/615 |
| Steel, sheet | Vitreous enameling | 1400–1650/1030–1170 |
| Steel skelp | Welding | 2550–2700/1673–1755 |
| Steel slabs | Rolling | 1750–2275/1228–1519 |
| Steel spikes | Heating | 2000–2200/1365–1480 |
| Steel springs | Annealing | 1500–1650/1090–1170 |
| Steel strip, cold rolled | Annealing | 1250–1400/950–1033 |
| Steel, tinplate sheet | Box annealing | 1200–1650/920–1170 |
| Steel, tinplate sheet | Hot mill heating | 1800–2000/1250–1365 |
| Steel, tinplate sheet | Lithographing | 300/420 |
| Steel tubing (see Steel skelp) | | |
| Steel wire | Annealing | 1200–1400/920–1030 |
| Steel wire | Baking | 300–350/420–450 |
| Steel wire | Drying | 300/422 |
| Steel wire | Patenting | 1600/1144 |
| Steel wire | Pot annealing | 1650/1170 |

(*continued*)

**TABLE 1.1 (*Continued*)**

| Material | Operation | Temperature, F/K |
|---|---|---|
| Steel, alloy, tool | Hardening | 1425–2150/1050–1450 |
| Steel, alloy, tool | Preheating | 1200–1500/920–1900 |
| Steel, alloy, tool | Tempering | 325–1250/435–950 |
| Steel, carbon | Hardening | 1360–1550/1010–1120 |
| Steel, carbon | Tempering | 300–1100/420–870 |
| Steel, carbon, tool | Hardening | 1450–1500/1060–1090 |
| Steel, carbon, tool | Tempering | 300–550/420–560 |
| Steel, chromium | Melting | 2900–3050/1867–1950 |
| Steel, high-carbon | Annealing | 1400–1500/1030–1090 |
| Steel, high-speed | Hardening | 2200–2375/1478–1575 |
| Steel, high-speed | Preheating | 1450–1600/1060–1150 |
| Steel, high-speed | Tempering | 630–1150/605–894 |
| Steel, manganese, castings | Annealing | 1900/1311 |
| Steel, medium carbon | Heat treating | 1550/1117 |
| Steel, spring | Rolling | 2000/1367 |
| Steel, S.A.E. | Annealing | 1400–1650/1030–1170 |
| Steel, stainless | Annealing[3] | 1750–2050 (3)/1228–1505 |
| Steel, stainless | Annealing[4] | 1200–1525 (4)/922–1103 |
| Steel, stainless | Annealing[5] | 1525–1650 (5)/1103–1172 |
| Steel, stainless | Austenitizing[5] | 1700–1950(5)/12001339 |
| Steel, stainless | Bar and pack heating | 1900/1311 |
| Steel, stainless | Forging | 1650–2300/1172–1533 |
| Steel, stainless | Nitriding | 975–1025/797–825 |
| Steel, stainless | Normalizing | 1700–2000/1200–1367 |
| Steel, stainless | Rolling | 1750–2300/1228–1533 |
| Steel, stainless | Sintering | 2000–2350/1366–1561 |
| Steel, stainless | Stress relieving[6] | 400–1700/478–1200 |
| Steel, stainless | Tempering (drawing) | 300–1200/422–922 |
| Steel, tool | Rolling | 1900/1311 |
| Tin | Melting | 500–650/530–615 |
| Titanium | Forging | 1400–2160/1033–1450 |
| Tungston, Ni-Cu, 90-6-4 | Sintering | 2450–2900/1616–1866 |
| Tungston carbide | Sintering | 2600–2700/1700–1755 |
| Type metal | Stereotyping | 525–650/530–615 |
| Type metal | Linotyping | 550–650/545–615 |
| Type metal | Electrotyping | 650–750/615–670 |
| Varnish | Cooking | 520–600/545–590 |
| Zinc | Melting[1] | 800–900/700–760 |
| Zinc alloy | Die-casting | 850/730 |

[1]Refer to appendix for typical pouring temperatures.
[2]Includes gray and ductile iron.
[3]Austenitic stainless steels only (AISI 200 and 300 series).
[4]Ferritic stainless steels only (AISI 400 series).
[5]Martensitic stainless steels only (AISI 400 series).
[6]Austenitic and martensitic stainless steels only.
All RJR 5-26-03 are by permission from *reference 52.*

Industrial heating operations encompass a wide range of temperatures, which depend partly on the material being heated and partly on the purpose of the heating process and subsequent operations. Table 1.1 lists ranges of temperatures for a large number of materials and operations. Variations may be due to differences in the material being heated (such as carbon contents of steels) and differences in practice or in measuring temperatures.

Rolling temperatures of high quality steel bars have fallen from about 2200 F (1200 C) to about 1850 F (1283 C) in the process of improving fine-grain structure. The limiting of decarburization by rolling as cold as possible also has reduced rolling temperatures.

In any heating process, the maximum furnace temperature always exceeds the temperature to which the load or charge (see glossary) is to be heated.

## 1.2. CLASSIFICATIONS OF FURNACES

### 1.2.1. Furnace Classification by Heat Source

Heat is generated in furnaces to raise their temperature to a level somewhat above the temperature required for the process, either by (1) combustion of fuel or by (2) conversion of electric energy to heat.

Fuel-fired (combustion type) furnaces are most widely used, but electrically heated furnaces are used where they offer advantages that cannot always be measured in terms of fuel cost. In fuel-fired furnaces, the nature of the fuel may make a difference in the furnace design, but that is not much of a problem with modern industrial furnaces and combustion equipment. Additional bases for classification may relate to the place where combustion begins and the means for directing the products of combustion.

### 1.2.2. Furnace Classification by Batch (Chap. 3) or Continuous (Chap. 4), and by Method of Handling Material into, Through, and out of the Furnace

*Batch-type furnaces and kilns*, termed “in-and-out furnaces” or “periodic kilns” (figs. 1.1 and 1.2), have one temperature setpoint, but via three zones of control—to maintain uniform temperature throughout, because of a need for more heat at a door or the ends. They may be loaded manually or by a manipulator or a robot.

Loads are placed in the furnace; the furnace and it loads are brought up to temperature together, and depending on the process, the furnace may or may not be cooled before it is opened and the load removed—generally through a single charging and discharging door. Batch furnace configurations include box, slot, car-hearth, shuttle (sec. 4.3), bell, elevator, and bath (including immersion). For long solid loads, crosswise piers and top-left/bottom-right burner locations circulate for better uniformity.

Bell and elevator kilns are often cylindrical. Furnaces for pot, kettle, and dip-tank containers may be fired tangentially with type H flames instead of type E shown.

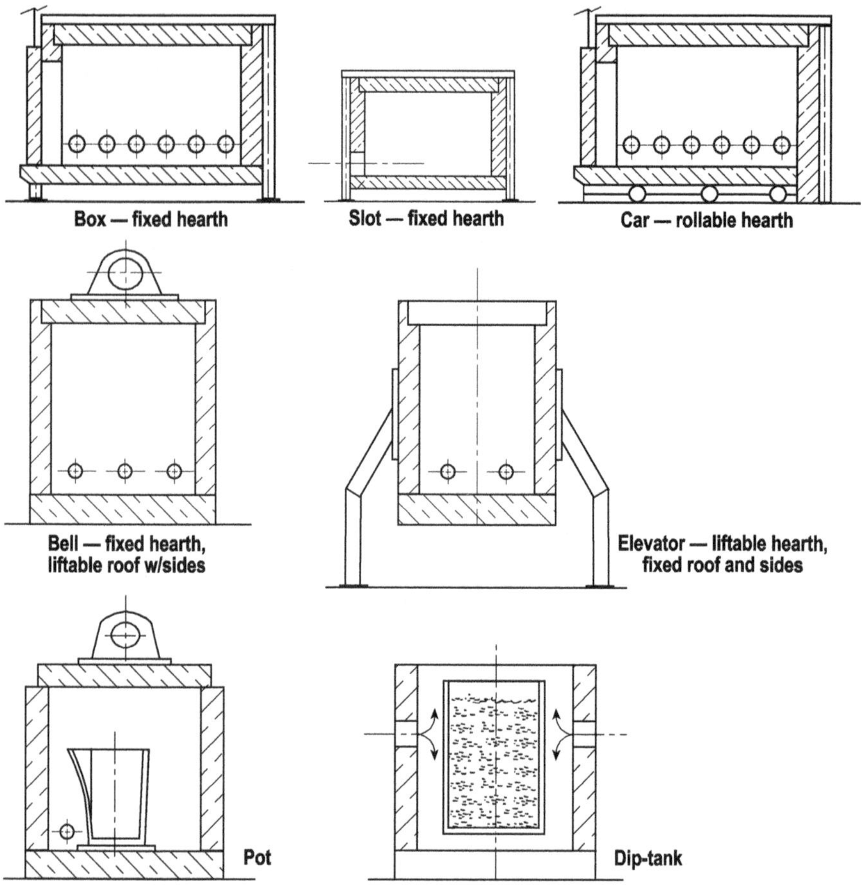


**Fig. 1.1.** Seven (of many kinds of) batch-type furnaces. (See also shuttle kilns and furnaces, fig. 4.8; and liquid baths in fig. 1.12 and sec. 4.7.)

(For flame types, see fig. 6.2.) Unlike crucible, pot, kettle, and dip-tank furnaces, the refractory furnace lining itself is the 'container' for glass "tanks" and aluminum melting furnaces, figure 1.2.

*Car-hearth (car type, car bottom, lorry hearth) furnaces,* sketched in figure 1.1, have a movable hearth with steel wheels on rails. The load is placed on the car-hearth, moved into the furnace on the car-hearth, heated on the car-hearth, and removed from the furnace on the car-hearth; then the car is unloaded. Cooling is done on the car-hearth either in the furnace or outside before unloading. This type of furnace is used mainly for heating heavy or bulky loads, or short runs of assorted sizes and shapes. The furnace door may be affixed to the car. However, a guillotine door (perhaps angled slightly from vertical to let gravity help seal leaks all around the door jamb) usually keeps tighter furnace seals at both door-end and back end.*

*See suggested problem/project at the end of this chapter.

**Fig. 1.2.** Batch-type furnace for melting. Angled guillotine door minimizes gas and air leaks in or out. Courtesy of Remi Claeys Aluminum.

Sealing the sides of a car hearth or of disc or donut hearths of rotary hearth furnaces is usually accomplished with sand-seals or water-trough seals.

*Continuous furnaces* move the charged material, stock, or load while it is being heated. Material passes over a stationary hearth, or the hearth itself moves. If the hearth is stationary, the material is pushed or pulled over skids or rolls, or is moved through the furnace by woven wire belts or mechanical pushers. Except for delays, a continuous furnace operates at a constant heat input rate, burners being rarely shut off. A constantly moving (or frequently moving) conveyor or hearth eliminates the need to cool and reheat the furnace (as is the case with a batch furnace), thus saving energy. (See chap. 4.)

Horizontal *straight-line continuous furnaces* are more common than rotary hearth furnaces, rotary drum furnaces, vertical shaft furnaces, or fluidized bed furnaces.

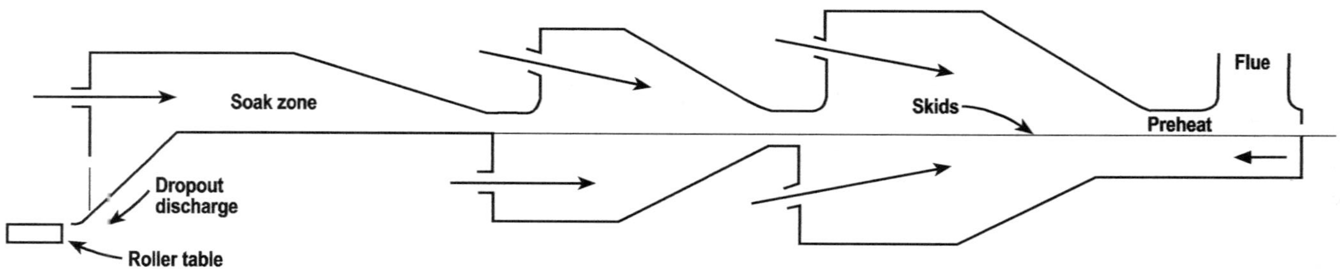


**Fig. 1.3.** Five-zone steel reheat furnace. Many short zones are better for recovery from effects of mill delays. Using end-fired burners upstream (gas-flow-wise), as shown here, might disrupt flame coverage of side or roof burners. End firing, or longitudinal firing, is most common in one-zone (smaller) furnaces, but can be accomplished with sawtooth roof and bottom zones, as shown.

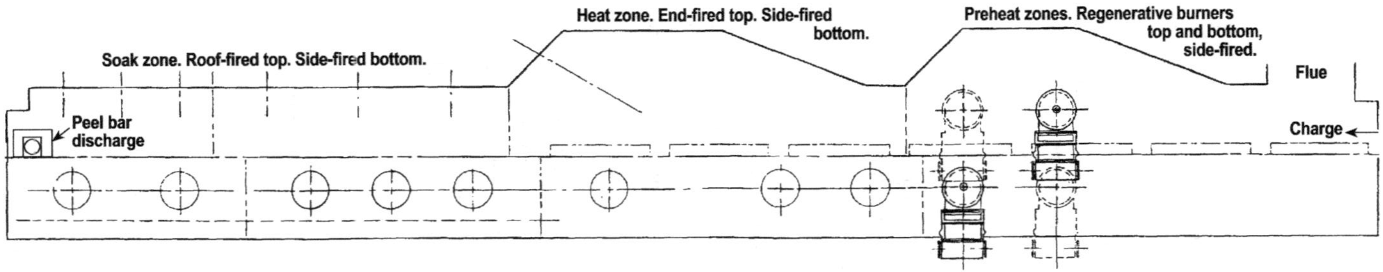


**Fig. 1.4.** Eight-zone steel reheat furnace. An unfired preheat zone was once used to lower flue gas exit temperature (using less fuel). Later, preheat zone roof burners were added to get more capacity, but fuel rate went up. Regenerative burners now have the same low flue temperatures as the original unfired preheat zone, reducing fuel and increasing capacity.

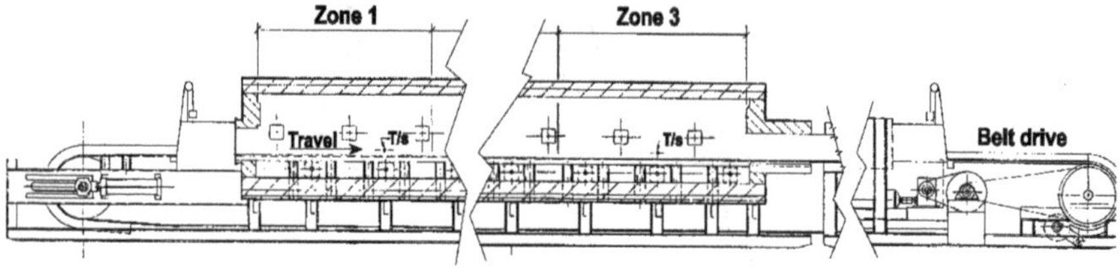


**Fig. 1.5.** Continuous belt-conveyor type heat treat furnace (1800 F, 982 C maximum). Except for very short lengths with very lightweight loads, a belt needs underside supports that are nonabrasive and heat resistant—in this case, thirteen rows, five wide of vertical 4 in. (100 mm) Series 304 stainless-steel capped pipes, between the burners of zones 2 and 4. An unfired cooling one is to the right of zone 3.

Figures 1.3 and 1.4 illustrate some variations of steel reheat furnaces. Side discharge (fig. 1.4) using a peel bar (see glossary) pushing mechanism permits a smaller opening than the end (gravity dropout) discharge of figure 1.3. The small opening of the side discharge reduces heat loss and minimizes uneven cooling of the next load piece to be discharged.

Other forms of straight-line continuous furnaces are woven alloy wire *belt conveyor furnaces* used for heat treating metals or glass "lehrs" (fig. 1.5), plus alloy or ceramic *roller hearth furnaces* (fig. 1.6) and *tunnel furnaces/tunnel kilns* (fig. 1.7).

Alternatives to straight-line horizontal continuous furnaces are rotary hearth (disc or donut) furnaces (fig. 1.8 and secs. 4.6 and 6.4), inclined rotary drum furnaces (fig. 1.10), tower furnaces, shaft furnaces (fig. 1.11), and fluidized bed furnaces (fig. 1.12), and liquid heaters and boilers (sec. 4.7.1 and 4.7.2).

*Rotary hearth or rotating table furnaces* (fig. 1.8) are very useful for many purposes. Loads are placed on the merry-go-round-like hearth, and later removed after they have completed almost a whole revolution. The rotary hearth, disc or donut (with a hole in the middle), travels on a circular track. The rotary hearth or rotating table

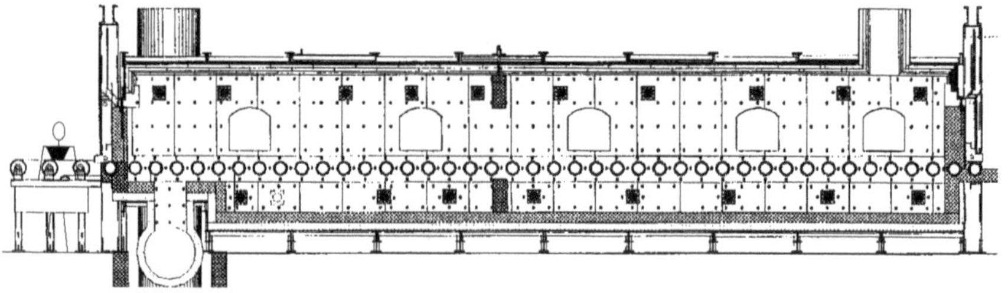

**Fig. 1.6.** Roller hearth furnace, top- and bottom-fired, multizone. Roller hearth furnaces fit in well with assembly lines, but a Y in the roller line at exit and entrance is advised for flexibility, and to accommodate "parking" the loads outside the furnace in case of a production line delay. For lower temperature heat treating processes, and with indirect (radiant tube) heating, "plug fans" through the furnace ceiling can provide added circulation for faster, more even heat transfer. Courtesy of Hal Roach Construction, Inc.

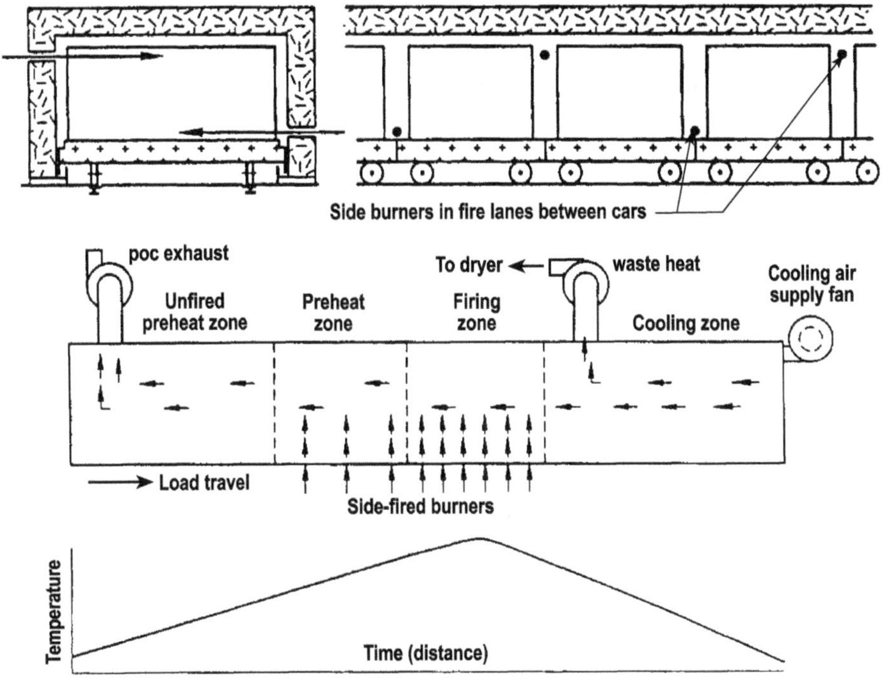


**Fig. 1.7.** Tunnel kiln. *Top row,* end- and side-sectional views showing side burners firing into fire lanes between cars; *center,* flow diagram; *bottom,* temperature vs. time (distance). Ceramic tunnel kilns are used to "fire" large-volume products from bricks and tiles to sanitary ware, pottery, fine dinnerware, and tiny electronic chips. Adapted from and with thanks to *reference 72*.

furnace is especially useful for cylindrical loads, which cannot be pushed through a furnace, and for shorter pieces that can be stood on end or laid end to end. The central column of the donut type helps to separate the control zones. See thorough discussions of rotary hearth steel reheat furnaces in sections 4.6 and 6.4.

*Multihearth furnaces* (fig. 1.9) are a variation of the rotary hearth furnace with many levels of round stationary hearths with rotating rabble arms that gradually plow granular or small lump materials radially across the hearths, causing them to eventually drop through ports to the next level.

*Inclined rotary drum furnaces, kilns, incinerators, and dryers* often use long type F or type G flames (fig. 6.2). If drying is involved, substantially more excess air than normal may be justified to provide greater moisture pickup ability. (See fig. 1.10.)

*Tower furnaces* conserve floor space by running long strip or strand materials vertically on tall furnaces for drying, coating, curing, or heat treating (especially annealing). In some cases, the load may be protected by a special atmosphere, and heated with radiant tubes or electrical means.

*Shaft furnaces* are usually refractory-lined vertical cylinders, in which gravity conveys solids and liquids to the bottom and by-product gases to the top. Examples are cupolas, blast furnaces, and lime kilns.

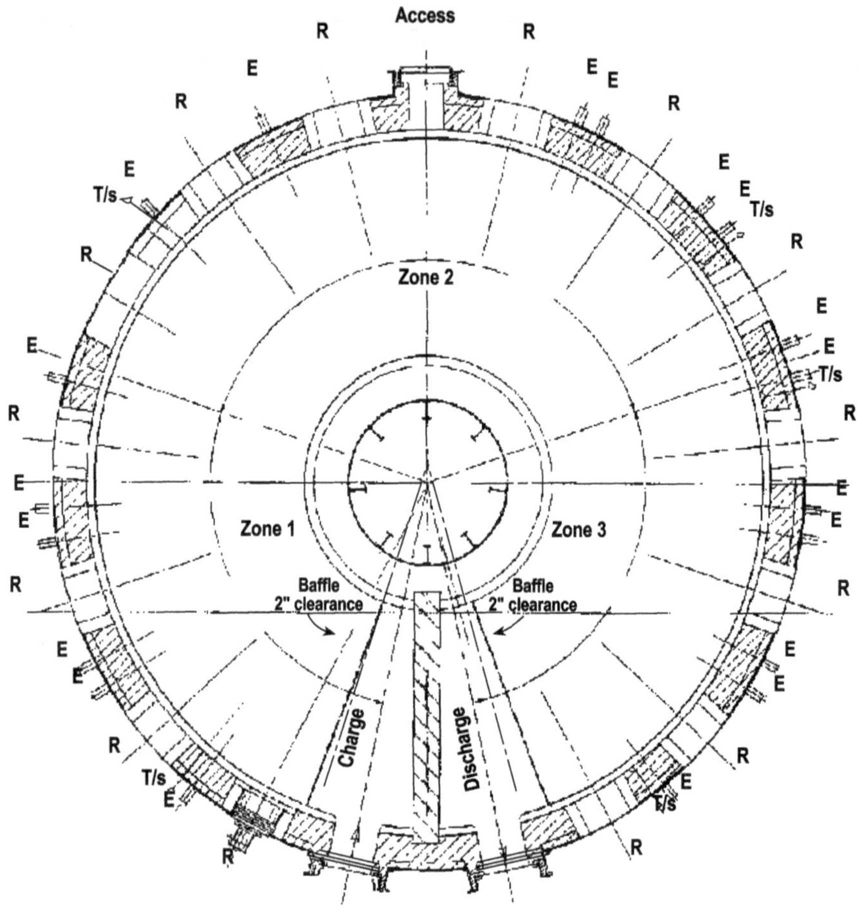


**Fig. 1.8.** Rotary hearth furnace, donut type, sectioned plan view. (Disk type has no hole in the middle.) Short-flame burners fire from its outer periphery. Burners also are sometimes fired from the inner wall outward. Long-flame burners are sometimes fired through a sawtooth roof, but not through the sidewalls because they tend to overheat the opposite wall and ends of load pieces. R, regenerative burner; E, enhanced heating high-velocity burner. (See also fig. 6.7.)

*Fluidized bed furnaces* utilize intense gas convection heat transfer and physical bombardment of solid heat receiver surfaces with millions of rapidly vibrating hot solid particles. The furnaces take several forms.

1. A refractory-lined container, with a fine grate bottom, filled with inert (usually refractory) balls, pellets, or granules that are heated by products of combustion from a combustion chamber below the grate. Loads or boiler tubes are immersed in the fluidized bed above the grate for heat processing or to generate steam.

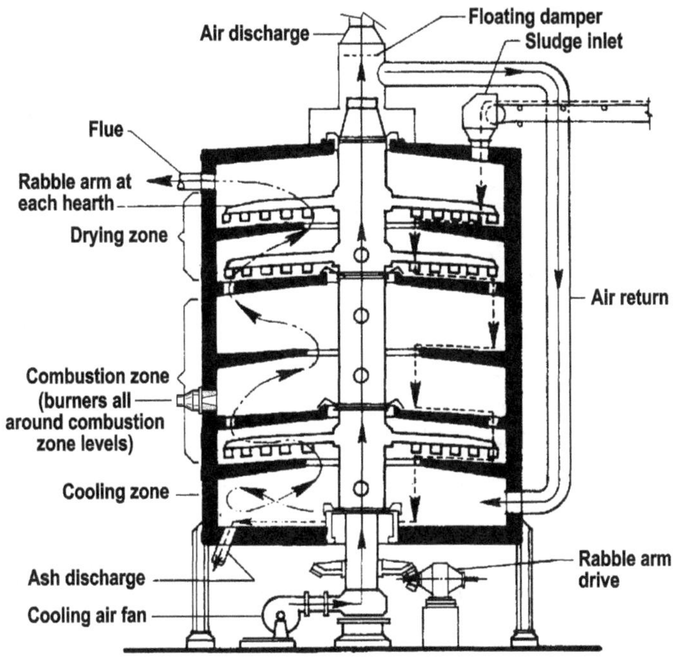


**Fig. 1.9.** Herreshoff multilevel furnace for roasting ores, calcining kaolin, regenerating carbon, and incinerating sewage sludge. Courtesy of *reference 50.*

2. Similar to above, but the granules are fuel particles or sewage sludge to be incinerated. The space below the grate is a pressurized air supply plenum. The fuel particles are ignited above the grate and burn in fluidized suspension while physically bombarding the water walls of the upper chamber and water tubes immersed in its fluidized bed.
3. The fluidized bed is filled with cold granules of a coating material (e.g., polymer), and loads to be coated are heated in a separate oven to a temperature above the melting point of the granules. The hot loads (e.g., dishwasher racks) are then dipped (by a conveyor) into the open-topped fluidized bed for coating.

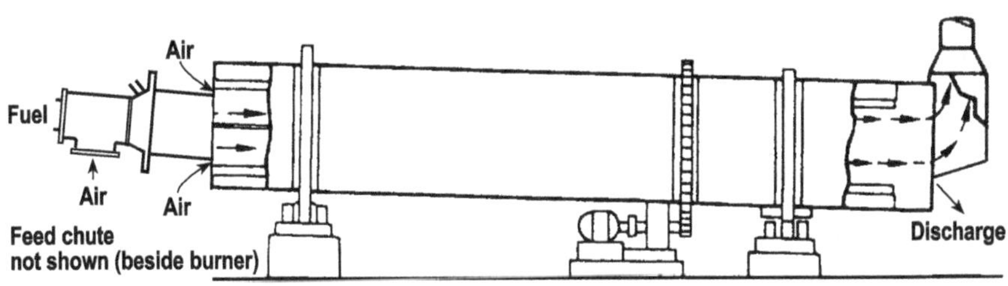


**Fig. 1.10.** Rotary drum dryer/kiln/furnace for drying, calcining, refining, incinerating granular materials such as ores, minerals, cements, aggregates, and wastes. Gravity moves material co-current with gases. (See fig. 4.3 for counterflow.)

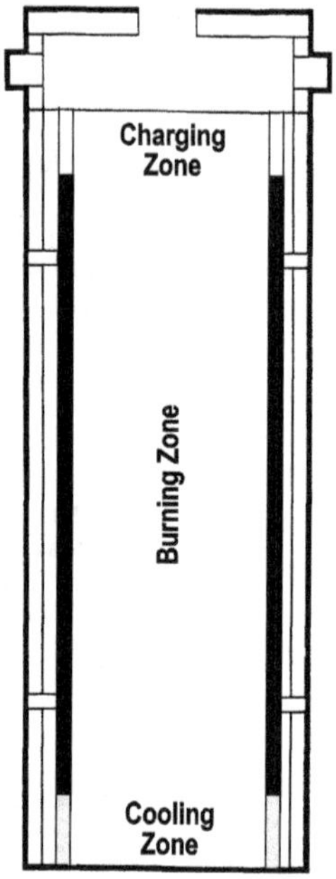


**Fig. 1.11.** Lime shaft kiln. Courtesy of *reference 26,* by Harbison-Walker Refractories Co.

*Liquid heaters.* See Liquid Baths and Heaters, sec. 4.7.1, and Boilers and Liquid Flow Heaters, sec. 4.7.2.

### 1.2.3. Furnace Classification by Fuel

In fuel-fired furnaces, the nature of the fuel may make a difference in the furnace design, but that is not much of a problem with modern industrial furnaces and burners, except if solid fuels are involved. Similar bases for classification are air furnaces, oxygen furnaces, and atmosphere furnaces. Related bases for classification might be the position in the furnace where combustion begins, and the means for directing the products of combustion, e.g., internal fan furnaces, high velocity furnaces, and baffled furnaces. (See sec. 1.2.4. and the rotary hearth furnace discussion on baffles in chap. 6.)

*Electric furnaces* for industrial process heating may use resistance or induction heating. Theoretically, if there is no gas or air exhaust, electric heating has no flue gas loss, but the user must recognize that the higher cost of electricity as a fuel is the result of the flue gas loss from the boiler furnace at the power plant that generated the electricity.

Resistance heating usually involves the highest electricity costs, and may require circulating fans to assure the temperature uniformity achievable by the flow motion of the products of combustion (poc) in a fuel-fired furnace. Silicon control rectifiers have made input modulation more economical with resistance heating. Various materials are used for electric furnace resistors. Most are of a nickel–chromium alloy, in the form of rolled strip or wire, or of cast zig-zag grids (mostly for convection). Other

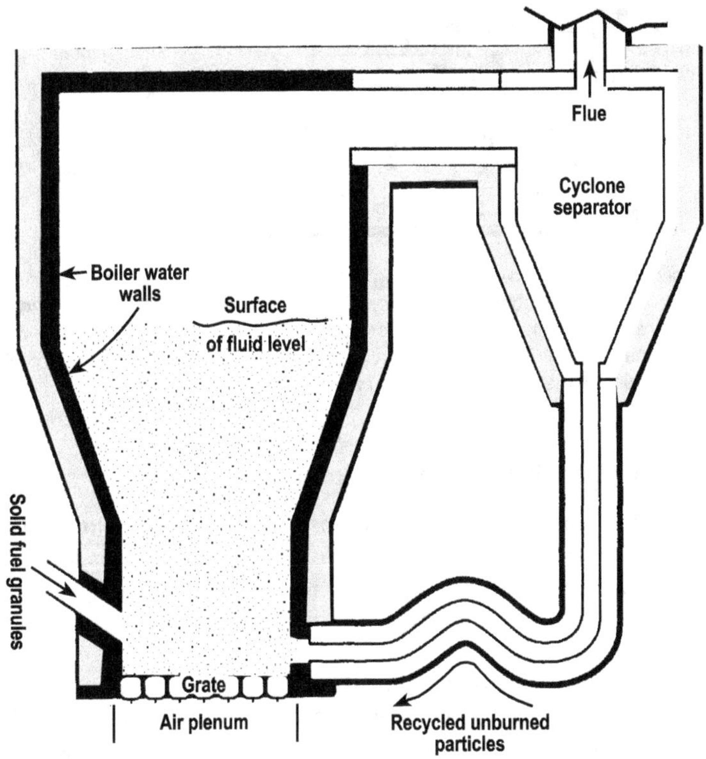


**Fig. 1.12.** Circulating fluidized bed combustor system (type 2 in earlier list). Courtesy of Reference 26, by Harbison-Walker Refractories Co.

resistor materials are molten glass, granular carbon, solid carbon, graphite, or silicon carbide (glow bars, mostly for radiation). It is sometimes possible to use the load that is being heated as a resistor.

In induction heating, a current passes through a coil that surrounds the piece to be heated. The electric current frequency to be used depends on the mass of the piece being heated. The induction coil (or induction heads for specific load shapes) must be water cooled to protect them from overheating themselves. Although induction heating usually uses less electricity than resistance heating, some of that gain may be lost due to the cost of the cooling water and the heat that it carries down the drain.

Induction heating is easily adapted to heating only localized areas of each piece and to mass-production methods. Similar application of modern production design techniques with rapid impingement heating using gas flames has been very successful in hardening of gear teeth, heating of flat springs for vehicles, and a few other high production applications.

Many recent developments and suggested new methods of electric or electronic heating offer ways to accomplish industrial heat processing, using plasma arcs, lasers, radio frequency, microwave, and electromagnetic heating, and combinations of these with fuel firing.

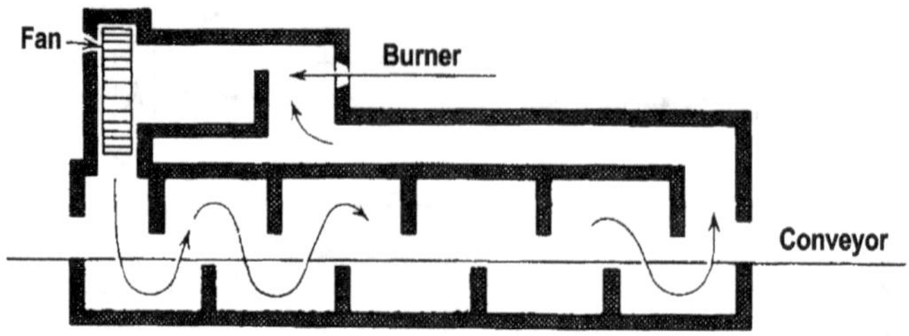


**Fig. 1.13.** Continuous direct-fired recirculating oven such as that used for drying, curing, annealing, and stress-relieving (including glass lehrs). The burner flame may need shielding to prevent quenching with high recirculating velocity. Lower temperature ovens may be assembled from prefabricated panels providing structure, metal skin, and insulation. To minimize air infiltration or hot gas loss, curtains (air jets or ceramic cloth) should shield end openings.

### 1.2.4. Furnace Classification by Recirculation

For medium or low temperature furnaces/ovens/dryers operating below about 1400 F (760 C), a forced recirculation furnace or recirculating oven delivers better temperature uniformity and better fuel economy. The recirculation can be by a fan and duct arrangement, by ceiling plug fans, or by the jet momentum of burners (especially type H high-velocity burners—fig. 6.2).

Figure 3.17 shows a batch-type direct-fired recirculating oven, and figure 1.13 illustrates the principle of a continuous belt direct-fired recirculating oven. All require thoughtful circulation design and careful positioning relative to the loads.

### 1.2.5. Furnace Classification by Direct-Fired or Indirect-Fired

If the flames are developed in the heating chamber proper, as in figure 1.1, or if the products of combustion (poc) are circulated over the surface of the workload as in figure 3.17, the furnace is said to be direct-fired. In most of the furnaces, ovens, and dryers shown earlier in this chapter, the loads were not harmed by contact with the products of combustion.

Indirect-fired furnaces are for heating materials and products for which the quality of the finished products may be inferior if they have come in contact with flame or products of combustion (poc). In such cases, the stock or charge may be (*a*) heated in an enclosing muffle (conducting container) that is heated from the outside by products of combustion from burners or (*b*) heated by radiant tubes that enclose the flame and poc.

***1.2.5.1. Muffles.*** The principle of a muffle furnace is sketched in figure 1.14. A pot furnace or crucible furnace (fig. 1.15) is a form of muffle furnace in which the container prevents poc contact with the load.

A double muffle arrangement is shown in figure 1.16. Not only is the charge enclosed in a muffle but the products of combustion are confined *inside* muffles called *radiant tubes*. This use of radiant tubes to protect the inner cover from uneven heating

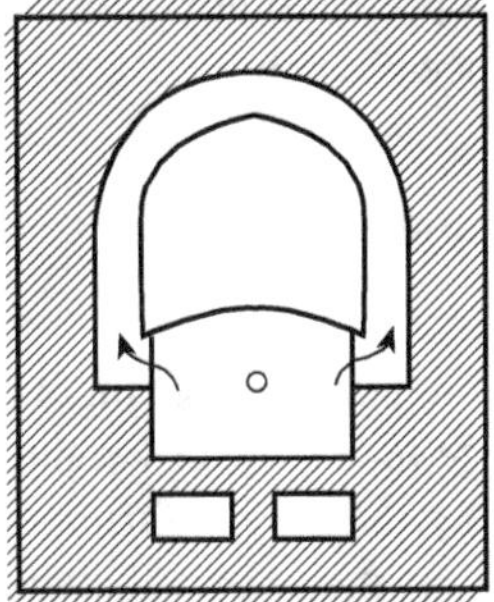

Fig. 1.14. Muffle furnace. The muffle (*heavy black line*) may be of high temperature alloy or ceramic. It is usually pumped full of an inert gas.

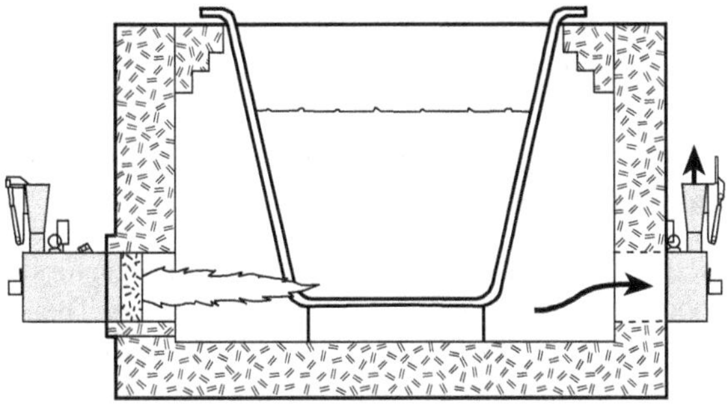

**Fig. 1.15.** Crucible or pot furnace. Tangentially fired integral regenerator-burners save fuel, and their alternate firing from positions 180 degrees apart provides even heating around the pot or crucible periphery. (See also fig. 3.20.)

is being replaced by direct-fired type E or type H flames (fig. 6.2) to heat the inner cover, thereby improving thermal conversion efficiency and reducing heating time.

***1.2.5.2. Radiant Tubes.*** For charges that require a special atmosphere for protection of the stock from oxidation, decarburization, or for other purposes, modern indirect-fired furnaces are built with a gas-tight outer casing surrounding the

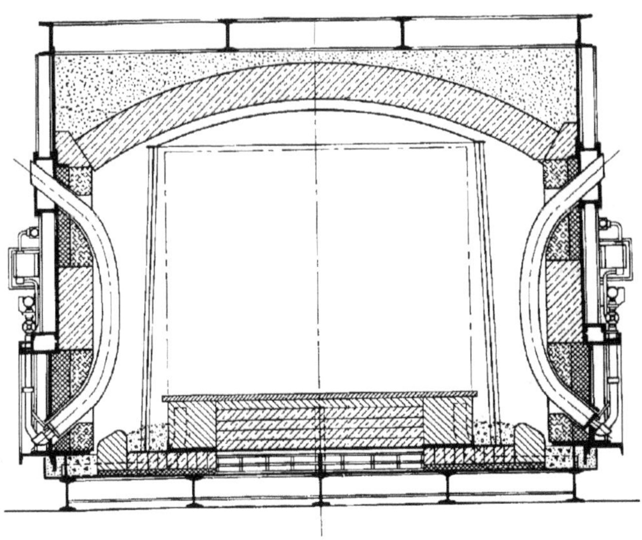

**Fig. 1.16.** Indirect-fired furnace with muffles for both load and flame. Cover annealing furnaces for coils of strip or wire are built in similar fashion, but have a fan in the base to circulate a prepared atmosphere within the inner cover.

refractory lining so that the whole furnace can be filled with a prepared atmosphere. Heat is supplied by fuel-fired radiant tubes or electric resistance elements.

### 1.2.6. Classification by Furnace Use (including the shape of the material to be heated)

There are *soaking pits* or ingot-heating furnaces, for heating or reheating large ingots, blooms, or slabs, usually in a vertical position. There are *forge furnaces* for heating whole pieces or for heating ends of bars for forging or welding. *Slot forge furnaces* (fig. 1.1) have a horizontal slot instead of a door for inserting the many bars that are to be heated at one time. The slot often also serves as the flue.

Furnaces named for the material being heated include bolt heading furnaces, plate furnaces, wire furnaces, rivet furnaces, and sheet furnaces. Some furnaces also are classified by the process of which they are a part, such as hardening, tempering, annealing, melting, and polymerizing. In *carburizing furnaces*, the load to be case-hardened is packed in a carbon-rich powder and heated in pots/boxes, or heated in rotating drums in a carburizing atmosphere.

### 1.2.7. Classification by Type of Heat Recovery (if any)

Most heat recovery efforts are aimed at utilizing the "waste heat" exiting through the flues. Some forms of heat recovery are air preheating, fuel preheating, load preheating (Fig. 1.17), recuperative, regenerative, and waste heat boilers—all discussed in chapter 5.

*Preheating combustion air* is accomplished by recuperators or regenerators, discussed in detail in chapter 5. *Recuperators* are steady-state heat exchangers that transmit heat from hot flue gases to cold combustion air. *Regenerators* are non-steady-state devices that temporarily store heat from the flue gas in many small masses of

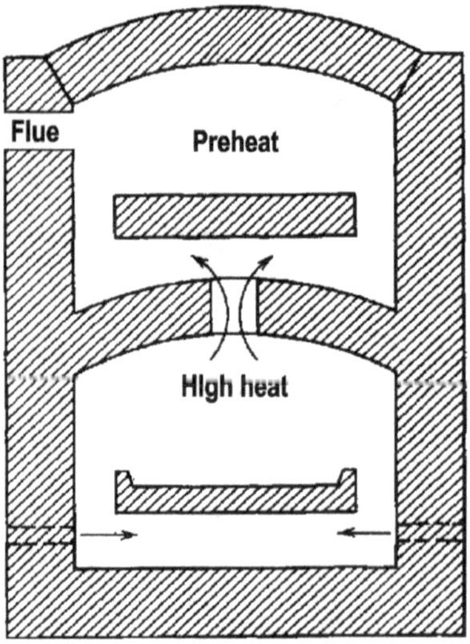


**Fig. 1.17.** Tool heating furnace with heat-recovering load preheat chamber.

> Regenerative furnaces were originally called "Siemens furnaces" after their inventors, Sir William Siemens and Friedrich Siemens. Their objective, in the 1860s, was a higher flame temperature, and therefore a higher glass melting furnace temperature from their gaseous fuel (which was made from coal and had low heating value), but they also saved so much fuel that they were soon used around the world for many kinds of furnaces.

refractory or metal, each having considerable heat-absorbing surface. Then, the heat-absorbing masses are moved into an incoming cold combustion air stream to give it their stored heat. Furnaces equipped with these devices are sometimes termed *recuperative furnaces* or *regenerative furnaces*.

Regenerative furnaces in the past have been very large, integrated refractory structures incorporating both a furnace and a checkerwork refractory regenerator, the latter often much larger than the furnace portion. Except for large glass melter "tanks," most regeneration is now accomplished with integral regenerator/burner packages that are used in pairs. (See chap. 5.)

Boilers and low temperature applications sometimes use a "heat wheel" regenerator—a massive cylindrical metal latticework that slowly rotates through a side-by-side hot flue gas duct and a cold combustion air duct.

Both preheating the load and preheating combustion air are used together in steam generators, rotary drum calciners, metal heating furnaces, and tunnel kilns for firing ceramics.

### 1.2.8. Other Furnace Type Classifications

There are stationary furnaces, portable furnaces, and furnaces that are slowly rolled over a long row of loads. Many kinds of continuous "conveyor furnaces" have the stock carried through the heating chamber by a conveying mechanism, some of which were discussed under continuous furnaces in section 1.2.2. Other forms of conveyors are wire-mesh belts, rollers, rocker bars, and self-conveying catenary strips or strands. (See sec. 4.3.) In porcelain enameling furnaces and paint drying ovens, contact of the loads with anything that might mar their surfaces is avoided by using hooks from an overhead chain conveyor. For better furnace efficiency and for best chain, belt, or conveyor life, they should return within the hot chamber or insulated space.

"Oxygen furnace" was an interim name for any furnace that used oxygen-enriched air or near-pure oxygen. In many high-temperature furnaces, productivity can be increased with miniumum capital investment by using oxygen enrichment or 100% oxygen ("oxy-fuel firing"). Either method reduces the nitrogen concentration, lowering the percentage of diatomic molecules and increasing the percentage of triatomic molecules. This raises the heat transfer rate (for the same average gas blanket temperature and thickness) and thereby lowers the stack loss.

Oxygen use reduces the concentration of nitrogen in a furnace atmosphere (by reducing the volume of combustion air needed), so it can reduce NOx emissions. (See glossary.)

Such oxygen uses have become a common alteration to many types of furnaces, which are better classified by other means discussed earlier. See part 13 of *reference 52* for thorough discussions of the many aspects of oxygen use in industrial furnaces.)

"Electric furnaces" are covered in section 1.2.3. on fuel classification.

The brief descriptions and incomplete classifications given in this chapter serve merely as an introduction. More information will be presented in the remaining chapters of this book—from the standpoints of safe quality production of heated material, suitability to plant and environmental conditions, and furnace construction.

## 1.3. ELEMENTS OF FURNACE CONSTRUCTION (see also chap. 9)

The load or charge in a furnace or heating chamber is surrounded by side walls, hearth, and roof consisting of a heat-resisting refractory lining, insulation, and a gas-tight steel casing. All are supported by a steel structure.

In continuous furnaces, cast or wrought heat-resisting alloys are used for skids, hearth plates, walking beam structures, roller, and chain conveyors. In most furnaces, the loads to be heated rest on the hearth, on piers to space them above the hearth, or on skids or a conveyor to enable movement through the furnace. To protect the foundation and to prevent softening of the hearth, open spaces are frequently provided under the hearth for air circulation—a "ventilated hearth."

Fuel and air enter a furnace through burners that fire through refractory "tiles" or "quarls." The poc (see glossary) circulate over the inside surfaces of the walls, ceiling, hearth, piers, and loads, heating all by radiation and convection. They leave the furnace flues to stacks. The condition of furnace interior, the status of the loads, and the performance of the combustion system can be observed through air-tight peepholes or sightports that can be closed tightly.

In modern practice, hearth life is often extended by burying stainless-steel rails up to the ball of the rail to support the loads. The rail transmits the weight of the load 3 to 5 in. (0.07–0.13 m) into the hearth refractories. At that depth, the refractories are not subjected to the hot furnace gases that, over time, soften the hearth surface refractories. The grades of stainless rail used for this service usually contain 22 to 24% chromium and 20% nickel for near-maximum strength and low corrosion rates at hearth temperatures.

Firebrick was the dominant material used in furnace construction through history from about 5000 B.C. to the 1950s. Modern firebrick is available in many compositions and shapes for a wide range of applications and to meet varying temperature and usage requirements. High-density, double-burned, and super-duty (low-silica) firebrick have high temperature heat resistance, but relatively high heat loss, so they are usually backed by a lower density insulating brick (firebrick with small, bubblelike air spaces).

Firebrick once served the multiple purposes of providing load-bearing walls, heat resistance, and containment. As structural steel framing and steel plate casings became more common, furnaces were built with externally suspended roofs, minimizing the need for load-bearing refractory walls.

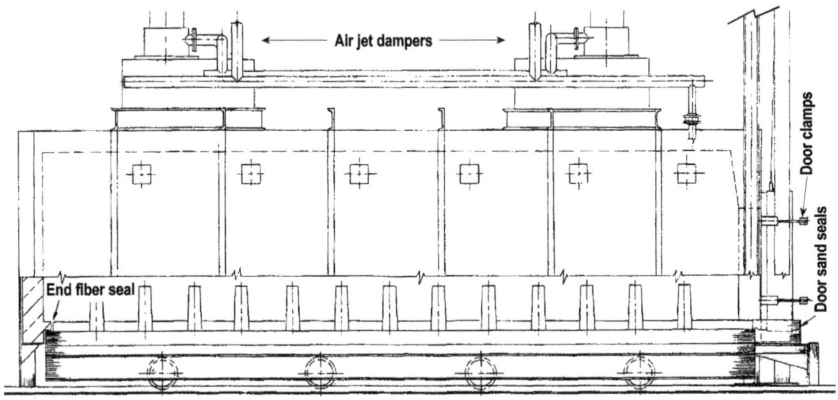


**Fig. 1.18** Car-hearth heat treat furnace with piers for better exposure of bottom side of loads. The spaces between the piers can be used for enhanced heating with small high-velocity burners. (See chap. 7.) Automatic furnace pressure control allows roof flues without nonuniformity problems and without high fuel cost.

Continuing improvements in monolithic refractories, particularly in bonding, have resulted in their steadily increasing usage—now substantially over 60% monolithic.

More detailed information on furnace structures and materials is contained in chapter 9, figure 1.18, and *reference 26.*

## 1.4. REVIEW QUESTIONS AND PROJECTS

**1.4Q1.** How can furnace loads be heated without scaling (oxidizing)?

**A1.** Heat loads inside muffles with prepared atmosphere inside, or heat loads in a prepared atmosphere outside of radiant tubes or electric elements.

**1.4.Q2.** How can loads be moved through a continuous furnace?

**A2.** By using a rotary hearth, a roller hearth, overhead trolleys suspending the load pieces, a pusher mechanism, a walking mechanism, or by suspending continuous strip or strands between rollers external to the furnace (catenary).

**1.4.Q3.1.** "Very high temperature furnaces" are operated above what temperature?

**A3.1.** Above 2300 F (1260 C).

**1.4.Q3.2.** Furnaces considered "high temperature" are operated in what range?

**A3.2.** Between 1900 F (1038 C) and 2300 F (1260 C).

**1.4.Q3.3.** Furnaces considered "midrange temperature" are operated in what range?

**A3.3.** Between 1100 F (593 C) and 1900 F (1038 C).

**1.4.Q3.4.** Furnaces considered "low temperature" are operated below what temperature?

**A3.4.** Below 1100 F (593 C).

**1.4.Q4.** When rolling high quality fine-grained steel, what range of furnace exit temperatures is now used, and why?

**A4.** Temperature of 1850 F (1010 C) to 1950 F (1066 C), to hold grain growth to a minimum after the last roll stand.

**1.4.Q5.** Why is it more difficult to successfully operate a rotary continuous furnace than a linear continuous furnace?

**A5.** Because in a rotary furnace, the furnace gases move in two opposite directions to the flue(s) or to a flue and to the charge and discharge doors.

**1.4.Q6.** In what ways is electric energy used in industrial heat processing?

**A6.** *By resistance*, using heating elements to provide convection and radiation, or using the load piece as a resistor itself, but this is very limited. *Or by induction* heating, in which an induced current agitates the load molecules, thereby heating them. The flux lines are concentrated near the load piece surfaces, so this does some internal heating whereas convection and radiation are surface phenomena.

**1.4.Q7.** What kinds of loads can be processed in shaft furnaces?

**A7.** Limestone to remove the $CO_2$ to make lime (lime kiln); iron ore, to remove oxygen, reducing the ore to iron (blast furnace); pig iron, to melt it for casting in a foundry (cupola).

## 1.4. PROJECTS

### 1.4.Proj-1.

Are you familiar with all the terminology relative to industrial furnaces? If not, you will find it helpful to set yourself a goal of reading and remembering the gist of one page of the glossary of this book each day. You will find that it gives you a wealth of information. Start now—read one page of the glossary each day.

### 1.4.Proj-2.

Build rigid models of car-hearth furnaces with (*a*) the door affixed to the car and (*b*) a slightly longer hearth so that a guillotine door closes against the car hearth surface. Decide which door arrangement will maintain tighter gas seals at BOTH front and back ends of the car through many loadings and unloadings. (See fig. 1.18.)

# 2

# HEAT TRANSFER IN INDUSTRIAL FURNACES

## 2.1. HEAT REQUIRED FOR LOAD AND FURNACE

To evaluate the input required for a process, one must first determine the heat required into the load, which is discussed in sections 2.1.1. and 2.1.2. below. The means by which the load is heated is usally a furnace, kiln, or oven, but these 'means' themselves require some heat over and above what they deliver to the load.

$$\text{Energy input to a furnace} = \frac{\text{'heat needs' for load \& furnace}}{\%\text{available heat}/100\%} \tag{2.1}$$

Find flue gas exit temperature from figure 5.3, then %available heat from figure 5.1 or 5.2. Heat first must be generated (liberated, released) in the furnace, then transferred to the load (stock, charge, ware), and finally, distributed in the charge to meet the specifications of the metallurgical or ceramic engineer. These specs usually cover final temperature of the charge, temperature uniformity of the charge, and time at temperature. Rates of heating and cooling are often specified.

For a clear understanding of the heating process, it is advisable to begin with the physical properties of the material to be heated. The heat to be imparted to the load is Weight × Specific Heat × Temperature Rise, or by use of figures 2.1 and 2.2.

$$Q = w \times c \times \Delta T = w \text{ (change in heat content)} \tag{2.2}$$

### 2.1.1. Heat Required for Heating and Melting Metals

Handbooks (such as *reference 52*) list the mean specific heats of metallic and non-metallic materials.

Figure 2.2 is a graph of the heat contents of irons and steels, illustrating the effect of varying percents of carbon. Addition of the usual small amount of alloying elements, such as nickel, chromium, or manganese, changes the heat content of steel by only a negligible amount. The specific heat of "Inconel" (79.5% nickel, 13% chromium, 6.5% iron) differs by only 1% from the specific heat of mild steel.

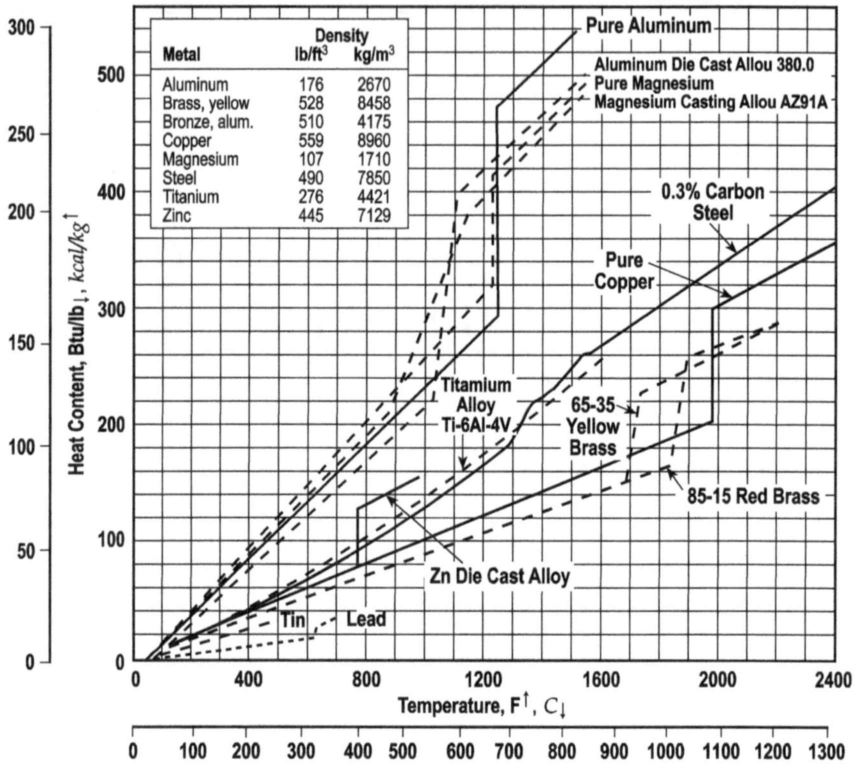


**Fig. 2.1.** Heat contents of metals at industrial processing temperatures.

Use of the heat content graph data and equation 2.2 are demonstrated in example 2.1 to determine the amount of heat absorbed by a material as it is heated through a prescribed temperature range.

**Example 2.1**: A 250-lb bar of 0.30% carbon steel is to be heated from 100 F to 2200 F. From figure 2.2, the heat content (above 0 F), when the bar is put into the furnace is 11 Btu/lb. When it is taken out of the furnace, if uniformly heated to 2200 F, its heat content will be 369 Btu/lb. By equation 2.1, $Q = 250\,(369 - 11) = 89\,500$ Btu, absorbed by the bar.

## 2.1.2. Heat Required for Fusion (Vitrification) and Chemical Reaction

If, as in burning lime or fusing porcelain enamel, the purpose is used to cause chemical reactions, specific heats and reaction heats should be obtained from chemical and ceramic engineering handbooks, such as *references 16*, *46*, and *82*. In the "firing" of ceramic materials, much heat also is required for "driving out" and evaporating moisture.

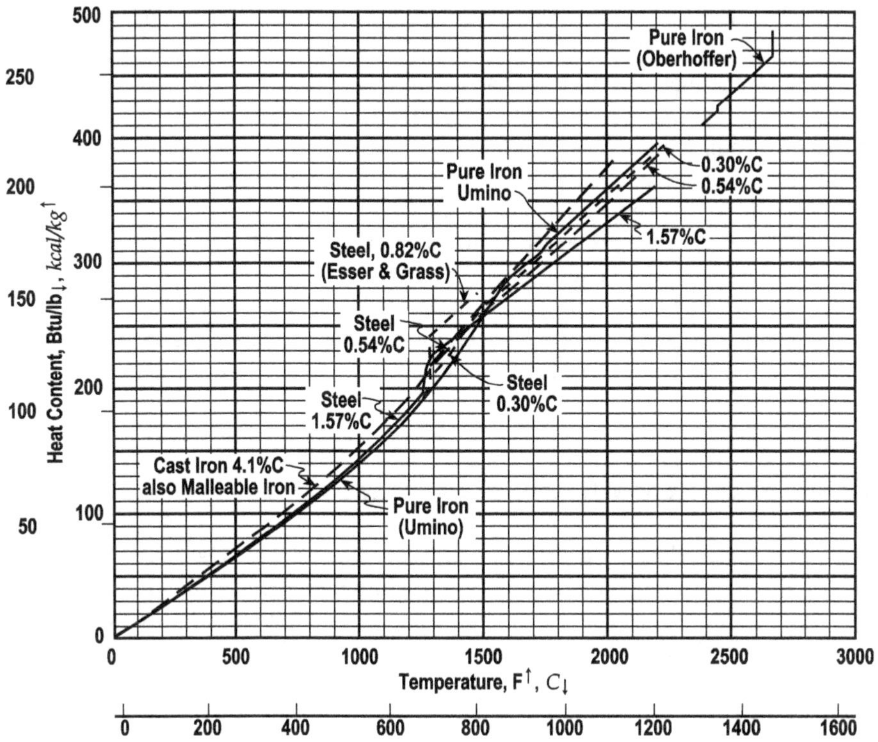


**Fig. 2.2.** Heat contents of irons and steels, showing the small effects of carbon content on heat contents of pure iron, cast iron, and malleable iron with 4.1% carbon; steels from 0.3 to 1.57% carbon. Compare this with fig. 2.5 showing effects on thermal conductivity over a narrower temperature range.

In addition to imparting sensible heat, *enameling* requires heat of fusion (vitrification) and chemical reactions. The metal on which the enamel is deposited requires a large part of the total heat, so some information on enameling is furnished next.

The porcelain enamel batch, composed of borax, quartz, feldspar, soda, cryolite, and metallic oxides, is first melted to form a glass, which is then disintegrated by pouring it into water, forming "frit." For typical batch mixtures of grip coat or ground coat of enamel, the heat absorbed in its formation is 1540 Btu/lb. of frit. This includes sensible heat in raising it to 2000 F, heat of fusion, and heat absorbed by chemical reactions. The corresponding number for the cover coat frit is 1309 Btu/lb of frit.

The frit is ground to powder with the addition of about 12% of its weight of clay and quartz or tin oxide, mixed with water (45% by vol.). This mixture is coated on the metal to be porcelain enameled, and dried before it enters an enameling furnace. The heat absorbed by the enamel itself when heated to 1650 F, but not including drying, is 395 Btu/lb of grip-coat enamel and 370 Btu/lb of cover-coat enamel. The weight of enamel applied is usually about 0.077 pounds per square foot (psf) for the grip coat and 0.108 psf for the cover coat, on each side of the metal.

The heat absorbed by the enamel, in heating to 1650 F, is 61 Btu/ft$^2$ for the grip coat, two sides, and 61 + 80 = 141 Btu/ft$^2$ for the grip plus cover coat. The heat absorbed by the metal itself, if 24-gauge sheet steel (0.025 in. thick), is about 280 Btu/ft$^2$; hence, the heat absorbed by the enamel is about 22% as much as the heat to the metal during the grip-coat heating and 50% during the cover-coat heating. For thicker metal, the percentage of heat absorbed by the enamel will be less, and far less for castings. The supports that carry the ware through the furnace may absorb as much heat as the metal plus coatings, although efforts have been made to reduce the weight of the fixtures by better design.

In many heating operations, additional heat is needed for containers, trays, or supports. Water-cooled skids absorb heat. If the furnace and its loads are to be heated together from cold conditions, the furnace walls may absorb almost as much heat as the loads.

## 2.2. FLOW OF HEAT WITHIN THE CHARGED LOAD

If a load is heated electrically—by actually using the load as a resistance in a circuit or by induction heating—the flux lines will concentrate just inside the surface. In fuel-fired heating processes, heat enters the load through its surface (by radiation or convection) and diffuses throughout the piece by conduction. This heat flow requires a difference in temperature within the piece. Steady heat flow through a flat plate is described by:

$$q = (k/x)\,(A)\,(\Delta T), \tag{2.3}$$

where

- $q$ = heat flow rate, in Btu/hr,
- $k$ = the load's thermal conductivity, in Btu/ft$^2$hr°F/ft, from figure 2.3,
- $x$ = the maximum thickness through which the heat travels (half thickness if heated from two sides),
- $A$ = the cross-sectional area of the load, perpendicular to the heat travel direction within the load, and
- $\Delta T$ = the maximum temperature difference within a load piece.

For other than flat plates, heat flux lines are seldom parallel, rarely steady. In transient heat flow, determination of the temperature at a given time and point within the load necessitates use of the finite element method.

Elevating the furnace temperature (a high "thermal head") or "high-speed heating" often results in nonuniform heating, which necessitates a longer soak time, sometimes defeating the purpose of high-speed heating.

### 2.2.1. Thermal Conductivity and Diffusion

Figure 2.3 shows the great variation in thermal conductivities of various metals, which has a direct bearing on the ability of heat to flow through or diffuse throughout

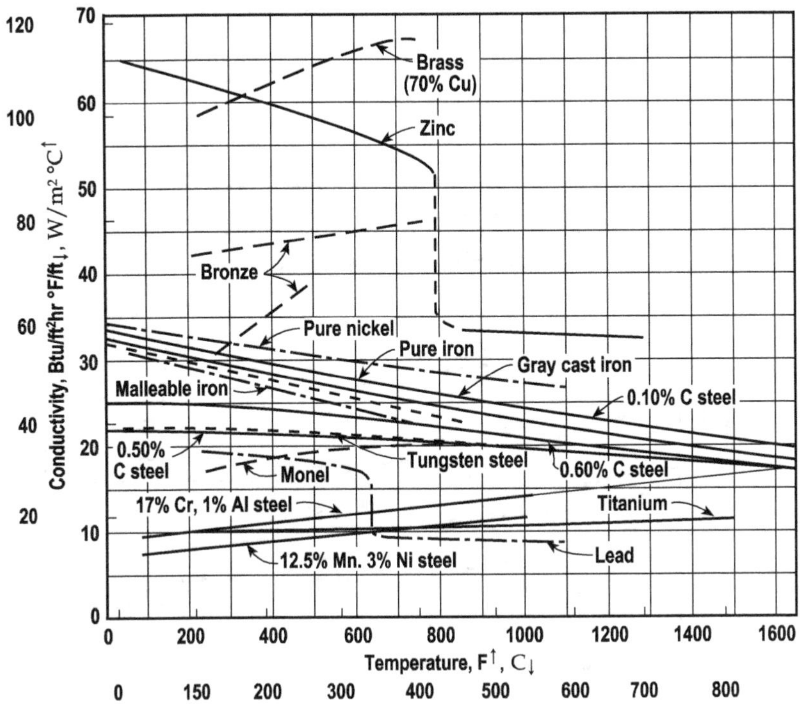


**Fig. 2.3** Thermal conductivities of some metals. Not shown is copper for which thermal conductivity ranges from 215 Btu ft/ft$^2$hr°F at 200 F to 200 Btu ft/ft$^2$hr°F at 1300 F. (See also figs. 2.4 and 2.5.)

them, and therefore has a very strong effect on temperature distribution or uniformity in solids. The whole factor that affects temperature distribution is *thermal diffusivity*, which is thermal conductivity divided by the volume specific heat of the solid material, or

$$\text{Thermal diffusivity}, \sigma = \frac{\text{thermal conductivity}, k}{(\text{specific heat}, c)\ (\text{density}, \rho)}. \tag{2.4}$$

In equation 2.4, the numerator is a measure of the rate of heat flow into a unit volume of the material; the denominator is a measure of the amount of heat absorbed by that unit volume. With a higher ratio of numerator to denominator, heat will be conducted into, distributed through, and absorbed.

Figures 2.3 to 2.5 and table 2.1 list conductivity and diffusivity data for many metals. Figure 2.5 exhibits surprisingly great variations of thermal conductivity for steels of various compositions. At 60 F (16 C), the conductivity, $k$, of steel #2 is more than five times that of steel #13.

Thermal conductivities and diffusivities of solids vary greatly with temperature. Specific heats and densities vary little, except for steels at their phase transition point. The thermal conductivities of solid pure metals drop with increasing temperature, but the conductivities of solid alloys generally rise with temperature.

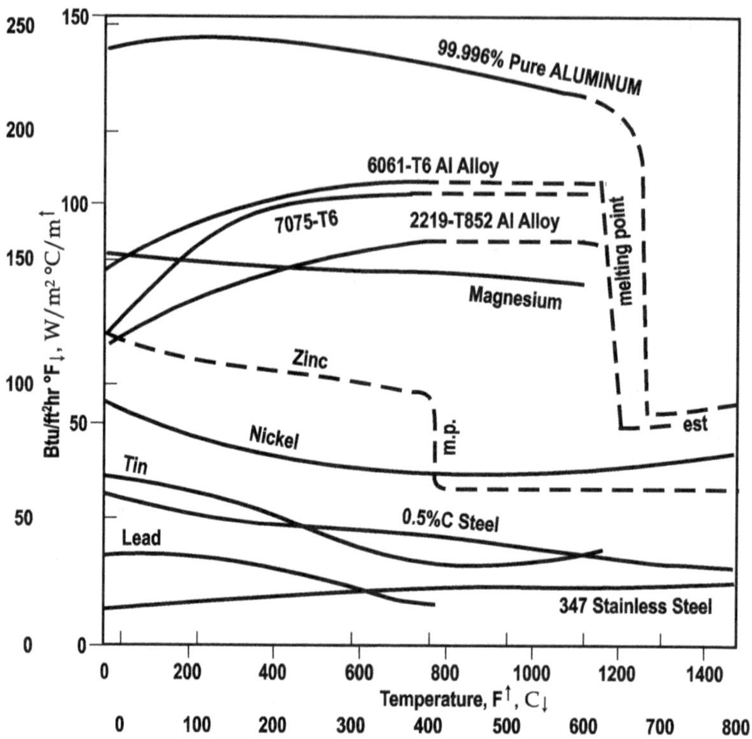


**Fig. 2.4** Thermal conductivities of more metals. (See also figs. 2.3 and 2.5.)

### 2.2.2. Lag time

The effect of thermal conductivity on heat flow and internal temperature distribution is shown in figure 2.6 for three same-size bars or slabs of ferrous alloys #1, #6, and #13 (from fig. 2.5) heated from two sides. The surface temperatures of all three will rise very quickly, but the interior temperatures of #6 and #13 will rise more slowly because of their poorer diffusivities. The #13 bar will take the longest time to come to thorough equilibrium with furnace temperature.

Solid material that is heated in industrial furnaces is not necessarily continuous. Very often, the charge consists of coiled strip material or separate pieces piled to various depths or close side by side. In such cases, heat only can flow from one piece to the adjacent piece through small contact points on their surfaces, or through gas-filled spaces—the thermal conductivity of which is very small. A pile of crankshafts is an example of low overall conductance, but high-velocity burners may be able to blow some gases between the pieces.

A stack of supposedly flat plates is an example of very low conductance. Even gaps thinner than a page of this book constitute much more thermal resistance than solid metal. Some people erroneously think a stack can be treated as a solid, but thin

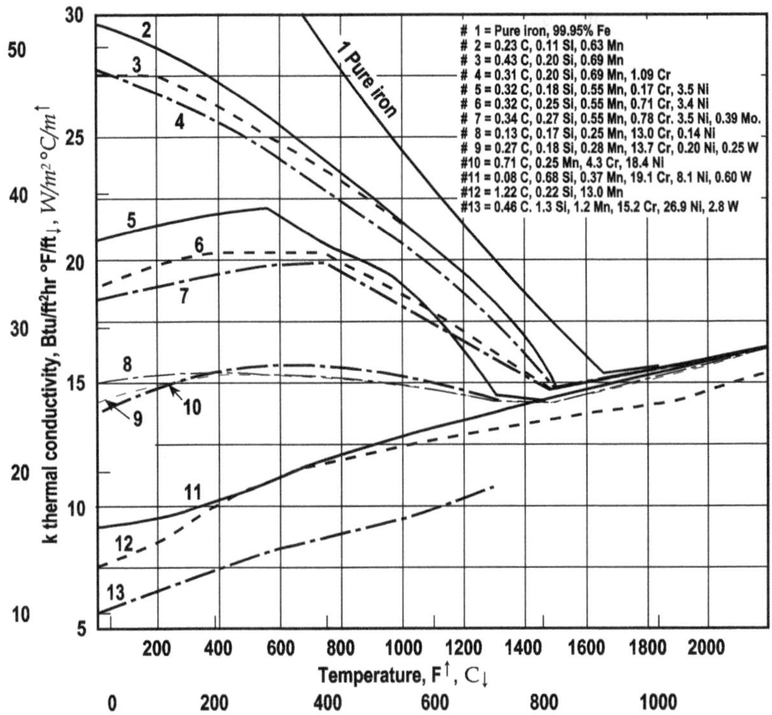


**Fig. 2.5** Thermal conductivity of pure iron and some ferrous alloys.

air spaces are insulators. If the plates are not perfectly flat, or identically dished, the differing air gaps will result in bad nonuniformities in temperatures and warping, probably resulting in junking of the whole stack.

Rapid heat flow in each piece of a piled charge is obtained only by circulation of hot gases through the piled material by convection and gas radiation. Those gas masses must be constantly replaced with new hot gas because they have low mass, low specific heat, and thin gas beam thickness, so they cool quickly without delivering much heat to the loads. For uniform heating and precise reproducibility, piling of pieces must be avoided. Use piers, piles, kiln furniture, or some other form of spacers generously; better yet, load pieces only one-high, but spaced above the hearth. Do not allow crumbs of refractory, scale, or anything else to accumulate on the furnace or oven floor because they impede circulation, choke flues, and may contaminate load surfaces.

## 2.3. HEAT TRANSFER TO THE CHARGED LOAD SURFACE

In furnace practice, heat is transferred by three modes—conduction, convection, and radiation. This book discusses only those essentials of heat transfer that are helpful to

**TABLE 2.1. Conductivity, specific heat, and diffusivity of metals at 100 F (37.8 C) (from *reference 85* and others, see also tables 4.2a, b of *reference 51*)**

| Metal | Thermal conductivity (Btu ft/ft$^2$hr°F) | Density (lb/ft$^2$) | Specific heat (Btu/lb°F) | Diffusivity (ft/hr) |
|---|---|---|---|---|
| ALUMINUMS: Cast | 108 | 165 | 0.248 | 2.6 |
| Drawn and annealed | 126 | 168 | 0.248 | 3.0 |
| Alloy, 92% Al, 8% Cu | 88 | 180 | | |
| COPPERS: Copper | 220 | 558 | 0.104 | 3.8 |
| Brass | 58 | 530 | 0.092 | 1.2 |
| Bronze | 42 | 510 | 0.086 | 1.0 |
| Manganese bronze | 42 | | | |
| Phosphor bronze | 33 | 554 | 0.087 | 0.68 |
| IRONS: Pure | 33 | 490 | 0.110 | 0.61 |
| Cast, gray | 31 | 442 | 0.122 | 0.55 |
| Malleable | 31 | 458 | 0.122 | 0.55 |
| LEAD: Solid | 19 | 708 | 0.031 | 0.87 |
| Molten | 9.5 | 650 | 0.034 | 0.43 |
| NICKELS: Nickel | 33 | 537 | 0.103 | 0.60 |
| Monel metal | 16 | 555 | 0.13 | 0.22 |
| STEELS: Chrome, 3% Cr | 21 | | | |
| (Varies with 10% Cr | 13 | 483 | 0.120 | 0.22 |
| heat treatment) 20% Cr | 10 | | | |
| Machinery steel | 30 | 488 | 0.115 | 0.54 |
| Manganese steel, 10% Mn | 7.2 | 498 | 0.125 | 0.12 |
| Nickel steel, 5% Ni | 18 | 492 | | |
| 15% Ni | 15 | | | |
| 30% Ni | 5 | 500 | 0.119 | 0.09 |
| Tool steel | 23 | 481 | 0.120 | 0.40 |
| ZINCS: Zinc | 63 | 446 | 0.094 | 1.5 |
| Die-cast metal, Zn base | 54 | 432 | | |

designers and operators of industrial furnaces. Most industrial furnaces, ovens, kilns, incinerators, boilers, and chemical process industry (cpi) heaters use combustion of fuels as their heat source.

Combustion, as used in industrial furnaces, comes from rapid and large chemical reaction kinetics—conversion from chemical energy to sensible heat (thermal) energy. Increasing fuel and oxidant (usually air) mixing surface area or increasing temperature of the reactants can cause faster combustion reactions, usually resulting in higher heat source temperatures. Fuel oxidation reactions are exothermic, so they can develop into a runaway condition (e.g., thermal energy being released faster than it can be carried away by heat transfer). This positive feedback can cause an explosion.

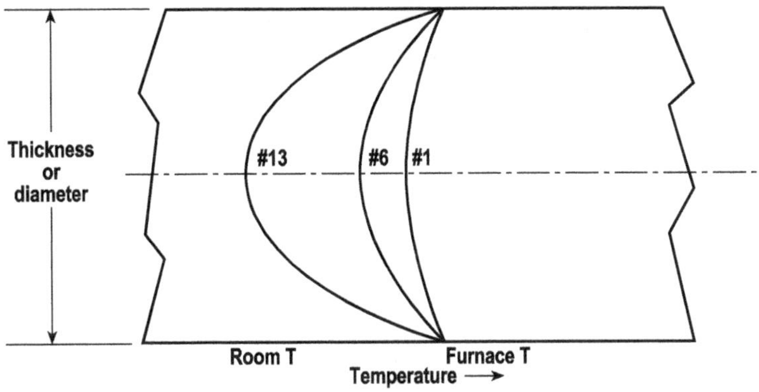


**Fig. 2.6** Transient temperature distributions in three same-size metal bars shortly after being simultaneously put in a hot furnace. Numbers are from fig. 2.5.

A flame is a thin region of rapid exothermic chemical reaction, small examples of which are a candle flame and a Bunsen burner flame. In a Bunsen burner, a thoroughly premixed laminar stream of fuel gas and air is ignited by an external heat source, and a cone-shaped reaction zone (flame front) forms. Turbulence increases the thickness and surface area of the reaction zone, resulting in higher burning velocity. Laminar burning velocity for natural gas is about 1 fps (0.305 m/s); turbulent burning velocity may be two to ten times faster.

In a laminar flame, thermal expansion from chemical heat release may combine with increased reactivity caused by higher temperatures, resulting in acceleration to a turbulent flame. Except for long luminous flames, most industrial flames are turbulent. (See fig. 6.2 for descriptions of a number of generic industrial flame types; see also *references 51 and 52*.)

If a flame is confined, it may suddenly become a detonating flame, the velocity of which may increase from a normal flame velocity of 1 fps (0.305 m/s) for natural gas to 4,400 mph (7,080 km/h). This results in the pressure behind the flame front increasing from 1 atmosphere to 15 atmospheres, and that increase drives the flame front to sonic velocity. This shock wave releases energy in the form of sound (a boom or thunderclap). Many small-scale thermal expansions within a burner flame may cause flame noise or (in extreme cases) combustion roar, which may be harmful to human ears or considered to be noise pollution. Fortunately, most industrial furnaces are well insulated, thermally and soundwise, so flame noise in not usually harmful to workers nor bothersome to neighbors. This and thermal energy conservation are good reasons to keep furnace doors and other openings closed. Burner manufacturers can usually offer less noisy burner options.

### 2.3.1. Conduction Heat Transfer

Conduction heat transfer is molecule-to-molecule transfer of vibrating energy, usually within solids. Heat transfer solely by conduction to the charged load is rare in

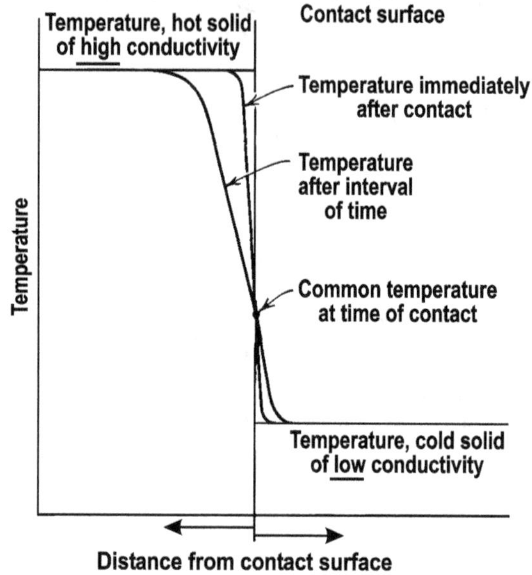


**Fig. 2.7** Effect of conductivity and time on temperature gradients in two solids of different temperatures and conductivities, in firm contact with one another.

industrial furnaces. It occurs when cold metal is laid on a hot hearth. It also occurs, for a short time, when a piece of metal is submerged in a salt bath or a bath of molten metal.

If two pieces of solid material are in thorough contact (not separated by a layer of scale, air, or other fluid), the contacting surfaces instantly assume an identical temperature somewhere between the temperatures of the contacting bodies. The temperature gradients within the contacting materials are inversely proportional to their conductivities, as indicated in figure 2.7.

The heat flux (rate of heat flow per unit area) depends not only on the temperatures of the two bodies but also on the diffusivities and configurations of the contacting bodies. In practice, comparatively little heat is transferred to (or abstracted from) a charge by conduction, except in the flow of heat from a billet to water-cooled skids (discussed in chap. 9).

When a piece of cold metal is suddenly immersed in molten salt, lead, zinc, or other molten metal, the molten liquid freezes on the surface of the cold metal, and heat is transferred by conduction only. After a very short time, the solid jacket, or frozen layer, remelts. From that time on, heat is transferred by conduction and convection. For that reason, discussion is postponed to the next section. Experimental determination of the heat transfer coefficient for heating metal solids in liquids is difficult, so practice is to record "time in bath for good results" as a function of thickness of strip or wire, as shown in section 4.7.1. on liquid bath furnaces.

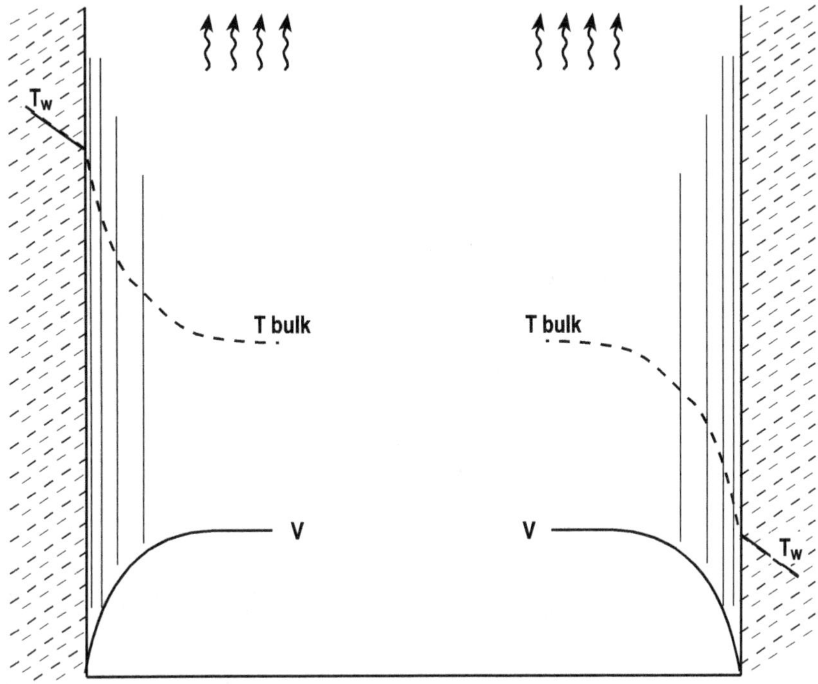


**Fig. 2.8** Convection film theory. Temperature and velocity profiles. *Left*, hot solid wall heating cooler turbulent fluid stream; *right*, Warm turbulent fluid stream heating cooler solid surface.

### 2.3.2. Convection Heat Transfer

Convection heat transfer is a combination of conduction and fluid motion, physically carrying heated (or cooled) molecules to another surface. If a stream of gaseous fluid flows parallel to the surface of the solid, as indicated in figure 2.8, the vibrating molecules of the stream transfer some thermal energy to or from the the solid surface.

A "boundary layer" of stagnant, viscous, poorly conducting fluid tends to cling to the solid surface and acts as an insulating blanket, reducing heat flow. Heat is transferred through the stagnant layers by conduction. If the main stream fluid velocity is increased, it scrubs the insulating boundary layer thinner, increasing the convection heat transfer rate. The conductance of the boundary layer ($h_c$, or film coefficient) is a function of mass velocity (momentum, Reynolds number).

For convection heat transfer with flow parallel to a plane wall,

$$Q_c/A = q = h_c(T_s - T_r) = (7.28)\,(\rho)\,(V^{0.78})(T_s - T_r) \tag{2.5}$$

where $h_c$ = convection film coefficient in Btu/ft$^2$hr°F, $\rho$ = density in lb/ft$^3$, and $V$ = velocity in ft/s.

The coefficient and exponent vary with the fluid, temperature level, and configuration. For turbulent flow, the exponent on velocity, $V$, is about 0.52 to 0.61 for flow across a single cylinder, 0.67 for flow across a bank of cylinders, 0.75 for flow parallel to a flat surface, and 0.80 for flow inside a pipe.

Figure 2.9 shows some convection (film) coefficients, $h_c$. Table 4.5 of *reference 51* lists many specific values for $h_c$.

In furnaces that operate below 1100 F, heat transfer by convection is of major importance because radiation is weak there. Modern high-velocity (high-momentum) burners give $h_c$ convection heat transfer coefficients as high as 6 Btu/ft$^2$hr°F (34 W/°Km$^2$). High velocities often provide more uniform temperature distribution around a single piece load, or among multiple piece loads, because more mass flow carries additional sensible heat at more moderate temperatures. At low furnace/oven temperatures, high rates of total heat transfer can be obtained only by high gas velocities because heat transfer by radiation at 1000 F is less than one-tenth of what it is at 2200 F. High-velocity (high momentum) burners are widely used to fill in where radiation

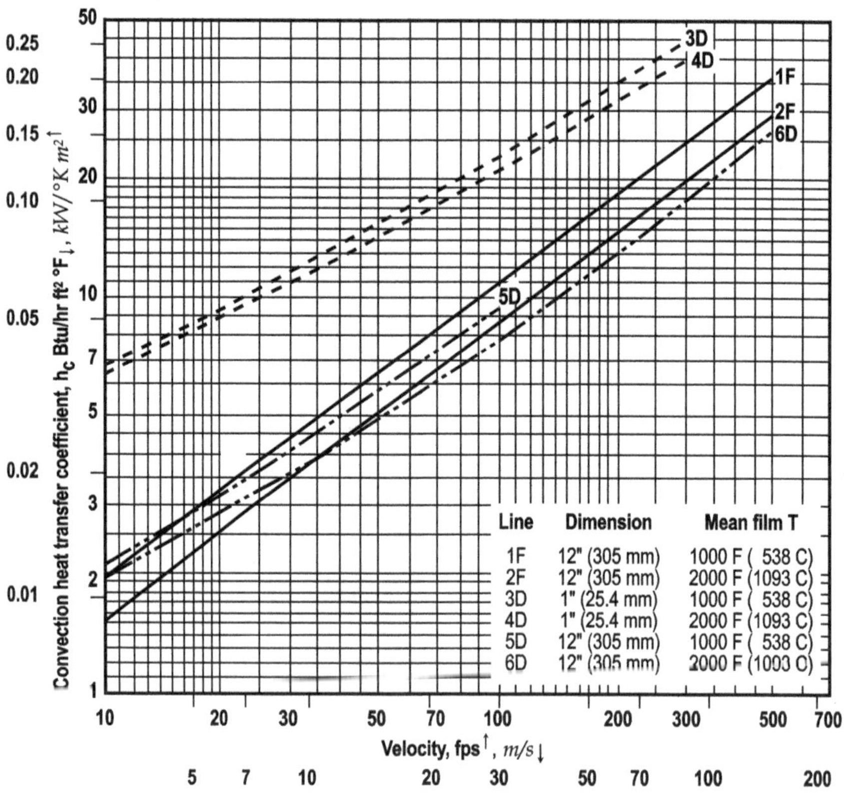


**Fig. 2.9** Convection (film) coefficients, $h_c$, for hot air or poc. F = flow parallel to a flat surface of length F; D = flow across a cylinder of diameter D. Courtesy of North American Mfg. Co. (See also table 3.2.)

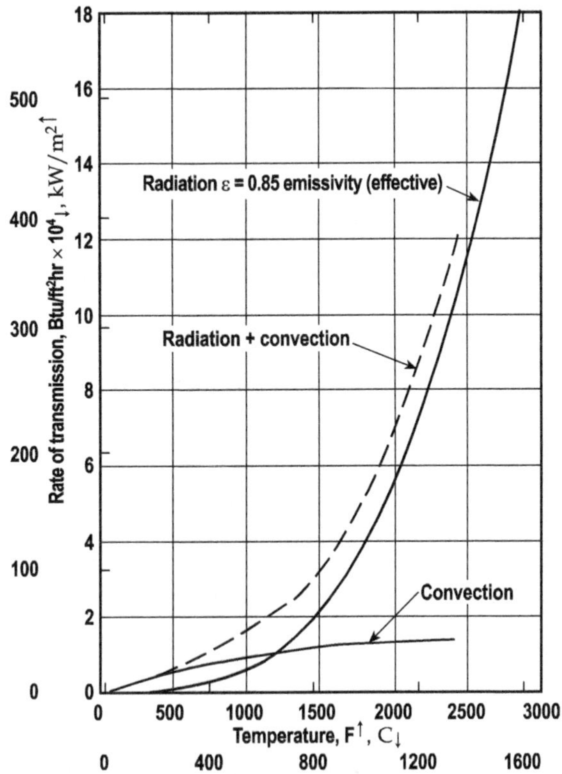


**Fig. 2.10** Comparison of relative power of radiation and convection in various temperature ranges, based on a typical emittance of 0.85. Radiation is dominant in high-temperature processes, convection in low-temperature heating. Adapted with permission from North American Mfg. Co.

cannot reach because of shadow problems. (See fig. 2.10.) This situation is discussed in the following section. Page 99 of *reference 22* analyzes radiation versus convection.

### 2.3.3. Radiation Between Solids

Solids radiate heat, even at low temperatures. The *net radiant heat* actually transferred to a receiver is the difference between radiant heat received from a source and the radiant heat re-emitted from the receiver to the source. The net radiant heat flux between a hot body (heat source) and a cooler body (heat receiver) can be calculated by any of the following Stefan-Boltzmann equations.

$$\text{Radiation heat flux} = Q_r/A = q_r, \text{ in Btu/ft}^2\text{hr} = \tag{2.6}$$

$$= 0.1713\ F_e F_a \left[(T_s/100)^4 - (T_r/100)^4\right]$$

if $T_s$ and $T_r$ are in degrees rankine.

$$\text{Radiation heat flux} = Q_r/A = q_r, \text{ in kcal/m}^2\text{h} = \tag{2.7}$$
$$= 4.876\left[(T_s/100)^4 - (T_r/100)^4\right] F_e F_a$$
if $T_s$ and $T_r$ are in degrees Kelvin, or

$$\text{Radiation heat flux} = Q_r/A = q_r, \text{ in kW/m}^2 = \tag{2.8}$$
$$0.00567\left[(T_s/100)^4 - (T_r/100)^4\right] F_e F_a$$
if $T_s$ and $T_r$ are in degrees Kelvin, or

$$\text{Radiation heat flux} = Q_r/A = q_r, \text{ in } MJ/\text{m}^2\text{h} = \tag{2.9}$$
$$0.02042\left[(T_s/100)^4 - (T_r/100)^4\right] F_e F_a$$
if $T_s$ and $T_r$ are in degrees Kelvin.

Equations 2.6 to 2.9 are correct for radiation through vacuum or transparent gases that do not absorb heat (gas mixtures that do not contain tri-atomic or heavier molecules). Table 2.2 explains the units in these equations. Table 2.3 lists $F_e$ and $F_a$ values. Figure 2.11 gives a visual study of the 4th power effect of absolute temperature on radiation heat transfer.

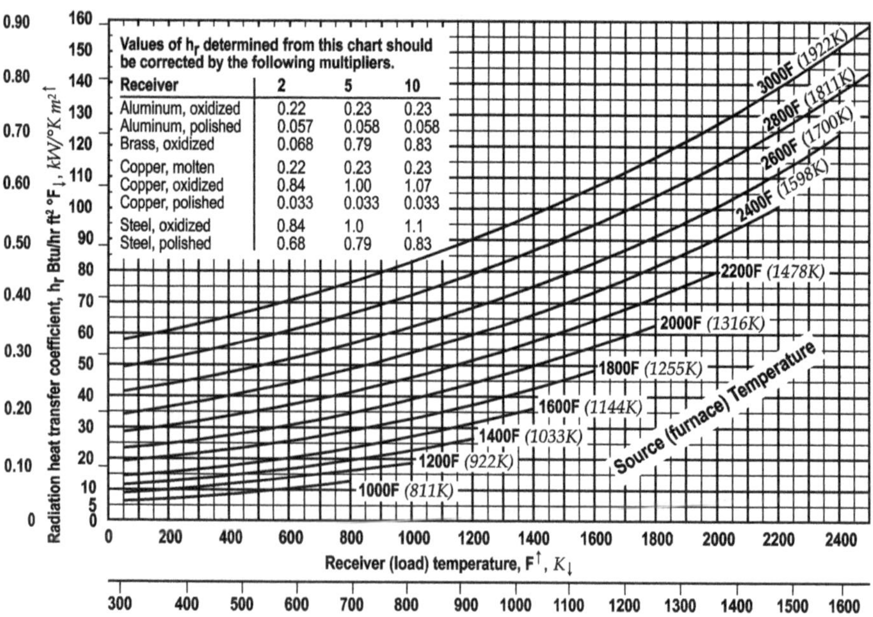


**Fig. 2.11** Radiation heat transfer coefficients from refractory wall materials (emissivity = 0.52). Multipliers (box) correct for emissivity of oxidized aluminum, copper, or steel. Column headings 2, 5, and 10 = (refractory area/metal area). Courtesy of North American Mfg. Company.

**TABLE 2.2. Heat transfer units, in order per preceding equations**

| Symbol/Explanation | US units | SI units |
|---|---|---|
| $Q$ = heat | Btu | kcal, Wh |
| $q = Q/t$ = heat flow rate | Btu/hr | kcal/h, W |
| $t$ = time | hour, hr | h |
| $A$ = area | $ft^2$ | $m^2$ |
| $q/A$ = heat flux | $Btu/ft^2hr$ | $kcal/m^2h$, $W/m^2$ |
| $F_e$ = emittance factor | | (See table 2.3) |
| $F_a$ = arrangement factor | | (see table 2.3) |
| $e = \epsilon$ = emissivity | | (1.0 is perfect, black body) |
| $T_s$ = source temperature | F or R | C or K |
| $T_r$ = receiver temperature | F or R | C or K |
| $h_c$ = convection coefficient or film coefficient | $Btu/ft^2$ hr°F | $kcal/m^2h°C$, W/°C $m^2$ |
| $h_r$ = radiation coefficient | $Btu/ft^2$ hr°F | $kcal/m^2h°C$, W/°C $m^2$ |
| $q_r$ from Equations 2.6–2.9 | $Btu/ft^2hr$ | $kcal/m^2h$, $W/m^2$ |
| $U = (h_c + h_r)$ = *overall coefficient of heat transfer for convection and radiation side-by-side in parallel* | $Btu/ft^2$ hr°F | $kcal/m^2h°C$ |
| $1/U = (1/h_c) + (1/h_r)$ = *overall coefficient of heat transfer for layered series, one after the other* | $Btu/ft^2$ hr°F | $kcal/m^2h°C$ |

The emissivities of some metals are listed in table 2.4; other materials are in *reference 51*. Values of emissivity and absorptivity of most materials are close to the same. *Emissivity* is the radiant heat emitted (radiated) by a surface, expressed as a decimal of the highest possible (black body) heat emission in a unit time and from a unit area. Emittance is the apparent emissivity of the same material for a unit area of *apparent surface* that is actually much greater, due to roughness, grooving, and so on. Absorptivity is the radiant heat absorbed by a surface per unit time and unit area, expressed as a decimal of the most possible (black body) heat absorption.

Engineers have used $F_e = 0.85$ in conventional refractory furnaces, but table 2.4 shows that temperature, surface condition, and alloy can make considerable difference. If stainless-steel strip is heated in less than three min. in a catenary furnace, the emissivity may not change even though the temperature increases from ambient to 2000 F. By measuring both strip surface temperature and furnace temperature, it has been possible to revise heating curve calculations, assuming that oxidation has not changed the emissivity nor absorptivity during the heating cycle.

Tables 2.3 and 2.4 can be used to determine values of $h_r$ for practical furnace situations. These can be compared directly with $h_c$ from figure 2.9 or table 3.2. The $h_r$ and $h_c$ can be added together as specified in the last four lines of table 2.2.

Even when $T_s$ and $T_r$ are not far apart, the difference between the fourth powers of temperature is very large. This is shown by the top right (elevated temperature) portion of figure 2.16, where even small temperature differences result in high heat transfer rates. For instance, 1°F temperature difference at 2200 F causes about 5.5 times as much heat transfer as 1°F temperature difference causes at 1000 F. The

**TABLE 2.3. Emittance factors $F_e$ for various configurations, applicable with equations 2.6 to 2.9 and where radiation is through a vacuum or through transparent gases that do not absorb heat (gas mixtures that do not contain triatomic or heavier molecules).**

| Configuration | Factor $F_e^*$ |
|---|---|
| Surface with emittance $e_1$ surrounded by a larger surface with emittance $e_2$. | $e_1$ |
| Surface with emittance $e_1$ surrounded by a smaller surface with emittance $e_2$. | $\frac{1}{(1/e_1) + (1/e_2) - 1}$ |
| Parallel planes with emittances $e_1$ and $e_2$ and with the space between the planes much smaller than either plane. | $\frac{1}{(1/e_1) + (1/e_2) - 1}$ |
| Concentric spheres or long cylinders, With the ratio of surface areas of inner to outer sphere or cylinder being $(S_1/S_2)$ and with inner surface emittance of $e_1$ and outer surface emittance of $e_2$. | With mirror reflection: $\frac{1}{(1/e_1) + (1/e_2) - 1}$ With diffuse reflection: $\frac{1}{(1/e_1) + (S_1/S_2)(1/e_2) - 1}$ |

*Factors for finding radiation per unit area of the smaller surface, $S_1$. The arrangement (or configuration) factor, $F_a$, for all the above is 1.0. For other shape factors, see *reference 74.*

coefficient of heat transfer by radiation, $h_r$, in Btu/ft$^2$hr°F, varies widely with the temperatures of the heat exchanging source and receiver. This $h_r$ = (Eq. 2.6 to 2.9) divided by $(T_s - T_r)$ can be used in equation 2.10.

$$Q_r/A = q_r = h_r(T_s - T_r). \tag{2.10}$$

(For appropriate units, see eqs. 2.6 to 2.9.)

The extent to which this radiation heat transfer coefficient varies is readily seen from the nest of curves in figure 2.11, where the coefficient appears as ordinate while the heat exchanging temperatures appear as abscissae and curve parameter labels. The heat transfer coefficients in figure 2.11 are for black body radiation, so they must be multiplied by an emittance factor, $F_e$, and by an arrangement factor, $F_a$, from table 2.3. Tables 4.6, 4.7, and 4.8 of *reference 51* list many emittances.

**Example 2.2**: Oxidized copper 3" × 3" billets are being heated in an electrically heated furnace that has an average heat source temperature of 1600 F. The refractory area is five times the exposed metal area. The loading arrangement is such that the equivalent exposure to furnace radiation is only 6 in. of the 12" periphery of each billet. The billet weight is 34.9 lb/ft of length.

*a.* What is the rate of heat transfer to the billets when their surface temperature has reached 1400 F? *b.* How fast will the billet temperature rise?

**Solution a**. The heat absorbing surface for each foot of length is one-half of the 1 ft$^2$ surface per foot of length = 0.5 ft$^2$/ft. From figure 2.11, the coefficient of radiant

**TABLE 2.4. Total hemispheric emittances (and absorptances) of metals and their oxides, selected from *references 42, 51,* and *70.* Emittances of refractories and miscellaneous nonmetals are listed in chapter 4 of *reference 51.***

| Metal, condition | Temp F/C | Emittance |
|---|---|---|
| **Aluminum,** polished | 71/23 | 0.04 |
| | 1067/575 | 0.057 |
| oxidized at 1110 F | 392/200 | 0.110 |
| | 1112/600 | 0.19 |
| molten, clean skimmed | | 0.12–0.33 |
| alloy 1100-0 | 200-800/93-427 | 0.05 |
| alloy A3003 Oxidized | 600-900/316-482 | 0.4 |
| alloy 6061-T6, chemically cleaned, rolled | 140/60 | 0.07 |
| alloy 6061-T6, forged | 140/60 | 0.10 |
| alloy 7075-T6, polished | 980/527 | 0.14 |
| **Brass,** oxidized | 372/200 | 0.61 |
| | 1112/600 | 5.59 |
| **Cadmium** | 77/25 | 0.02 |
| **Chromium,** polished | 100/38 | 0.08 |
| | 1000/538 | 0.26 |
| **Copper,** polished | 212/100 | 0.05 |
| oxidized | 536/280 | 0.5 |
| | 1400/760 | 0.855 |
| molten | 1970/1077 | 0.16 |
| | 2330/1279 | 0.13 |
| **Iron,** oxidized | 390/200 | 0.64 |
| | 1110/600 | 0.78 |
| (see also steel) | 1700/927 | 0.87 |
| | 2040/1116 | 0.95 |
| molten | 2550/1400 | 0.29 |
| **Lead,** polished | 260/127 | 0.056 |
| oxidized | 392/200 | 0.63 |
| **Magnesium** | 500/268 | 0.13 |
| oxide | 1880/1027 | 0.16 |
| **Molybdenum** | 1000/538 | 0.82 |
| **Monel,** oxidized | 1110/600 | 0.46 |
| **Nickel,** oxidized | 392/200 | 0.37 |
| | 1112/600 | 0.48 |
| | 2000/1093 | 0.86 |
| Inconel X-750, buffed | 140/60 | 0.16 |
| Inconel X-750, oxidized | 600/316 | 0.69 |
| Inconel X-750, oxidized | 1800/982 | 0.82 |
| Inconel B, polished | 75/24 | 0.21 |
| Inconel sheet | 1400/760 | 0.58 |
| Haynes alloy C, oxidized | 600/316 | 0.9 |
| | 2000/1093 | 0.96 |
| Haynes alloy 25, oxidized | 600/316 | 0.86 |
| | 2000/1093 | 0.89 |
| Haynes alloy X, oxidized | 600/316 | 0.85 |
| | 2000/1093 | 0.88 |
| **Platinum,** oxidized | 500/260 | 0.07 |
| | 1000/538 | 0.11 |
| **Steel,** mild, oxidized | 77/25 | 0.8 |
| | 1112/600 | 0.79 |
| c, molten | 2910/1600 | 0.28 |
| c, plate, rough | 104/40 | 0.94 |
| | 752/400 | 0.97 |
| 304A stainless, balck oxide | 80/27 | 0.3 |
| 304A, stainless, machined | 1000/538 | 0.15 |
| 304A, stainless, machined | 2140/1444 | 0.73 |
| 310 stainless, oxidized | 980/527 | 0.97 |
| 316 stainless, polished | 450/232 | 0.26 |
| 316 stainless, oxidized | 1600/871 | 0.66 |
| 321 stainless, polished | 1500/816 | 0.49 |
| 347 stainless, grit blasted | 140/60 | 0.47 |
| 347 stainless, oxidized | 600/316 | 0.88 |
| 347 stainless, oxidized | 2000/1367 | 0.92 |
| **Tin,** commercial plated | 212/100 | 0.08 |
| **Titanium,** polished | 60/16 | 0.12 |
| | 1900/1038 | 0.24 |
| oxidized | 60/16 | 0.18 |
| oxidized gray | 1040/560 | 0.55 |
| alloy A-110A7, polished | 225/107 | 0.18 |
| alloy A-110A7, polished | 1400/760 | 0.46 |
| alloy A-110A7, oxidized | 225/107 | 0.17 |
| alloy A-110A7, oxidized | 1375/746 | 0.63 |
| alloy C110M, oxidized | 800/427 | 0.61 |
| alloy Ti-95A, oxidized | 800/427 | 0.48 |
| **Tungsten,** filament, aged | 5000/2760 | 0.35 |
| **Uranium** oxide | 1880/982 | 0.79 |
| **Zinc,** commercial 99.1% | 500/260 | 0.05 |
| oxidized | 1000/538 | 0.11 |
| galvanized sheet | 100/38 | 0.28 |

heat transfer, $h_r$, is found to be $36 \times 1.0 = 36$ Btu/ft$^2$hr°F. Therefore, the transfered radiation $= Q_r = h_r A(T_s - T_r) = 36 \times 0.5 \times (1600 - 1400) = 3600$ Btu/hr ft of length.

**Solution b**: From *reference 52*, table A16, the specific heat of copper is 0.095 Btu/lb°F, and the density is 559 lb/ft$^3$. The weight of copper per foot of length is therefore (559 lb/ft$^3$) × (3/12) (3/12) (12/12) = 34.9 lb per lineal foot. The heat transferred per hour to each lineal foot, from solution *a*, divided by the heat absorbed per degree temperature rise and per lineal foot will give the degrees rise per unit time:

$$\frac{(3600 \text{ Btu/hr ft of length})}{(0.095 \text{ Btu/lb°F})\ (34.9 \text{ lb/ft of length})} = 1086\text{°F/hr, or } 18.1\text{°F/min.}$$

The emittance factors in tables 2.3 and 2.4, and in figure 2.11 do not include triatomic gas radiation and absorption, which leads to the next section.

### 2.3.4. Radiation from Clear Flames and Gases

There are two origins of radiation from products of combustion to solids: (1) radiation from clear flame and from gases and (2) radiation from the micron-sized soot particles in luminous flame.

Radiation from clear gas does not follow the Stefan-Boltzmann fourth-power law. The only clear gases that emit or absorb radiation appreciably are those having three or more atoms per molecule (triatomic gases) such as $CO_2$, $H_2O$, and $SO_2$. An exception is diatomic carbon monoxide (CO), which gives off less radiation. The other diatomic gases, such as $O_2$, $N_2$ (and their mixture, air), and $H_2$ have only negligible radiating power.

Gaseous radiation does not follow the 4th-power law because gases do not radiate in all wavelengths, as do solids (gray bodies). Each gas radiates only in a few narrow bands, as can be seen on a spectrograph in figures 2.17 and 2.18.

In figure 2.12, the whole area under each curve represents black body radiation from solid surfaces (per Planck's Law). Two shaded bars show the narrow radiating bands for carbon dioxide gas. Similar but shorter bands for the other common triatomic gas, $H_2O$, are shown in figures 2.17 and 2.18.

Radiation from clear gases depends on their temperature, on the partial pressure or %volume of each triatomic gas present, and on the thickness of their gas layer. Calculation of the heat transfer from radiating clear gases to solids is possible by use of figures 2.13 and 2.14, derived from data in *reference 42* and corrected for each triatomic gas being slightly opaque to radiation from the other, and for 0.9 receiver surface absorptivity. The curve labels are the arithmetic mean of bulk gas and solid receiver surface temperatures. The coefficients of radiant gas heat transfer from figures 2.13 and 2.14 should not be used for temperature differences greater than 500°F (278°C). No correction need be made for the peculiar behavior of water vapor if the mean temperature is above 1200 F (649 C). To calculate the heat flux rate in Btu/ft$^2$hr, multiply $h_{gr}$ (the reading from the vertical scale) by $F_a$ and by the $\Delta T$ between gas source and solid receiver surface, as in equation 2.11.

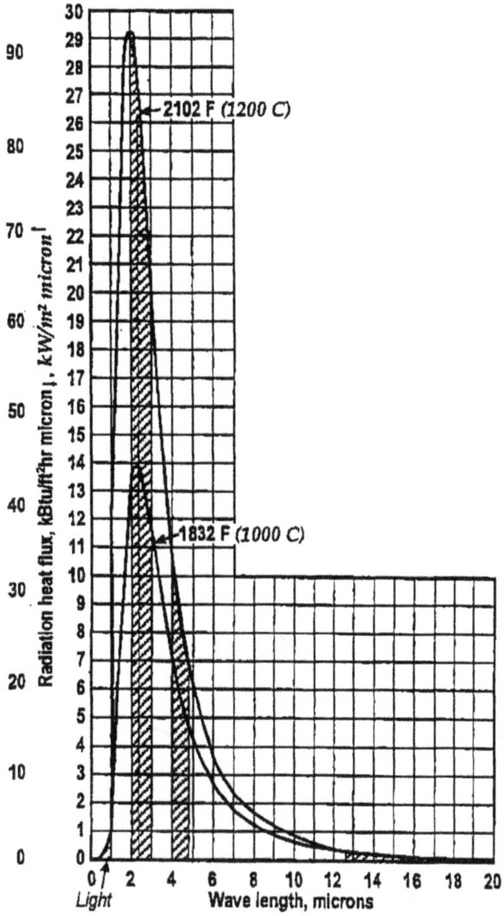


**Fig. 2.12** Comparison of radiation intensity of a "black body" solid at two selected temperatures. Superimposed on this plot are two shaded bands of carbon dioxide gas radiation and a small corner of a band for sunlight. (See also fig. 2.18.)

$$q_{gr} = Q_{gr}/A = (h_{gr} \text{ or } F_e)(F_a)(T_g - T_r) \tag{2.11}$$

wherein $gr$ = gas radiation, $g$ = gas (source), and $r$ = receiver. For a cloud of radiating gas, $F_a$ can be assumed equal to 1.0.

**Example 2.3**: A reverberatory batch melting furnace, fired with natural gas, has a 36" high gas blanket between the molten bath surface and the furnace roof. The absorptivity of the 1500 F molten bath surface is estimated to be 0.3.* When the poc are at 2000 F, calculate the radiant heat flux from the poc gases to the load.

*Absorptivities (usually close to the same as emissivities, from *reference 51*) are typically 0.9 for clean refractory or rough iron or steel, or 0.7 for glazed refractory.

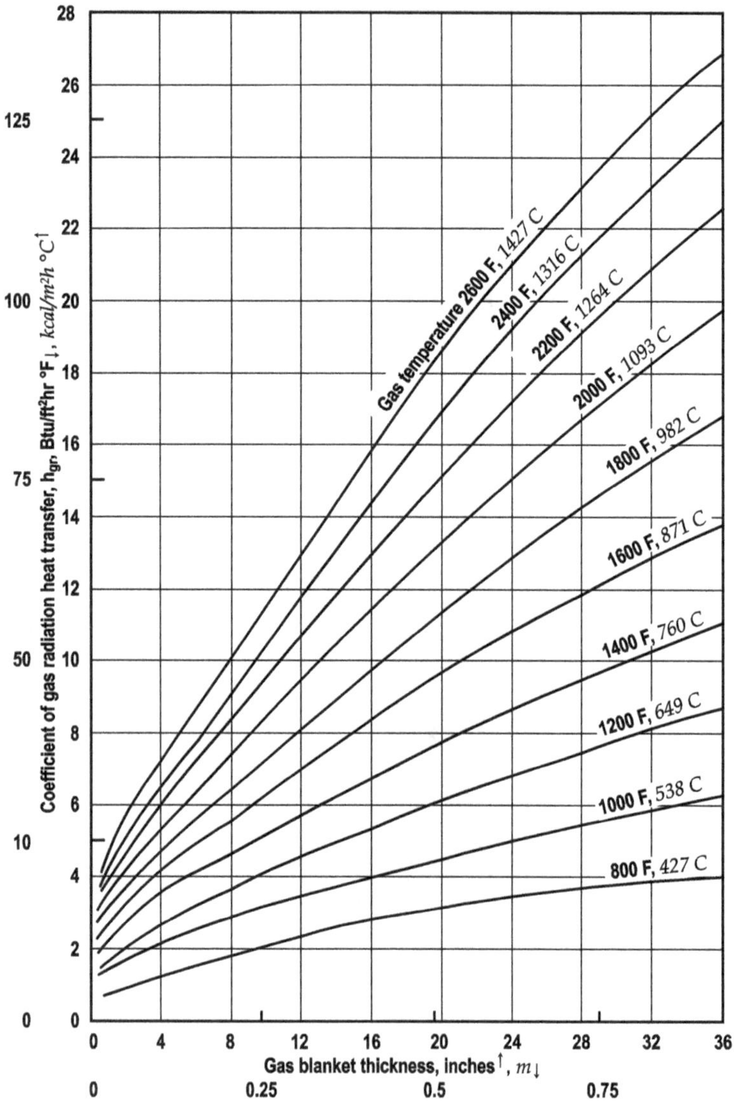


**Fig. 2.13** Triatomic gas radiation heat transfer coefficients for 1 to 36 in. (0.3–0.9 m) thick gas blankets with poc having 12% $CO_2$ and 12% $H_2O$ (products of a typical natural gas with 10% excess air) at average gas temperatures [(surface + gas)/2] of 1400 F to 2400 F (760–1316 C). (Continues on fig. 2.14.)

From figure 2.13, for a 2000 F source temperature, read $h_{gr} = 19.5$ Btu/ft$^2$hr°F. By equation 2.11, $q_{gr} = 19.5\ (0.3)\ (2000 - 1500) = 2925$ Btu/hr ft$^2$. Measuring or estimating temperatures in a high-temperature stream of poc is difficult. (See sec. 2.4 and 5.1.) In contrast to convection formulas, radiation formulas contain no velocity factors. However, velocity of radiating gases is important because hot gases cool in

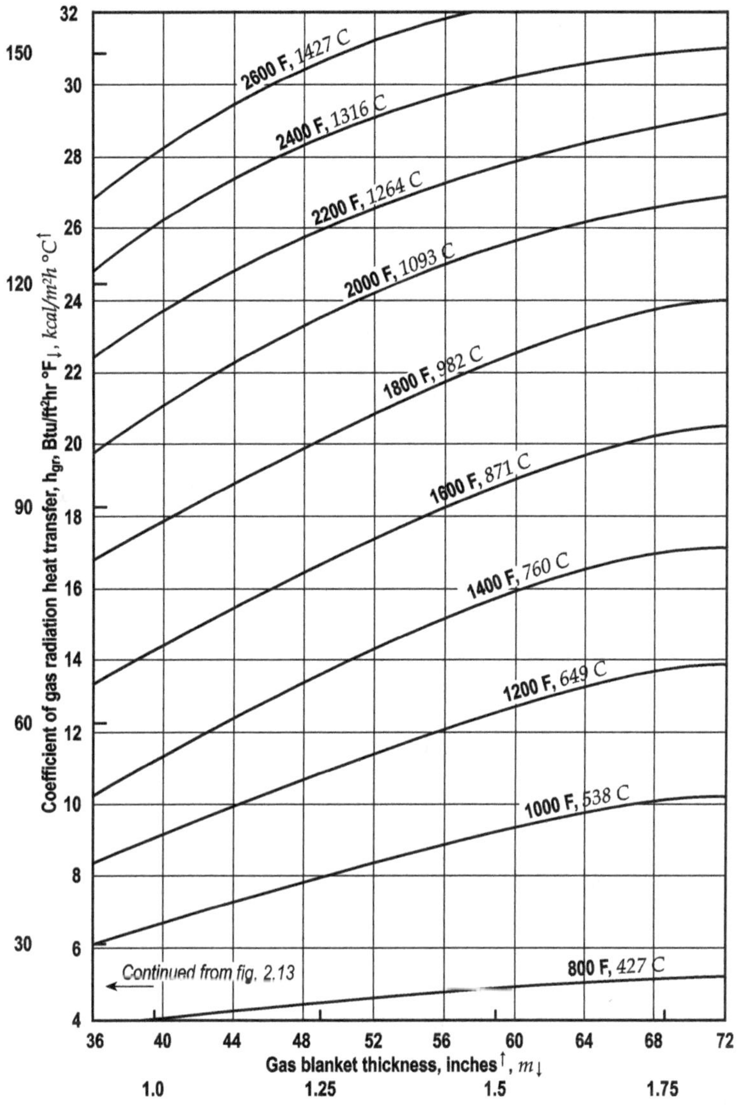


**Fig. 2.14** Triatomic gas radiation heat transfer coefficients for 36 to 72 in. (0.91–1.83 m) thick gas blankets with poc having 12% $CO_2$ and 12% $H_2O$. The data of figs. 2.13 and 2.14 are for gas blankets of 12% $CO_2$ and 12% $H_2O$, but most natural gases produce about 12 $CO_2$ and 18% $H_2O$, so the actual radiation will be somewhat higher. (Continued from fig. 2.13.)

the process of radiating to colder surfaces (walls and loads). The temperature of a radiating gas gets lower in the direction of gas travel. To maintain active gas radiation, the gas must be continually replaced by new hot gas, which also improves convection. Higher gas feed velocities reduce the temperature drop along the gas path. This book shows how critical this factor is to maintaining good temperature uniformity in high-temperature industrial furnaces.

Furnace builders have generally designed furnaces on the basis of refractory radiation heating the load, with usually reasonable results, but some situations cannot be explained by refractory radiation alone.

Author Trinks' early editions made it clear that direct radiation from furnace gases delivered 62% (±2%) of the heat to the load, and refractories transferred the remaining 38% (±2%). His calculations (*reference 83*) showed that gas temperatures required to transfer the heat to refractory and load are generally much higher than assumed. Engineers are encouraged to continue use of the familiar refractory furnace calculations, but to use gas radiation calculations as a "go/no go" gauge to check on the results. Coauthors Shannon and Reed believe that future furnace designers will calculate combined gaseous and refractory heat transfer rates as soon as sufficient experimental data become available.* Accuracy may then be improved by using a dynamic three-dimensional computer iteration of the 4th power effect over the actual range of varying poc temperatures.

**Example 2.4**: A proposed natural-gas-fired furnace will need a heat transfer coefficient of 16 Btu/ft$^2$hr°F. (a) Determine the needed mean furnace gas temperatures with 18", 36", 54", and 72" heights of the furnace ceiling above the tops of the load pieces (gas blanket thicknesses). (b) Compare probable NOx emissions.

From figures 2.13 and 2.14, read the second line of the following table:

| Gas thickness, "/m | 18" 0.46 m | 36" 0.91 m | 54" 1.8 m | 72" 1.8 m |
|---|---|---|---|---|
| Mean furnace gas T, F/C | 2440 F 1340 C | 1760 F 960 C | 1480 F 805 C | 1340 F 721 C |
| NOx emissions | Very high | High | Medium | Lower |

Figure 2.16 compares magnitudes of gas-to-load radiation and gas-to-refractory-to-load radiation for a specific furnace/flame configuration.

A study of a 7' (2.13 m) high steel reheat furnace versus a 9' (2.74 m) high similar furnace (using the Shannon Method explained in chap. 8) showed that the 7' furnace required a higher average gas temperature than the 9' to heat the same load at the same rate—because of its shorter gas beam height.

### 2.3.5. Radiation from Luminous Flames

If a fuel-rich portion of an air/fuel mixture is exposed to heat, as from a hotter part of the flame, the unburned fuel molecules polymerize or suffer thermal cracking, resulting in formation of some heavy, solid molecules. These soot particles glow when hot, providing luminosity, which boosts the flame's total radiating ability.

This can be witnessed in a candle flame by immersing a cold dinner fork or piece of screenwire in the yellow part of the flame. It will quench the flame and collect soot. Without it, however, enough oxygen will eventually be mixed with the wax vapor to complete combustion of the soot.

*Suggested research project, described at the end of this chapter. No convection, conduction, or particulate radiation are included in Shannon Method calculations for steel reheat furnaces.

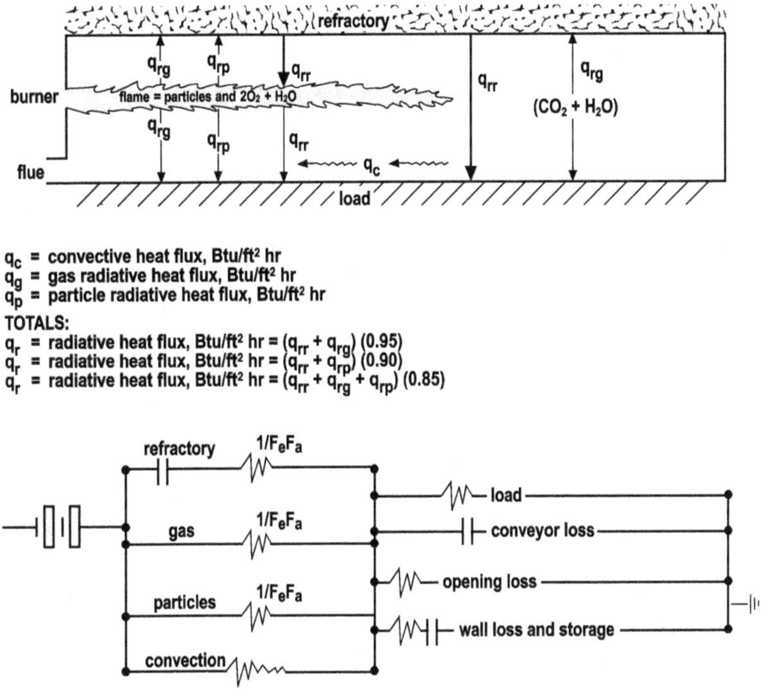


**Fig. 2.15** Combining of concurrent heating modes in a refractory-lined furnace, kiln, incinerator, or cpi heater, with suggested formulas and electrical analogy.

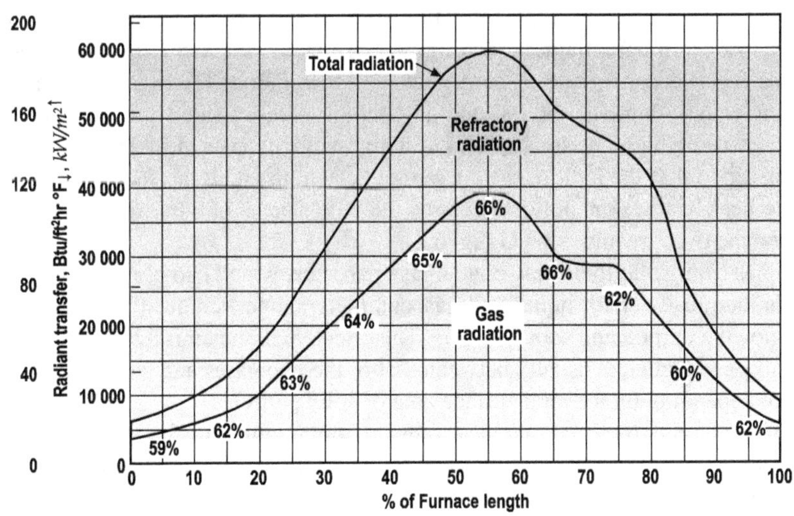


**Fig. 2.16** Comparison of direct gas radiation from gases *to load (lower curve)* with radiation from gases *to refractory to load (gray area between curves).* At the peaks, 66% is direct gas radiation and the remaining 34% is gas radiation to rofractory that is then re-radiated to the load. (See also fig. 5.5.)

It is possible to prevent the polymerization by aerating the lower part of a candle flame by blowing through a thin cocktail straw, thus converting the entire candle flame to blue flame (no soot, less total radiation, higher poc temperature immediately beyond the flame tip). (See *reference 19*, "The Chemical History of a Candle," by Michael Faraday, 1861.)

Let us now switch from the candle analogy to a real-world burner. If fuel and air are not thoroughly mixed promptly after they leave the burner nozzle, they may be heated to a temperature at which the hydrocarbons crack (polymerize). Further heating brings the resulting particles to a glowing temperature. As oxygen mixes with them, they burn. As the flame proceeds, formation of new soot particles may equal the rate of combustion of previously formed particles. Farther along the flame length, soot production diminishes, and all remaining soot is incinerated. This series of delayed-mixing combustion processes should be complete before the combustion gases pass into the vent or flue. If the flame were still luminous at the flue entry, smoke might appear at the stack exit. (Smoke is soot that has been cooled [chilled, quenched] below its minimum ignition temperature before being mixed with adequate air.)

The added radiating capability of luminous flames causes them to naturally cool themselves faster than clear flames. This is performing their purpose—delivering heat. The cooling phenomenon might negate some of the gain from the higher luminosity (effective emissivity).

Luminous flames often have been chosen because the added length of the delayed-mixing luminous flames can produce a more even temperature distribution throughout large combustion chambers. As industrial furnaces are supplied with very high combustion air preheat or more oxy-fuel firing, luminous flames may enable increases in heat release rates.

Fuels with high carbon/hydrogen ratios (most oils and solid fuels) are more likely to burn with luminous flames. (See fig. 2.17.) Fuels with low C/H ratios (mostly gaseous fuels) can be made to burn with luminous flames (1) by delayed mixing, injecting equally low-velocity air and gas streams side-by-side (type F, in fig. 6.2), and (2) by using high pressure to "shoot" a high-velocity core of fuel through slower moving air so that the bulk of the air cannot "catch up" with the fuel until after the fuel has been heated (and polymerized) by the thin 'sleeve' of flame annular interface between the two streams (type G, fig. 6.2).

Flames from solid fuels may contain ash particles, which can glow, adding to the flame's luminosity. With liquid and gaseous fuels, flame luminosity usually comes from glowing carbon and soot particles. The effective flame emissivity, as measured by Trinks and Keller, is usually between that of the poc gases and a maximum value of 0.95, depending on the total surface area of solid particles.

It is common experience that heat transfer from a luminous flame is greater than that from a clear flame having the same temperature. The difference in the rate of heat transfer is quite noticeable in furnaces for reheating steel and for melting glass or metals. The difference becomes more pronounced at high temperature, where the radiating power of each triatomic gas molecule increases, but the gain is partially canceled by the decreasing density of radiating molecules per unit volume.

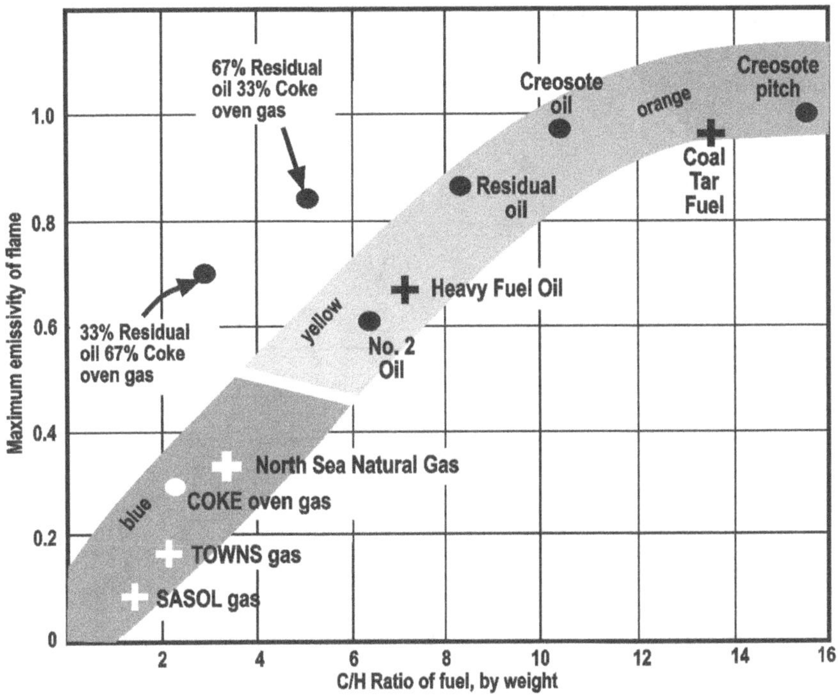


**Fig. 2.17** Effect of fuel C/H ratio on flame emissivity. (From *reference 78b* and *reference 85.*)

In another phenomenon, the bands of gaseous radiation (fig. 2.18) hold their wavelengths regardless of temperature. At higher temperatures, however, the area of high intensity of solid radiation (glowing soot and carbon particles) moves toward shorter wavelengths (away from the gas bands). In higher temperature realms, radiation from clear gases does not increase as rapidly as radiation from luminous flames.

Flame radiation is a function of many variables: C/H ratio of the fuel, air/fuel ratio, air and fuel temperatures, mixing and atomization of the fuel, and thickness of the flame—some of which may change with distance from the burner. Fuels with higher C/H ratio, such as oils, tend to make more soot, so they usually create luminous flames, although blue flames are possible with light oils. Many gases have a low C/H ratio, and tend to burn clear or blue. It is difficult to burn tar without luminosity. It is equally difficult to produce a visible flame with blast furnace gas or with hydrogen.

Sherman's data on flame radiation (*reference 80*) give peak values of 200 000 Btu/ft$^2$hr for flames from tar pitch or residual oil, but the radiation from the average for the whole flame length may be half as much. When comparing luminous and nonluminous flames, it is important to remember (a) Soot radiation (luminous) usually ends where visible flame ends because soot is most often incinerated at the outer "surface" or "skin" of the flame, where it meets secondary or tertiary air; and (b) gas radiation (nonluminous) occurs from both inside and outside the visible flame

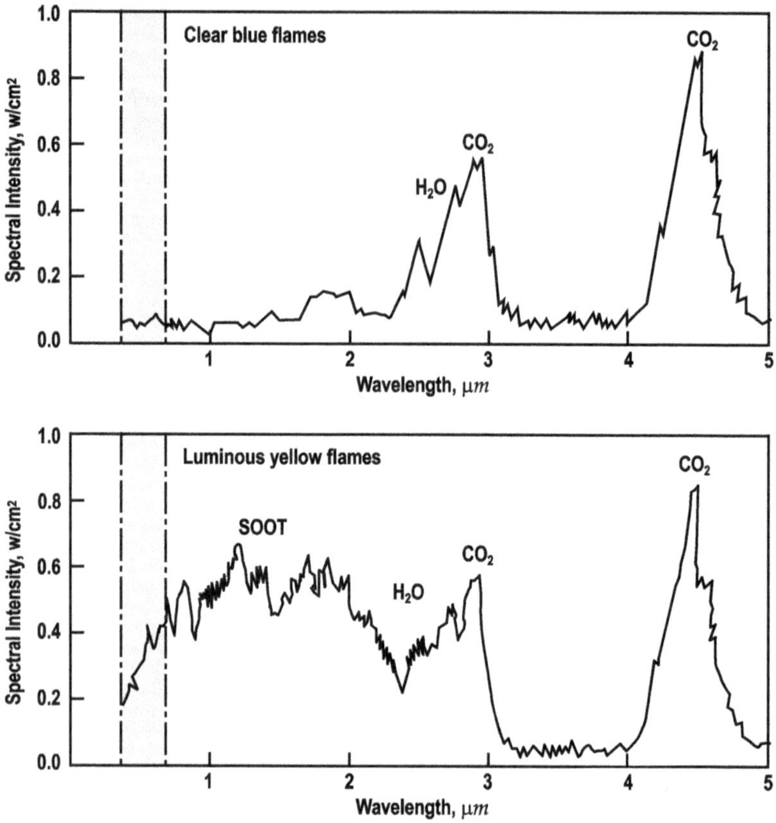


**Fig. 2.18** Spectographs of radiation from clear and luminous flames. Nonluminous flames (*top graph*) are blue; luminous flames (*lower graph*) are yellow and emit soot particle radiation. Both luminous and nonluminous flames and invisible poc gases emit triatomic gas radiation. Courtesy of *Ceramic Industry* journal, Feb. 1994, and Air Products & Chemicals, Inc. (*reference 13*).

envelope, greatly increasing the uniformity and extent of its coverage, although gas radiation within the flame is somewhat shadowed by any surrounding soot particles or triatomic gases, and gas radiation outside the flame may be from cooler gases.

The effect of excess fuel on flame radiation is considerably greater than the effect of less excess air. The effects of fuel-air mixing on luminosity, and the means for adjusting the mixture, are discussed in *reference 52*.

The merits and debits of clear flames versus long luminous flames have been debated by engineers for years. Modified burners and control schemes are helping to utilize the best of both. A problem common to many burner types is change of the flame characteristic as the burner input is turned down.

Problems with some clear flame burners are (1) movement of the hump in the temperature profile closer to the burner wall as the firing rate is reduced and (2) at lower input rates, temperature falls off more steeply at greater distances from the burner wall (e.g., the temperature profile of a burner firing at 50% of its rated capacity

Trinks' and Mawhinney's 5th Edition mentions heating more load per unit of hearth area "by alternating short-flame and long-flame burners." Prior to that, one of Professor Trinks' countrymen, Dipl. Ing. Otto Lutherer, Chief Engineer of North American Mfg. Co., dreamed of being able to increase the heat flux to a furnace load by alternating luminous and clear flames in furnaces.

Mr. Lutherer reasoned that the opaque soot particles in luminous flames would increase radiation to furnace loads and refractory crown, and that if clear flames then momentarily replaced them, that would allow the refractory to radiate to the load and "dump" its accumulated high-thermal-head heat on the load.

Otto must be smiling now, with the development of adjustable thermal profile flames and of 20-sec-on and 20-sec-off regenerative burner flames, both of which fulfill his dream as well as Prof. Trinks' and Matt Mawhinney's idea of alternating flame patterns (with respect to time) for better overall transfer.

or below is at its peak temperature [maximum heat release] at or near the burner wall, falling off further from the burner wall). At lower firing rates, the temperature drop-off gets worse. At higher firing rates, the burner wall temperature decreases as the peak temperature moves away from it. In some steel reheat furnaces at maximum firing rate, the temperature difference between the burner wall and the peak may be 300°F (170°C).

The problem of a temperature peak at the far wall during high fire is exacerbated by inspiration of furnace gases into the base of the flame, delaying mixing of fuel with oxygen. If the burner firing rate is increased, the inspiration of products of complete combustion increases exponentially. Resulting problems are many. When side-firing a furnace at low firing rate, the peak temperature is at the burner wall, but at maximum firing rate, the peak temperature may be at the furnace center or the opposite wall. Thus, the location of a single temperature control sensor is never correct.

If the temperature sensor were in the burner wall, low firing rates would have peak temperature hugging the furnace wall and driving the burner to low fire rate; thus, the rest of the furnace width would receive inadequate input. At high firing rates, a sensor in the burner wall will be cool while the temperature away from the burner wall would be very high, perhaps forming molten scale on the surfaces of the load pieces at the center and/or far wall. To remedy this problem, inexperienced operators may lower the set point, reducing the furnace heating capacity.

Another example of the effect of the problem occurs with the bottom zone of a steel reheat furnace when fired longitudinally counterflow to the load movement, and with the control sensor installed 10 to 20 ft (3–6 m) from the (end-fired) burner wall. At low-firing rates, with the zone temperature set at 2400 F (1316 C), the burner wall may rise to more than 2500 F (1371 C). At that temperature, scale melts and drips to the floor of the bottom zone where it may later solidify as one big piece. At high firing rates, the peak temperature may move beyond the bottom zone T-sensor,

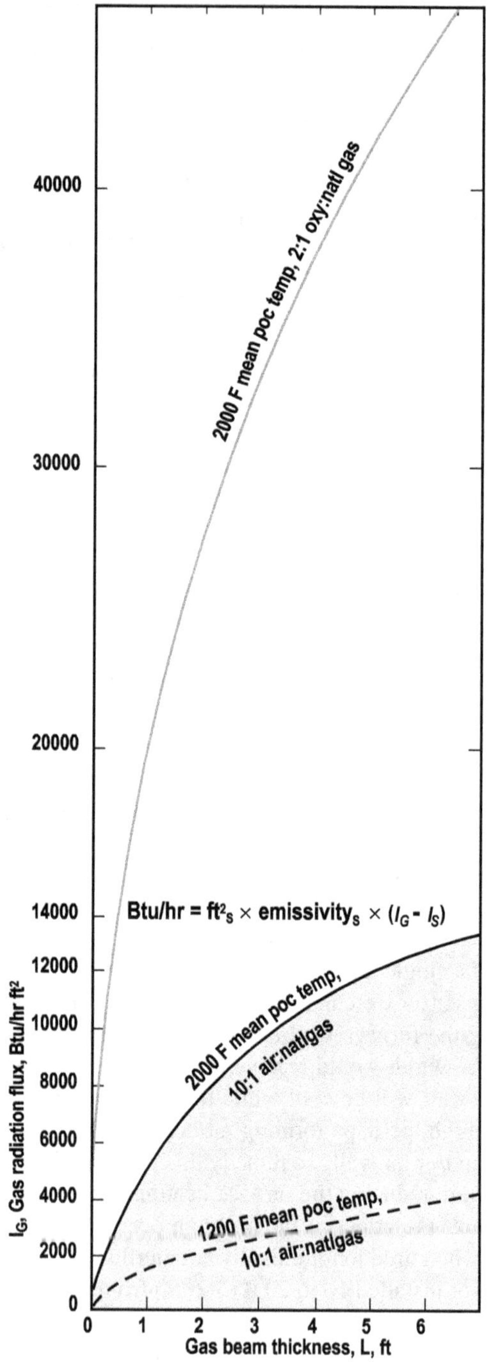


**Fig. 2.19** Comparisons of gas radiation intensity for three situations. A three-fold increase with oxy-fuel firing is caused of elimination of diluting $N_2$.

possibly melting scale some distance toward the charge end of the furnace. Again, to avoid the problem, operators may lower temperature control settings, reducing the furnace capacity.

Control of the aforementioned problems requires an additional temperature sensor in each zone and a means for changing the mixing rate characteristic of the burner in response to the temperature measurements. Burners with adjustable spin (swirl) can be set to prevent much of the problem, especially if combined with a low-fire, forward-flow gas or air jet through the center of the burner. Such a jet is typically sized for 5% of maximum gas or air flow.

Long, luminous flames, either laminar type F or turbulent type G (fig. 6.2), tend to have much less temperature hump and do not change length as rapidly when input is reduced. They can be great "levelers," providing better temperature uniformity. The change from air-directed to fuel-directed burners, using 5 to 15 psi (35–105 kPa) natural gas, usually available at no extra cost, has solved many nonuniformity problems.

This information on in-flame soot radiation and triatomic gas radiation has been known for some time, but recent developments may be changing the picture:

(a) Use of oxy-fuel (100% oxygen), both of which elevate flame turndown (see fig. 2.19). The major gain from oxy-fuel firing is from more intense radiation heat transfer because of the higher concentration of triatomic gases, due to the elimination of nitrogen from the poc. This also decreases the mass of gas carrying heat out the flue (reducing stack loss).

(b) Some lean premix gas flames (designed for low NOx emissions) make a ubiquitous flame field (seemingly transparent) through much of the chamber (see "flameless combustion" in the glossary).

## 2.4. DETERMINING FURNACE GAS EXIT TEMPERATURE

Improving energy use in furnaces requires knowledge of the flue gas exit temperature. Many studies and articles oversimplify the measurement of furnace gas exit temperature or simply assume it to be the temperature of the furnace (refractory wall) at the flue entry—neither of which is correct.

Measurement of flue gas exit temperature is difficult because the radiation rates to a measuring device are greater from solids than from the gases, the temperature of which is to be measured. Accurate measurement of poc gas temperature requires: (1) a low mass sensor with multiple radiation shields, and (2) a suction device to induce a high sample gas velocity over the sensor. The velocity should be increased until no higher signal can be detected. A practical rule of thumb has been that the velocity energy source should be capable of accelerating the flue gas across the temperature sensor to 500 fps (152 m/s). Table 2.5 shows that to fill only a single 0.5" ID (13 mm ID) radiation shield with this rule-of-thumb velocity would require pump suction and flow rates necessitating a cumbersome suction pumping apparatus.

**TABLE 2.5. Pumping requirements for 500 fps (152 m/s) sample gas velocity**

| Estimated sample flue gas temperature | Required suction pressure drop* | Required or volume flow rate |
|---|---|---|
| 1000 F = 538 C | 53"wc = 1270 mm | 40.9 cfm = 69.5 $m^3$/h |
| 1500 F = 816 C | 40"wc = 1016 mm | 40.9 cfm = 69.5 $m^3$/h |

*static pressure (sp) measured in water column height on a manometer.

Because actual measurement of the flue gas temperature may be difficult, an estimated or calculated gas temperature is often used. Our peers have been estimating flue gas exit temperature as either (Guess #1) the furnace temperature, or (Guess #2) the furnace temperature plus 200°F or plus 111°C (Celsius). Guess #1 violates the fact that heat flows from a high-temperature source to a low-temperature receiver, and therefore makes the unlikely assumption that the poc path through the furnace has been so long that the gases have cooled to the furnace wall temperature, in which case they would no longer transfer heat to the furnace walls. In guess #1, the thermal efficiency (available heat) would be higher than actual.

A shortcut method for estimating furnace gas exit temperature is offered by the graph of figures 2.20 and 5.3, adapted by coauthor Shannon from radiant tube data, and extrapolated above 1800 F (1255 C). Also refer to "Estimating Furnace temperature profile for calculating heating curves" in chapter 8.

*NOTE: The convention used in this book is to omit the* <u>*degree mark (°)*</u> *with a temperature level (e.g., water boils at 212 F or 100 C), and to use the degree mark only with a temperature difference or change (e.g., the difference,* ΔT, *across an insulated oven wall was 100°F, or the temperature changed 20°F in an hour).*

In contrast to the formulas for heat transfer by convection, gas radiation formulas contain no velocity factor. Yet, *gas velocity is important in gas radiation*, as follows. If a stationary hot gas radiates to a colder surface, the gas necessarily loses temperature and finally becomes just as cold as the surrounding surfaces. To maintain active

**TABLE 2.6. Effective radiation beam length, s, of clear gas flames.** From *reference 27* (H. C. Hottel and R. B. Egbert: "The Radiation of Furnace Gases," ASME Transactions, May 1941). Those authors comment that for the range of P × s encountered in practice, the actual value is always less than these figures, and suggest that a satisfactory approximation consists in taking 85% of the limiting value, which is 4 × volume/total inside area.

| Shape of radiating gas volume | Beam length, s |
|---|---|
| Cube, sphere, or right circular cylinder with height = diameter, radiating to a spot at the center of its base | 0.6 × diameter or edge |
| Same, radiating to whole surface | 0.9 × diameter |
| Infinitely long cylinder | 0.9 × diameter |
| Space between infinite parallel planes | 1.8 × distance between planes |
| 1 × 2 × 6 rectangular parallel piped, radiating to any of its faces | 1.06 × shortest edge |

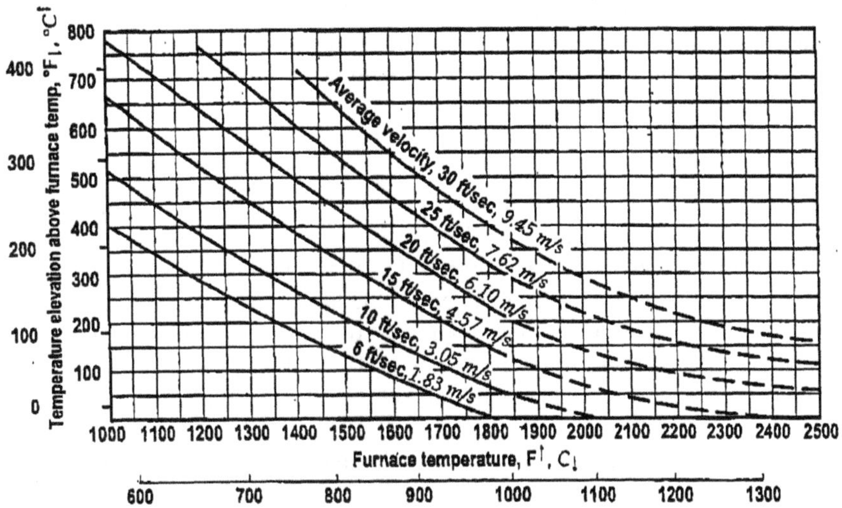


**Fig. 2.20** Elevation of flue gas exit temperature above furnace temperature, for a variety of velocities (average across-the-furnace cross section in the vicinity of the flue). (Same as fig. 5.3.)

radiation, the radiating gas must be replaced continually by fresh hot gas. A gas that radiates to a cold surface becomes colder and colder in the direction of the gas travel. With higher gas velocity (and therefore higher gas mass flow), the radiating gas stream's temperature will drop more gradually along the path of travel.

### 2.4.1. Enhanced Heating

The aforementioned path of gas travel is usually through a "tunnel" formed by piers on each side, the load above, and the hearth below. With less poc gas temperature drop because of higher total flow as they traverse the "tunnel" length, the lengthwise tunnel temperature uniformity will be improved. Control of the bottom "pumping" burners should be separate from control of the top (main) burners, thus effectively maintaining a small temperature drop between firing end and exit end of the tunnels. This may increase the bottom zone firing rate, but it will be well worth it if uniformity (product quality) is improved, and particularly if it reduces the total firing time for a uniformly heated load.

It has been common practice to try to increase the clearance under the load in forge and heat treat furnaces, but the opposite has been found to be better in view of the phenomena described in the previous paragraph, especially when one becomes aware of the poor life-to-cost ratio of tall piers.

This apparent enigma warrants a philosophical discussion* because it may seem that product quality (temperature uniformity) and fuel economy (efficiency) might be at odds. First, there is terrible economic loss in producing rejects because one must expend a duplicate quantity of fuel to redo the load properly, plus added labor, material, and machine time. Second, even on a continuous furnace, which naturally has a temperature differential from charge end to discharge end, those arguments for cross-wise temperature uniformity do not contradict conventional measures for fuel economy.

### 2.4.2. Pier Design

For this discussion, "piers" refer to supports, posts, pillars, skid rails, kiln furniture, stanchions—any devices used in a furnace, oven, or kiln to allow radiation and convection circulation under the load(s), and to avoid chilling of the bottoms of load pieces by direct contact with (conduction to) the hearth, which is often colder. Tall or high piers may be 30 in. (0.75 m) high or more to accommodate underfiring with large burner flames. Short or low piers may be 10 in. (0.25 m) high or as needed to accommodate underfiring with small high-velocity burners ("pumping, circulating, or enhanced heating burners").

Ideally, piers should be of low weight so that they do not add appreciably to the furnace load nor slow heat-up time. They should be narrow at the point of contact with the bottom surface of the load to minimize "shadowing" dark streaks or "striping" of the load. Using old reject billets is not recommended because of their weight and because they make scale that accumulates in the gas passageways between piers. High alloy or refractory piers are preferred if it is practical for them to support the weight of the load.

In batch-type furnaces, reducing underload clearance, reducing triatomic gas concentrations, and using high-velocity burners to inspirate furnace gases for increased mass flow under the load has reduced cross-wise load-bottom temperature differentials to less than 15°F (8°C). It is important to remember that the high-velocity underpass gases do not exit the furnace at the end of their pass, but circulate around the load(s) several times, and that they enhance radiation and convection in other parts of the furnace.

**Case Study**

In a batch forge furnace, the space above the load(s) was held at 2250 F, wall to wall. High-velocity stirring burners were fired between the 8 in. tall piers supporting the load(s). The burners were operated with fuel turndown only to minimize the concentration of triatomic molecules while inducing a high mass of inert gas from above the load. The wall-to-wall temperature drop under the product was very low—a maximum of 6°C (3.3°C). Chapter 8 discusses temperature uniformity in more detail.

*Suggested furnace design and operating policy priorities:

1st—Safety.

2nd—Product Quality.

3rd or 4th—Productivity.

4th or 3rd—Fuel Economy, conservation, and cost reduction.

Improved fuel economy can result in gains in many aspects. Pollution minimization may rank anywhere in this order, depending on local conditions.

## 2.5. THERMAL INTERACTION IN FURNACES

The many modes of heat transfer (heat flow) in a fuel-fired furnace are shown in figures 2.15 and 2.21. Some radiation usually accompanies high-velocity convection jet flames; some convection may accompany luminous and gas-radiating flames. Heat is transferred from high-temperature heat sources to lower temperature heat receivers, or heat sinks.

### 2.5.1. Interacting Heat Transfer Modes

Heat flows from the flame and products of combustion (poc) to the load(s) via six routes:

1. Direct gas (and clear flame) radiation from triatomic gas molecules (mainly $CO_2$ and $H_2O$) to surfaces of loads and walls that they can "see"*
2. Direct particulate radiation from soot particles within the flame to surfaces of the charged loads and walls that they can "see"
3. Direct convection from any poc molecules that flow across the surfaces of loads and walls

4. to 6. Indirect re-radiation from walls (already heated by routes 1, 2, or 3 to the surfaces of loads that they can "see"

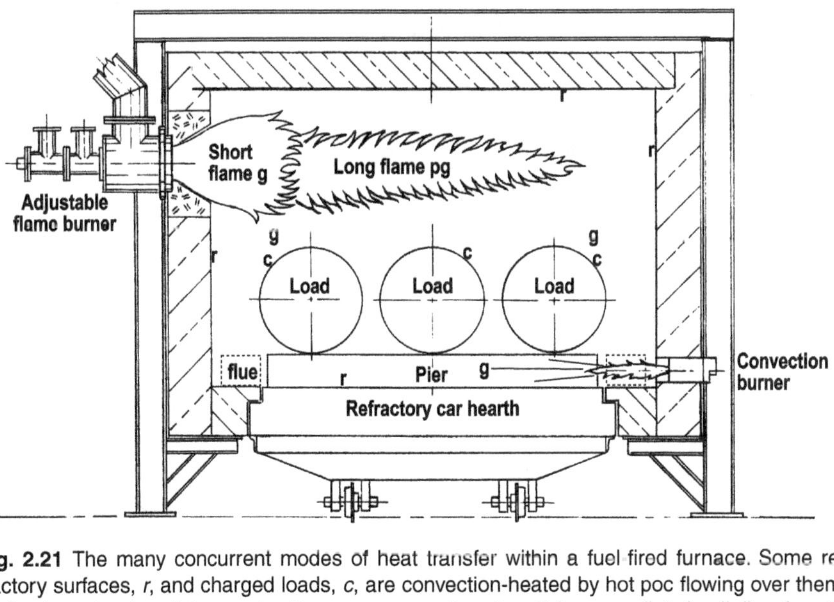


**Fig. 2.21** The many concurrent modes of heat transfer within a fuel fired furnace. Some refractory surfaces, *r*, and charged loads, *c*, are convection-heated by hot poc flowing over them. Triatomic molecules of the combustion gases, *g*, and soot particles, *p*, radiate in all directions to refractories, *r* and loads, *c*. The surfaces of *r* and *c* in turn radiate in all possible directions, such as *r* to *r*, *r* to *c*, *c* to *c*, and *c* to *r*.

Radiation and convection are surface phenomena. Only conduction, induction, and electrical resistance heating through the load itself can transmit heat beneath the surfaces of solid opaque objects. Induction flux lines tend to crowd just below the surface of large solid load pieces, so they, too, rely on conduction to deliver heat to the centers of large pieces. The molecules of triatomic combustion gases and the particles of soot radiate in all directions (spherically), but the surrounding 'cloud' of other molecules or particles can absorb (filter out) some of their radiation. Every unit of flat surface of a load or wall radiates throughout the hemisphere that it can "see." Both the re-radiation and absorption of these large solid surfaces may be slightly diminished by the aforementioned filtering effect of soot particles and triatomic molecules.

The soot particles are confined within the visible flame. The triatomic molecules are everywhere within the furnace, but can absorb and emit radiation only within narrow wavelength bands. Interference among the several modes of heat transfer can make calculation of net heat transfer in a fuel-fired furnace difficult. Some of the many variables that must be considered are composition, velocity, temperature, and beam thicknesses of the poc and well as emissivities, absorptivities, conductivities, densities, and specific heats of the refractory wall and load materials.

A technique for calculating *steel heating curves*, using the lag time theory, is explained in Chapter 8. That theory states that the center temperature of a piece of steel will follow the surface temperature of the piece by a given time-lag, irrespective of the rate at which the steel is being heated, if the rate of heating is nearly constant. With this theory, average core temperature and/or bottom surface temperature of a metal piece can be predicted accurately using a graph of apparent thermal conductivities of the metal throughout the expected temperature range. (Fig. 2.22 for steels.) The internal temperatures of the metal during transition may not be known, but that will not be defeating if the heating curves for before-and-after situations are known. Time-lag for a piece of steel is calculated by equation 2.12.

$$\text{Time-lag, minutes} = \frac{(\text{thickness, inches})^2}{10}\,(\text{exposure factor})\,(\text{conductivity factor}) \tag{2.12}$$

where the *exposure factors* are 1 for four-side heating, 2 for two-side heating, and 8 for one-side heating. The exposure factor for other configurations and spacings can be read from figures 8.2 and 8.4. The conductivity factor for a steel containing a specific percent carbon can be determined from figure 2.22.

Calculation of a furnace heating curve using the Simplified Time-Lag Method uses a trial-and-error solution that deals with furnace temperature, steel surface temperature, and firing with less than 20% excess air. This method results in only slight errors. If oxygen enrichment or air preheating is involved, as much as 15% added heat transfer may occur as indicated by higher heat transfer coefficients inferred in

*The word "see" implies a direct straight line of sight. Radiation that "hits" triatomic gas molecules, soot particles, piers, or kiln furniture may be absorbed by those "receivers," diminishing the heat that reaches the surfaces of the loads.

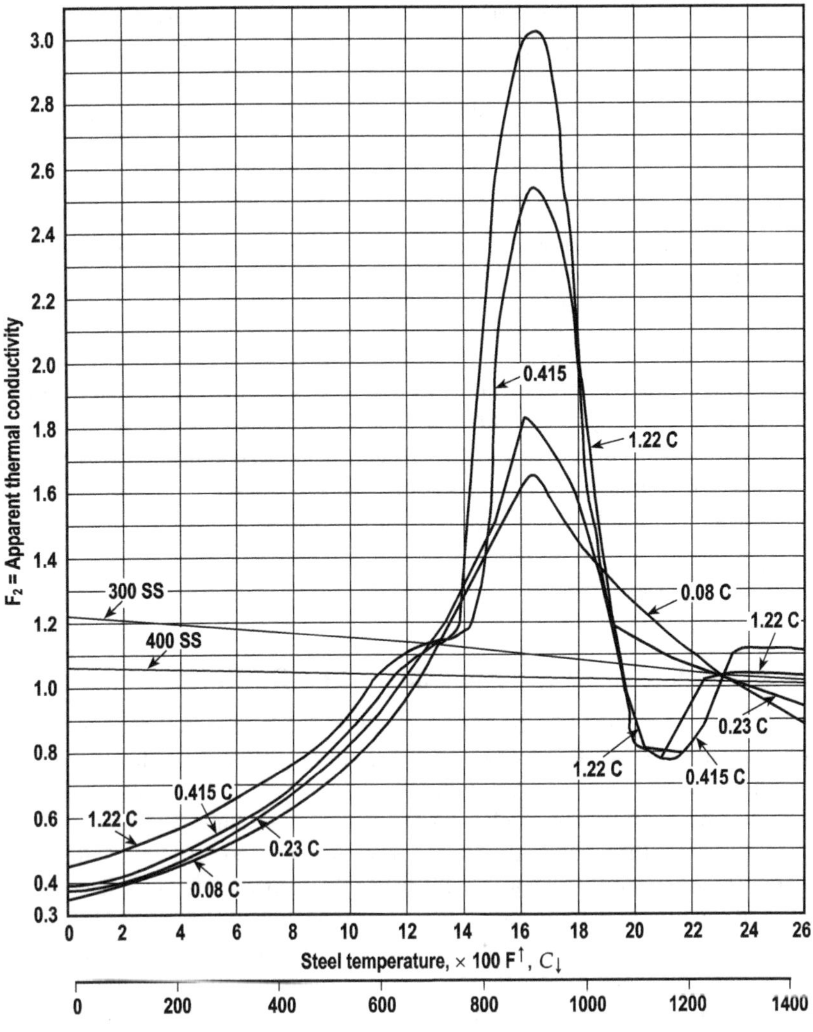


**Fig. 2.22 and 8.4** Effect of carbon content in various steel grades on heat absorption is shown by these "grade factors" used in the last steps of table 8.7 (worksheet) for the Shannon' Method for plotting steel heating curves. The peaks in this graph show the effect of the dramatic increase in heat absorption for steels containing various percentages of carbon, C, during the crystalline phase changes between 1200 F and 1900 F (650 C and 1038 C). SS = stainless steel.

figures 2.13 and 2.14 at higher air temperatures and higher partial pressures of $CO_2$ and $H_2O$.

Radiation heat transfer, as used in the simplified time lag method for creating furnace heating curves (temperature vs. time) is really an average condition of the gas blanket temperature, gas blanket thickness, and vapor pressure of triatomic gases. With high excess air, the heat transfer will be less due to lower percentages of the diluted triatomic gases and a lower average gas blanket temperature. Other "average" conditions assumed in the simplified time lag method are a 3 ft (0.9 m) gas beam and 3450 F (1900 C) adiabatic flame temperature.

To increase the rate of heat transfer above that determined by the simple time-lag methods:

1. Increase the gas blanket thickness
2. Increase the percentage of triatomic gases in the products of combustion—by using less excess air or by enriching the combustion air with oxygen
3. Increase the gas blanket temperature
   a. with preheated combustion air
   b. with higher flame temperature fuel (e.g., coal tar theoretical flame temperature is 4100 F versus natural gas theoretical flame of 3800 F)
   c. With fuel-directed burners, which will increase combustion speed and reduce recirculation of products of combustion that normally dilute the flames with inert and lower temperature furnace gases
4. By reducing air infiltration
5. By reducing all heat losses

### 2.5.2. Evaluating Hydrogen Atmospheres for Better Heat Transfer

Below is a summary of calculations that coauthor Reed made for coauthor Shannon to help a customer evaluate improving heat transfer by substituting hydrogen (better gas conductivity) for air as a recirculating medium in a furnace. This was a very special case because (1) the stock being annealed was stainless steel at 1750 F—higher temperature than that used in most cover annealers and (2) no inert atmosphere, and therefore no inner cover, was used because the load was stainless steel. Radiant tubes were used for indirect firing instead of an inner cover.

Coauthor Shannon warned that the safety hazard from fire or explosion with hydrogen requires that a hydrogen–inert gas mix be used only below the lower limit of flammability. The lower explosive limit is 4% hydrogen in a hydrogen–air mix. The upper limit is 74.2% hydrogen in an $H_2$–air mix.

Thinking ahead, however, to the fact that others may want to explore the possibility of enhancing heat transfer through the use of hydrogen, it was decided that an evaluation of the heat transfer gain was in order. The following comparison procedure is outlined for those who might want to consider applying it to their processes in the future.

#### *2.5.2.1. Calculating Comparable Heat Transfer Rates.* 

See the section on forced convection heat transfer coefficients, $h_{cf}$, in any heat transfer text.

$$\text{Nusselt number, } Nu = h_{cf} L/k = C Re^x Pr^y \tag{2.13}$$

The *Nusselt number, Nu*, is a dimensionless number wherein $C$, $x$, and $y$ are constants determined by experiment or experience for specific fluids, configurations, and temperatures. Values for all fluid properties, including Prandtl number, $Pr$, should be evaluated at an estimated mean film temperature—mean between bulk stream temperature and wall surface temperature. The Nusselt number, $Nu$, is a dimensionless ratio of convection to conduction capabilities of the fluid, wherein $h_{fc}$ is the forced convection film coefficient, in Btu/ft²hr°F, and $L$ is length of the surface parallel to the gas flow if less than 2 ft (0.61 m). If more than 2 ft and turbulent flow, use 2 ft (0.61 m), $k$ is the thermal conductivity of the gas, in Btu ft/ft² hr°F (See table.)

$$\text{Reynoldsnumber, } Re = \rho V L/\mu \tag{2.14}$$

*The Reynolds number, Re,* is a dimensionless ratio of momentum to viscous forces in the heating or cooling fluid, wherein $\rho V$ = momentum, in which density is in lb/ft³ and velocity is in ft/hr, and absolute viscosity is in lb/hr ft, all at mean film temperature.

$$\text{Prandtl number, } Pr = c\mu/k \tag{2.15}$$

*The Prandtl number, Pr,* is a dimensionless ratio of fluid properties that affect heat flow, wherein $c$ = specific heat, Btu/lb °F, $\mu$ = absolute or dynamic viscosity in lb/hr ft, and $k$ = thermal conductivity in Btu ft/ft² hr°F. Values of $Pr$ range from 0.65 to 0.73 for most gas mixtures based on hydrogen or nitrogen. When raised to the suggested $y = 0.43$, the last term of the Nusselt equation ranges from 0.83 to 0.87, so use of 100% hydrogen instead of air would improve the forced convection heat transfer coefficient, $h_{fc}$, by a small amount, but other parts of the Nusselt equation raise it more. Some engineers simplify the Nusselt equation by substituting the average value 0.85 for Pr when dealing with these gases.

**TABLE 2.7. Properties of hydrogen, $H_2$, at one atmosphere**

| | TEMPERATURE | | | | | |
|---|---|---|---|---|---|---|
| | 60 F<br>15.6 C | 500 F<br>260 C | 900 F<br>482 C | 1200 F<br>649 C | 1750 F<br>954 C | 1850 F<br>1010 C |
| Specific heat, $c_p$, Btu/lb °F and cal/gm °C | 3.405 | 3.469 | 3.494 | 3.548 | 3.714 | 3.712 |
| Thermal conductivity, $k$, Btu ft/ft²hr°F | 0.101 | 0.159 | 0.214 | 0.238 | 0.286 | 0.303 |
| Density, $\rho$, lb/ft³ | 0.00443 | 0.00289 | 0.00203 | 0.00166 | 0.00125 | 0.00120 |
| Viscosity absolute, $\mu$, lb/hr ft | 0.0210 | 0.0318 | 0.0401 | 0.0459 | 0.0560 | 0.0571 |
| Prandtl number, $c\mu/k$ dimensionless | 0.71 | 0.69 | 0.66 | 0.70 | 0.73 | 0.70 |

**TABLE 2.8. Properties of air at one atmosphere**

| | TEMPERATURE | | | | | |
|---|---|---|---|---|---|---|
| | 60 F<br>15.6 C | 500 F<br>260 C | 900 F<br>482 C | 1200 F<br>649 C | 1750 F<br>954 C | 1850 F<br>1010 C |
| Specific heat, $c_p$, Btu/lb °F and cal/gm °C | 0.240 | 0.247 | 0.260 | 0.269 | 0.281 | 0.283 |
| Thermal conductivity, $k$, Btu ft/ft$^2$hr°F | 0.0148 | 0.0250 | 0.0338 | 0.0402 | 0.0502 | 0.0517 |
| Density, $\rho$, lb/ft$^3$ | 0.0763 | 0.0413 | 0.0292 | 0.0239 | 0.0180 | 0.0172 |
| Viscosity absolute, $\mu$, lb/hr ft | 0.0440 | 0.0670 | 0.085 | 0.0970 | 0.116 | 0.118 |
| Prandtl number (dimensionless) $c\mu/k$ | 0.71 | 0.66 | 0.65 | 0.65 | 0.65 | 0.65 |

Pages 549 to 551 of *reference 36* (Karlekar and Desmond's'heat Transfer , 2nd ed.) give refinements on "Flow over Flat Plates," using recommendations of *reference 88*, wherein the constants in the Nusselt equation, above, should be: $C = 0.29$, $x = 0.8$, and $y = 0.43$.

For "large temperature differences," Whitaker recommends $Nu_{av} = 0.036$, $Pr_{av} = 0.43$, $Re_L = 9200$, $\mu_s/\mu_w = 0.25$, where the last term is the ratio of viscosities at

**TABLE 2.9 Summary comparison of convection heat transfer rates**

| *100% Hydrogen vs. 100% Air, at 80 fps gas velocity* | | Cycle Start | Midcycle | Cycle End |
|---|---|---|---|---|
| Load surface temp | | 60 F | 900 F | 1750 F |
| Mean gas film temp | | 500 F | 1200 F | 1850 F |
| Temp difference, gas to load | | 440°F | 300°F | 100°F |
| *With 100% Hydrogen* | Re | 56 604 | 22 676 | 13 104 |
| | Pr | 0.691 | 0.692 | 0.695 |
| | Nu | 216 | 114 | 60.9 |
| Film coefficient, $h_c$, Btu/ft$^2$hr°F | | 16.6 | 13.0 | 9.22 |
| Heat flux, Btu/ft$^2$hr | | 7304 | 3888 | 922 |
| *With 100% Air* | Re | 356 654 | 141 922 | 83 929 |
| | Pr | 0.66 | 0.65 | 0.65 |
| | Nu | 925 | 409 | 261 |
| Film coefficient, $h_c$, Btu/ft$^2$hr°F | | 11.6 | 8.22 | 6.75 |
| Heat flux, Btu/ft$^2$hr | | 5122 | 2466 | 675 |

free stream temperature and at wall temperature. Reed interprets Whitaker's '9200' as based on the transition from laminar to turbulent flow for air or products of combustion, estimated at $Re = 10\,000$. However, with hydrogen, the density is so small that the laminar-to-turbulent transition $Re$ may be $< 9200$, resulting in a negative answer; thus Reed omitted the '−9200' term from all his calculations, to give comparable results.

***Conclusions:*** For the state of the art at this writing, and with the previous set of conditions, the listed gains look promising. They must be weighed against the costs of precautions to minimize the risks of handling hydrogen.

## 2.6. TEMPERATURE UNIFORMITY

In most heating applications, temperature uniformity is a major player in product quality. Furnace users have insisted that temperature differences from thermocouples in gridlike racks should be within ±25°F, or 10°F with no loads in the furnace. After the loads are placed in a furnace, the thermocouple grid uniformity check should be replaced by T-sensors strategically attached to the loads because the following heat transfer variables become dominant.

### 2.6.1. Effective Area for Heat Transfer

With a load placed in a furnace or oven, its effective area for heat transfer is determined by its location relative to other loads, the sidewalls, and the end walls.

**Situation a**: For products loaded in a two-high configuration on 12" high piers, the effective heat transfer area of the top load(s) would be their full projected top surface area. Because of the thinner gas cloud or "blanket" adjacent to the lower row of load pieces, their effective heat transfer area would be less. (See fig. 4.7.)

**Situation b**: For two ingots placed end-to-end in a furnace, the active heat transfer area would be in the range of 70 to 80%, with top and bottom firing, depending on the load width relative to the furnace width. Ingots loaded side-by-side with top and bottom firing would have active areas of 40 to 80%, depending on the ratio of load spacing and furnace width.

**Situation c**: With products loaded in three-high rows, the top and bottom rows are similar to situation a except that they must supply heat to the middle row. The effective area of the middle row can only be estimated by experience with the specific configuration.

**Situation d**: When loads are elevated on lightweight supports at least 3 ft. high, the effective area for heat transfer from below may be increased from the 30% of situation a to as high as 100%. This might raise the total circumferential effective area of a single piece from 73 to 86%. In a two-high configuration with tall supports, the effective heat transfer area of the bottom rows would be a mirror image of the top minus the shadow effects of the supports. Tall supports with two side-by-side ingots might increase their effective heat transfer areas from 40 or 50% to 80%.

Positioning the loads to raise their effective heat transfer area not only improves heat transfer rates but also reduces the lag time (time it takes for the core or lowest %exposed area side to reach the temperature of the hottest surfaces). This benefit reduces thermal stresses in the product, resulting in shorter cycles (less fuel and higher productivity) plus higher quality products.

### 2.6.2. Gas Radiation Intensity

Gas radiation intensity depends on: (*a*) thickness of the gas radiation blanket or cloud, (*b*) concentration of triatomic molecules in the gas radiation cloud, and (*c*) average temperature of the gas cloud, including the flame.

### 2.6.3. Solid Radiation Intensity

Solid radiation intensity depends on: (*a*) projected areas "seeing" other hotter or colder solids and gases, (*b*) solid particles in the flames (luminous flames), and (*c*) temperature differences between interacting solids.

### 2.6.4. Movement of Gaseous Products of Combustion (See also chap. 7.)

Furnace gas movement enhances convection, but it also causes mixing in downstream zones, raising or lowering the gas cloud temperature and thereby affecting the load temperature. Slower moving poc gases have more contact (cooling) time, but are less vigorous in viscously thinning the stagnant boundary layer, which acts as an insulator.

***Roof flues*** should generally be used only when there is bottom firing. Otherwise, hot gases will not flow to the bottom to maintain a hot gas blanket temperature, so bottom heat losses will take heat from the load(s) via solid radiation and conduction. The resultant nonuniformity in load temperature will be intolerable.

***Bottom flues*** are preferred to keep temperature differences low. When a furnace is top-fired only, bottom flues bring hot gases to the hearth, partially balancing bottom heat losses and load heat requirements. If flues are placed in the centers of the side walls of a long furnace at hearth level, flue gases will move toward the center flues, reducing the flow of hot gas to the door and back end. Wise positioning of flues (elevationwise, lengthwise, crosswise) requires much experience.*

In higher temperature furnaces, the interradiation from hotter solid surfaces to cooler surfaces tends to self-correct minor nonuniformities. For example, in batch furnaces and ovens, the door end and back end incur the greatest heat losses. In one instance it was found that in an 1100 F (593 C) oven, a 150°F (83°C) differential was sufficient to level out the temperatures from center to each end. However, in a 2250 F (1232 C) furnace, only a 70°F (39°C) difference was necessary to level out the temperatures (because of the 4th power effect in the Stefan-Boltzmann radiation

*Revered old-time furnace designer, Lefty Lloyd, exaggerated this point, saying: "You can put the burners anywhere you want, but just let me locate the flues."

> Downdrafting vs. Updrafting. A similar situation can occur inside stacks of loads in a furnace, kiln, or oven. Ceramic kiln operators learned this the hard way long ago. In a top-flued kiln (updraft), if one vertical space between loads happens to get a little hotter than the other gas columns, its lower density will cause its gases to rise faster, pulling more hot gas into itself. This quickly rachets its temperature so much above the rest of the kiln that all adjacent load pieces became rejects. If the kiln were "downdrafted" (burners at top, flues at the bottom), an overheated column of gas would be bucking the general flow pattern and receive less gas flow, and therefore automatically cool itself until at the same uniform temperature as the rest of the load.

equation). In many situations, the 70°F (39°C) differential is an unacceptable nonuniformity of temperature.

Personnel working around hot furnaces must be protected from burns near hot flues. Best practice is to position lightweight, insulated, vertical ducts (open at both ends with a 1 ft high gap between their open bottom ends and the floor to admit cooling air) so that all poc exiting the furnace are drawn up into these ducts by their own "chimney effect." This "barometric damper" also tends to minimize excessive "draw" by flues that get too hot, which could otherwise "snowball" into a very uneven temperature situation within the furnace chamber. Likewise, failure to clean scale or other blockages from flue entrances can cause uneven heating because nonblocked flues will get hotter and pull more "draft" by natural convection.

Modern practice tends to use a single large flue instead of multiple small flues because of the difficulty in balancing multiple flues for even heating. Undersized flues may be very difficult to enlarge, but oversized flues can be partially reduced in size quite easily.

An "ell" (90-degree turn) is recommended in a flue line to prevent straight-line furnace radiation out the flue, wasting fuel, and chilling part of the load. This is particularly important if there is cleanup or heat recovery equipment beyond the flue because of possible radiation damage to that equipment.

### 2.6.5. Temperature Difference

To have temperature uniformity within each load piece and among the pieces, furnace gases and solids must have low temperature differences. All heat supplied by the combustion reaction flows either (1) directly from the hot poc gases to the load or (2) from the poc gases to the refractory, and is then re-radiated to the load. Heat transfer is a form of 'potential flow,' moving from high temperature to low temperature. Thus, the flame and poc gases must be hotter than the refractory, and the refractory must be hotter than the load.

Until recently all intrafurnace heat transfer was erroneously thought to be via solid-to-solid radiation or by convection, ignoring gas radiation. Many cases have led engineers to realize that radiation heat transfer directly from gases to load may be as much as 60% of the total heat transferred in a 2400 F furnace. Therefore, to have uniform product temperature, uniform gas and refractory temperatures are essential. To hold ±15°F (±8°C) load temperature, the gas cloud[6] temperature must not drop more than 30°F (17°C) while passing the load. Limiting gas cloud* temperature drop to this very small quantity requires changing heat release to the poc,* heat transfer from the poc, and/or mass of flowing poc.

***Change the heat release rate*** (chemical reaction rate), which depends on the energies and directions of the air and gas streams, and shape of the burner tile. In each of these reaction variables, a fixed pattern of poc temperature profiles can be generated if no dynamic flow rate adjustments are made. Generally, higher inputs will drive the peak heat release point farther away from the burner wall. Conversely, the point of peak heat release will be closer to the burner wall at firing rates less than 30% of maximum. Adjustable Thermal Profile (ATP-type) burners were conceived to provide dynamic adjustment, producing a near-flat thermal profile.

With an ATP-type burner, the heat release pattern of the flame can be automatically adjusted by the difference in temperatures sensed at two points in the furnace. One of those temperatures also can limit energy inputs so that both ends of the load(s) will be controlled to raise or lower their temperatures together. If ATP-type burners cannot be fitted to spaces that are too narrow, other means (discussed later) must be used to avoid load temperature nonuniformities. This is usually done by designing for no more than a 30°F (16°C) poc temperature drop as the gases pass from one end of the load to the other.

*Change the heat transfer* from the poc gases: when firing between piers, lower the pier height to reduce the thickness of the radiating gas cloud or use a higher level of excess air to dilute the triatomic gases with oxygen and nitrogen. Excess air also lowers flame and gas cloud temperatures.

*Use enhanced heating*: Operate with very high velocity burners to inspirate great quantities of furnace gas into the tunnels between the piers. With this high mass flow of gas between the piers and between the load and the hearth, the burner poc temperature is nearly uniform, resulting in a more uniform load temperature (reflecting the more uniform poc temperature).

*Taking advantage of adjustable thermal profile type burners* above and below the loads will give the best uniformity, productivity, and economy. With the recommended control system, they can actually hold temperature dfferentials near zero. For maximum adjustability, ATP burners should flue through bottom ports or through the center of the zone roof. An ATP system will be capital intensive, but low in operating costs. If ATP-type burners do not fit, high-velocity burners with or without thermal turndown (excess air) are the next best choice for improved temperature uniformity, but this may increase operating cost.

*Incorporate pulse firing*, which takes advantage of all the energy of high fire velocity (momentum) in limited time firings instead of throttling burners to low

*gas cloud = gas blanket = gas beam = poc = furnace gases, which may include pic.

fire where their circulating ability would be decreased. This method for moving masses of gas is already widely used with burners of 2.5 million Btu/hr (2640 MJ/hr) capacity and less, doing a helpful job in this size range where ATP burners are not yet available. Stepfire operates burners in sequence at maximum firing rates to move large masses of gas, thereby supplying the transferred heat with minimum gas temperature drop (minimum temperature differential from end to end of each gas flow path). This, combined with a control based on an individual model, will provide near-best uniformity with greatly reduced energy cost.

## 2.7. TURNDOWN

Turndown is the ratio of maximum to minimum firing rate without having to provide a change in air/fuel ratio. For example, on a soaking pit, the maximum firing rate might be 35 kk Btu/hr at 5% excess air with 10 in. of water column air pressure to reach the desired pit temperature of 2400 F as soon as possible, with the available 1000 F combustion air.. After 1 to 5 hr, this firing-rate requirement might drop to a minimum of 3 kk Btu/hr.

The turndown ratio in this case would be 35/3 = 11.7 without changing the air/fuel ratio. The pressure (energy) will drop as the square of the flow, so the air pressure at the burner will drop from 10" of water to $10/(11.7)^2 = 0.073$" of water. $G$ (specific gravity relative to stp air) for 1000 F air $= (60 + 460)/(1000 + 460) = 0.356$; so from equation 5/6 of *reference 51*, the 0.073"wc air pressure will provide only an air velocity at the diverter in the burner of $66.2 \times \sqrt{(0.073/0.356)} = 30$ fps. This will be too low to mix the air and fuel thoroughly, so at about 5 kk Btu/hr, a turndown of 7:1, the air/fuel ratio can be changed from 5 to 50% excess air (1.5 times stoichiometric air flow) or an air flow of 30 (1.5) = 45 ft/sec to increase the air energy to mix the fuel and the air.

There are other ways to increase mixing energies and mass flows. For example, 5 to 10% of the maximum airflow can be in a jet down the center of the fuel tube of the burner. This will allow the use of the pressure upstream of the air control valve to provide 10" of water column to accelerate the air to mix with the fuel: $66.2\ (10/0.356)^{0.5} = 350$ fps.

The use of excess air to achieve temperature uniformity costs more fuel, but so does holding the furnace in a soak mode for a long time to achieve uniformity. An alternative to high excess air is to use pulse firing so that the desired high mass flow is either high or off.

## 2.8. REVIEW QUESTIONS AND PROJECT

**2.8.Q1.** Which mode of heat transfer travels only in straight lines? Which can go around corners?

**A1.** Radiation travels straight, like light; therefore has a shadow problem. Convection can go anywhere that a moving gas stream can.

**2.8.Q2.** How does 'enhanced heating' benefit heat transfer to load pieces that can be separated by spaces on a furnace hearth or by piers and spaces between the loads and the hearth?

**A2.** Furnace gas flowing between the loads not only helps convection heat transfer but also continually passes and replaces hot triatomic gas molecules (with high radiating capability) through the "'tunnels" between or under the loads.

**2.8.Q3.** What kind of gases radiate appreciable amounts of heat?

**A3.** Triatomic gases, of which $CO_2$ and $H_2O$ are the most common in furnace gases.

**2.8.Q4.** Use the following blank table to check off what heat sources use which heat transfer methods. Use a 1 for primary sources and a 2 for secondary sources.

| | HEAT TRANSFER METHODS | | | | |
|---|---|---|---|---|---|
| HEAT SOURCES | Conduction | Convection | Gas radiation | Solid* radiation | Induction |
| Electric resistor | | | | | |
| Electric induction | | | | | |
| Clear (blue) flame | | | | | |
| Luminous flame (soot particles) | | | | | |
| Refractory walls and roof | | | | | |
| Refractory hearth, furniture, piers | | | | | |

| HEAT SOURCES | HEAT TRANSFER METHODS | | | | |
|---|---|---|---|---|---|
| | Conduction | Convection | Gas radiation | Solid* radiation | Induction |
| Electric resistor | 2 | 1 | | 1 | |
| Electric induction | 2 | | | | 1 |
| Clear (blue) flame | 2 | 1 | 1 | | |
| Luminous flame (soot particles) | 2 | | 1 | 1 | |
| Refractory walls and roof | 2 | 2 | 2 | 1 | |
| Refractory hearth, furniture, piers | 2 | 2 | 2 | 1 | |

## 2.8. PROJECT

Refer to the "need for experimental test data" mentioned in section 2.3.4 just before example 2.4. Check with Gas Technology Institute, Chicago, IL, Massachusetts Institute of Technology, Cambridge, MA, and International Flame Research Foundation, Ijmuiden, the Netherlands, for past and future research.

# 3

# HEATING CAPACITY OF BATCH FURNACES*

## 3.1. DEFINITION OF HEATING CAPACITY

The heating capacity of a furnace is usually expressed by the weight of charged load† that can be heated in a unit of time to a given temperature, for the coldest part of that load, without overheating the rest of the charge. Because the cost of a furnace is approximately proportional to its size, heating capacity per unit of size is important. This "specific heating capacity" is expressed as: weight heated per hour, and per unit of furnace volume, OR weight heated per hour, and per unit of hearth area. The latter is more frequently used. Neither ratio is a perfect measure of heating capacity, as is shown by the following examples.

When annealing huge tanks, the furnace must be large enough to house the tank and to leave room for circulation of products of combustion around the tank, so the weight capacity per unit of volume seems small. If a long shaft is suspended in a vertical cylindrical annealing furnace, the annealing capacity per unit of hearth area would appear to be very great.

Furnace heating capacity depends on factors such as rate of heat liberation, rate of heat transfer to the load surface, and rate of heat conduction (diffusion) to the coldest point in the load.

## 3.2. EFFECT OF RATE OF HEAT LIBERATION

In *electric heating furnaces,* the heat release rate is expressed in kW. In both direct resistance and induction heating, the heat is generated within the material of the

*Many parts of chapter 4 on continuous furnaces contain useful information that also applies to batch furnaces, but they are not included here (to keep this book compact). Readers are advised to study both chapters 3 and 4.

†The terms "load," "charge," "product," "work," and "stock" are used interchangeably in this book and in industry. (See the glossary.)

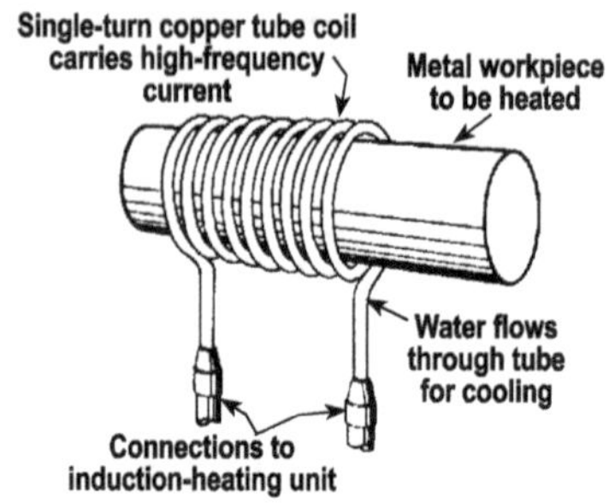


**Fig. 3.1.** Heating by induction. The part of the load surrounded by the coil is inductively heated. Some heat may "stray" to adjacent areas by conduction.

heated load. In electric *resistance heating,* the rate of heat release per unit of (element covered) wall area depends on economic life of the elements, element material, design and spacing of the elements, furnace temperature, and furnace atmosphere.

*Induction heating* uses a medium- or high-frequency electric coil (water cooled) to induce a current in a metal load. (See figs. 3.1 and 3.2.) The flux lines are most concentrated just below the surface of the load. Conduction distributes the heat across the load. The heat flow is not reduced by surface resistances as with convection and radiation.

In *fuel-fired furnaces,* heat release rate is usually expressed in heat units liberated per unit of furnace volume in unit time, commonly in Btu/ft$^3$hr or MJ/m$^3$hr. Closely related to rate of "furnace heat release" is the combustion volume or flame volume. Generally, the furnace volume should be at least equal to the sum of the maximum flame volume and the maximum load volume. The volume of the flame is a function of the "combustion intensity condition" discussed with table 3.1 subsequently. and where $Fc$ is a configuration factor to assure that all of any one flame's volume is contiguous.

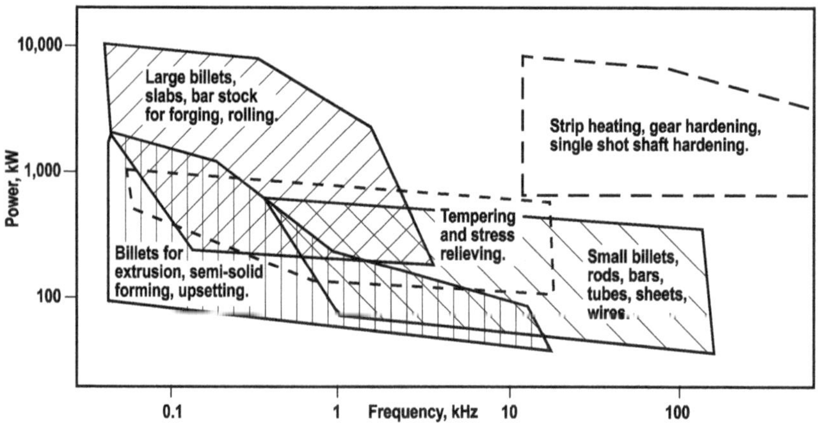


**Fig. 3.2.** Induction heating application parameter ranges. Courtesy of Inductoheat, Inc., Madison Heights, MI.

**TABLE 3.1. Generalized descriptions of six "combustion intensity conditions" for use in equation 3.1, and in example 3.1**

| Combustion Condition | Description | Approx. Max. Gross Btu/ft$^3$hr* | MJ/m$^3$hr* |
|---|---|---|---|
| 1 | Very poor fuel and air mixing, coarse fuel, cold air, inclusion of space in which no combustion takes place in what might be considered "combustion volume." Cold air. | 5 400 | 208 |
| 2 | Fair (to poor) fuel and air mixing, fair utilization of combustion chamber volume, coarse fuel, cold air. Similar to condition 1, except 500 F (260 C) air. | 21 600 | 800 |
| 3 | Good fuel–air mixing, good use of combustion space, fine atomization or powdered fuel, cold air. Same as condition 2, but 500 F (260 C). | 36 000 | 1 300 |
| 4 | Thorough fuel and air mixing or premixing, perfect utilization of combustion space, fine atomization or powdered fuel, 500 F (260 C) air. Same as condition 3, but 1000 F (538 C) air. | 64 800 | 2 400 |
| 5 | Thorough fuel and air mixing or premixing, perfect utilization of combustion space, fine atomization of fuel, 1000 F (538 C) air. Also, the discharge from many small burners. | 118 800 | 4 400 |
| 6 | Premixed fuel and air from closely spaced, small orifices firing against refractory surfaces to speed combustion. In the combustion space proper, as much as 3 600 000 Btu/ft$^3$hr* or 134 000 MJ/m$^3$hr* are released. Space is needed between burners and load to avoid overheating. | 1 800 000 | 67 000 |

*Reference 18* lists $10^4$ to $10^6$ Btu/ft$^3$hr (373 MJ/m$^3$hr to 37 300 MJ/m$^3$hr) with nozzle-mix burners, and $10^6$ to $10^7$ Btu/ft$^3$hr (37 300 to 373 000 MJ/m$^3$hr) with industrial premix burners.

If air and fuel are premixed upstream of a burner nozzle, mixing (and therefore combustion) may occur more rapidly than with nozzle mixing, and surely more thoroughly than with delayed mixing (perhaps with a detached flame) out in the furnace. Presumably, faster mixing and combustion will require less furnace volume, but the aerodynamics and the directions of the velocity vectors can influence flame shape to the point where flame volume may be less dependent on air or fuel momentum.

Most premix burners have been removed from industrial use for the following reasons:

(a) Nozzle-mix burners remove the hazard of flammable mixtures inside burner feed pipes, ducts, valves, plenums, headers, and burner bodies.

(b) Nozzle-mix burners have wider lighting windows and broader stability limits. Burning can be maintained from 40% rich to more than 2000% excess air, improving safety and operating flexibility.

(c) With nozzle-mix burners, combustion air can be preheated, causing combustion to proceed even more rapidly and saving fuel.

A few premix burners and their flames plus many nozzle-mix burners and their flames are shown throughout pt 6 of *reference 52*. Special premixing arrangements with low flashback hazard are now being used in some low NOx industrial burners.

Figure 3.3 shows geometrically similar burners and flames. If a single large long flame was installed in the center of a large furnace wall, some space surrounding the flame might be wasted. On the other hand, many small short flames might better utilize the wall area and permit reduced furnace volume. However, there are large modern burners that can hold a whole burner wall as hot as the point of traditional maximum heat release. With these burners, controlling spin of the poc can produce a nearly level temperature profile from burner wall to far wall. Automatic furnace pressure control makes possible the use of roof flues without nonuniformity problems and high fuel cost.

Using many small burners to utilize the whole wall area is a way to achieve good temperature uniformity. (See figs. 3.4 and 3.5, and sec. 7.4.) There are large burners that can hold the burner wall as hot as the point of conventional maximum heat release. These adjustable thermal profile burners (fig. 6.1) can automatically hold a desired temperature profile by controlling the spin of the products of combustion. Optimum use of furnace space and overall refractory wall radiation usually favors the hottest possible burner wall (maximum flame spin, minimum flame length). In

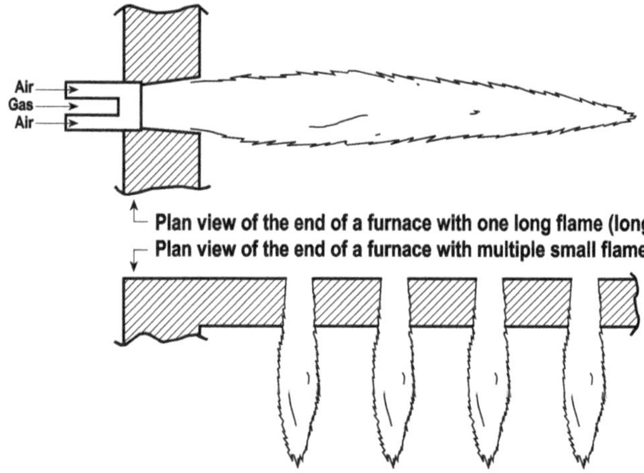


**Fig. 3.3.** A side-fired arrangement makes better use of the combustion space, giving better temperature uniformity. The best, described later, uses spin to adjust their heat release pattern. (See also discussions on circulation in chap. 7.)

**Fig. 3.4.** Car-hearth heat treat furnace with piers, ceramic fiber walls, and high-velocity burners (*top left* and *bottom right*). Courtesy of Horsburgh and Scott Co., Cleveland, OH.

longitudinally fired furnaces, very hot burner walls can reduce fuel rates by 10% while increasing productivity by 10%.

It is difficult to predict the volume needed for complete combustion. Table 3.1 gives broad generalizations that require judgment in their use.

**Example 3.1**: Find the rate of heat liberation needed to heat 0.4% carbon steel to 2200 F on a hearth. A loading rate of 80 lb/ft$^2$hr is very good for a single zone batch furnace. From figure 2.2, interpolate the gain in steel heat content from 60 F to 2200 F as 365 Btu/lb, so 80 × 365 = 29 200 Btu/ft$^2$hr, which is 8.11 Btu/s for each square foot of hearth. From an available heat chart for natural gas (*reference 51*), the best possible efficiency for an estimated 2400 F flue gas exit temperature with 10% excess air would be 31.5%, so the rate of heat liberation required = 29 200 Btu/ft$^2$hr output divided by (31.5 useful output/100 gross input) = 92 700 gross Btu/ft$^2$hr.

With good fuel and air mixing, combustion condition 3 in table 3.1 suggests about 36 000 gross Btu/ft$^3$hr as the volumetric heat release intensity. Thus, for the situation in example 3.1, the required combustion space would be 92 700/36 000 = 2.58 ft$^3$ psf of hearth, or 2.58 ft of inside furnace height. For some load configurations (e.g., large thin-walled shapes), such a low furnace roof might endanger product quality with flame impingement, and would be difficult for access for repair. Yielding to these practical considerations with a higher roof would reduce the required combustion heat release intensity, which is on the safe side.

*Flame temperature* affects heat transfer to the load(s), and therefore affects the furnace capacity. In gaseous heat transfer, it is the average temperature of the gas blanket that transfers the heat. Neither the flame temperature nor the poc temperature

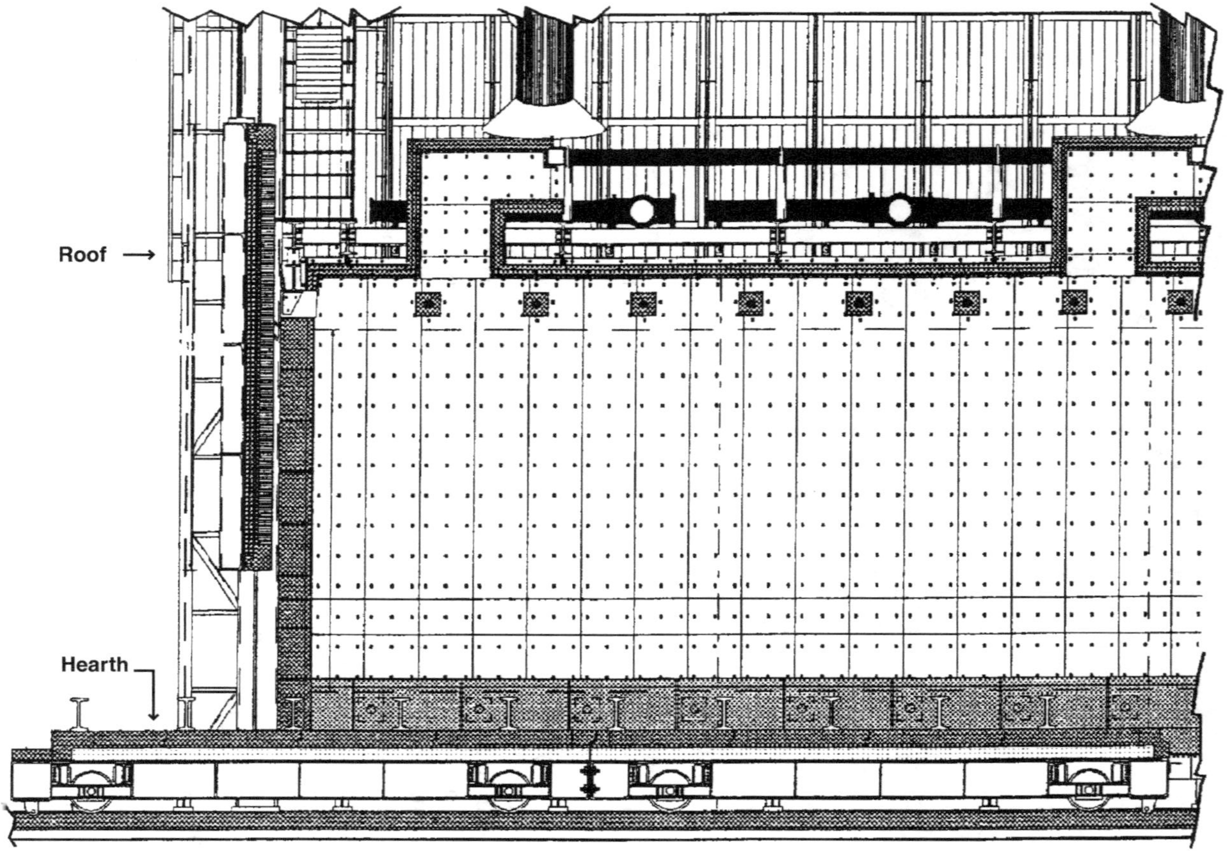


**Fig. 3.5.** Large car-hearth furnace such as used for stress-relieving large vessels. The fiber-lined 90° flues avoid "black hole" cold spots in the furnace roof preventing uneven load temperature. Courtesy of Hal Roach Construction Co.

should ever drop lower than the temperature of the adjacent load(s). This rarely happens except (1) with 'lean' fuel gases[‡] or very long heat transfer time or distance (2) with high burner turndown resulting in insufficient sensible heat in the poc to make up for heat losses, (3) with cold air infiltration, or (4) with poor furnace gas circulation [e.g., poor flue port location(s). (See chap. 7).]

Whereas each fuel molecule burns at the ideal (adiabatic) flame temperature, the reaction heat is transferred to surrounding gases, liquids, and solid objects as combustion proceeds. Only by infinitely rapid combustion, or by combustion in a perfectly insulated chamber, can the adiabatic flame temperature be reached.

Values for adiabatic flame temperatures can be read from the $x$-intercepts of available heat charts or from *reference 51*. With lean fuels, high temperatures can be obtained only by preheating the air, the fuel, or both, or by using oxygen-enriched air or oxy-fuel firing.

## 3.3. EFFECT OF RATE OF HEAT ABSORPTION BY THE LOAD

Because ample heat can usually be released at sufficiently high temperatures in industrial furnaces, the next problem to be studied in calculation of furnace capacity should be heat transfer to the furnace load and temperature equalization within the load. With adequate heat release at sufficiently high temperature assured, note the following factors that affect furnace capacity.

### 3.3.1. Major Factors Affecting Furnace Capacity

1. Exposure of the load to heat transfer
2. Temperature of the furnace walls when cold load is charged
3. Temperature to which the load is to be heated
4. Temperature of the products of combustion
5. Emissivity of the products of combustion
6. Absorptivity and emissivity of the walls (Absorptivity are emissivity are nearly the same for most materials)
7. Absorptivity of the load to be heated
8. Degree to which excess air, or excess fuel, is to be used
9. Thickness of the cloud of products of combustion
10. Load thermal conductance (conductivity including effects of voids)
11. Required temperature uniformity within the load
12. Thickness of load(s) to be heated
13. Furnace configuration, including dimensions, volume, and hearth

[‡]Lean fuel gases, such as blast furnace gas and some producer gases, have low hydrogen/carbon ratios, and therefore have low calorific or heating value.

14. Locations of temperature control sensors
15. Number of furnace control zones
16. Temperature uniformity within the furnace
17. Quantity of infiltrated air (furnace pressure control)
18. Velocity of the poc passing over the load surfaces
19. Thickness of the gas blanket (beam)
20. Fuel carbon/hydrogen ratio
21. Burner location and flame type

It is difficult to combine all the preceding variables into a single equation, model, or computer program for furnace design. Engineers have calculated tables, drawn charts, and developed spreadsheets for combinations of the variables that fit the types of furnaces and loads that frequently occur in their practice. This reference book cannot furnish procedures for every conceivable combination. Instead, a generalized method will be developed that will suffice for many practical purposes.

Generally, (*a*) the rate of heat transfer to the load determines the best possible heating rate for *thin loads* whereas (*b*) temperature equalization within the load(s) determines heating capacity rates for *thick loads*, especially those having low thermal conductivity.

See chapter 2 for more about heat transfer phenomena. Heat flux, $q = Q/A$, heat transfer rate per unit of exposed area, is the product of the average coefficient of heat transfer ($U$) and the temperature difference ($\Delta T$) between the heat source (flame, refractory, poc) and heat receiver (load):

$$q = Q/A = U \times (\Delta T) = (h_r + h_c) \times (\Delta T) \tag{3.1}$$

where $Q$ is heat transfer rate in Btu/hr or MJ/hr, and $U$, $h_r$, and $h_c$ are heat transfer coefficients in Btu/ft$^2$hr°F or MJ/m$^3$hr°C; where $h_r$ varies with $[(T^4_{abs,s})-(T^4_{abs,r})]/(T_s - T_r)$, source emissivity, receiver absorptivity, and configuration, and $h_c$ is a function of Re (velocity = a major factor).

In batch-type furnaces, temperatures of poc and refractories must be controlled to avoid overheating the load if a mill delay or other problem requires the load to stay in the furnace an unusually long time. This necessitates that the temperature of the poc be no more than about 5% (from 0 F, not absolute) above the prescribed final surface temperature of the load. The excess temperature may be 8% above final load temperature if occasional overheating causes no serious damage to the load.

The data available on emissivities of refractories at high temperatures indicate that they are generally lower than 0.9. When cold stock is put into a furnace, the refractory temperature drops temporarily by radiation to the cold load and through open doors. Some parts of the refractories may have lower temperatures than indicated by the temperature sensors.

The following summary of observations was gleaned from time versus temperature profile graphs in *reference 85*, where they were intended to give the reader a "feel" for how temperature of a load rises. A 2 ft thick steel plate was heated from the top side only, with a 2 ft thick gas beam above, as follows: (a) heated to within 100 F of refractory temperature in 13% less time with 2800 F refractory than with 2400 F refractory; (b) heated to 60% of its final temperature in the first half of heating time; and (c) The time–temperature path was almost a straight line for the first half of the heating time, and then like a half-hyperbola (similar to the trajectory of a ball thrown up at an angle).

Current practice requires engineers to have more than a "feel" for load heating patterns (time–temperature profiles). They must acquire an ability to determine the effects of many operating and design variables on various loads' time–temperature curves. The Shannon Method, which enables one to calculate specific time–temperature curves, is discussed briefly several places in this book and then detailed in chapter 8. The reader is encouraged to adapt the Shannon Method for processes other than the steel reheat and forging cases illustrated here.

Figure 3.5 shows a 40 ft (12.2 m) long car-hearth in a 17.5 ft (5.3 m) high fiber-lined furnace with high-velocity burners at top and between the piers. Automatic furnace pressure control makes it possible to use top flues. Drilled square air manifolds shoot curtains of air across the flue exits as throttleable "air curtain dampers" for furnace pressure control.

## 3.4. EFFECT OF LOAD ARRANGEMENT

In batch-type furnaces, two questions arise: (*a*) What is the effect of arrangement of individual pieces on furnace capacity? (*b*) What is the effect of thickness of the pieces on furnace capacity? Obviously, space must be provided between the pieces for the manipulating tongs or other loading and unloading equipment. Unless the spaces between the pieces are inordinately large or small, the heating capacity is not noticeably affected because the bare spots of the hearth receive radiation from the gases as well as the roof and the side walls. The heat received by the hearth is then re-radiated to the work and assists in heating it. For reasonable heat transfer exposure (temperature uniformity and fuel economy), a minimum spacing ratio, $C/W =$ (center-to-center)/$W$ of figure 3.7, is 1.6. Somewhere above a spacing ratio of 2.0, the loss of furnace capacity (because wider spacing permits fewer pieces across the furnace) usually necessitates adding furnace capacity to reach an optimum combination of product quality and productivity.

The square billets in figure 3.6 were laid on a hearth so that the width of each empty space between them equaled the width of each billet (spacing ratio, $C/W = 2/1 = 2$),

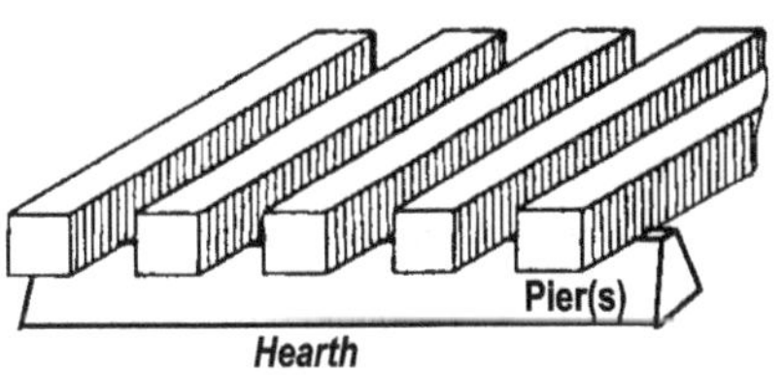


**Fig. 3.6.** Three steps to better heat access: loads spaced out, loads elevated on lightweight piers, and enhanced heating between piers.

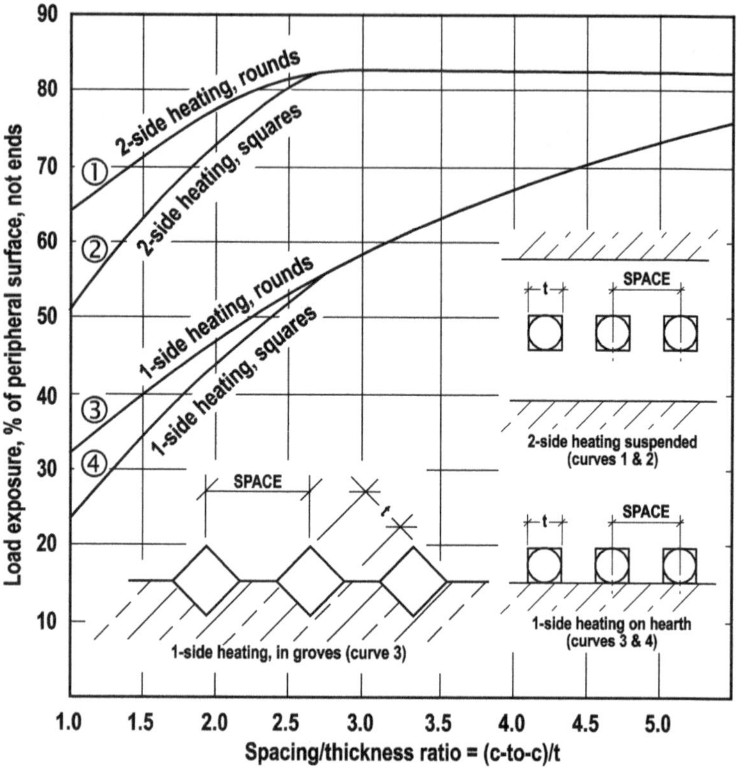


**Fig. 3.7.** %Exposure versus workpiece spacing ratio. Billet "spacing ratio" = centerline to centerline distance, *C*, divided by billet width or diameter, *W*. Use a centimeter scale for interpolating.

the weight per square foot of hearth would be the same as if the same area were covered by a plate or slab half as thick. The heating surface of the billets would be 50% larger than the heating surface of the plate. However, the vertical heating surfaces are not as effective as the horizontal heating surfaces. Radiation from the hearth (which would not be as hot as the roof) increases the transfer of heat to the vertical surfaces. The net result would be that the weight of billets heated in unit time would be about equal to the rate at which the half-as-thick plate could be heated, except for added time-lag of the thicker pieces. The curves of figure 3.7 give exposure data for a variety of arrangements.

**Example 3.2**: Heat a load of three steel rounds, 24" (0.61 m) diameter, for forging in a furnace 8.5 ft (2.6 m) wide × 6 ft (1.83 m) high inside. Loads are on piers with centerlines 3.2 ft (0.98 m) apart. High-velocity burners fire through "alleys" between the pieces-enhanced heating). The center piece is the most difficult to heat because outer pieces shield it from side radiation and convection; thus, it will govern the heating time required.

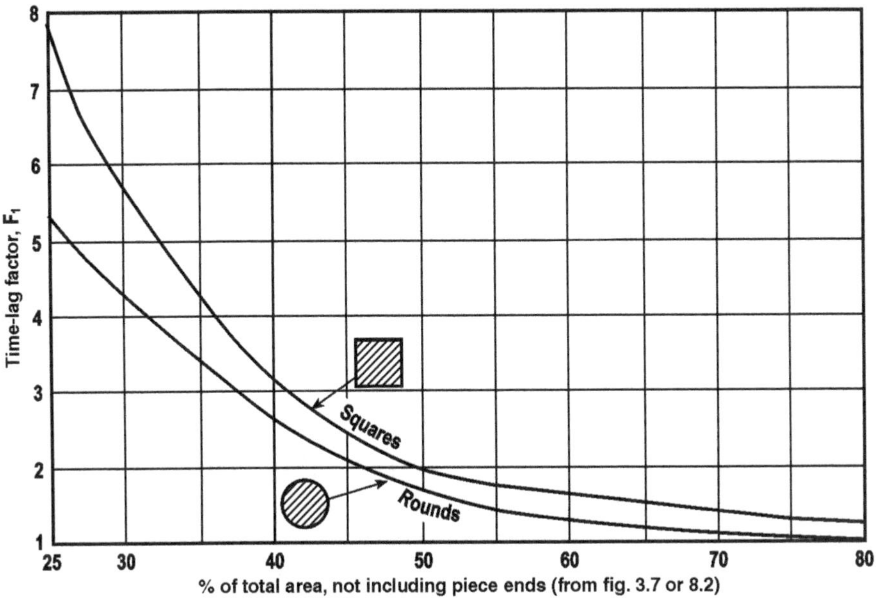


**Fig. 3.8.** Time-lag factors, for squares and rounds with various sides exposed, or various percents of total area exposed. Use a centimeter scale for interpolation (see example 3.1). Lag time, minutes = (0.1) ($F_1$) (thickness in inches)$^2$ = (155) ($F_1$) (thickness in meters)$^2$

Dividing the circumference of the center load into four quarters, each of which should theoretically receive 25% of the heat to that piece. (See figure 3.9.) Small numerals are the authors' estimate of the true % received by each quadrant, totaling 60% with enhanced heating. (If enhanced heating had not been applied, the bottom quadrant would probably have received almost none, totaling only about 46%.) From fig. 3.8, for 60% exposure on a cylindrical shape, read a time-lag factor, $F$, of 1.25; thus, the time-lag will be 0.1 (1.25) (24) (24) = 72 min.

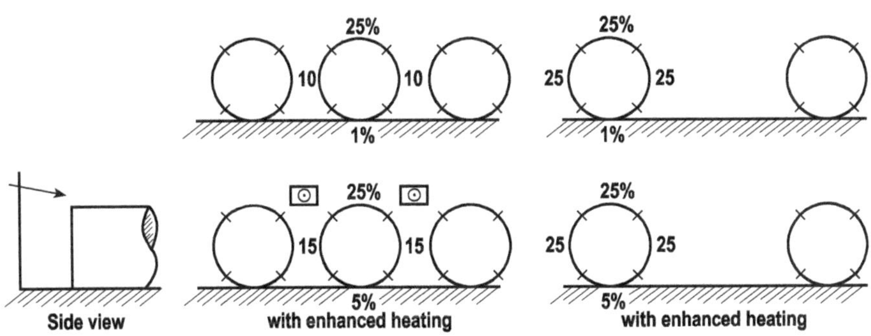


**Fig. 3.9.** Two loading and two firing situations for example 3.2.

| TABULATED SUMMARY for EXAMPLE 3.2 | Exposure (%) | Factor (*F*) | Lag (min) | Total (hr) | Average (hr/piece) | Benefits |
|---|---|---|---|---|---|---|
| 3 pieces at once[a] | | | | | | Fewer hours & less fuel per piece. |
| w/o enhanced heating | 46 | 1.75 | 101 | | | |
| w/ enhanced heating | 60 | 1.25 | 72 | 23.5 | 7.8 | |
| 2 pieces at once[b] | | | | | | Fewer hours per load. More even temp. |
| w/o enhanced heating | 76 | 1.09 | 63 | | | |
| w/ enhanced heating | 80 | 1.06 | 61 | 20.0 | 10.0 | |

[a]Center-to-center spacing = 2.3 feet = 0.7 m.
[b]Center-to-center spacing = 4.6 feet = 1.4 m.

By the Shannon Method explained in Chapter 8, a temperature-versus-time heating curve was calculated for the center piece, and the total heating time was found to be 23.5 hr. If the center piece were removed to give the two outer pieces better heat transfer exposure, the heating time for the two remaining pieces would be 20 hr.

In figure 3.10, pieces in row 1 lean against row 2. Sidewise stacking is almost as bad as vertical stacking because the $\Delta T$s so created within the pieces cannot be tolerated for high quality. The side of piece 1 facing piece 2 will be 50° to 100°F (28° to 56°C) below the right face of piece 1, which faces the hot furnace. If piece 1 is press forged, it will curl ("banana"—see glossary) toward its cold surface and may crack, causing the piece to be scrapped. After piece 1 has been removed, piece 2 will have an even colder side (facing the back wall), with more problems.

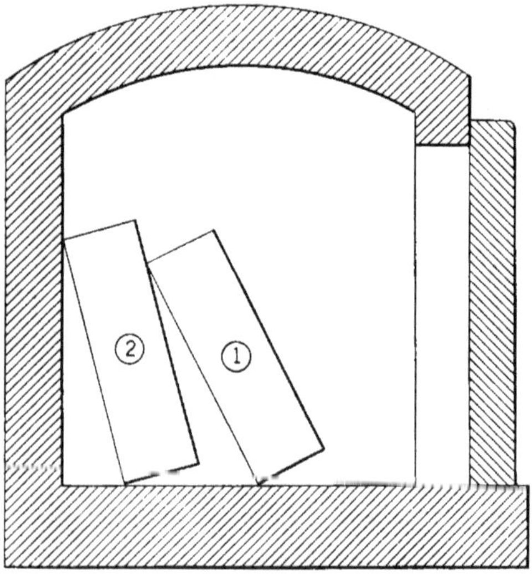


**Fig. 3.10.** Box furnace, in-and-out furnace, or soak pit with two rows of slabs.

The solution is to place the pieces on piers, preferably 12" (300 mm) high, and fire very high velocity burners between the piers, controlling the turndown of the burners with temperature sensors through the wall opposite those burners by reducing fuel input while holding the combustion air flow constant. In forge shops, each press is best surrounded by four furnaces: #1 furnace being charged, #2 heating up, #3 soaking, and #4 furnace being worked out.

### 3.4.1. Avoid Deep Layers

Some think that stacking loads three or more layers high is efficient use of furnace space, but it causes nonuniform heating, which reduces productivity per furnace, per man-hour, and per unit of fuel. It takes more than three times as long to heat a three-high stack than it takes to heat a single layer. (See fig. 3.11.) Putting the bottom row of load pieces on piers will allow one-side heating from below by radiation from the hot combustion gas and from the refractory hearth. The top row of loads will get one-side heating from above by radiation from hot gas and refractory. Without vertical and horizontal spacers, load pieces between the top and bottom rows will be heated at unknown rates depending on unknown quantities of gas moving between the layers. Read about bottom-fired furnaces in chapter 7.

When heat treating is performed on multiple layers, the cycle time needed to achieve the required grain size will be unpredictable. For best results with minimum time, heat one layer at a time, with over- and underfiring. Increasing need for tighter temperature control in rolling, forging, and heat-treating operations is forcing more careful integration and control of radiation patterns and high-velocity gas circulation techniques.

In ceramic kiln firing, similar problems are discussed by Mr. Chris Pilko of Eisenmann Corp. on pp. 32–35 of the Dec. 2000, *Ceramic Industry.*

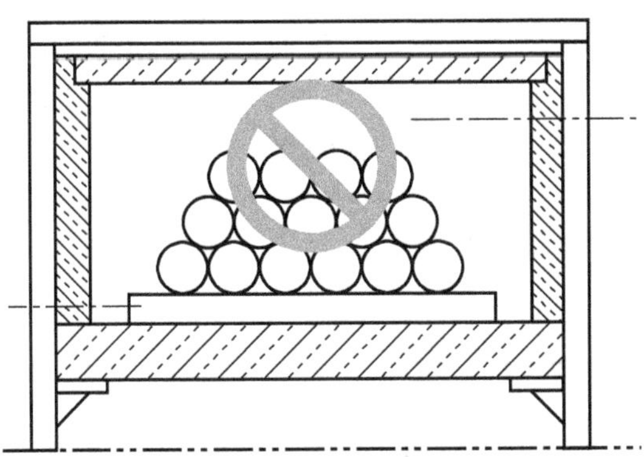

**Fig. 3.11.** Do not stack loads unless separated by horizontal spacers to allow gas flow between layers.

## 3.5. EFFECT OF LOAD THICKNESS

Many charts have been developed for predicting the time it takes to heat steel. (See figs. 3.12 and 4.21a.) The industry now has better methods for predicting required heating times (e.g., the Shannon Method, in chap. 8). It combines (*a*) the radiation heat transfer equation for the time it takes to transfer the required heat to the load, with (*b*) lag time theory. Together, (*a*) and (*b*) predict how fast and how uniformly a product can be heated, knowing the size and nature of the pieces to be heated and their location relative to the furnace gases and the refractory.

The lag time theory uses the following equations and factors to determine the extra time required for the center of a load piece to catch up with its surface temperature. The time necessary for a piece to reach a required temperature with uniformity throughout depends on the conductivity, density, and thickness of the material, and the number of sides exposed for heat transfer. Equations 3.1 and 3.2, for heating steel, show that the lag time increases as the square of the thickness. (See fig. 3.8.)

$$\text{Lag time, minutes} = (0.01)\,(F_1)\,(\text{thickness in in.})^2 \tag{3.1}$$

$$\text{Lag time, minutes} = (15.5)\,(F_1)\,(\text{thickness in m})^2 \tag{3.2}$$

where $F_1 = 8$ for one-side heating, $F_1 = 2$ for two-side heating,
$F_1 = 1.25$ for three-side heating, $F_1 = 1$ for four-side heating.

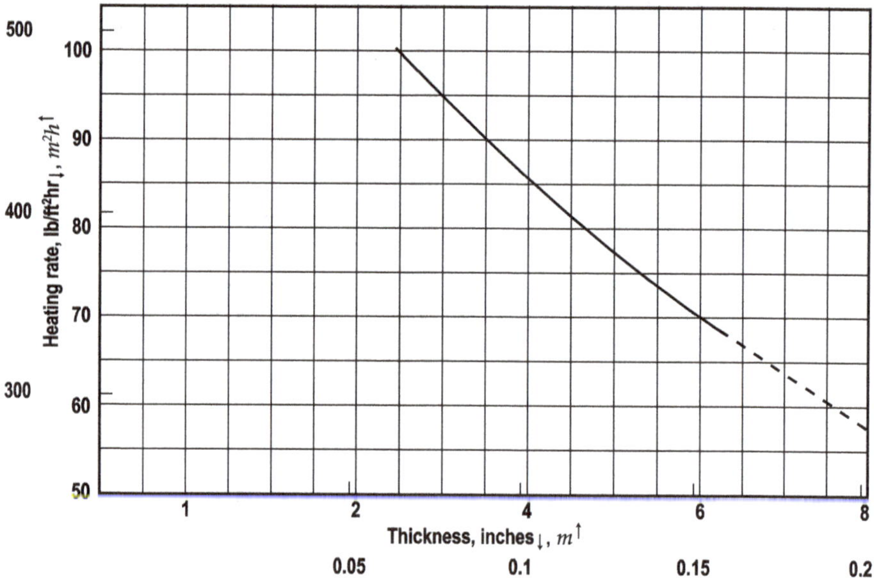


Fig. 3.12. Typical heating rates for various steel thicknesses in a batch reheat furnace. The dashed lower end of the curve indicates that greater than 6" (0.15 m) steel thickness is not recommended for one-side heating. (See also fig. 4.21.)

Large steel objects of certain compositions must be heated slowly to avoid steep temperature differentials across their thickness, which can produce strains in the metal. These are usually harmless in mild steel, but can cause cracks in tender steels and brittle metals. The cracking is accompanied by a peculiar noise that is called "the clink." Obviously, the slow and careful heating of large objects reduces the heating capacity of a furnace. A furnace operator should use a heating curve (chapter 8) for the specific metal analysis being heated to determine a safe rate of furnace temperature rise to prevent the metal from being damaged. When the temperature differential in a piece exceeds 400°F, trouble will likely occur.

## 3.6. VERTICAL HEATING

If long objects are heated to high temperatures, they may sag under their own weight. For that reason, they are usually heated suspended in a tall vertical furnace. The usual rules about lb/hr $ft^2$ of hearth, or kg/hr $m^2$ of hearth are meaningless in this case. Vertical dimensions range from 4 ft (1.3 m) to > 60 ft (18 m). Engineers may use the product of the vertical dimension and the larger horizontal dimension in place of the hearth area to use their rules of weight heated per unit of area. However, this "laying the furnace on its side" does not help for ingots or slabs in soaking pits nor for stack coil annealing furnaces.

A practical loading limitation for ingots in *soaking pits* is to keep the total ingot cross-sectional area between 30 and 40% of the total pit plan view area at a level above the burner. Greater than this percentage of hearth coverage will result in larger temperature differentials (top to bottom) of each ingot.

A second major criteria for soaking pits is *firing rate*. To calculate the maximum firing rate in *US units*, multiply the pit's Length × Width × 125 000$^+$ Btu/$ft^2$hr for cold air to a maximum of 200 000$^+$Btu/$ft^2$hr if using 700 F combustion air. Then, with cold air, add 30%$^+$ to the firing rate. Corresponding numbers for calculating firing rate in SI units are multiply pit hearth area by 33 800$^+$kcal/$m^2$h with cold air to a maximum of 54 100$^*$kcal/$m^2$h if using if using 370 C air. Then with 15 C air, add 30% to the firing rate.

To estimate the *fuel use when charging cold ingots, in US units,* multiply the charged tons by 2$^*$ kk Btu/ton when using cold air, or by 1.6$^*$kkBtu/ton when using 700 F air. To estimate the *fuel use when charging cold ingots, in SI units*, multiply the charged tons by 0.56$^*$ kcal/metric ton with cold air, or by 0.448$^*$kkBtu/metric ton with 350 C air.

**Example 3.3**: Find the maximum firing rate necessary for a 9-hr heating cycle for heating 80 short tons of steel from 60 F to 2250 F, with a flue gas exit temperature of 2400 F during the maximum firing rate period. The steel is to be heated with natural gas in an 8′ × 22′ × 15′ deep soaking pit. A recuperator produces 700 F preheated air during the maximum rate period. A Shannon Method heating curve (sec. 8.1 to 8.3) predicts the total heating time from 60 F to 2250 F will be 9 hr. Charge and draw time

$^*$experience factor.

may add 1 hr. The soak time from the burners' automatic cutback until the first piece is drawn may add 2 hr. Wall and gap losses total 1.3 million Btu/hr.

**Solution 3.3**: From figure A-14 in the appendix of *reference 52* at 2250 F, find that the heat content of steel (from base 60 F) is 355 Btu/lb. Thus, the load requires (80 ton/hr) (2000 lb/ton) (355 Btu/lb) = 56.8 kk Btu per hour. For wall and gap losses, add 1.3 kk Btu/hr. Therefore, the total *'heat need'* (required available heat) = 56.8 + 1.3(9) = 68.5 kk Btu/hr.

From an available heat chart for natural gas (such as fig. 5.1 in chap. 5), at 2400 F flue gas exit temperature with 700 F air preheat, read 42% available heat; thus, the required gross input = 68.5/0.42 = 163 kk gross Btu/hr. That 163 gross divided by (9 − 1 − 2) hr = 27.2 gross kk Btu/hr as the *required burner firing rate* during the 6 hr of firing. The *heating capacity of the pit* will be 80 tons/9 hr = 8.88 tph of cold steel.

In one-way, top-fired soaking pits, complications stem from large temperature differentials from burner wall to wall opposite the burner. With burners that produce straight ahead poc[†] gas flow lines, the temperature differential in the space above the ingots can be 140 to 300 °F (78 to 167 °C),with the highest temperature near the wall opposite the burner.

Spinning the products of combustion helps greatly. Sometimes there is too much spin, but more often there is not enough. Even with the degree of spin controlled to give a flat temperature profile in the combustion chamber, the pit bottom temperature may be 100 to 200 °F (55 to 110 °C) hotter at the opposite end than at the burner end.

To correct this problem, three controlling temperature sensors are needed: two in a sidewall above the height of the bridgewall, 18" in from each end wall, and one below the burner The sensor near the opposite wall controls the energy input and provides a setpoint for cascade control of the degree of poc spin (by the burner), which is sensed by the thermocouple near the burner wall. The third temperature sensor (below the burner but above the ingots) limits the maximum temperature of the pit, thereby preventing washing[‡] the top surfaces of the ingots.

With this soaking pit control system, ingots are all heated alike in much shorter time, and with no greater temperature differential ($\Delta T$) from top to bottom of the ingots than 40 °F (22 °C) with a hearth coverage of 35%. Greater density of hearth coverage increases the $\Delta T$.

## 3.7. BATCH INDIRECT-FIRED FURNACES

The principal purpose of indirect firing is to protect the furnace load from corrosion, oxidation, carbon and/or hydrogen absorption, or other reactions with the poc. The protection is accomplished by placing a solid barrier wall between the poc and the load, and by pumping an inert atmosphere into the chamber on the side of the wall where the load is located. The barrier wall may be refractory or metal, but it must

[†]poc = products of combustion.

[‡]melting the oxide (surface slag).

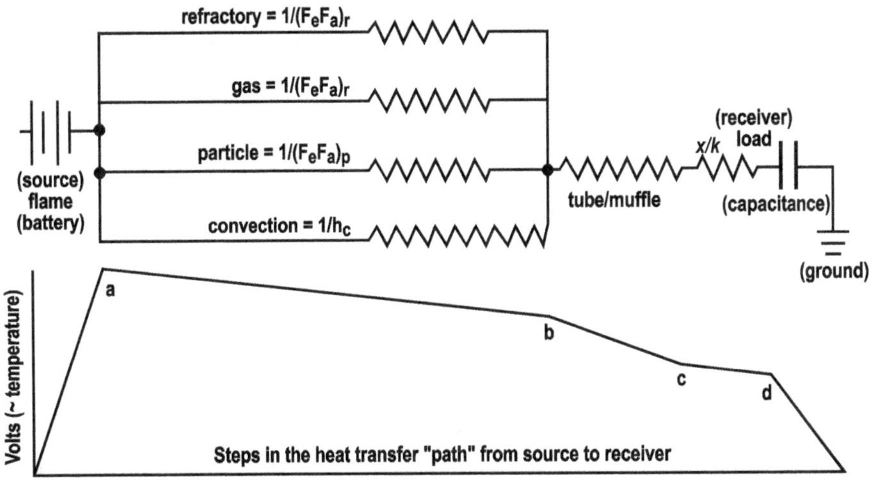


**Fig. 3.13.** Electrical analogy and accompanying graph of the temperature (voltage) profile from energy source to receiver.

be a gas-tight separation between the load and the flames and poc. The poc are then vented via a sealed exhaust through the outer wall. If the barrier wall appears to be a container for the loads, it is termed a *muffle*. A barrier wall wrapped around a flame is a *radiant tube*. Before controllable-flame-shape burners were developed, muffles and radiant tubes also were used to even out temperature irregularities in the load. In those cases, non-gas-tight "semi-muffles" were acceptable.

Both radiant tubes and ceramic muffles have higher flue gas exit temperatures than direct-fired furnaces, which means lower available heat and higher fuel cost; thus, electric heating may be able to compete with them. The muffle or tube wall acts as another resistance in the energy flow path from flame to load. Figure 3.13 is a modification of the electrical analogy of figure 2.15, showing the added resistance of the tube and the heat transfer "path" from source to receiver for indirect firing. The downhill slide from b to c represents the effect of three resistances in series: tube inner surface resistance, tube wall thickness resistance ($x/k$), and tube outer surface resistance (including the poor-conducting boundary layers on tube inner wall, tube outer wall, and load surfaces). For a direct-fired situation (no tube), the flame and poc would probably have cooled all the way from a to c, delivering much more heat to the load and less out the flue. For this reason, heat recovery devices such as recuperators or regenerators are often used with indirect firing. (See *reference 86* and figs. 3.14 and 3.16.)

There always will be a considerable temperature drop across a muffle wall or a radiant tube wall. Forced circulation on the load side of the wall helps reduce the resistance of the stagnant film clinging to the wall surface and minimize temperature nonuniformities within complex loads.

The heating capacity of furnaces that are equipped with flame-in-tube muffles (radiant tubes) is limited by the heat that can be radiated from the tubes. The heating capacity of an indirect-fired furnace is less than that of a direct-fired furnace having

**Fig. 3.14.** Heat treating furnace with radiant U-tubes on the roof and back wall. The return legs (2nd and 4th from the hearth) are less radiant than the burner legs (1st and 3rd from the hearth). Tumbling around the bends completes gas–air mixing so the renewed delayed-mixing flame (type F, fig. 6.2) causes a glow in the second leg. Courtesy of Rolled Alloys, Temperance, MI.

the same wall temperature because radiating and convecting poc that are hotter than the furnace wall cannot "see" nor "touch" the load, and because of the temperature drop through the muffle or tube. Radiant tubes are often used in continuous furnaces (chap. 4).

The input to muffles or radiant tubes is limited by the strength, durability, and conductivity of their wall materials. The great temperature difference across a muffle or tube wall not only reduces its useful life but also causes the products of combustion to exit at a very high temperature, raising the fuel bill. For both reasons, muffle and tube walls are made as thin as practical, using a material that has both high thermal conductivity and resistance to heat. Alloy steels and silicon carbide are the most suitable materials for muffles and radiant tubes. Silicon carbide radiant tubes can withstand higher temperatures and are more resistant to oxidation than nickel–chrome alloy steel tubes, but the latter are less brittle and cheaper.

Muffles are prone to leak, especially in furnaces above 1800 C (982 C), where most have been replaced by radiant tubes. For lower temperatures,electrically heated furnaces or furnaces with radiant tubes and forced circulation have largely replaced muffle furnaces, except for cover annealing furnaces.

**Radiant-tube-fired furnaces** are most popular in the steel heat treating industry. Depending on the loading density, uniform heating often requires "covering the walls" with tubes as shown in figures 3.14 and 3.16. In lightly loaded furnaces, small (3" or 76 mm) diameter tubes may line the side walls, often with pull-through eductors and pilots on the top (flue) ends. Most batch and continuous furnaces, however, use 4" to 10" (104 to 253 mm) diameter tubes.

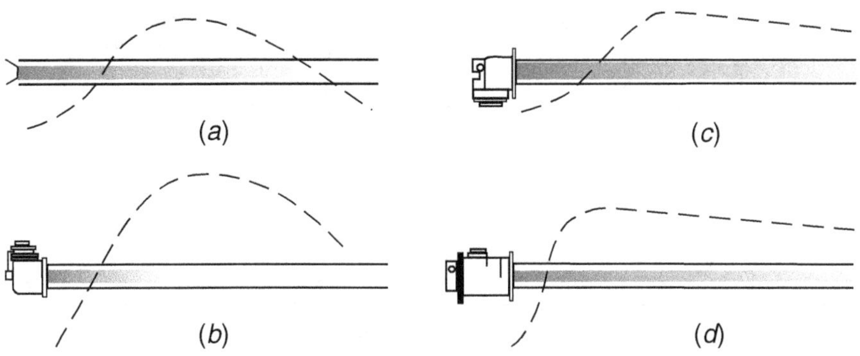


**Fig. 3.15.** Evolution of gas-fired radiant tube flames. a = premix flame, open burner. b = nozzle-mix flame, sealed-in burner. c = long, laminar, delayed-mix flame (type F) sealed-in. d = partial premix, followed by long, laminar, delayed-mix flame, sealed-in.

Aluminum heat treating (aging, homogenizing), uses indirect-fired air heaters, with a bank of radiant tubes positioned across an air duct. Circulation rates are typically at 8 to 10 air changes per minute. The process temperature levels are well below 1000 F (538 C).

As users of gas-fired radiant tubes realized that they had to invest in better materials to avoid frequent tube replacement, they demanded flames that would provide more even temperature distribution along the tube length, and that would assure that every part of the expensive tube length would be used for a high rate of heat transfer. Figure 3.15 shows the growth from simple to sophisticated.

Radiant tubes can be straight (fig. 3.15), U (fig. 3.14), W (fig. 3.16), or trident (three-legged, with burners at both ends and a common flue leg in the middle to give higher convection and less gas temperature in this last pass to compensate for its reduced interior radiation). Single "bayonet" radiant tubes have two concentric passes with a turnaround cap on the end opposite the burner, and with exhaust through the burner. In all cases, consideration must be given to support for the tube, and allowance for expansion and contraction. Vertical tube arrangements reduce hot tube sagging, but upfiring risks problems with falling scale interfering with the nozzle flow pattern. With downfiring, it is difficult to keep a tight seal to prevent outleakage around the burner.

Regenerative radiant tube burners are installed in pairs, each with a bed of heat storing media, usually alumina pellets or balls. While the burner on the right of each W-tube in figure 3.16 is firing, the bed of regenerative pellets in the left burner's body is being reheated by the exit gases from that tube. In about 20 sec, the bed will be as hot as it can get. At the same time, the bed in the right burner, which has been preheating air from energy stored in a previous cycle, will have cooled to the point where its delivery temperature of preheated combustion air is dropping below the design level. At that point, the positions of both air and gas valves on both burners are switched (air and gas on the left burner open, air and gas on the right burner close,

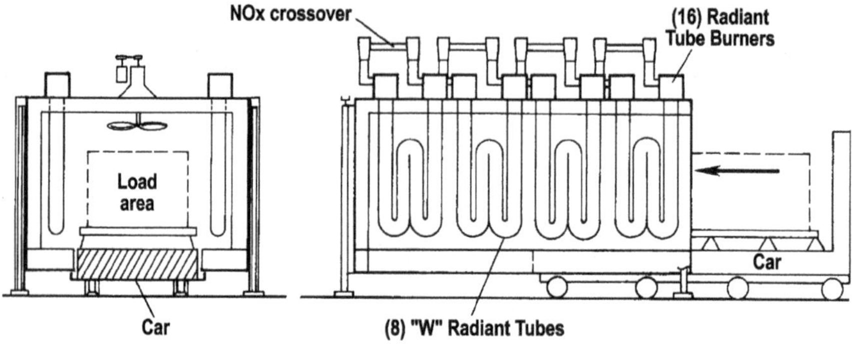


**Fig. 3.16.** A heat treating car-hearth (batch) furnace. Both sides of the furnace are heated by four W-radiant-tubes with a total of eight pairs of regenerative burners. "Plug fans" through the roof drive recirculation down between the load pieces.

and the right burner's air eductor opens to pull exhaust poc gas through its bed). Cycle times longer than about 20 sec (for this bed depth) result in less available heat. The NOx crossover allows flue gas recirculation to minimize NOx emission.

Regenerative radiant tube burners have the following advantages over recuperative radiant tube burners: (1) the regenerative beds extract heat more effectively from the tube exit gases than is usually possible with recuperators, thus assuring better fuel economy, (2) the final throw-away gas is so much cooler that it is no longer necessary to pay double time to those working around the recuperators because of terribly hot working conditions, and (3) the aforementioned alternating firing of each tube (right to left, then left to right) keeps the radiant tube more evenly heated, prolonging the tube life and giving a more even distribution (lengthwise and timewise) to the radiant input from the tubes to the furnace loads.

Point 3 of the previous paragraph is confirmed by the following data comparing a W-tube fired by a recuperative one-way burner versus a pair of regenerative burners alternatively firing both ways.

| | Recuperative | Regenerative |
|---|---|---|
| Maximum tube temperature | 1850 F 1010 C | 1850 F 1010 C |
| Minimum tube temperature | 1329 F 721 C | 1641 F 893 C |
| Average tube temperature | 1657 F 903 C | 1793 F 978 C |
| Furnace temperature | 1610 F 877 C | 1750 F 954 C |
| Typical thermal efficiency | 55–60% | 75–80% |

In any furnace, the time required to get the bottom center load piece to specified temperature determines heating cycle time (or for a continuous furnace, the furnace length divided by the conveyor speed). Attaching a temperature sensor to the most difficult-to-heat part of the load (and to the least difficult-to-heat part of the load) will make it easier to estimate the cutback time in the firing cycle.

**Example 3.4**: *Data* for a furnace such as shown in fig. 3.16. Inside dimensions = 18' × 12' × 10' high. Load = 12 000 pounds of steel weldments to be stress relieved at 1100 F.

*Find:* Gross heat input rate for the burners to match the tubes' radiating capability.

*Design estimates*: 6" diameter tubes with 9' of height and 0.6 of circumference exposed on the outer two legs, and 7' of height and 0.5 of circumference exposed on two inner legs (224 ft$^2$ effective surface for eight W-tubes). From tube supplier recommendations, operating tube temperature to heat a load to 1100 F should be 1600 F. From p. 94 of *reference 51,* tube emissivity = 0.66 and load absorptivity = 0.97.

**Solution to Example 3.4**: For parallel planes, third case on p. 97 of *reference 51*, find the emissivity factor, *Fe*, to use with an arrangement factor of *Fa* = 1.0 in formula 4/1a on p. 81 and with a black body radiation rate from the table on page 82, as follows:

$$1/Fe = 1/e_1 + 1/e_2 - 1 = 1/0.66 + 1/0.97 - 1 = 1/1.546;$$

$$\text{so } Fe = 0.647 \text{ with } Fa = 1.0.$$

For 1600 F tube temperature and 1100 F load temperature, find that the black body radiation rate is 20 700 Btu/ft$^2$hr.

$$\text{Radiation heat flux} = \text{Black body radiation rate} \times F_e \times F_a = 20\,700 \times 0.647 \times 1.0$$

$$= 13\,393 \text{ Btu/ft}^2\text{hr.}$$

Total radiation heat transfer rate for eight W-tubes = 13 393 × 224 ft$^2$ = 3 000 000 Btu/hr, or for one W-tube = 375 000 Btu/hr. The reader can estimate that the flue gas exit temperature with an average tube outside surface of 1600 F will be 1800 F. From an available heat chart for natural gas, at 1800 F and 10% excess air, read 48% available heat. Therefore, each of the sixteen regenerative burners should have a gross input capacity of 375 000 / 0.48 = 781 000 gross Btu/hr.

## 3.8. BATCH FURNACE HEATING CAPACITY PRACTICE

Heat transfer in batch-type furnaces is limited by the same variable factors as in all other furnaces (e.g., furnace temperature, refractory radiation, gas radiation, convection, scale on the load, hearth heat loss, and location of the control temperature measurement). See also the list of improvements that can help furnace productivity in sections 4.6.1, 4.6.1.2, and 4.6.1.3. Tables B.3 and B.4 in *reference 52* give heat requirements for drying.

Reducing temperature difference within the load pieces can sometimes nearly double furnace capacity by reducing the need for long holding periods. It is important to remember that the longer the heating cycle, the longer the fuel meter is turning. Exposing all possible surface area of each load piece to be heated is a cardinal rule.

Loading patterns must be rethought with each new size and shape of load. If load pieces are thicker than 4 in. (100 mm), at least 8-in. (200 mm) spacers are needed to permit heating from two or more sides. Engineers should take advantage of hollow pieces by trying to aim hot gas streams into their interiors.

Giving all parts of every load the most practical $\Delta T$ (heat-driving force) is logical, but often overlooked. To facilitate this, hot gas temperature across a hearth should be controlled to a flat (not drooping) temperature profile by maintaining high gas flow volume all the way across the whole loading area. Temperature profile control is a crucial part of modern burner technology. It not only reduces nonuniformities in the heated product (fewer rejects, which cost double fuel, labor, machine time, and sometimes material) but also minimizes holding time (fuel meter running time, operators' time-clock time).

Guides for good heating results in weight production per unit of hearth area or per unit of furnace volume are useful for judging normal needs for good heating (ball-park planning) (see thumb guides in the appendix). However, there are so many specific variables that affect each particular situation that the only safe way to engineer a good design is to plot time–temperature heating curves for each product, process, and furnace. (See chap. 8.)

### 3.8.1. Batch Ovens and Low-Temperature Batch Furnaces

Batch ovens and low-temperature batch furnaces (400–1400 F, 200–760 C) are in a range where convection capability may exceed radiation capability. (See fig. 2.10 in chap. 2.) Convection is used for effective heating in this temperature range where radiation is weak or has a "shadow problem" because it travels only in straight lines.

**Example 3.5**: Compare radiation to a 100 F (38 C) load in a 1000 F (538 C) oven with a 2200 F (1205 C) furnace. From a black body radiation table such as p. 82 or 83 of *reference 51*, the furnace would transfer only 7.6/85.5 = 0.89 or 8.9% as much radiation heat transfer as the oven. The heat needed to be imparted to the 100 F (38 C) load to bring it to 900 F (480 C), compared to the heat to be imparted to the same 100 F (38 C) load to bring it to 2100 F (1150 C) is (900 – 100)/(2100 – 100) = 0.40 or 40%. Therefore, if the heat were to be transferred by radiation only, the low-temperature oven would have to be 40/8.9 or 4.5 times as large as the high-temperature furnace.

Increasing the convection heat transfer rate is accomplished by using circulating fans, by using high-velocity burners, by judicious load placement and spacing as advised in chapter 7, and by enhanced heating. At one time, use of more excess air also was advocated to help circulation and convection, but as fuel costs have gone up, that method has been largely abandoned in the higher temperature ranges.

Circulation and flow concerns of chapter 7 require that boundary layers of stagnant poc gases be swept away, or thinned down, by high velocity. The magnitude of velocity is often indicated by momentum; hence, the interchangeable terms *high-velocity burners* and *high-momentum burners.* Momentum is Velocity × Density, but the gain from slightly higher density at low temperatures is almost insignificant.

The true measure of convection effectiveness is *Re*.* The higher density of low-temperature gases provides a very small gain in both *Re* and heat transfer.

Convection heat transfer can be helped by exterior recirculating fans as in direct-fired recirculating ovens (fig. 3.17), or internal recirculating fans, usually in the oven or furnace ceiling, blowing down into the load. Protection of fan motors on top of the furnace may be a maintenance problem. The velocity and volume of circulating fans are limited by the reduction of furnace size, cost, and increased temperature uniformity on one hand, and the cost of fan power on the other. The optimum varies with the cost of power, the openness of the loading, and the absorptivity of the load. (A brighter load justifies a higher velocity because its radiation reception is poorer.) The power delivered to the fan is converted to heat.

In figure 3.17, the hot recirculating gases being blown from left to right deliver some of their heat to the loads and are therefore cooler as they exit at the right. Mixing the hot products of combustion with the cooler recirculated gases that have already passed over the loads is accomplished by a circulating fan capable of withstanding the temperature of the stream between the burner and the oven. Those cooler recirculated gases produce a cooler "hot mix temperature" in a manner similar to (but less effective than) that of using excess air (see figs. 3.17, 3.18, 7.6, and 7.7). Control for this case should involve at least two T-sensors. In a batch oven or furnace, the sensors can be placed in contact with a piece of the load, one at the center of the load, heightwise, one on the incoming gas side (left, high limit), and one on the returning gas side (right, input control).

While the furnace gases pass along or through the material that is to be heated, they lose temperature, raising two questions: (1) When the load piece at the point of first contact with furnace gases has reached the desired temperature, what is the temperature of the last load piece at the point where the gases leave? (2) When the coldest part of the load has reached the desired temperature, how much is the hottest part of the load overheated?

The preceding two questions cause one to wonder how to evaluate a log mean temperature difference for the purpose of calculating the heat transfer to the load. There is a practical answer to this and to how to get the most even temperature distribution within the load: Use enough blower power and velocity to assure a temperature drop in the gas stream less than the allowable temperature difference within the load, in which case use a simple average temperature drop for the calculation (see table 3.2).

**Example 3.6**: A forced convection oven, 5 ft wide × 10 ft from front to back, with 1100 F hot recirculated gases, is to heat 1500 lb/hr of steel disks, 2 ft in diameter and

*Reynolds number, a ratio of momentum forces to viscous forces, *Nr* or $Re = (\rho)(V)D/\mu$, where $\rho$ is fluid density, $V$ is fluid velocity, $\mu$ is fluid viscosity (absolute), and $D$ is some significant dimension such as the diameter of a pipe. Units used must all cancel out, that is, make *Re* a dimensionless number. Example: $Re = (\text{lb/ft}^3) \times (\text{ft/hr}) \times \text{ft/(lb/hr ft)}$. Try canceling out the same units in numerator and denominator, and you have no units left—a dimensionless number. As an example, the change from laminar to turbulent flow inside a pipe (where $D$ is the inside diameter of the pipe) is in the range $Re = 2100$ to 3000, no matter what units are used.

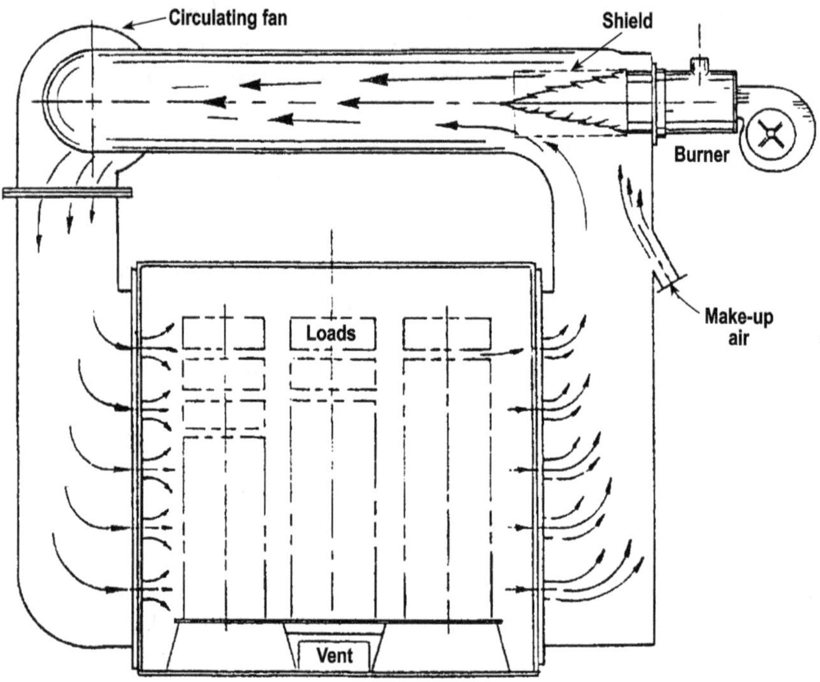


**Fig. 3.17.** Batch recirculating oven passes gases through the loads many times, saving fuel. The circulating gases have burner poc, and thus help uniformity.

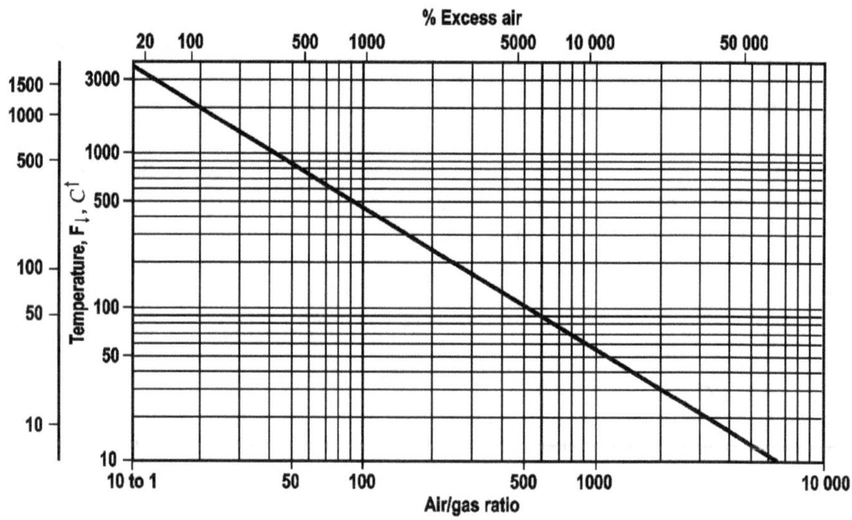


**Fig. 3.18.** More excess air and more recirculated gases reduce the temperature rise of the oven gases, lowering the hot-mix temperature. Courtesy of Dick Bennett's "Energy Notes" in the Sept. 1999 issue of *Process Heating*.

**TABLE 3.2. Heat transfer coefficients, $h_r^*$ for ovens and low-temperature furnaces with gas temperature 100°F (55.6°C) higher than final load temperature**

| | | Radiation coefficient, $h_r^*$, in Btu/ft$^2$hr°F, kW/°C m$^2$ | | | |
|---|---|---|---|---|---|
| Gas Temp (F, C) | Area ratio, load/wall | Oxidized steel or copper | Bright steel or copper | Oxidized aluminum | Bright aluminum |
| 800, 427 | 0.4 | 5.6, 32 | 2.8, 16 | 1.1, 6.2 | 0.4, 2.3 |
| 800, 427 | 0.7 | 4.0, 22 | 2.0, 11 | 0.7, 2.8 | 0.3, 1.7 |
| 800, 427 | 1.0 | 2.4, 13 | 1.2, 6.8 | 0.4, 2.3 | 0.2 1.1 |
| 1000, 538 | 0.4 | 8.1, 46 | 4.1, 23.3 | 1.6, 9.1 | 0.5 2.9 |
| 1000, 538 | 0.7 | 5.8, 33 | 2.9, 16.5 | 1.1, 6.2 | 0.4, 2.8 |
| 1000, 538 | 1.0 | 3.5, 20 | 1.8, 10.2 | 0.7, 4.0 | 0.2, 1.2 |
| 1200, 649 | 0.4 | 12.0, 68 | 6.0, 34 | 2.3, 13 | 0.7, 4.3 |
| 1200, 649 | 0.7 | 8.6, 49 | 4.3, 24 | 1.6, 9.1 | 0.5, 3.1 |
| 1200, 649 | 1.0 | 5.2, 30 | 2.6, 15 | 1.0, 5.7 | 0.3, 1.8 |
| 1400, 760 | 0.4 | 16.2, 92 | 8.1, 46 | 3.1, 17.6 | 1.0, 5.7 |
| 1400, 760 | 0.7 | 11.6, 66 | 5.8, 33 | 2.2, 5.7 | 0.7, 3.9 |
| 1400, 760 | 1.0 | 7.0, 40 | 3.5, 19 | 1.4, 7.9 | 0.4, 2.5 |

*For convection at 20 fps, add about 2.5 Btu/ft$^2$hr°F, 14 W/°C m$^2$; at 40 fps, add about 4.0 Btu/ft$^2$hr°F, 23 W/°C m$^2$.

0.20-in. thick and weighing 25 lb each to 1050 F. If the oven is charged with ten disks at a time, what hot gas velocity is required?

**Procedure:** Solve Equation 3.1 for the required $h_c$; then use equation 2.3 to calculate the required velocity, or work backwards through table 3.2 to find a velocity that will provide the required $h_c$. From the required velocity and flow area of the oven, the required circulation volume can be calculated.

**Solution:** Calculate the required $q$. The time required in the oven will be $t = (10$ disks $\times$ 25 lb)/1500 lb/hr) = 0.167 hr or 10 min for each batch of disks. The exposed steel surface area for each batch = $A = 10$ disks $\times$ 6.28 ft$^2$ (both sides) = 62.8 ft$^2$. The weight in the oven will be $w = 10$ disks $\times$ 25 lb = 250 lb. The average specific heat of steel in the 60 F to 1100 F range is $c_p = 0.135$ Btu/lb°F, the initial receiver temperature, $T_{ri} = 100$ F; $T_{rf} = 1050$ F; the initial source temperature, $T_{si} = 1100$ F. (A guideline might be that the system should provide sufficient convection so that source temperature "droop" $(T_{si} - T_{sf})$ will be less than the $\Delta T$ tolerance in the final temperature throughout the load.)

From the specific heat equation, the required heat input for each batch of 10 disks will be

$$\begin{aligned} Q &= w\, c_p \text{ (temperature rise or } T_{sf} - T_{si}) \\ &= (250\text{ lb}/0.167\text{ hr}) \times 0.135\text{ Btu/lb°F} \times (1050 - 100) \qquad (3.3) \\ &= 192\,000\text{ Btu/hr (available heat, not gross).} \end{aligned}$$

Interpolate the mean $h_r$ (the mean coefficient of radiant heat transfer from figure 3.16 for somewhat oxidized steel and a load/wall area ratio of about 0.8) as about 5 Btu/ft$^2$h°F.

$$\text{Log mean } \Delta T^* = \frac{[(1100 - 100) - (1100 - 1050)]}{\text{Ln}(1000/50)} = (1000 - 50)/3.0 = 317°\text{F} \tag{3.4}$$

The required overall coefficient of heat transfer, $U$, can now be calculated by solving equation 3.5 for $U$ (dividing both sides of equation 3.1 by $\Delta T$).

$$U = \frac{Q/A}{\Delta T} = \frac{192\,000 \text{ Btu/hr}}{(6.28 \text{ ft}^2) \times 317} = 9.6 \text{ Btu/ft}^2\text{hr°F}. \tag{3.5}$$

$U = hr + hc = 9.6$. From above $h_r = 5$, so $h_c$ must be $9.6 - 5 = 4.6$ Btu/ft$^2$hr.

Solve equation 2.5 from chapter 2 for velocity, $V$. The density of the boundary layer, ρ, at 600 F mean film temperature, from table A2.a of *reference 51* is 0.0375, therefore, $hc = 4.6 = 7.28(\rho)(V)^{0.78} = 7.28(0.0375)(V)^{0.78}$, and using an engineering pocket calculator, $V = 37.8$ fps bulk stream velocity required.

Alternatively, by interpolation in table 3.2 find that an $h_c$ of 4.6 will be attainable with a bulk stream velocity of about 40 fps. The oven and its loading configuration must provide a circulation pattern to assure at least 38 fps hot gas flow across all the load surface. If the flow is end to end with baffles arranged for 10 sq ft of cross-sectional area, the fan will need a capacity of 10 ft$^2$ × 38 ft/sec = 380 cfs at 1100 F. The temperature of the loads at the cooler end of the furnace will depend on the method of loading. To attain a minimum temperature difference between the loads at the two ends, the loads should be charged at the cool end first and removed from the hot end last. Good control practice is to drop the circulating gas temperature to 1050 F as soon as the loads at the hot end reach 1050 F.

### 3.8.2. Drying and Preheating Molten Metal Containers

Drying and preheating molten metal containers—crucibles, pots, ladles—must be performed slowly and evenly to avoid damaging their refractory lining. These dryout and preheat jobs involve low temperature inputs to refractory-lined chambers built for high temperature. After initial or relining, these vessels must be dried out very slowly (*a*) to avoid trapping vapor below the finished surface and (*b*) to properly cure the refractory minerals. That requires high air circulation to carry away the evaporated liquid vehicle, that is, mass transport. (See sec. 4.2.)

*Logarithmic mean temperature difference (LMTD) is described on pp. 126–128 of *reference 51*. It corrects for the curvature of the temperature lines from beginning to end of the heat process whether over time as in batch furnaces or over distance in continuous furnaces. A rough method uses a "$\frac{2}{3}$ rule" that estimates the mean receiver (load surface) temperature will be the initial load temperature plus $\frac{2}{3}$ of the receiver load surface temperature rise, $T_{rf} - T_{ri}$, or in Example 3.6, LMTD $= 100 + (\frac{2}{3})(1050 - 100) = 733$°F.

The dangers in these jobs are overheating the surface and undercuring the interior of the wall-lining material. Use of excess air and much recirculation to maintain low hot mix temperatures (see glossary) are common practices. This might suggest using high-velocity (high-momentum) burners to induce more carrier air to evacuate the evaporated liquid, but care must be taken to avoid impingement hot spots in target areas and sidewall areas too close to the burners. Because drying and preheating burners must often be positioned in pouring openings, the design engineer may be confronted with little choice of burner flame configuration and position for optimum drying or preheating.

With thick rigid refractory linings, there is danger of fracture from shock thermal expansion when they are cold and suddenly filled with molten liquid; thus, they are usually preheated before every filling. The dryout burners also are usually used for preheating, but a different time-versus-input program should be used. It is wise to seek the advice of the refractory supplier or both dryout and preheat cycle timing.

The need to do the preheating before every use forces most installations to build a dry/preheat station convenient to the operation. For very large ladles, this "station" may be a vertical wall of folded ceramic fiber, with a burner installed in the center of the wall, firing horizontally. The ladle is laid on its side on a platform on wheels on rails so that the ladle can be rolled snugly against the fiber wall. The poc flue through leaks between the ladle and the wall, mostly at the top. Different controlled/timed cycles are advised for various sizes, materials, and thicknesses.

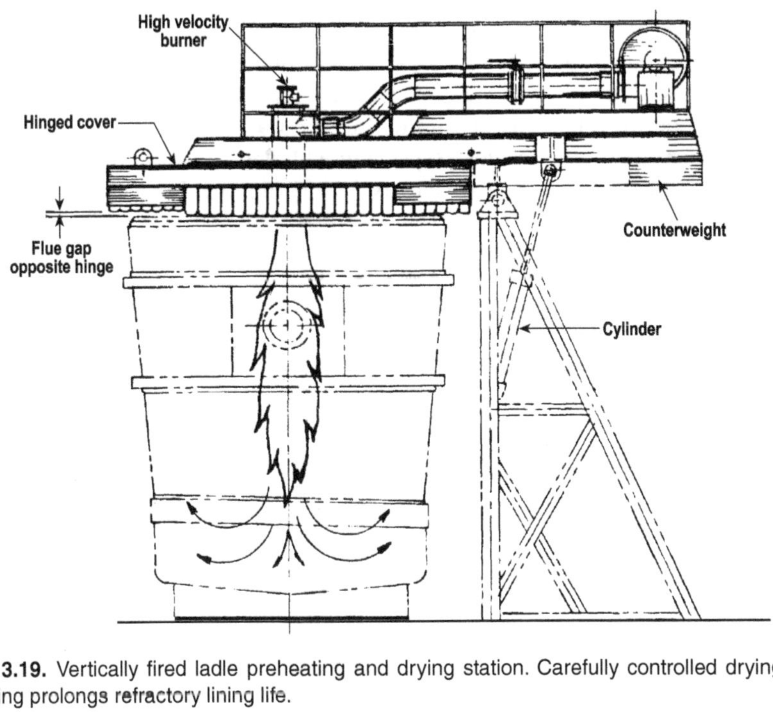


**Fig. 3.19.** Vertically fired ladle preheating and drying station. Carefully controlled drying and heating prolongs refractory lining life.

Another configuration is shown in figure 3.19, wherein the ladle is kept right side up. In both vertically and horizontally fired arrangements, it is necessary to provide a burner/flame configuration that reaches to the bottom of the ladle with sufficient velocity and excess air to provide the vehicle for both convection and mass transport, especially during drying. A high-momentum flame is preferred to drive heat to the ladle bottom, assuring hotter gate and porous plug areas.

### 3.8.3. Low Temperature Melting Processes

Lead, solder, and other materials that melt at temperatures below 1000 F (537 C) are melted in a variety of steel alloy containers, usually in small batches. Carefully positioned, small premix type A flames or nozzle-mix type E or H flames (fig. 6.2) are used within fiber-lined furnaces. Figure 3.20 shows the use of pairs of tangentially fired regenerative burners around a melting container to take advantage of the alternating firing of regenerative burners to even out temperatures around the periphery, prolonging container life.

*Galvanizing tanks or kettles* (batch or continuous) may contain tons of liquid zinc or alloy into which steel articles are dipped to give them a protective coating to inhibit rusting. Small to large units handle items from fasteners to pipe to highway guardrails. A refractory furnace surrounds the sides of the liquid holding tank (alloy steel), but the top is open for access for dipping the articles to be coated manually, by crane, or by conveyor.

In figure 3.21, careful choice of burner type, size, and position is essential to avoid hot spots on the tank wall, which shorten the tank life. When one of these fails, a pit full of solidified zinc is an expensive and time-consuming recovery operation. Type E (fig. 6.2) swirled flat-flame burners are excellent for spreading heat sideways in the narrow space between the tank and inside furnace wall. However, long tanks need many such burners, raising the cost, especially with flame monitoring devices. This problem has forced the use of high-velocity type H (fig. 6.2) burners at two corners

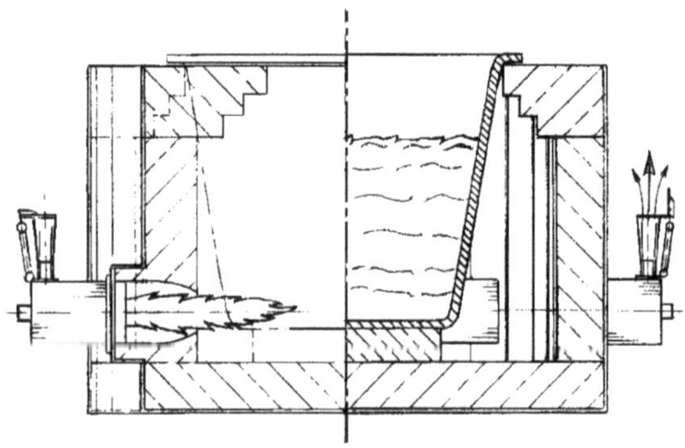

**Fig. 3.20.** Large metal melting pot furnace. With large containers, tangential heating minimizes nonuniformity around the periphery. More small type E or type H burners usually help. (See also fig. 1.15.)

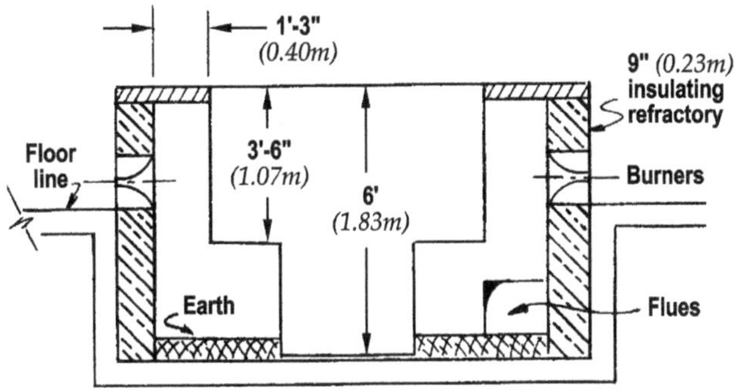


**Fig. 3.21.** Sectional view through a galvanizing tank or kettle.

of the tank, firing horizontally along the long sides of the tank. The size and position of such burners are crucial to avoid hot spots, with their devastating effect on tank life. A recent large galvanizing tank was designed for a net sidewall input of 9500 Btu/ft$^2$hr.

### 3.8.4. Stack Annealing Furnaces

Stack annealing furnaces are bell-type furnaces in which stacked coils of steel wire or strip are heated to about 1250 F (680 C), copper heat treated at 500 to 900 F (2.60 to 480 C) (see figure 3.12). They may be direct fired or indirect fired, depending on the materials being annealed. "Cover annealing furnaces" have a gas-tight inner cover or "bell" within the bell furnace in which a prepared atmosphere is circulated by a base fan. Radiant tubes may be used instead of an inner cover. (See fig. 3.22.)

If the properties of the material being heated could be adversely affected by slight overheating, the difference between furnace gas temperature and final load temperature must be kept small, especially if the heated material has poor thermal conductance. This combination of two requirements is encountered in the annealing of thick coils of thin strip steel.

Most cover annealers are single stack furnaces, but there are some multistack annealers with three, four, six, or eight stacks, each with a bell cover, all within one rectangular furnace. (Radiant tubes were used in addition to the inner covers in the past because of poor heating between the inner covers.) Now, type H high-velocity burners are fired down or up between the inner covers.

Although the strip is coiled under tension, successive wraps do not have continuous contact with one another because the apparently smooth surface of the strip has microscopic irregularities. These thin spaces are filled with trapped air, which has very poor thermal conductivity. The result is that the heating time may be more than 2 hr per inch of coil radial thickness.

For annealing commercial-quality steel strip, the goal is no more variation than 70 F (39 C); for deep-drawing quality, no more than 34 F (19 C). Cooling times under the inner cover may be almost as long as the heating cycle. With wider and

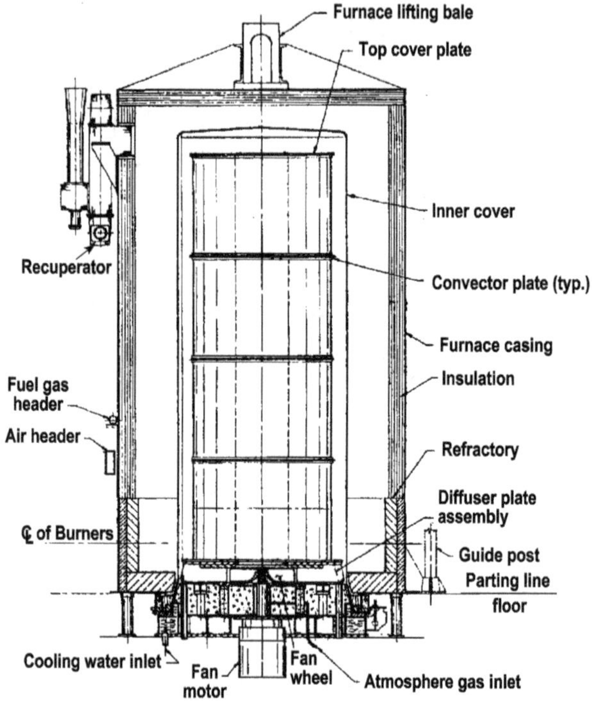


**Fig. 3.22.** Single stack cover furnace with four-coil load. Recuperator with suction Venturi is the size of a person. Circulating fan in base drives prepared atmosphere through coiled strip under alloy cover. Bell-type furnace is lowered over a loaded inner cover. One or two circles of high-velocity, tangentially fired burners fire between the inner bell cover and the and outer bell furnace.

longer coils, total time may be one week. This is the reason why there are acres and acres of these furnaces needed to keep up with growing automobile needs.

As wider strip needs to be annealed, there is greater heat soak distance to the center of each coil. Delivering heat to the innermost laps has become the governing factor determining production rate. Higher power fans enhance internal convection. Tests by Lee Wilson Engineering Co. found that heating time was about 1.2 hr/axial inch from each coil end to the coil's midwidth for commercial quality strip, and 1.6 hr/axial inch for deep-draw quality (or about 0.47 hr/axial cm for commercial quality or 0.63 hr/axial cm for deep-draw quality).

Various methods have been used to promote faster heating and cooling of large coils, such as (*a*) using hydrogen (an excellent conductor) within the cover, (*b*) loosely winding coils to allow more gas to be forced between the laps, (*c*) adding convector plates to let hot gases flow between the stacked coils, and (*d*) placing a large solid "star" (fig. 3.24) in the hard-to-heat middle of the coil (1) to force hot gases to "convect" faster along the inner surface of the coil, and (2) to absorb heat from the hot circulating gases and then re-radiate that heat toward the inner surface of the coil.

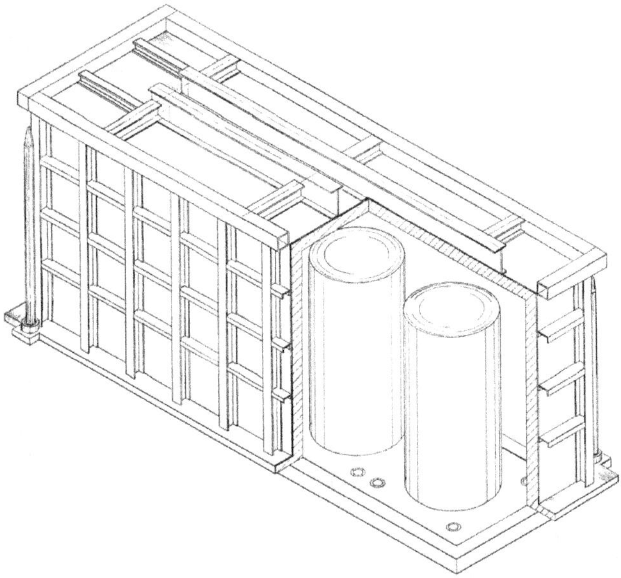

**Fig. 3.23.** A multistack annealer can be difficult to heat uniformly. Bottom-up firing (shown) or top-down firing is recommended.

### 3.8.5. Midrange Heat Treat Furnaces

Midrange heat treating, steel and glass, 1200 to 1800 F (650 to 980 C), includes glass annealing lehrs and steel heat treating furnaces (hardening, annealing, normalizing, etc.). Batch heat treating furnaces may be direct fired or indirect fired (usually with a prepared atmosphere and radiant tubes). Their sizes and shapes are numerous and governed by the necessary method for handling the loads. Simple box furnaces and car-hearth, lorry-hearth, or car-bottom batch heat treat furnaces are some of the most common configurations.

Bottom flueing is preferred, but in-the-wall vertical flues have been found too costly, and they pull a harmful negative pressure at the hearth level. With top firing, the best arrangement is hearth-level flues with automatic furnace pressure (damper) control. If fired with top and bottom burners, use of a roof flue with automatic furnace pressure control is suggested. The flue location should be determined to enhance the design circulation pattern. (See chap. 7.)

The heating capacity of furnaces that operate within this temperature range can be determined in the same manner as that used for high-temperature furnaces. (See sec. 3.8.8.) Although this midtemperature level needs less heat to be imparted to each unit weight of load, the heating time is longer and heating capacity is lower because heat transfer by radiation is weaker than it is at higher temperatures, as shown in figure 2.16. The coefficient of heat transfer from 1600 F to 1200 F is about 40% of the coefficient for the same 400°F difference between 2200 F and 1800 F, but that decrease is counterbalanced by the lower amount of heat required.

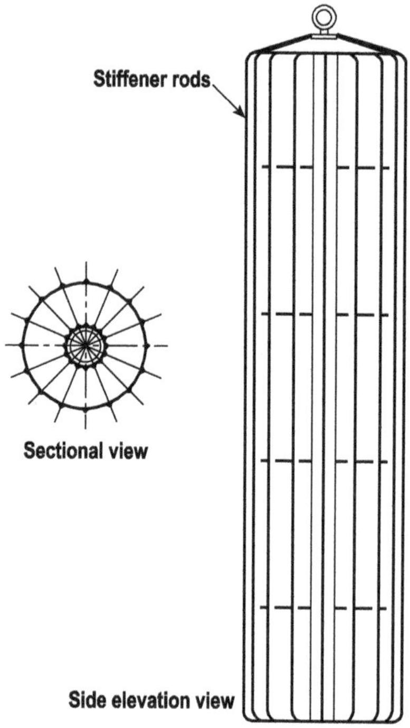


**Fig. 3.24.** Shannon Star, for placement in the center hole of a strip coil, breaks up the center core gas stream, forcing the center space gases to wipe away the stagnant boundary layer on the inner lap of the coil. The stainless-steel central post and radial fins do more than a convection "corebuster" because they also absorb heat from the core gases and then provide a lot of re-radiating surface that heats the inner surface of the coil.

If there is an operation bottleneck because of lack of heating capacity of a furnace in this temperature range, control techniques are available to increase capacity by raising the temperature of the furnace above the final product temperature. If bright metals such as stainless steel or titanium are to be heated, the rate of radiation will be low because of their lower emissivity (eq. 2.6); therefore, convection velocity should be increased. An excess of furnace or gas temperature over the desired final load temperature is permissible with steel provided the hottest location has a T-sensor to automatically control heat head. A flue gas temperature somewhat higher than the final load temperature can be used with aluminum because of its lower absorptivity and higher thermal conductivity.

For **heat treatment of railway wheels**, see sec. 7.4.5.1.

### 3.8.6. Copper and Its Alloys

Copper and its alloys are often heated to temperatures within this midrange and above (see figure 3.25.)

To compare heating (soak) times and production rates of copper alloys with those of steel, use equations 3.6 and 3.7, both based on the ratio of diffusivities. (See also eq. 3.2a and 3.2b and fig. 3.25.) Thermal diffusivity (see glossary), $\alpha$ = thermal conductivity divided by volume specific heat, $k/c(\rho)$.

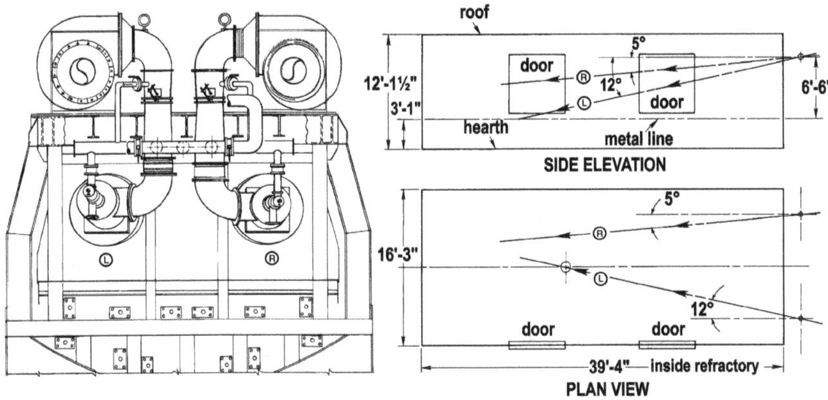


**Fig. 3.25.** Tilting copper remelt furnace operated as high as 2600 F (1427 C) with dual-fuel, fuel-directed, ATP burners, using retractable atomizers and up to 4% oxygen enrichment. 400 tons per day.

$$\text{Soak time for material b} = (\text{known soak time for material a})\,(\alpha_a)/(\alpha_b) \qquad (3.6)$$

The productivity, weight heated-through per unit time, is directly proportional to the ratio of the diffusivities:

$$\text{Weight/time for material b} = (\text{weight/time})_a\;(\alpha_b/\alpha_a) \qquad (3.7)$$

Judging from the previous formulas and the difference in temperature levels, a *guideline* might be to allow about two times as much time for copper to be heated psf exposed. As for steel, see equations 3.1 and 3.2, and figure 2.11.

## 3.8.7. High Temperature Batch Furnaces, 1990 F to 2500 F (for forging steel pieces 12" [0.3l m] and smaller, see sec. 6.10)

To increase the capacity of high-temperature batch furnaces such as those used for forging and rolling large thick loads, the major objective should be to heat the whole load uniformly from charge to draw time, by observing the following general recommendations. Applying these recommendations will improve product quality, raise productivity, and lower fuel use. If heating rates are to achieve (and continue at) high levels, the air/fuel ratio controls, furnace pressure controls, and temperature controls must be kept in good operating condition. "Controls" include controllers, sensors, and actuators.

Use two-side heating by placing the load(s) on piers and firing above and below them. Any load more than 4" (0.1 m) thick should be placed on piers in the furnace so that the loads do not have cold bottoms. The piers should be a minimum of 8" high (0.2 m) so that underfiring can be used to heat the pieces from below (and traditional overfiring to heat from above). If the load pieces must be placed in the furnace in several layers (not good for good surface area exposure), they should be spaced apart to allow convection and radiation to reach all surfaces. More than two layers is unwise, unless horizontal spacers are used with forced circulation between layers.

Piers and spacers themselves can add to the mass of the load and absorb useful heat that should have gone to the load; therefore, make them light and open to encourage convection and radiation through the interstices. Admittedly, lightweight spacers may not last as long as massive reject billets or highway-divider-like refractory shapes, but the lightweight spaces will not stretch the cycle time while the gas meters and the time clocks spin.

Load the furnace with piece-to-piece centerline distance about twice the piece thickness. (See the first paragraph of sec. 3.4.) No load should be closer to a furnace wall than one-half of the thickness of the piece.

Use adjustable thermal profile burners above the load on one side of the furnace. Control these burners by two temperature sensors, each at the level of the top of the load—one in the burner wall and one opposite. Bring the two temperatures up as one by controlling the spin of the air through the burner. Follow the fuel input until minimum fuel input is registered in all zones. Add 1 hr for thin loads and 2 hr for thick loads, then draw the first piece.

Divide the furnace into lengthwise zones, two very small end zones, with the center space as one or, preferably, two zones.

Enhance furnace bottom temperature with many small high-velocity (high-momentum) burners, firing with constant air, variable fuel, that is, excess air as they turn to low fire, to hold the same temperatures below the load(s) as above. Install fuel meters on each zone. When the fuel flows in all zones reach their minimums, hold as long as necessary for the required minimum temperature differential between surface and core, as determined from time–temperature heating curves. Then remove and process the loads.

***3.8.7.1. Certification*** To sell their products, forging suppliers must meet their customers high-quality standards by holding to increasingly tight temperature tolerances. Often, a furnace temperature uniformity test must be performed and certified on an empty furnace. Certification without loads in a furnace may be an improvement over no testing, but putting loads in the furnace changes firing rates, gas movement, and heat transfer at nearly all locations in the furnace. Temperature uniformity within each zone from charge to draw saves time, often 25%. Production benefits accrue from the shorter time cycles. If uniform product temperature is to be achieved, better means of internal furnace temperature control must be developed for use both above and below the loads, for example, adjustable thermal profiling and step-firing.

***3.8.7.2. Control Above the Load(s)*** With the advent of the fuel-directed burner, two temperature locations in a longitudinal direction can be held at the same or a constant difference in temperature. Therefore, firing across the width of a furnace above the product can be controlled to a nearly flat temperature profile regardless of the product size or location.

In addition to the two-point temperature control, other temperature measurements and control loops in each zone can be added to act as control monitors. Through low select devices on the output signal, these monitors can automatically take control of energy input to prevent hot spots. With sufficient monitors, overshooting of product temperature can be eliminated.

With this type of control system and burners, the temperature control above the product can be excellent if sufficient zones are installed. The minimum number of zones should be three: one for each end wall and one for the main body of the furnace. If there are two side-by-side doors, five zones are desirable: one for each sidewall, two for furnace body, and one behind the doorjambs in the furnace center.

Control below the load(s) depends on the load location. If the product is placed on the hearth, the temperature difference top to bottom will never be uniform and will depend on the following:

1. Product thickness. Greater thickness will increase temperature differences.
2. Product shape. Rectangular pieces are a greater problem than round pieces.
3. Hearth heat loss. Reducing hearth heat loss reduces temperature nonuniformities in the product.
4. Scale thickness. More scale on the hot faces of the product means poorer temperature uniformity and slower heat transfer. As loose scale accumulates in the spaces between the piers, it will disrupt the flow of gases through that tunnel, further upsetting temperature distribution. High-pressure air-jet pipes at one end of each tunnel and operated when there is no load in the furnace will help keep the tunnels clean, but the end spaces need frequent manual cleanout.
5. Number of sides exposed to heat transfer. More are better. Under no circumstance should loads be piled on top of one another.

Every effort should be made to provide space between the loads and the hearth, particularly for loads more than 4 in. (100 mm) thick. Loads more than 6 in. (150 mm) thick should not be placed on a hearth unless their center-to-center distance is at least twice their thickness.

Load height above the hearth (pier height) should be sufficient to avoid overheating of the undersides of the load by flame impingement from the underfiring burners; therefore, the burner supplier should be consulted. (See enhanced heating by circulation in chap. 7.) If the management cannot be convinced to fire under the loads, 4 in. (100 mm) clearance (pier height) will be better than nothing, but the clearance must be maintained by periodic removal of scale or all the gain will be lost.

For truly uniform temperature across the bottoms of the load pieces, approximately equal clearances under and above the loads must be provided, plus equal firing. Equal firing treatment above and below may not be practical in many high-temperature jobs. The following provides some "judgment numbers" for installation of enhanced heating "pumping burners" firing between the piers. Such burners not only add their own products of combustion but induce three to five times their own poc mass from the furnace gases above. The clearance (pier height) should accommodate the flame of a small, very high velocity burner with at least 150% excess air flame stability. Generally, satisfactory temperature uniformity across the furnace wll be attained if the burners are spaced 30 in. (0.76 m) apart or less, firing across an 8 ft (2.4 m) hearth, each with one million gross Btu/hr (1.055 GJ/h) input or less, each with maximum velocity of combustion products leaving the burner tile of 200 mph (322 km/h), or a tile pressure of at least 4 in. (102 mm) of water column.

To assure minimum bottom temperature difference across the furnace width of the load, two T-sensors should be installed, one on each side of the furnace (arrows #3 and #4 in fig. 3.26). The #4 T-sensors should be positioned 1 to 3 in. (25 to 75 mm) above the pier top in the wall opposite the high-velocity burners, controlling the fuel input (with combustion air flow held constant). The #3 T-sensor should be at the same elevation as the #4 sensor, on the same side as the high-velocity burners. In a heavily loaded furnace at forging temperature, the two opposite lower sensors should be within ±6°F (3.3°C) of one another.

To keep the temperature differences small within the load(s) across the furnace, heat transfer beneath the load from the gas blanket to piers and product must be kept relatively low. To minimize heat transfer from the gas stream, the thickness of the stream must be very small (8 to 12 in., or 200 to 300 mm), and the percentage of triatomic gases in the products of combustion must be low. Excess air will lower the percentage of triatomic gases and reduce the temperature drop of the gas stream under the load from the burner wall to the opposite wall.

Pier mass should be kept to a minimum to reduce the need for extra fuel to heat the piers. That heat would have to be supplied by the gases moving below the load, adding to the temperature loss of those gases, and therefore adding to the temperature nonuniformity of the undersides of the load(s) along the length of the pier tunnel. The underfiring tunnels must be kept clear of scale to avoid impeding the gas flow.

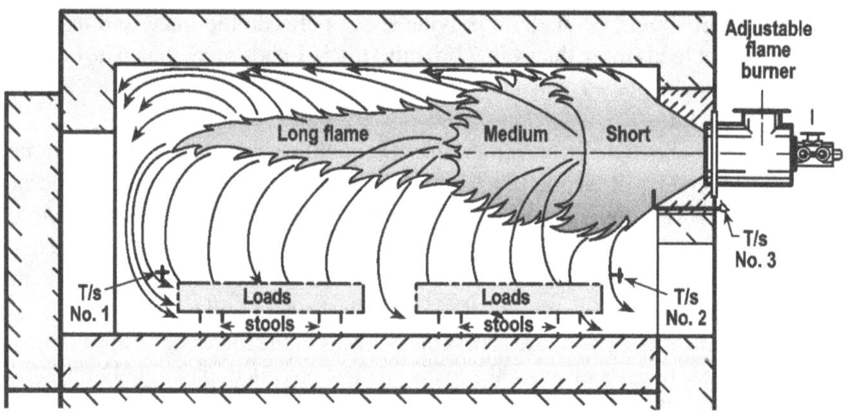


**Fig. 3.26.** Batch furnace for good uniformity control, with top backwall fired by adjustable thermal profile burners and bottoms of sidewalls fired by high-velocity burners; multiple T-sensors on both sides. Flow lines show the sweeps of gases of the ATP burners' spinning short mode flames, medium length flames, and long mode flames. (See also figs. 2.21, 6.1, and 6.23.)

Good temperature uniformity requires that flues be positioned to minimize interaction between zones. With the above "enhanced heating" scheme, the temperature profiles above and below the loads will be very flat, providing very low temperature differences within the product even with a variety of loads and loading patterns.

The above enhanced heating and controls cannot provide uniform temperatures if the charge is not logically placed on the piers. For example, untrained operators may pile loads on top of one another, restricting heat transfer to one or more pieces, which may then have less than one side exposed to radiation and/or convection. The result will be that their cores will be too cold to forge or roll. Care also must be exercised to avoid placing load pieces too close to a sidewall where very little hot gas moves, causing one side of the piece to be very cold. Persons who load furnaces must be made aware of the importance of their work in maintaining quality products.

*Increasing high-temperature batch furnace capacity.* Most of the wasted production capacity of batch furnaces comes from uneven heating that requires sitting and soaking out the temperature irregularities. The gas meter is usually still spinning during this temperature-evening-out period. Thus, whatever improves production rate usually improves fuel economy as well. The principal improvement in productive capacity of high-temperature batch furnaces can be made by heating the whole load uniformly, charge-to-draw, by the following general means:

1. Two-side heating with the load on piers and firing above and below the load.
2. Charge the furnace with the load centerline distance between pieces at least twice the thickness of the pieces. In addition, no load pieces should be closer to the walls than one-half the piece thickness.
3. Install adjustable profile burners above the load on one side only. Control these burners by two thermocouples, one on each side of the furnace and each at the height of the top of the load. Bring the two temperatures up as one. Follow the fuel input until minimum fuel input is registered in all zones. Add an hour or two, then draw the first piece.
4. Divide the furnace lengthwise in a minimum of three zones. Four zones is an even better approach. Construct the furnace into two very small end zones with the large center space divided into one or two zones.
5. Control the furnace bottom temperatures with many small, high-velocity burners firing with constant air to hold the same temperatures below the load as above it. Install fuel meters on each zone. When the fuel flows reach minimum in all zones, hold for several hours, then remove the load from the furnace for processing. The benefits will accrue from shorter cycles, many times by 25% because uniformity of zone temperatures is held from charge-to-draw requiring minimum soak time.

An alternative to adjustable thermal profile burners above the loads for topside crosswise temperature uniformity might be staggered opposed regenerative burners because the alternate firing from right then left would help develop "level" temperature patterns, as is done with regenerative burners on both ends of a long radiant tube. However, this would require a similar concurrent alternating of the small high-velocity tunnel burners below, which could be done with pulsed firing.

To achieve ongoing high production rates, low fuel rates, and good temperature uniformity, everyone—management, operators, maintenance people—must be aware of sensible loading practice, and that there are many other furnace items that need constant care. These include air/fuel ratio control, furnace pressure control, and temperature (input) control—all of which must be maintained in top operational order if heating rates are to be held at high levels. "Control" does not just mean the controller, but the whole control system—sensor, controller, actuator, and all connections among them.

### 3.8.8. Batch Furnaces with Liquid Baths

Heating solids by immersion in liquid baths happens by convection. For viscous liquids (liquid salts and liquid metal), motion is so minor that conduction is the primary heating mode. Conduction transfers heat to the load pieces so much more rapidly than from flame to bath liquid that the conduction resistance between liquid and solid surface often can be ignored. Soak time from the solid surface to solid core might be a consideration in salt baths or liquid metal baths if the load pieces are of very heavy cross section.

Factors affecting liquid bath heating capacity are: (1) the surface transferring heat to the bath must be large enough to permit required heat flow without damaging the container or the liquid, and (2) a good practice consensus is that the volume of the bath must be large enough that immersion of the load(s) will not reduce the bath temperature by more than 25 F or 14 C, which translates to equations 3.8 and 3.9, based on the specific heat equation, $Q = w\,c\,\Delta T$, where $Q$ is Btu or kcal, $w$ is weight in pounds or kg, $c$ is specific heat, $\Delta T$ is temperature change in °F or °C:

$$(\text{wt} \times \text{sp ht} \times 25)_{\text{bath}} \text{ must} = \left[\text{wt} \times \text{sp ht} \times (T_{out} - T_{in})\right]_{\text{load}}. \tag{3.8$_{\text{US}}$}$$

$$(\text{wt} \times \text{sp ht} \times 14)_{\text{bath}} \text{ must} = \left[\text{wt} \times \text{sp ht} \times (T_{out} - T_{in})\right]_{\text{load}}. \tag{3.9$_{\text{SI}}$}$$

Weight of the "load" includes any containers, hooks, and conveyors that might be immersed in the bath.

In addition to the heat to be imparted to the total load during immersion (right side of eq. 3.8 and 3.9), heat input is needed to make up for loss from an uncovered bath surface by radiation and convection. Emissivity ($e$) of a *salt bath* is approximately 0.9. *Lead baths* are purposely covered with lead oxide ($e = 0.63$) and with charcoal (estimated mean $e = 0.7$) to reduce radiation and convection heat loss and to minimize oxidation.

Crucible or pot furnaces are used for *melting and alloying brass and other nonferrous alloys* in small foundries. They need very uniform heating around the container periphery to prolong pot life. Container replacement cost is a major item for small foundries. Alternate firing of centrifugally aimed regenerative burners greatly lengthens container life.

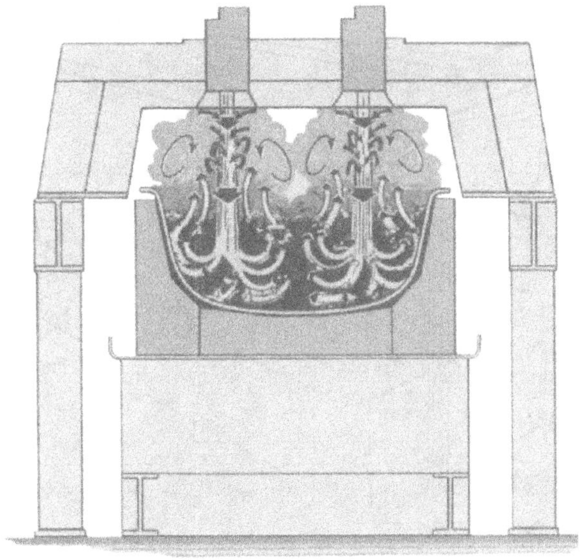

**Fig. 3.27.** Scrap preheater with high-momentum flames driving through the interstices of iron scrap, to preheat it prior to big ladle melting, and to incinerate paint and oil on the scrap.

Small liquid bath furnaces, including foundry pot furnaces and small salt bath furnaces, are sometimes heated electrically by resistors or by induction. Resistors may be positioned between the container and a surrounding insulator or refractory furnace wall, or they may be inserted into the bath from above. In larger units, such as *scrap iron preheating* prior to melting in a large mill ladle, high-velocity flames are directed vertically into the scrap batch. (See fig. 3.27.) All figures in this section 3.8.8 are courtesy of the North American Manufacturing Co.

Molten zinc for galvanizing (surface oxide emissivity 0.1) is contained in open-topped, rectangular steel "tanks" or "kettles," with walls of 1" to 2" boiler plate or firebox steel. Test data on the tank shown in figure 3.28 (*reference 49*) showed that the container wall temperature was more uniform with four type H flames than with 18 type E flames (fig. 6.2), but such comparisons are highly dependent on burner spacing, burner size, and distance from container to wall.

If the heat is transferred through the metallic tank sidewalls, the surface area through which heat is transferred must be large enough to avoid injury to the kettle by overheating (oxidation, warping). The tank walls can be corroded quickly by the zinc if the kettle wall temperature gets too high. Such corrosion is very costly because of danger of a breakout if the steel wall temperature exceeds 900 F (462 C) or if heat transfer to the container wall exceeds 14 000 Btu/ft$^2$hr. Designers aim for 10 000 Btu/ft$^2$hr, hoping that the rate of heat transfer at the hottest spot will not exceed the danger point. Temperature uniformity is very important. Flames must not impinge upon nor be aimed toward the kettle. Burners should have their closest flame surface at least 15 in. (380 mm) from the tank wall.

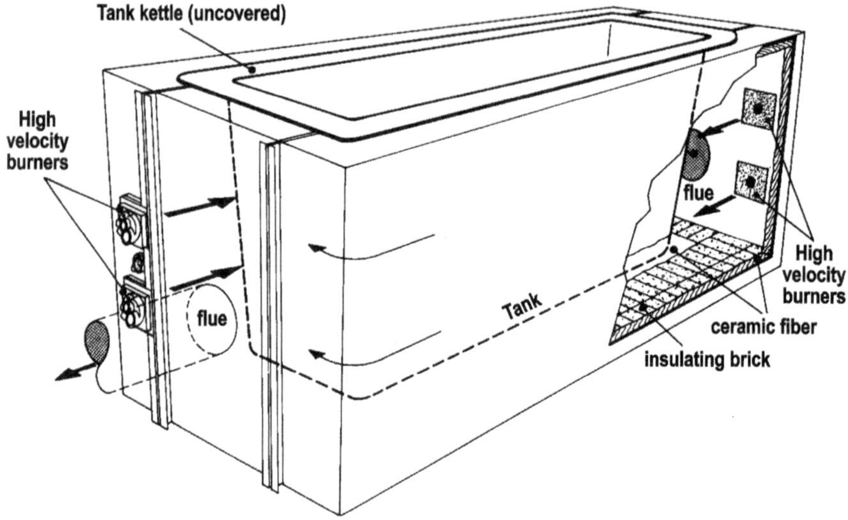


**Fig. 3.28.** Galvanizing tank rebuilt with high-velocity end firing replacing side firing for better tank life and to use fewer burners.

Galvanizing gurus Larry Lewis and Jim Bowers recommend 14 tons of molten zinc in the tank for each ton of load to be galvanized per hour. Others recommend as high as 20:1. Because dross settles to the bottom of the kettle, the kettle should be deep enough that articles to be galvanized will be at least 1 ft (305 mm) above the kettle bottom. For the same reason, heat should be applied no closer to the outside bottom of the tank sidewall than 1 ft or preferably 1.7 ft (0.5 m).

The term *reverberatory* originated because the thermal radiation seemed to vibrate, reflect, bounce, or reverberate around the inside of the furnace. Radiation *is* a vibrating wave phenomenon, but it does not cause noise as "reverberatory" may imply. Maybe Granddad's burner was unstable and therefore noisy, especially with the echo effect of the then-typical high roof (crown), which was probably built that way for easy access by humans for loading or for making repairs.

Unfortunately, the high space above the bath later came to be used to pile a high load of metal pigs, sows, scrap, or "batch," the sandlike raw material in glass melters. The high pile of solid load interfered with refractory radiation and reduced the beam for gas radiation. When told of this problem, some people not only lowered the pile but lowered the roof, diminishing the sidewall refractory radiating capability and the gas beam radiating capability.

Maybe Granddad's way with the high crown and the name "reverberatory" was pretty good after all!

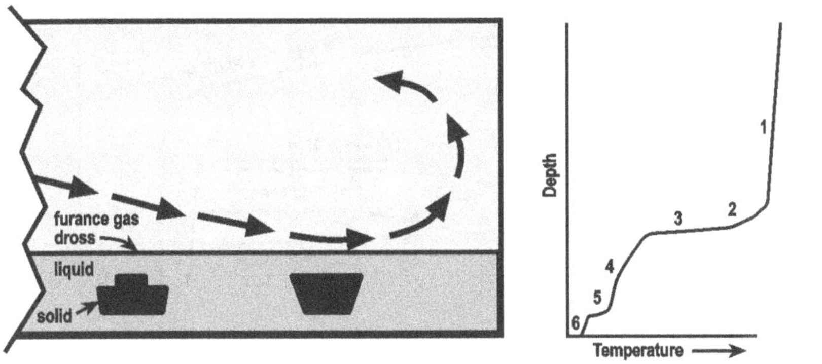


**Fig. 3.29.** Immersed metal solids are hard to heat. Temperature profile (*right*) shows $\Delta T$s through (1) furnace gas, (2) boundary resistance, (3) dross, (4) liquid, (5) sediment, and (6) base.

Most *aluminum melters* and molten *aluminum holding (alloying) furnaces,* as well as *glass melting 'tanks'* and *frit smelters* are refractory-lined 'reverberatory' furnaces. Heat is transferred to the bath from above by radiation and convection. The bath surface must have enough surface area to accept the needed heat transfer rate, right side of equations 3.8 and 3.9, and to avoid harm to the bath/load or refractories above the combustion space.

In a liquid bath used for melting, there may be slow melting of submerged metal solids because of poor liquid-to-solid heat transfer. (See fig. 3.29.) Heating from the top down in a liquid bath depends on conduction or convection. Some stirring or pumping velocity can be supplied to add forced convection heat transfer. The pumping equipment can be expensive to buy and to maintain.

A higher furnace space temperature simply aggravates the steep temperature gradient in the first few millimeters below the bath surface, which with aluminum, lowers the conductivity of the liquid further. (The thermal conductivity of liquid aluminum is much lower than that of solid aluminum—see fig. 3.30.) Raising the furnace space temperature or impinging poc on the bath surface can aggravate the problem by accelerating oxide (dross) formation, which then becomes an insulating blanket between the furnace space and the molten load. Thorough draining of the molten batch helps minimize the effect of the old liquid heel in covering part of the next solid batch, thereby shielding it from exposure to furnace radiation. (See fig. 3.31.)

To better expose solid loads for melting, it is preferable not to cover them with molten liquid, but of course that is the ultimate objective of the furnace! A step in the direction of faster, more productive melting is to completely drain the furnace before charging new solid loads—in other words, to leave no "heel" either liquid or solid. A tilting melter or holding furnace such as shown in figure 3.31 is very helpful in this effort.

**Quality control problems** with melting aluminum and its alloys include oxide (dross) formation and hydrogen absorption. These two phenomena can have a bad effect on product quality by making oxide inclusions or porosity.

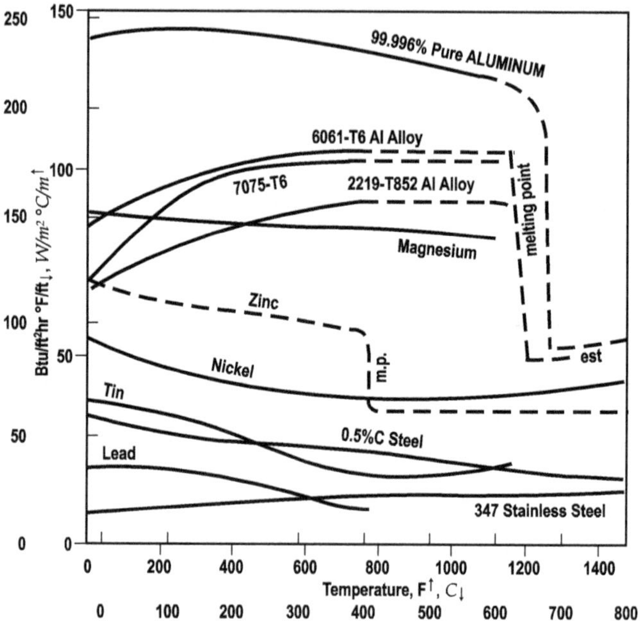


**Fig. 3.30.** Effect of temperature on thermal conductivity of metals. Note the major loss in thermal conductivity of aluminum when it is melted.

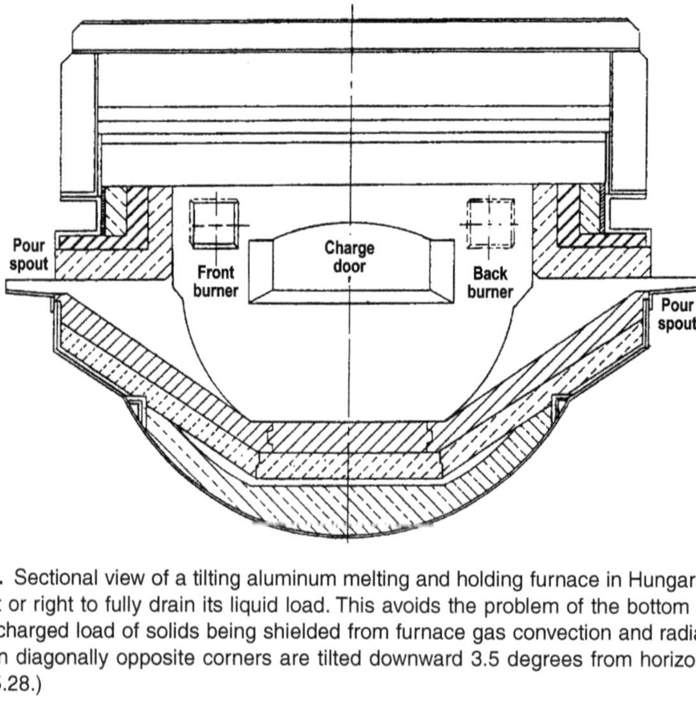


**Fig. 3.31.** Sectional view of a tilting aluminum melting and holding furnace in Hungary that tips either left or right to fully drain its liquid load. This avoids the problem of the bottom portion of the next charged load of solids being shielded from furnace gas convection and radiation. Two burners in diagonally opposite corners are tilted downward 3.5 degrees from horizontal. (See also fig. 5.28.)

Some ways to reduce these problems are:

1. Maintain a leak-tight furnace, with minimal opening of door and peep sights
2. Use an automatic furnace pressure control with the set point at +0.02" wc (0.05 mm water gauge) to prevent air inflow
3. Use a quality air/fuel ratio controller set as close to stoichiometric as practical, but erring on the oxidizing side (because dross is easier to remove than absorbed hydrogen)
4. Avoid flame or hot poc impinging directly on the molten bath surface
5. Do not use a liquid metal circulating device that sucks in air or poc along with the metal

## 3.9. CONTROLLED COOLING IN OR AFTER BATCH FURNACES

After heat treating, some materials need to be cooled slowly, sometimes more slowly than they would cool if just left in the furnace with the doors closed. This requires use of in-furnace recirculating fans and/or excess air. On the available heat chart of figure 5.1, the $x$-intercept of the curves is the theoretical flame temperature (adiabatic flame temperature), also termed "hot-mix temperature" in high excess air (lower temperature) realms. Examples for average natural gas: 3450 F (1899 C) with 5% excess air, 2700 F (1482 C) with 50% excess air, 1810 F (988 C) with 150% excess air, 1290 F (691 C) with 275% excess air, 985 F (530 C) with 400% excess air. Gradually increasing excess air to 400% will slowly cool the load to 985 F. Programmed control of excess air provides programmed temperature control for cooling.

For faster cooling, with no fuel, example 3.7 is a possible compromise cooling method midway between cooling with excess air burners and convection cooling with cooling tube banks and high air circulation.

**Example 3.7**: Design radiation cooling U-tubes positioned across the ceiling of a chamber for cooling 38 000 lb/hr of cast iron pieces from 1800 F to 800 F. Usually a minimum tube spacing ratio of 2:1 is satisfactory. From figure A.7 in *reference 51*, iron has a heat content at 1800 F of 285 Btu/lb and at 800 F of 112 Btu/lb. Therefore, the cooling load will be (38 000 lb/hr) (285 Btu/lb − 112 Btu/lb) = 6 574 000 Btu/hr. With a 2% safety factor, design for 6.7 kk Btu/hr.

Assume the cooling air from a blower will enter the tubes at 100 F and be heated to 350 F (allowing it to get hotter will reduce the cooling capability of the tubes). Therefore, the average load (source) temperature = 1300 F, and the average cooling air (sink) temperature = 225 F. Interpolating from Table 4.1a in *reference 51*, the black body radiation from 1300 F loads to 225 F tubes will be 16 000 Btu/ft$^2$ hr. For an emissivity of 0.85, the loads' radiation to the cooling tubes = (16 000) (0.85) = 13 600 Btu/ft$^2$hr. Therefore, the total required tube surface will be 6 700 000 Btu/hr/13 600 Btu/ft$^2$hr = 493 ft$^2$. Adding a 15% security factor, use 570 ft$^2$.

For 11.5 ft long cooling U-tubes of 4" ID and 4.5" OD (23.59 ft equivalent length), the outside cooling surface area of *each tube* will be (23.59) ($\pi$) (4.5/12) = 27.8 ft$^2$.

Therefore, the number of U-tubes needed should be 570/27.8 = 20. The total *flow* area of the 20 U-tubes will be (20) (π) $(4/12)^2 = 7\ ft^2$.

In the temperature range below about 800 F (482 C), a hydrogen atmosphere might be considered, but air is safer and less expensive. Circulated air is the usual cooling medium. Air is made up of diatomic gases (oxygen and nitrogen) which do not receive nor emit radiation; thus, the cooling must be via the small amount of direct "solids radiation" from loads to cooling pipes and by convection. Fans are often used within these low-temperature furnaces to increase circulated air velocity next to the load surfaces and across cooling pipes for better convection cooling. Walls and ceiling of furnaces, ovens, or special cooling chambers can be covered with air-cooled or water-cooled pipes, and fan air streams should be designed to pass circulating air over their cooling surfaces and over the load surfaces.

It is often assumed that a 2 psi (32 osi) fan is the highest practical pressure for *in-pipe* cooling. From table 5.1 in *reference 51*, a 32 osi pressure drop can create 462 fps air velocity. It is rarely practical to raise the average *circulated* air velocity at the load surface above about 60 ft/s (18.3 m/s). Therefore, heat transfer is limited to low rates.

Constant exhausting of some of the resultant warmed circulating air is necessary to avoid reduction of the $\Delta T$ that is a major factor in the cooling heat transfer process. Any means for moving the circulating air to remove heat from the loads must be able to produce uniformly high velocity on all the product surfaces.

## 3.10. REVIEW QUESTIONS AND PROJECT

**3.10Q1.** List advantages of batch furnaces over continuous furnaces.

**A1.** Lower first investment cost. Less maintenance, because fewer moving parts. Save fuel if need is intermittent. Save fuel if new loads cannot be put in place promptly. Sometimes more versatile as to product size, shape, and temperature cycle. Easier to hold tight furnace pressure. Easier to hold a prepared atmosphere.

**3.10Q2.** How do shuttle furnaces and kilns overcome some of the disadvantages of batch furnaces?

**A2.** Less lost heat during unloading and reloading. Easier and safer to load and unload. Regularity for operators.

**3.10Q3.** List all the differences that must be considered when designing a furnace for a molten metal (including glass) as opposed to a furnace for heating solid pieces.

**A3.** Corrosive action of metal liquids, vapors, and oxides on refractories and metals used in furnace construction. Accumulation and removal of oxides (dross). Added weight of a liquid bath, compared with a rack of pieces. Charging and unloading problems. Safety and clean-up problems with liquid spills.

**3.10Q4.** If, in the case of example 3.7, you chose to use water cooling instead of air cooling, would the lower first cost of the cooler be enough to justify installing a cooling tower or cooling pond to avoid thermal pollution of a nearby stream?

**A4.** Answer depends on costs at the locality, but calculate for your specific situation.

**3.10Q5.** With loads 6" thick or greater, what separation between pieces is required for excellent uniformity?

**A5.** A space-to-thickness ratio of 2:1.

**3.10Q6.** Normally, how many zones should a 30 ft long car furnace have to handle a wide variety of product sizes?

**A6.** The minimum number of zones is three, but more zones will reduce cycle time and improve product uniformity. End zones should be smaller than zones between them. If the normal load has a mix of lengths, more zones are needed.

**3.10Q7.** Why is it advantageous to use hydrogen inside a bell furnace inner cover?

**A7.** Convection heat transfer often is limited by the conductivity of the boundary layer film on the product. Comparing the averge k values for hydrogen and air in tables 2.7 and 2.8, find that over a range of cover annealing temperatures the k of hydrogen is 6.25 as large as k of air.

**3.10Q8.** Why should load pieces not be piled more than two-high?

**A8.** Obviously, less surface area of the middle row of pieces is exposed to convection and radiation. Calculation of the cycle time required for the middle pieces would be very laborious and doubtful. The best way to judge when the middle pieces are heated to specification is by watching the curve of fuel input. (See A9.)

**3.10Q9.** With batch heating, what should a normal fuel input curve look like?

**A9.**

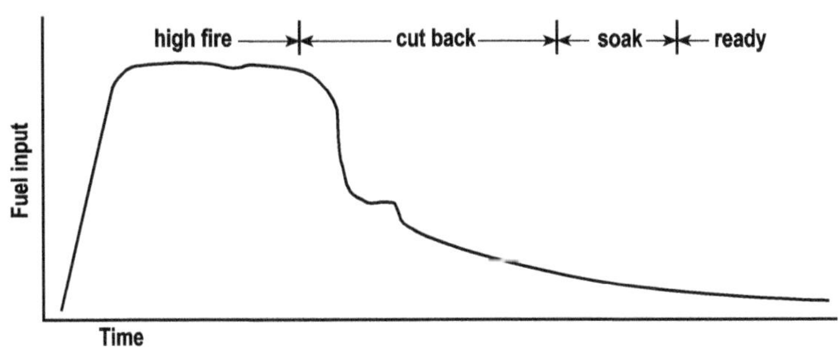

## 3.10. PROJECT

Search for or test for more data on heat and evaporation losses from open liquid tanks in all temperature ranges.

# 4

# HEATING CAPACITY OF CONTINUOUS FURNACES

## 4.1. CONTINUOUS FURNACES COMPARED TO BATCH FURNACES*

The loads move continuously or intermittently through continuous furnaces. They may be pushed, rolled, or walked through the furnace or they may rest on a rotating hearth or be suspended from a conveyor. Theoretically, the temperature versus length profile of a continuous furnace should be the same as the temperature versus time pattern for its batch predecessor that was found to be the optimum pattern for product quality and productivity. All too often, designers of continuous furnaces assume that the new furnace will operate continuously without interruptions or delays. That is rarely the case, especially with high-temperature furnaces used for heating large pieces having considerable time-lag before their core temperature catches up with their outer surface temperature.

Coauthor/Consultant Shannon often has been called to unravel serious problems resulting from the previous incorrect assumption, which continuous furnace buyers and sellers like because it lowers the first cost. That initial savings can turn out to be insignificant compared with operating costs resulting from unforeseen cyclic operations. It is much less expensive in the long run if the designer builds in ways to overcome the following problems that invariably happen after the constant delays: **Problem 1** = *Loads that have "sat" in a furnace during a delay will be overheated upon restart.* **Problem 2** = *Newly charged cold loads will not be able to catch up to acquire the required discharge temperature and uniformity*. These problems cause automatic control (or heater setpoint changes) that set up variable temperature wave patterns ("domino effects") down the length of the furnace, which this book calls "accordian effects." (See glossary.)

*Many parts of chap. 3 on batch furnaces may contain useful information that also applies to continuous furnaces, but is not included here (to keep this book compact). Readers are advised to study both chap. 3 and chap. 4.

### 4.1.1. Prescriptions for Operating Flexibility

Prescriptions for operating flexibility despite delays and interruptions:

(*a*) Install one or more burners in a previously unfired top preheat zone (preferably all the way to the charge entrance) with T-sensors to operate as a separate control zone—to sense the arrival of new cold loads sooner after a delay. If there is an unfired bottom preheat zone, add burner(s) there also, with controls to make them follow the lead of the top preheat zone. Some will say these actions defeat the fuel-saving feature of the unfired preheat zone, but regenerative burners can accomplish a similarly low flue gas exit temperature as without preheat zone burners.

(*b*) Replace the one or two heat zones with more smaller zones with controls

(*c*) and T-sensors to track the temperature changes from overheated loads right after a delay as they are replaced by underheated newly charged loads. Designers may decrease the number of control zones to lower the first cost of a furnace. *Increasing* the number of zones is necessary if the furnace and its operators are to improve capacity, increase operating flexibility, and lower fuel rate. For steel reheat furnaces, zone lengths may vary from 12 to 20 ft (3.66 to 6.1 m), but should not exceed 30 ft (9.1 m).

(*d*) If dilution air is used to protect recuperators or other equipment, both the fan pressure developed and its volume capacity may have to be increased to keep the diluted exit gas temperature below the danger level at the new maximum firing rate.

*The previous improvements will make a continuous furnace flexible and profitable.* The savings can be even more if done properly from the start. With industrial furnaces, it is usually true that *"Only the low bidder wins in a low-cost deal."* (See chap. 8 for sample heating curves illustrating these points.)

A continuous furnace may be heated so that the temperature of its zones is practically the same across the furnace. This temperature uniformity can be obtained by lengthwise firing in several zones (as illustrated by fig. 4.2), or by roof firing or side firing in several zones (as shown in fig. 4.3). In such furnaces, the heating capacity of a continuous furnace will equal or exceed the capacity of a same-size batch furnace.

Continuous furnaces are usually more fuel efficient than batch furnaces if their charge and discharge openings can be kept small and shielded from large radiation loss. Because they do not have to stop with doors open for loading and unloading, their walls, roof, and hearth stay at a nearly constant temperature with respect to time, thus avoiding repetitive storing and losing of heat from their refractory lining.

By eliminating the downtime for loading and unloading, continuous furnaces almost always can have better production capacity per unit time and per unit of hearth area than do batch furnaces. Of course, the cost of handling equipment to make possible the continuous loading and unloading raises the initial investment of continuous furnaces.

When fuel costs are high or fuel supply is a concern, continuous furnaces can be built and controlled with a graduated temperature profile from highest in the zones near the load-discharge end of the furnace to lowest in the load-charging end, and with the poc flowing counterflow to the load flow. This fuel-efficient configuration has often been modified to a "level" temperature profile when fuel costs have dropped and production demands have increased. Because new furnaces can be built shorter if planned for a level temperature profile, that has been done during low fuel cost eras. However, firing furnaces to produce a level temperature profile from end to end of the furnace has two very serious drawbacks:

**Drawback 1**: A reflective scale is generally formed when the preheat zone is held at temperatures at or above 2300 F (1260 C). The cause of the reflective scale is the normal softening of the scale above 2320 F (1271 C) and the lower conductivity of the surface. If a furnace has this problem, reducing the preheat zone temperatures and increasing the product discharge temperature will increase furnace productivity.

**Drawback 2**: The flue gas temperature is exceedingly high, resulting in very high fuel rates that have become intolerable. With conventionally fired furnaces, the preheat zone temperatures have been reduced by hundreds of degrees to save fuel. Furnace modeling by computer has been applied to reduce preheat zone temperatures as much as possible. A very effective way to correct delay problems and to reduce fuel rates is by installing a T-sensor (to control the first fired zone) in the sidewall of the flowing poc stream 6 ft (1.8 m) from the uptake (or downtake) flue.

Modern regenerator–burner packages permit low-end exit gas temperatures (400 to 500 F or 205 to 260 C) at every regenerator–burner anywhere in the furnace, and for process temperatures as high as 2500 F (1370 C), the high-productivity *level temperature profile* can be as efficient as a *graduated temperature profile*.

Modeling has had mixed results. For modeling to be effective, the furnace heating requirements must be nearly constant for the following reason. Picture a furnace operating in equilibrium at 70% capacity when the mill requirement increases to 90% capacity. To catch up, *all* the zones may be subjected to the 100% firing rate to accelerate to the new 90% rate. Newly charged pieces will be exposed to gas and refractory radiating powers equivalent to the 100% firing rate. When those newly charged pieces reach the midpoint of the furnace, they will be hotter than they should

***Scale*** (dross, oxide) forms if a load is subjected to ***too high temperature*** for ***too much time*** with ***excess oxygen*** in the furnace atmosphere. The presence of scale, and the extent of its formation, is difficult to determine within the furnace. Scale is usually obvious only after the damage is done.

A reflective-radiation sensor as a high limit might be practical. It is difficult to measure (detect) scaling, thus, it is not very practical to adjust for, or automatically prevent, its formation. Operators and supervisors must rely on knowledge and experience to anticipate scale problems and prepare to avoid or forestall them. (See sec. 8.3.)

be; thus, the model then must reduce firing rates and zone temperatures to some lower level such as 80%, which is below the actual need. This cycling is difficult to stop, especially when the mill requirements change frequently. With cyclical temperatures in various furnace zones, scale formation accelerates. Scaling increases as the 5th power of temperature, so it will increase with cycling or during high-input swings. Other variables involved in scale formation are time, atmosphere, and gas velocity, but temperature is the most predominant variable.

Regenerative burners have minimized the need for modeling, as long as the operator avoids reflective scale on the load. With the high thermal efficiency of regenerative beds, fuel efficiency and furnace productivity are practically two different problems—no longer closely interrelated. Operators can run with zone temperatures that can deliver furnace capacity whether the mill requires it or not. When the mill does need 100% output, the operator will be prepared, and the fuel rate will be barely higher than when controlling the furnace to exact mill needs.

The statements relating to batch type and continuous furnaces are for top-fired furnaces at a temperature corresponding to that of the batch type. The heating capacity of such furnaces is determined by hearth area, ceiling temperature, load absorptivity, time, and exposure of the load as well as composition and thickness of the load and of the poc.

The heating capacity of continuous furnaces usually exceeds that of batch type furnaces of the same hearth areas because:

1. Whereas batch furnace temperature must be held down to prevent overheating, temperature in the heating zone of a continuous furnace may be very high,

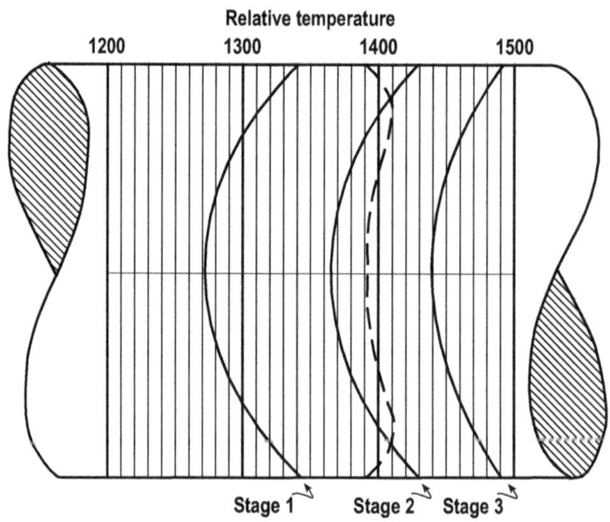


**Fig. 4.1.** Temperature patterns in a large, round load, showing changes with time in a batch or continuous furnace. The dashed line shows the temperature equalization (leveling) if there had been a delay (firing cutback) after stage 2.

if thin load temperature is carefully monitored and removed promptly. When heating thick pieces, the furnace should have a soaking zone for temperature equalization, as shown by the dashed curve in figure 4.1. For loads of high thermal conductivity, a soak zone may be omitted.

2. In a continuous furnace, the loads may be supported by skid rails, allowing more heat delivery to the load undersides (discussed later).

Continuous dryers, ovens, incinerators, and furnaces take any of a variety of forms such as rotary drum, tower, shaft, tunnel oven, multihearth (Herreshoff) kiln, and fluidized bed. As with all continuous furnaces, their design is very dependent on how the load(s) can be moved through the furnace (or occasionally, how the furnace can be moved over the loads).

## 4.2. CONTINUOUS DRYERS, OVENS, AND FURNACES FOR <1400 F (<760 C)

The reader should review section 3.8.1 on batch ovens and low-temperature batch furnaces because many of the ideas discussed there also apply to continuous dryers, ovens, and furnaces. Dryers and drying ovens usually release large quantities of water vapor or of solvents, the accumulation of which can have at least two bad effects: (1) an explosion hazard with flammable solvents and (2) a reduced rate of drying (mass transfer) with either water or solvent drying. Tables B.3 and B.4 of *reference 51* give heat requirements for drying.

### 4.2.1. Explosion Hazards

**Explosion hazards** develop as flammable vapors accumulate to a concentration that is within their flammable limits = explosion limits = lower explosive limit (LEL) and upper explosive limit (UEL). (See chap. 7 of *reference 47,* and *reference 48.*) Most codes and standards require built-in air dilution to keep the furnace atmosphere below one-fourth of the LEL, or one-half LEL with specific automatic control or alarm arrangements. The dilution changestemperature and mass transfer potentials (discussed later), and increases the convection velocity.

Many explosions in furnaces result from this sequence of events: (1) loss of combustion air flow (pressure); (2) so furnace atmosphere becomes fuel rich; (3) flame is extinguished because beyond its rich flammability limit; (4) someone shuts off the

> REMEMBER: *Safety is Job 1*, above quality, productivity, fuel economy, and pollution reduction. Explosions and the fires that follow not only cause loss of limbs and lives but loss of employees and employers (by death, incapacitation, layoff, or business failure).

fuel or opens a furnace door, either of which brings the furnace's %fuel in its air–fuel mixture back down into the flammable range; (4) creating a bomb awaiting ignition; and (5) which could be supplied by a constant (standing) pilot,* welding, an impact spark, or lighting a cigarette within a short distance of the furnace. For the reason shown by this scenario, it is recommended that fuel be controlled to the burner(s) only in response to, and in proportion to, the measured flow of air to the combustion chamber ("air primary" air/fuel ratio control). Then, if the air supply fails for any reason, the fuel flow will stop immediately, avoiding fuel accumulation.

### 4.2.2. Mass Transfer

The removal of water or solvents is a three-step process:

1. Heat is first transferred to the material that naturally contains water, such as milk, tobacco, carrots, or to which liquid water or solvent was added in a preceding process (such as for forming or coating). The heat is necessary to evaporate the liquid to a vapor form for easy removal (mass transfer).
2. The driving force that causes the liquid to migrate to the surface of the material or piece being dried is the difference in vapor pressure between the inside and the surface of the pieces being dried.
3. Similarly, the driving force causing the liquid to vaporize and causing the vapor to migrate away from the surface is the same difference in vapor pressure that caused (b).

The practical way to create and maintain an appreciable difference in vapor pressure to continually force rapid mass transfer is to move a stream of hot poc and air to constantly wipe the wet surface (i.e., convection heating). Neither radiant burners nor electric elements are as effective unless accompanied by circulating fans. Convection burners provide a circulating (wiping, mass transfer) effect.

Drying can be overdone if heat application is not carefully controlled. Overheating can cause a "skin" or "rust" to form on the surface, and that skin may impede further migration or evaporation. The pressure of the trapped vapor under the dried crust then rises from further heat application until it breaks the crust in a sort of steam explosion. Such small explosions may not be very damaging, like a furnace or oven explosion, but they may bloat or crack the load pieces so that they become rejects.

### 4.2.3. Rotary Drum Dryers, Incinerators

**Rotary drum dryers, calciners, kilns, and incinerators** tumble bulk material or pieces peripherally and lengthwise downhill, thus exposing all load surfaces, even

*A constant or standing pilot is prohibited by most insurers. (See *references 47* and *48*.) Many pilots are so stable that they can continue to operate when surrounded by a too-rich mixture. Flame monitors are often positioned to detect main *or* pilot flame. If the main flame goes out "on rich" but the pilot flame continues, the pilot flame may set off an explosion of an accumulated flammable mixture within the furnace or oven.

> Rejects are costly! Even if you can recycle the material, you cannot recover the cost of the labor, machine time, or fuel put into the rejected piece. All have to be bought again. If the job is on a rush delivery schedule, you cannot buy the lost time again. More than one business has gone down the drain because they let minor dips in product quality slip through to their customers, and the customers never came back; therefore, add "reputation" as another cost of rejects.

for small granules, to the poc and hot air which may be traveling counterflow or in parallel flow (co-current) through the rotating drum. (See fig. 4.2.)

In figure 4.2, the driving force that makes heat flow into the load is proportional to the height and area between the two temperature curves. Fuel consumption will be less with counterflow (lower final exit gas temperature). Increasing the counterflow drum length will save more fuel and heat the load to a higher final temperature whereas increasing the parallel flow drum length will "soak out" a more even temperature in the load and assure no overheating. (See fig. 4.3.)

Heat transfer in low-temperature rotary drums is largely by convection because radiation is naturally less intense in this temperature range. If the drum diameter is 5 ft (1.5 m) or more, radiation from triatomic gases can be helpful. However, many low-temperature rotary dryers use so much excess air (for moisture pickup) that the triatomic gas concentration is diluted significantly.

The granular material slides and rolls around in a long, narrow pile, the cross section of which is a segment of a circle, extending roughly from five o'clock to eight o'clock (0500 to 0800 hr) for clockwise rotation. Granules within the bottom segment slowly roll from the bottom to the top of the segment. Many rotary dryers

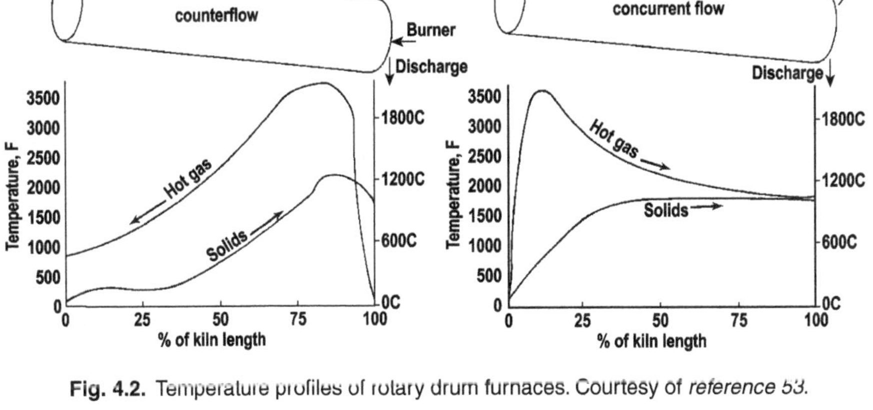


**Fig. 4.2.** Temperature profiles of rotary drum furnaces. Courtesy of *reference 53*.

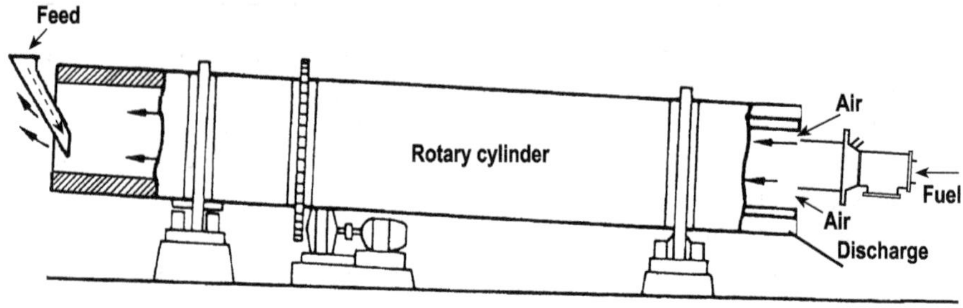


**Fig. 4.3.** A rotary drum dryer, kiln, incinerator, or furnace transports granular loads (*left to right*) by gravity and rotation, counterflow to the burner gases and induced air. Parallel flow or co-current flow (fig. 1.10) can be used with some load materials and processes.

have longitudinal shelves (lifters or flights) attached to the inner walls as shown in Figure 4.4. These scoop up some of the bottom segment granules and carry them up to near the top of the drum, where the granules pour across the hot gas stream, giving every granule excellent surface exposure to the hot gases—good convection contact—especially if the shelf lifters have an edge bent up in the direction of rotation. Some added rolling of granules occurs from pile bottom to top.

The lifters should not be used too close to the burner flame (1) because flame contact with the granules may be harmful and (2) because the life of the shelves would be shortened. Lifter flights have been as wide as 10% of drum inside diameter, but the greater widths require sturdier construction to carry a deeper pile, which obstructs gas flow. Many short, closely spaced flights make it difficult for maintenance persons to walk through the cold drum to inspect it. Parts 4 and 5 of figure 4.4 show the use of suspended chains to heat up when hanging across the hot gas stream, and then heat the load in the bottom of the drum by conduction (contact).

Care must be exercised in operating rotary drums so that the hot gas velocity is not too high relative to the size and weight of the granules, as that may cause carry-over into the exhaust (particulate emissions).

### 4.2.4. Tower and Spray Dryers

**Tower dryers and spray dryers** shower or cascade their liquids or granules down through a vertical tower with a horizontal burner (or air heater) at the bottom and off to the side so that the load pieces will not fall through the flame or into the burner. Considerable height, diameter, and precise control are required to assure that droplets have a free fall until they are thoroughly dried particles.

### 4.2.5. Tunnel Ovens

**Tunnel ovens** can be used for stress relieving and annealing copper and its alloys at 500 to 900 F (260 to 480 C). Tunnel ovens are so common for paint drying that they are often assembled from standardized fiber-lined, metal-encased sections that

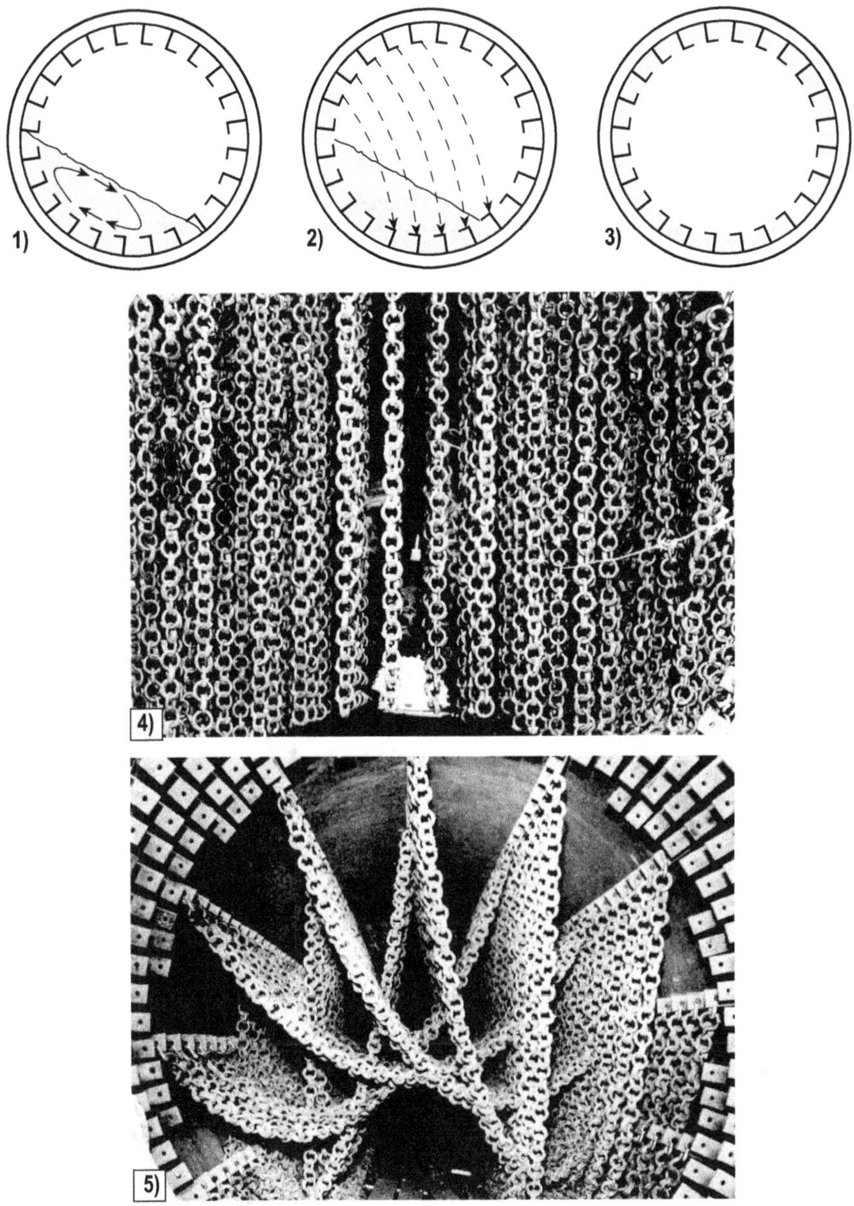


**Fig. 4.4.** Speed of drum rotation determines granules' fluid action. (1) Normal angle of repose of granules with no lifting shelves or with rotational speed too slow. Arrows in the segment cross section show the rolling effect that slowly exposes granules at the pile surface. (2) Optimum rotational speed with maximum cascading. (3) Excessive speed prevents cascading—centrifugal force holds the granules against the inner drum periphery. Curtain chains (4) and garland chains (5), attached around 360° of the inner periphery, absorb heat when suspended and give up heat when lying among the load granules. (Four and five are courtesy of Sept. 1980, *Pulp and Paper.*)

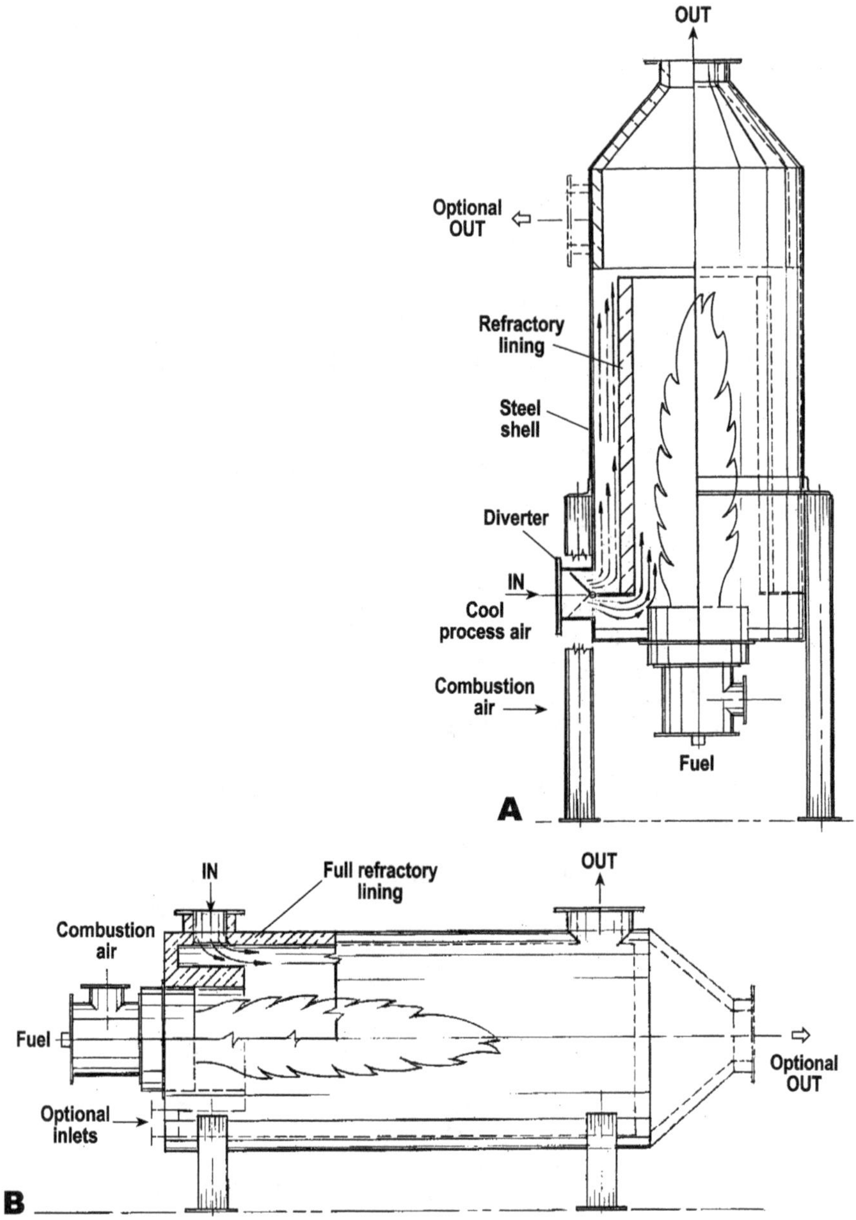


**Fig. 4.5.** Two of many configurations for direct-fired air heaters. Version A shows a parallel-flow arrangement with variable dilution, and a shield to prevent the air to be heated (the load) from quenching the flame. Version B has full counterflow and more insulation in the outer shell for higher in-and-out temperatures; thus, it is ideal for recirculation.

can be bolted together into a series of zones, each with its own circulating fan. Such a production line may have the same conveyor for preceding processes such as a spray washer, its dryer, and for applying paint. Surge or holding areas between these operations (often overhead to save floor space) provide flexibility and easier starting and stopping of the separate processes. Heat input controls of the zones must be coordinated or line delays may have "accordian" problems as described in sections 4.6, 6.4, after delays in multizone reheat furnaces.

Even though precautions have been taken to prevent explosions, fumes evaporating from the vehicles in coatings, binders, or adhesives may be volatile organic compounds to which pollution regulations apply. Carefully designed vent duct/fan systems are needed for the safety, health, and comfort of operators. Because it is difficult to operate "air locks" to keep hot air in and cold air out of a tunnel-type dryer with a continuously moving conveyor, it may have excessive end losses which may be minimized by air curtains or fiber rope curtains (which require carefulmaintenance). An advantange of open-ended ovens and furnaces is that they minimize the confinement that can turn a fire into an explosion.

### 4.2.6. Air Heaters

**Air heaters** to supply hot air for drying and other processes take many forms. Indirect air heaters are basically heat exchangers, which come in many forms. *Direct-fired air heaters* are less expensive and use less fuel, but they can be used only where no harm will be done to the process product by contact with poc. Thorough mixing and careful temperature control are necessary. Figure 4.5 shows some of the configurations possible.

## 4.3. CONTINUOUS MIDRANGE FURNACES, 1200 TO 1800 F (650 TO 980 C)

This section applies to all types of continuous furnaces operating in the stated temperature range, including furnaces for brazing, calcining, roasting, sintering, and the conventional "heat treating" operations such as annealing (metals and glass), normalizing, carburizing, hardening, and stress relieving. This section relates to conveyorized furnaces, tunnel kilns, pusher furnaces, and shaft furnaces. Rotary drum furnaces are covered in 4.2, catenary furnaces and strip-heating tower furnaces in 4.3, axial continuous (barrel) furnaces in section 4.5, and rotary hearth furnaces in section 4.6.1.

Some comments and warnings from chapter 3, sections 3.8.4 to 3.8.6 for batch-type furnaces operating in this temperature range may be applicable to continuous furnaces as well.

### 4.3.1. Conveyorized Tunnel Furnaces or Kilns

Conveyorized tunnel furnaces or kilns may be stretched versions of their batch equivalents, divided into several zones. Many types of conveyors are used. Figure 4.6 shows

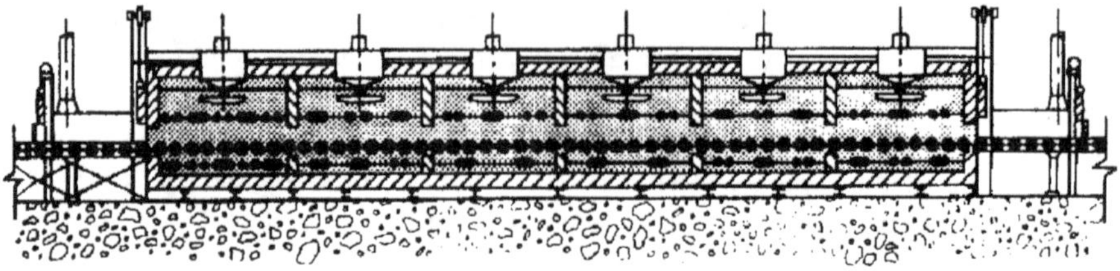

**Fig. 4.6.** Continuous roller hearth furnace, side-elevation sectional view. Through-the-roof plug fans drive circulation across radiant tubes above and below loads on rollers.

a continuous roller hearth furnace heated with radiant U-tubes above and below the loads on rollers instead of a conveyor. "Plug fans" through the furnace ceiling may be used to circulate prepared atmosphere gas over radiant tubes and the loads.

It is wise to return a conveyor within the furnace to save heat loss and to prolong its life by minimizing the amplitude and the frequency of the temperature cycle to which the conveyor materials are exposed. Many materials last longer if kept hot, rather than being constantly cycled between hot and cold. For flexibility during production line delays, it is advisable to provide a temporary storage area at each end of a conveyor furnace.

A common problem with many continuous furnaces is an "accordion" effect that occurs after line stoppages. Continuous furnaces are wonderful as long as they maintain steady-state operation. To envision the accordian effect, think of the changes with passage of time of the temperature pattern throughout the length of a furnace with temperature sensors located at the traditional positions near the ceiling of the furnace and near the load-exit-end of each zone.

After a delay, the temperatures of the walls and loads have tended to even out. Thus, the load in the zones 1 and 2 from the load entry will remain at a low fire-holding condition until those load pieces are worked out. By that time, new cold loads have started to fill the furnace, and have finally affected the sensors high at the ends of the zones, driving the burners to high fire. But the firing has begun much too late, so that the pieces are very cold entering the next zone. The loads, particularly those in the 1st and 2nd from entry zones, will have soaked under some residual wall heat during the delay and can quickly overheat before reaching a sensor that can turn down the high fire. The final zones have the same problem—a heat delay or cobbles, or both! Then, the overshooting will be followed by undershooting—the waves of an accordian hysteresis effect.

To prevent this problem, all control sensors should be close to the level of the tops of the loads. Input control sensors should be within about one-fourth of their zone length from the load entry end of their zones. Over-temperature sensors should be 5 to 10% of their zone length from the exit end of their zones, and set at the maximum furnace temperature allowed. With such a sensor-positioning arrangement, a modern quick-recovery temperature control has a chance to avoid a heat delay following a mill delay.

*Tunnel kilns,* widely used in firing ceramics and carbon shapes, use a long train of cars as a conveyor Each car may be similar to, but often narrower than, the car of a batch-type car-hearth furnace. Much of what is discussed in this book can apply to ceramic kilns, but the ceramic industries have so many publications on kiln construction and operation that this text will not dwell on them specifically.

*Roller-hearth conveyors* have an advantage over continuous belt and chain conveyors in that the conveying device can stay within the furnace all of the time (except for kiln furniture, saggers, or other containers that may ride on the rollers); thus, they do not carry as much heat out of the furnace. Rollers and their bearings can be maintenance problems. Recently, however, nickel aluminide (Ni3Al) steel rolls have proved better in a plate mill annealing furnace. These intermetallic alloys have higher strength and corrosion resistance at elevated temperatures than did formerly used alloys, and they are not as brittle as ceramic rolls or ceramic covered rolls.

The heating capacity of furnaces in this midtemperature range can be determined by calculating heating curves, as discussed in sections 4.6 and 8.2. The lower radiation intensity in this range warrants more attention to convection, surface exposure, and circulation (chap. 2 and 7).

### 4.3.2. Roller-Hearth Ovens, Furnaces, and Kilns

Some narrow and lightweight loads (such as tiles and dinnerware) permit the use of ceramic or alloy rollers instead of kiln cars. Warping of the rollers can cause tracking problems and may result in deformation of the loads. Rollers are made of high-temperature alloys, mullite, alumina, or silicon carbide, determined by the load, span, and temperature. Sometimes, rolls of several different materials are reused in the same furnace or kiln. Rollers are usually driven from one end only, usually by a chain or gear. Regular maintenance is required. Flat tiles are usually fired directly on the rollers; other types of loads in or on refractory setters, "kiln furniture." (See fig. 4.7.) One-high loads are common, but at lower temperatures there may be several levels traveling through a kiln or oven in series or in parallel.

The load pieces should be uniformly distributed across the rollers to permit uniform air flow and temperature distribution. With multiple roller levels, offsetting the load pieces can assure more uniform hot gas flow around all pieces.

### 4.3.3. Shuttle Car-Hearth Furnaces and Kilns

Shuttle car-hearth furnaces and kilns are hybrids between batch and continuous furnaces and kilns, combining the compact lower cost of a batch operation with the productivity and fuel economy of a continuous furnace or kiln. A shuttle furnace has doors at both ends and with two rolling hearths, permitting quick unloading and reloading of the furnace with minimum cooling during the switch-around. (See fig. 4.8.) The capital cost is only about 65% of two furnaces, but the production rate is almost doubled. The fuel economy per year and per ton heated is better because the doors are closed and the burners are in use more often.

Fig. 4.7. Roller kilns with top- and bottom-fired small, medium-velocity burners. Type E flat flames above the ware would permit a lower roof and assure more even sidewise heat spread. Upfired burners from below are not wise for fear of crumbs falling into the burners. Radiant tubes can be used above and below the rollers and ware to protect the loads from contact with poc. Courtesy of North American Mfg. Co.

### 4.3.4. Sawtooth Walking Beams

**Sawtooth walking beams** provide rollover action for round pieces. Figure 4.9 illustrates a pipe annealing furnace wherein the cold pipe is charged through a side opening on the rollers at right, then picked up by the sawtooth walking beam for intermittent stepping from right to left, and then discharged by the rollers at left through a side exit. Each time the walking beam returns a pipe to its next notch on the sawtooth, the pipe rolls down the incline of one tooth, exposing a different part of its periphery to flame, gas, and refractory radiation—like a chicken in a rotisserie.

Unlike most other conveyorized furnaces, walking beam furnaces accommodate top- and bottom-zone-firing. When used at lower temperatures (e.g., for annealing light sections such as pipe), the beam and supports may be of high-grade alloy without water cooling.

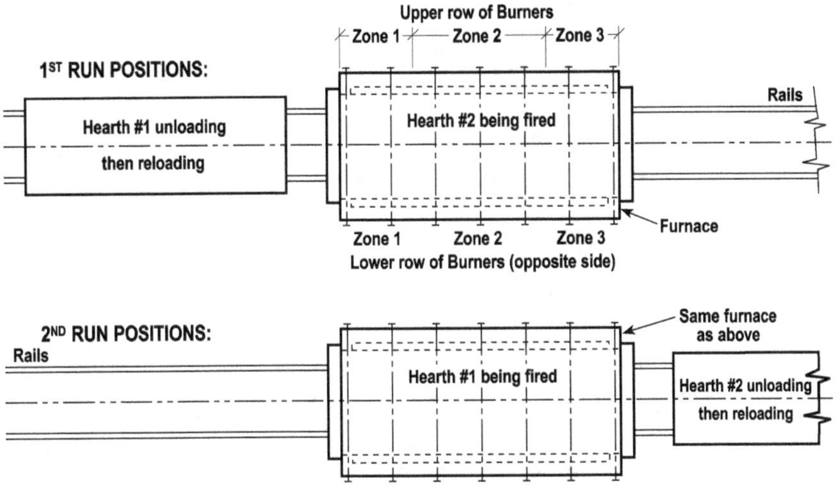


**Fig. 4.8.** Shuttle kiln or furnace. One furnace with two shuttle hearths and 33% longer rails can provide almost 100% more production with considerably less capital investment by heating loads a higher percentage of the time. To some extent, the shuttle arrangement also improves efficiency of personnel because there is less waiting around, and everyone is on a better schedule.

Furnaces for vertical strip* or strand (wire) do not have a conveyor, per se, because the strip or wire can be pulled over a series of rollers after it has been "threaded" through the furnace. A ***catenary furnace*** is a continuous horizontal furnace most often used for annealing stainless-steel strip. A long, thin load is supported by rollers at the entrance and exit, and therefore hangs in the shape of a catenary curve within the furnace. (See box on page 132 and fig. 4.10.)

With a light, thin load such as strip, heating capacity may be in the range of 100 to 300 psf of hearth. As with all furnaces, the authors recommend developing a heating curve for the specific load (chap. 8), and using that curve to determine necessary total furnace length. In this industry, a factor of 1.4 could be applied for needed future growth in production. To deliver the desired production rate, some plants use two to four furnace sections in series, with the supporting rollers out in the furnace room between sections. Hot strip may stretch with a long, deep catenary; therefore, a practical maximum section length is less than 60 ft (18 m).

Because of the low mass of a strip, the preheat zone may be operated at higher than maximum desired strip temperature, such as 2200 F (1200 C) to increase productivity (by perhaps 30%) above that possible with a preheat zone temperature at design strip exit temperature. Most of the strip running through the furnace will be below the design exit temperature, so no strip damage results from this practice. The discharge zone temperature must be close to the design maximum strip temperature to allow

*Vertical strip heating furnaces are sometimes called "tower furnaces," but should not be confused with tower dryers (sec. 4.2.4)

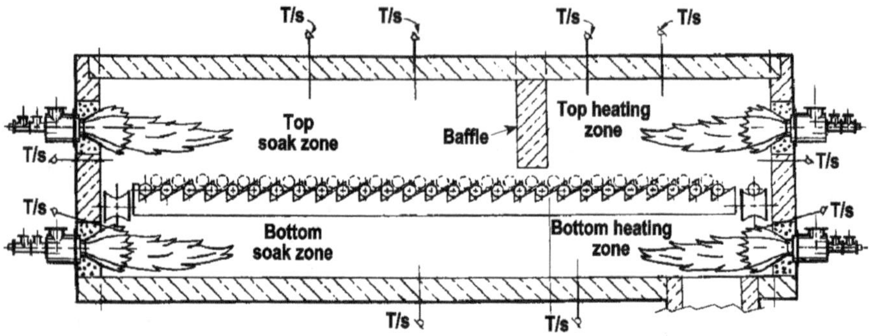


**Fig. 4.9.** Walking beam pipe annealing furnace. Bowing pipes (loads) had prevented smooth transfer of pipes with each "walk" of the beams. The original long flames concentrated too much radiation in the top segment of each pipe's periphery, causing bowing. Replacement with adjustable thermal profile burners and with Tc (temperature control) sensors has eliminated the pipe bowing that had prevented the conveyor from rolling the pipes over. The To (temperature observation) sensors help with manual control to avoid bowing close to the burners.

time at temperature for the desired physical changes to take place within the load material. With 300 series stainless steels, discharge zone temperatures are generally 1950 to 2050 F (1066–1121 C), but 400 series stainless steels are annealed at 1700 F ± 100°F (927 C ± 56°C).

If a line stop occurs, the 2200 F (1200 C) zone temperature can cause strip thinning or separation. Therefore, a protective control scheme is needed. (See temperature measurement and control discussions that follow.)

In the temperature range usually used for this process, the furnace walls, roof, and hearth provide excellent radiant heat transfer. The furnace height necessary to avoid flame impingement on the strip from lower burners also assures a good average beam for gas radiation to both top and bottom surfaces of the load.

*Catenary* = the graph of the hyperbolic cosine function = curve assumed by a heavy chain supported at two points not on the same vertical line (usually on the same horizontal line) = the curve of cables on a suspension bridge (left), or = the curve of a suspended string of beads all of same size and weight (center).

*Caterary arch* = a sprung arch in the shape of an inverted catenary curve, used in early refractory brick kilns and the St. Louis arch, "Gateway to the West."

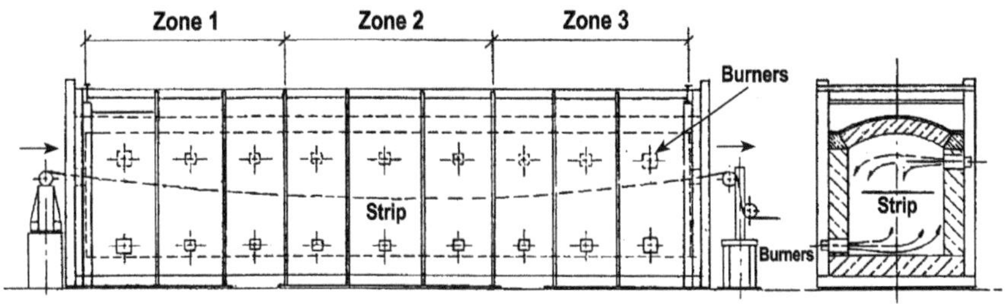


**Fig. 4.10.** Catenary furnace for heat treating metal strip. Careful strip tension control is needed to prevent strip sag to prevent strip contact with the flame. Better control can be achieved with the exit supporting roll water cooled and just within the exit end of the furnace and with a T-sensor near that roll and under the strip.

There are not very many catenary furnaces in the United States, so more capacity is needed. A need also exists for better communication between designers and operators of such furnaces to improve operation and productivity. The relatively light load in these furnaces requires a different approach to product temperature control. Caternary furnace design has often been a throwback to rules of thumb, such as 21 min/in. of strip thickness. Heating curves using reasonably correct emissivities, higher zone temperatures, and greater firing rates have predicted a possible 30% increase in productivity.

To attain an even more effective heat head control of a preheat zone, relocate the control measurement near the charge door, for example, 2 or 3 ft (0.7 to 0.9 m) into the zone. Such a measurement will require greater firing rates to achieve the same set points. The relocation will not be dangerous to the strip because the strip temperatures in preheat zones are several hundred degrees below final temperature. In addition, during a line stop, the relocated measurement will sense the rapid temperature rise and reduce energy input. (See "accordian effect" discussed earlier in this section.)

***4.3.4.1. Temperature Measuring Devices.*** Most furnace designers call for T-sensors with insulators on the wires in a 0.75 in. (19 mm) alumina protection tube, which, in turn, is in a 1.625 in. (41 mm) silicon carbide tube. Such a design causes far too much time lag to control a strip that may be in the furnace only 30 sec. There have been cases where the strip hardness varied down its length like a sine wave because of large time lags in control temperature measurement. To correct this problem, a 0.375 in. (9.5 mm) diameter alumina tube without a silicon carbide outer cover generally suffices. (A very small diameter, metal-encased thermocouple would have even less time lag, but its life would be shorter.)

An open-tube radiation temperature sensor at the furnace outlet has been found very useful by many operators. However, emissivity changes from coil to coil can erode confidence in strip temperature measurement. Their use inside the furnace may be even more variable.

A "K" thermocouple welded to the strip and pulled through the furnace to display a temperature profile is extremely effective in proving the thermal treatment of the strip. Such a temperature profile can be used immediately to adjust zone setpoints and to assure proper strip treatment. For the very best strip treatment, using a welded thermocouple on every coil seems appropriate for improving downstream processing.

*A control method variation* uses the output signal from a temperature control in a downstream zone as process variable for energy input in the next upstream zone, for example, soak zone temperature controls main heating zone input and/or heat zone temperature controls preheat zone temperature. Note that "zones" may sometimes be a series of closely spaced, separate catenary furnaces. If a very low setpoint for the output signal of the soak and/or heat zones is used to control the upstream zone, the soak time will be extended to allow the chrome carbides to dissolve into the strip and thereby produce a quality product.

The controllers for the preheat zone or zones should have an over-temperature loop to automatically assume control in case of difficulties. In case of a line stop, the output signal of the heat or soak zone temperature controller would be reduced, calling for lower firing rates in the preheat zones. To provide an additional means for reducing the fuel input quickly, push-button stations could be installed at the line control locations to shut off the fuel to the preheat zone or zones in less than one sec. Strip temperature is almost never the same as furnace temperature, following firing rate changes more closely than furnace temperature; thus, on/off control should not be used, and a rate bias triggered by soak zone firing rate may help. It is recommended that at least one roller should be within the furnace to allow a temperature sensor to be very near the strip. Sensors must have a surface-to-mass ratio similar to the strip. (Heavily encased sensors will have too much time delay.) Less protected sensors may have shorter life, but that is the cost of getting good control. (See fig. 4.11.)

Catenary furnaces are excellent candidates for fiber linings to reduce the refractory heat storage (flywheel) effects. With a lightweight lining, line stops are generally less of a problem.

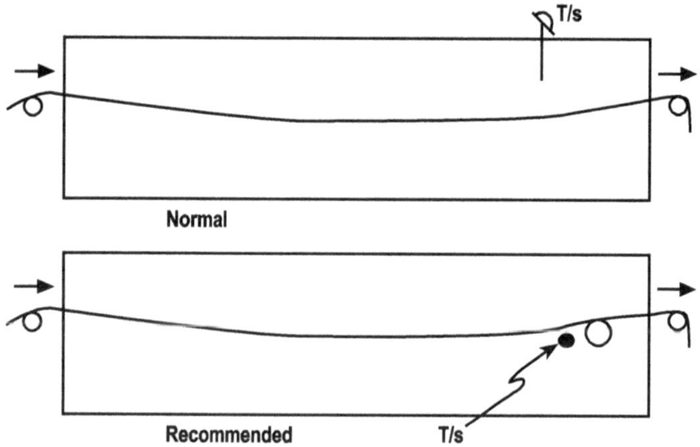


**Fig. 4.11.** Normal (*left*) and recommended (*right*) temperature sensor locations for catenary strip. The hollow shaft through the center of the added roll should be water cooled because the furnace temperature may be 2300 F (1260 C).

***4.3.4.2. Burners and Zones.*** Many past furnaces were built with burners staggered from side to side, omitting burners above the strip in some zones, and with some zones oversized and others smaller than they should have been. The primary difficulty with these early designs was lack of flexibility. There was no problem as long as the furnace was to operate at very slow strip speed, but because the operators' responsibilities were to achieve maximum throughput consistent with good quality, furnace problems often bottlenecked the process.

Burners should be about 2.5 ft (0.87 m) apart, above and below the strip. The burners above the strip should be on one side of the furnace and those below the strip on the other side, enhancing circulation velocity. The burners should have a near-flat heat-release pattern (preferably adjustable), providing a temperature profile across the furnace that is practically level. It is important to check the design and the actual operation to make sure that no bottom-row-burner flames impinge on the lowest part of the strip's catenary loop.

Zone lengths should not be longer than 15 ft (4.6 m) to allow adequate soaking times with various product requirements and maximum furnace lengths, taking advantage of additional heat heads for maximum furnace productivity. Regenerative burners can be used to reduce fuel input per ton of strip heated, with excellent results. Another means to save energy is a waste heat boiler, which can recover heat from a catenary furnace's flue gas—if there is a concurrent need for steam, such as for heating cleaning solutions.

### 4.3.5. Catenary Furnace Size

Heat transfer rate is a function of the gas blanket thickness, which should be 3 ft above and below the strip. For the strip hanging in the natural shape of a catenary curve with, for example, the low point of the strip 1.5 ft (0.5 m) below the top surface of the supporting rolls, the furnace bottom should be 4.5 ft (1.4 m) below the strip's highest level.

*Air/fuel ratio* should be on a burner-by-burner basis to nearly eliminate varying ratios throughout the furnace zones. (See fig. 4.12 and 4.13.) At low firing rates, burners should be run on high excess air to avoid exceeding zone temperature setpoints when the line speed is slow or stopped. The air/fuel ratio should be set by measuring gas and air flows to hold 15 to 25% excess air (about 3 to 5% excess oxygen) from maximum firing rate down to 30% of high fire input rate, where the ratio should be changed to about 200% excess air.

Most annealing of stainless-steel strip is done without a protective atmosphere in the furnace. However, combustibles must be avoided to prevent their effect on the surface chemistry of the strip. Likewise, high excess air at low fuel inputs may necessitate more aftercleaning, but some excess air protects the strip from a runaway furnace temperature condition. A simple cross-connected regulator with a low-flow tension spring (fig. 4.12) is ideal for this. Figure 4.13 shows a more accurate control.

**Warnings:** When designing a furnace, one should expect that eventually the process capacity will be furnace constrained, and that the furnace will be costly to upgrade or replace. Therefore, making the furnace somewhat larger than present needs, say 20% larger, will generally return the investment well.

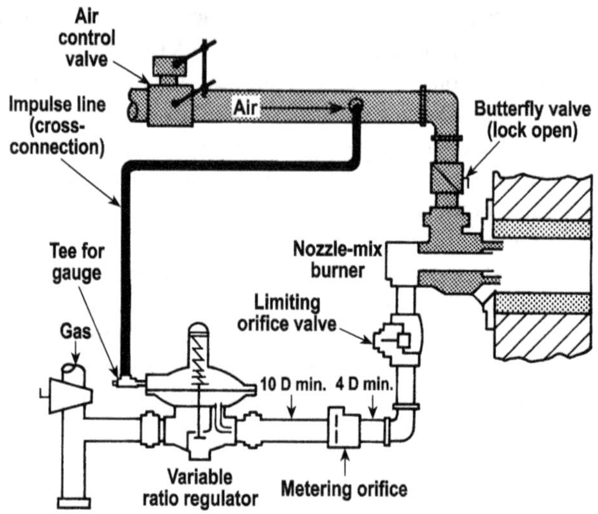


**Fig. 4.12.** Variable ratio gas regulator and piping. Extra spring length allows setting extra negative bias to gradually change air/fuel ratio from correct at high fire to a selectable lean air/fuel ratio at low fire. Courtesy of North American Mfg. Co.

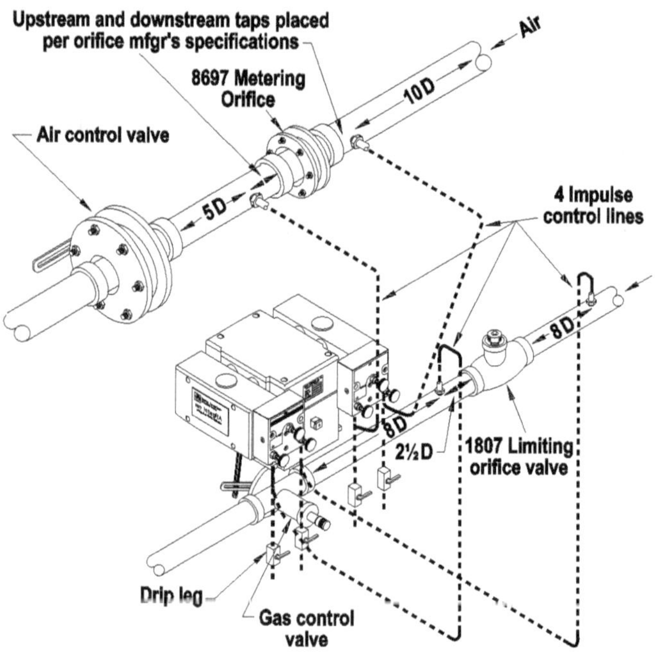


**Fig. 4.13.** Integrated ratio actuator controls air/gas ratio by comparing pressure drops across air and gas orifices. It automatically compensates for varying air temperature, thus providing mass flow control. An adjustment allows use of low-fire excess air for thermal turndown. Courtesy of North American Mfg. Co.

The reader is urged to reread the first $1\frac{1}{2}$ pages of this chapter concerning the inevitable discontinuous operation of continuous furnaces, the costly consequences thereof, and the necessary design corrections. Chapter 8 includes original and corrected time–temperature diagrams from an actual case.

## 4.4. SINTERING AND PELLETIZING FURNACES

Both sintering and pelletizing include induration* and are processes of ore beneficiation, including chemical and physical methods for enriching ores such as taconite-magnetite, hemitite, and geotite to less water and oxygen content, and strengthening the clinkers or pellets for less breakage and fines formation and to assure better hot gas passage through deep beds such as in blast (shaft) furnaces.

Sintering is a process of heat-agglomerating fine particles of naturally occurring fine ore, flue dust, ore concentrates, and other iron-bearing material into a clinkerlike material that is well suited for blast furnace use. (The term "sintering" also describes a process used in much *powder metallurgy*—a method for forming small metal shapes by a combination of heat and compression. Many such furnaces are batch type, and most are similar to heat treating furnaces such as those discussed in sec. 4.3.)

Sintering was originally used to provide a larger and more uniformly sized charge ore material for blast furnaces. In most cases, sintering also improved the ore charge chemically. Most of the raw ore was made up of very fine particles. In a blast furnace, the fine particles created increased resistance to the flow of reducing gases through the burden (ore, coke, and limestone). Fines would often create a "bridge" and leave voids. If these collapse, a relief valve opens, polluting the area with particulates and gases.

Air or highly oxidizing gas is passed through the bed, and the carbon and ore mixture is ignited by the hood. The heat from the burning coke raises the temperature of the pellets to 2300 F ± 100 F (1260 C ± 56 C), agglomerating the ore fines and forming irregularly shaped clinkers that are then screened for size. Any remaining fines are recycled. The air or oxidizing gas must be passed through the bed at a high enough rate to minimize the gas temperature drop so that the whole bed thickness is involved in the oxidizing process. If the flame progresses quickly down through the bed, the length of the traveling grate can be minimized.

In the continuous sintering process, a mixture of ore dust and coke breeze or anthracite coal is delivered to a traveling grate in a continuous bed about 18" (0.46 m) deep passing under an "ignition arch" or "ignition hood" of burners for induration. (See fig. 4.14.)

Blast furnace productivity increased by the use of sinter. In some parts of the world, nearly all ore is sintered. Sintering provides the charge sizing that iron melters had long wanted for their furnaces.

*Induration is a process of heating and agglomerating a clinker or pellet by grain growth and/or recrystallization.

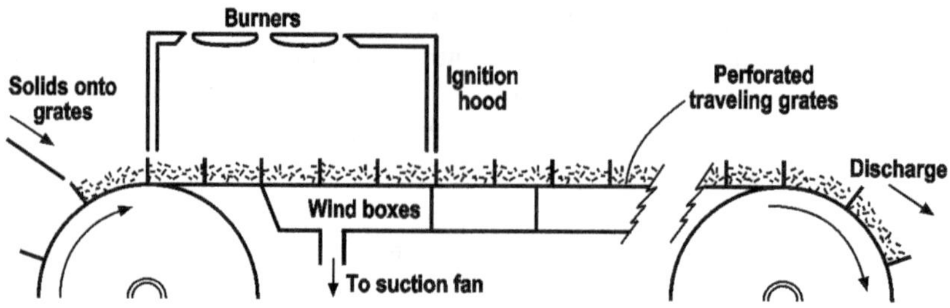


**Fig. 4.14.** Traveling grate furnace for roasting, sintering, or pelletizing ores. The ignition arch or hood may be fired with conventional type A flames or flat type E flames (shown, see fig. 6.2.)

### 4.4.1. Pelletizing

Converting the ore fines into pellets with more physical strength prevents them from being crushed, thereby avoiding obstruction of free flow of partially burned gases to reduce the ore. Continuous pellet-forming processes utilize heat recovery to minimize fuel cost. As the first step in the indurating process, pellets are formed on a large disc or in a rotary drum kiln, and then dried to prevent internal steam build-up.

Preheated air is used to burn oil or natural gas to form a gas stream (more than 10% $O_2$) to oxidize the ore at a very high temperature to make the pellets very hard and strong. These gases, still very hot when they leave the bottom of the pellet bed, are collected and used in updraft and downdraft drying of the bed and in pellet preheating. Further recycling of the hot gases may be justified as fuel costs rise.

The bed is then cooled enough to minimize damage to the belts used to convey the pellets from the plant. The portion of the cooling air that had been pumped up through the bed of pellets that gets to more than 1700 F (930 C) can be used as preheated combustion air.

Part of the warmed cooling air, at about 500 F (260 C), is used for a first zone of updraft drying of the pellets, but its temperature must be carefully controlled because pellets that are not suitably dried may explode, causing plugging and very dirty atmospheres in the vicinity of the machines.

A major problem with pelletizing plants is the NOx formed by the very high temperatures developed in the burners and heating chamber above the pellet bed. After the process reaches 1400 F (760 C), low NOx fuel injectors could be used above the beds to avoid the very high reaction temperature in the burners. To get the combustion chamber to 1400 F would require low NOx auxiliary burners. This technology has been used in many industries with excellent results. The NOx-forming temperature is lowered in the main combustion chamber by two major effects:

1. The reaction takes place within sight of both the product and the furnace refractories, both of which absorb some reaction heat (unlike a burner tile of quarl)
2. Inert molecules in the combustion chamber atmosphere join in the reaction because both the air and the fuel inspirate combustion chamber gases as they

are directed into the chamber by peripheral nozzles. The combustion chamber gases contain inerts that deter NOx formation absorbing heat, reducing the combustion reaction temperature, lowering NOx.

An additional means for reducing NOx would be to recycle some of the effluent bed gas into the suction of the cooling air fan. This will reduce the oxygen concentration in the combustion "air" to 13 to 17%, which along with fuel injection will reduce NOx by 50%.

## 4.5. AXIAL CONTINUOUS FURNACES FOR ABOVE 2000 F (1260 C)

### 4.5.1. Barrel Furnaces

Some hot forming processes such as continuous butt welding of tubes or pipes and sizing of tubes or pipes are facilitated by heating the stock ("skelp") as it travels axially through a furnace. Because such furnaces are long, there is a desire to shorten them by using very high temperatures. Supporting the load is a problem, solved by (*a*) a series of "barrel furnaces" with cooled rollers in the spaces between the barrels (see figure 4.15), or (*b*) one or more long furnaces with water-cooled pipes ("hairpins") or rollers within the furnace(s). (See fig. 4.16.)

Combustion gases are directed at the edges of the skelp to heat them to scale softening temperature (about 2320 F, or 1270 C). Temperatures in skelp-heating furnaces may reach 2600 F (1427 C), causing very high fuel bills unless recuperation or regeneration is used. A skelp-heating furnace may consume 2.5 kk Btu/US short ton or more (2,908 MJ/tonne or more). Regenerative burners have been applied to a few zones of this type of furnace with outstanding results. Steel slabs with 2.25" thickness (57 mm) have been heated for rolling in skelp furnaces at a rate of 165 lb/hr $ft^2$ of top- and bottom-load surfaces.

Water-cooled supports inside the furnaces should be reduced to a minimum for good fuel economy and furnace productivity. The high operating temperatures on these furnaces necessitate alert maintenance.

Skelp-heating furnaces sometimes exceed 150 ft (45 m) in length. For thick traveling stock, the last zone may be at a lower temperature soak zone for equalization within the stock thickness. Water-cooled rollers absorb more heat from the load, requiring extra bottom-side input. Barrels must be short enough to prevent sagging of the hot stock, especially at the load's leading edge. Fewer supports are needed for continuous bar, rod, or strip. Supports inside the furnace or between barrels absorb much heat.

For butt-welding skelp, the burners are often directed at the skelp edges so that these edges become hotter than the skelp body. When the edges reach scale softening temperature (2320 F, 1271 C), steel burning begins if the burners' poc has at least 1% $O_2$. The higher rate of burning sustains the reaction by virtue of its heat release of 2,850 Btu/lb of iron (1,583 kcal/kg). The iron is oxidized to $Fe_2O_3$, the most oxidized iron compound.

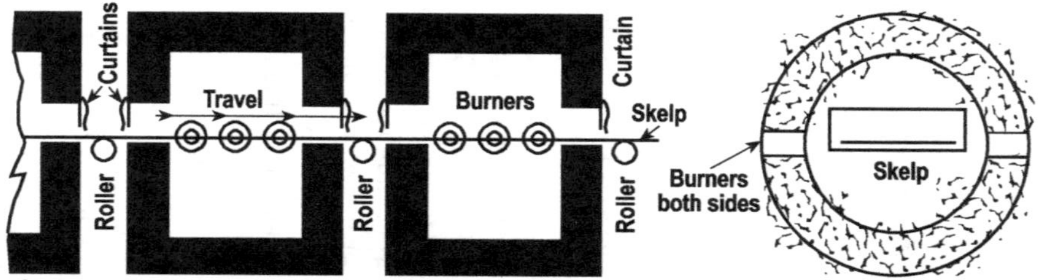


**Fig. 4.15.** Barrel furnaces for impingement heating of skelp edges—for welding into seamed pipe or tube. *left,* side view of three barrels; *right,* end view. Not shown, but necessary, are slag cleanout access doors in all sections.

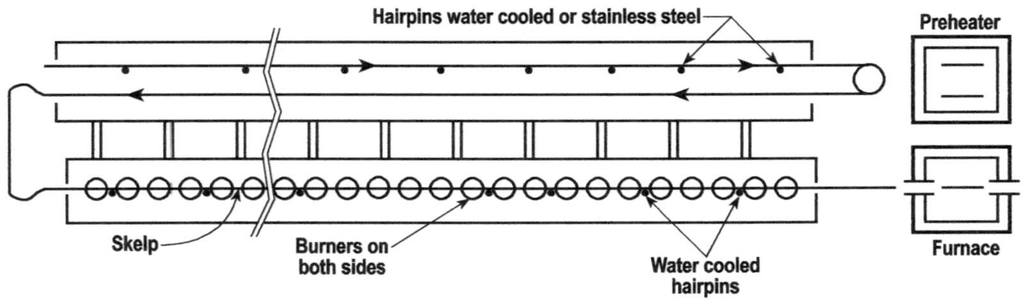


**Fig. 4.16.** Modern skelp-heating furnace with heat recovery by load preheating. Some furnaces use type H high-velocity impinging burners; others use refractory radiating burners similar to type E, but with concave refractory tiles. (See fig. 6.2 for these flame types.)

Butt-welding furnaces that use type E convex tile radiation burners instead of impingement are controlled by eye measurement of strip temperature. With impingement heating (type H burners), control is by observing the width of strip edge burning, a much more accurate way.

Calculating furnace size and firing rate can be accomplished by the Shannon Method detailed in chapter 8. The required furnace length = required heating time multiplied by stock feed speed. Heating times and cooling times between barrels should be figured and plotted alternately.

***4.5.1.1. Impingement Heating.*** This type of heating is sometimes used for operations at lower temperatures than the skelp welding process, such as heat treating and forging of pieces processed in long-run, mass-production equipment. Maintaining uniform surface temperatures with impingement heating requires many small burners; thus, temperature uniformity control and selecting a representative location for the T-sensor can be difficult.

***4.5.1.2. Unfired Preheat Section for Fuel Economy Versus Fired Preheat for Productivity.*** Unfortunately, a characteristic of impingement heating often is high flue gas exit temperature, which results in high fuel cost; thus, such cases are good candidates for addition of a heat recovery system. If an unfired preheat vestibule is selected as the vehicle for heat recovery, there may be a great temptation later to add burners to the preheat section for higher capacity. With any preheat section—unfired or fired—careful attention must be paid to gas flow patterns. Usually, little heat recovery is accomplished by simply passing flue gases through an insulated box holding some load pieces. The designer should have an understanding of heat flow (chap. 2) and fluid flow patterns (chap. 7).

Examples of nonuniform heating-control problems above 1000 F (538 C) are (1) nonuniform scale formation with carbon steels, (2) questionable completion of the combustion reaction (pic contact the load surface), (3) sticky scale with resultant rolled-in scale, (4) spotty decarburization of high carbon steels, (5) some stainless steels may not tolerate contact with the reducing atmosphere within the flames, and (6) using impingement heating for steel pieces of heavy cross section could cause formation of reflective scale with resultant reduction of heat transfer.

### 4.5.2. Shaft Furnaces

Shaft furnaces have been epitomized by blast furnaces and cupolas in the past, but those are being replaced by electric melters. Most use a solid fuel such as coke layered in with the load charge from the top. As the solid fuel burns, it heats the granular charged load to melting point, allowing the liquid metal to trickle down through the voids left by the coke. The only "burners" are gas or oxygen lances inserted through the sidewalls to hasten melting. Figure 4.17 illustrates a typical arrangement.

### 4.5.3. Lime Kilns

Lime kilns are sometimes built in a shaft-furnace configuration. Fuel and air are fed into the descending column of pebble-size limestone from burner beams across the

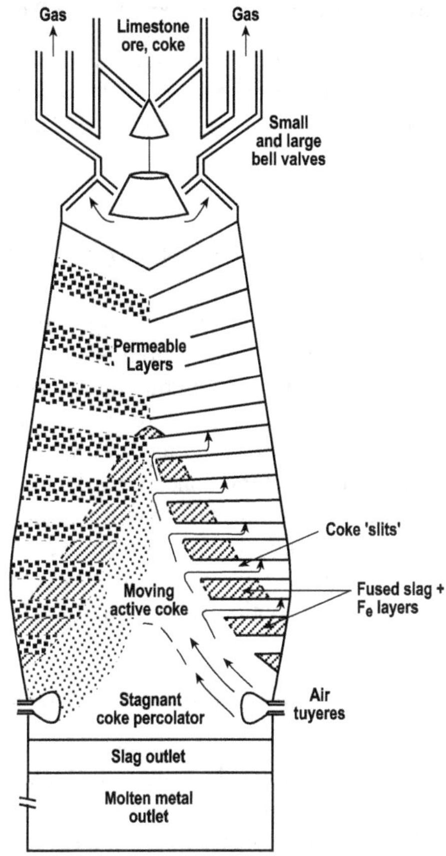


**Fig. 4.17.** Blast furnace, a shaft furnace. The fusion zone has alternate layers, 1.5 to 3 ft (0.5–1 m) thick of coke, then fused slag and iron. If cleaned, the off-gas (blast furnace gas) can be used as a fuel. Courtesy of *reference 11*.

shaft-furnace interior. The powderlike lime is extracted in a fluidlike form at the bottom. Lime kilns are more often built in rotary-drum configuration like cement kilns, mentioned later. (See pages 16, 124, 142, and 144.)

## 4.5.4. Fluidized Beds

Fluidized beds are similar to shaft furnaces. They contain a thick bed of inert balls, pellets, or particles through which are bubbled streams of hot poc rising through a grate or perforated plate from a combustion chamber below. The loads may be (*a*) the pellets or particles themselves, which need heat processing, (*b*) larger solid pieces needing some sort of heat treating, or (*c*) boiler tubes for generating steam (fig. 1.9), or tubes carrying liquids or solid particles that must be heated but protected from contact with poc.

The benefits of fluidized bed heating are (*a*) rapid heat transfer from the physical bombarding of the particles in the fluid bed and (*b*) more uniform heating of complex shapes because the load pieces are completely immersed in the heat transfer medium, which is the fluidized bed contacting all surfaces of each piece equally.

### 4.5.5. High-Temperature Rotary Drum Lime and Cement Kilns

High-temperature rotary drum lime and cement kilns are of similar configuration to rotary drum furnaces and dryers discussed in section 4.2, except that they are of higher temperature construction and longer. This is a very specialized field. (See Perry: "The Rotary Cement Kiln," *reference 64*.) A shaft-type lime kiln is shown in figure 1.11.

## 4.6. CONTINUOUS FURNACES FOR 1900 TO 2500 F (1038 TO 1370 C)

Thickness of heating stock does not limit heating capacity as much in continuous furnaces as it does in top-fired batch furnaces because heat can be imparted to the load from below. The limiting thickness depends on the thermal conductivity of the load and required temperature uniformity.

"Triple" firing of continuous furnaces refers to top heat, bottom heat, and separate firing of the soaking zone. When comparing heating capacities of such furnaces, statements regarding the hearth area of reference should be specific: whether top heating zone only, or top plus bottom area, or top plus bottom plus soaking zone, and finally whether based on load or hearth area. Hearth area is (effective hearth length in direction of motion) × (length of load piece across the hearth).

### 4.6.1. Factors Limiting Heating Capacity

Ideally, there should be no transfer of heat in soak zones, except the temperature equalization within the pieces. In fact, a slight loss of heat from the top speeds equalization. Temperature equalization between surface and interior is considered to be of less importance than elimination of dark *spots*. The soaking zone eliminates or reduces dark spots, but does not necessarily eliminate cold centers, which show up as greater thickness in the finished product (rejects).

Numerical values for the capacity of steel heating furnaces are based on uninterrupted operation throughout the work week. Delays in the mill or forge reduce the weight of steel heated in the furnace, but do not reduce the heating capacity of the furnace. Figure 4.21, later in this section, gives a good approximation of the weight of steel that can be heated per hour and per square foot of hearth, for various thicknesses, depending on the number of furnace zones. Specific heating curves must be developed to verify whether a particular product can be heated to a specified uniformity. Generally, steel pieces thicker than 6" (0.15 m) must be heated from both top and bottom.

Major factors in limiting heating capacity are the pounds heated per unit of hearth area, average gas cloud (blanket) temperature (with preheated air or oxygen enrichment, the average gas temperature rises), thickness of the gas cloud, number of zones, air/fuel ratio, and furnace heat losses. The heating capacities of all types of furnaces vary greatly with the nature and surface condition of the loads being heated. Another issue that must be addressed is fuels with low flame temperatures. These will result in high flue gas exit temperature, thus less heat transfer than with rich fuels because of lower $\Delta T$ between the flame and the load.

Capacity increases in direct proportion to the area exposed per unit weight and in proportion to the heat transfer coefficient, which increases with average gas temperature and gas blanket thickness (figs. 2.13 and 2.14). Obviously, heat transfer increases as zone temperature setpoints are raised, unless scale formation interferes—as it will do if the preheat or entry zone is raised above 2300 F (1260 C).

Other problems that limit production rates in either longitudinally fired or side-fired bottom zones are restricted gas passages in the bottom zones, and low-velocity luminous flame burners. Low-velocity luminous flames with their variable temperature profiles (hot at the burner wall at low firing rates. and hotter beyond the T-sensor at high firing rates) cause the melting of scale into the bottom zones. To counter this scale build-up problem, operators are prone to lower the bottom zone temperature by 100 F (56 C) or more.

In three- and five-zone furnaces, the clearance between the skid line and roof and between skid line and furnace bottom are usually designed equal to divide the gas flows equally between top and bottom. However, designers forget about the partial closure of the bottom gas passage by crossovers, which can cut the area by 33%, forcing the bottom gases into the top zones. In addition to the crossover restriction, scale drops off the incoming products partially filling the bottom zone gas passage further, forcing bottom gases into the top zone(s). Without hot gas and a thick gas blanket, heat transfer suffers greatly in the bottom zones. When these gases pass from the bottom zones to the top zones, they generally envelop the bottom zone temperature sensor, causing the bottom zone to be much colder than it should be, further reducing the furnace heating capacity. With modern burners, which can develop a profile to suit the conditions, the top and bottom zone temperatures can be nearly the same, increasing heat transfer and therefore furnace capacity.

Furnace heating capacity also is limited by the percentage of the hearth that is covered. For example, a pusher furnace 42 ft (12.8 m) wide and 80 ft (24.4 m) long may have a rated capacity of 200 tph. However, if it is loaded with slabs only 31.5 ft (9.60 m) long, then only 31.5/42 or 9.60/12.8 = 75% of the hearth is used; therefore, the heating capacity will be only 0.75 × 200 = 150 tph. Another factor in limiting furnace capacity is the shape of the furnace. If the roof is lowered in the charge end of the furnace and the bottom is raised, the quantity of radiant energy transferred from the gases in those areas is reduced because the thickness of the gas blanket is less, reducing the heat transfer from the gases. Reducing the cross-sectional area in the charge end of a furnace is generally a design error, lowering furnace capacity. If operators try pushing the furnace output, they will raise the fuel consumption.

The thickness of the product has a direct bearing on furnace capacity because the added time needed to raise the core or bottom to the heated surface temperature is proportional to the square of the thickness. To provide equalization (soaking) time at the furnace discharge with loads of larger cross section, heating must be started earlier; thus, the gas meter will be cranking up the fuel bill longer. A further problem arises from the fact that thicker load pieces will have a less steep temperature gradient from outside surface to core temperature, so heat transfer from the surface to the core will be slower. It is impossible to hurry this conduction heat transfer rate by raising the furnace temperature without raising the flue gas exit temperature, which raises the fuel bill.

In furnaces equipped with skid pipes, the soaking zone serves mainly for elimination of dark spots. If the greatest possible heating capacity in a given space is desired or necessary, the temperature in the heating zone is run up as high as circumstances permit (explained later) and some equalization of temperature, including elimination of dark spots, is obtained in a soaking zone. The length of the soaking hearth is determined by temperature difference between surface and core (in very thick sections) and by elimination of dark spots (in medium heavy sections). In the rolling of thin strip, micrometer measurements in the finished product reveal the location of the dark spots in the slab. For that reason, the length of the soaking zone depends upon the stringency of specifications on uniformity of thickness in the finished material.

In other words, the capacity of a furnace with a given soaking zone length depends on the required uniformity of gauge in the finished product. This fact explains the seemingly illogical practice of adding top heat in the soaking zone. Elimination of black spots is considered to be more important than top-to-bottom temperature uniformity.

Positioning of T-sensors should be thought through to provide temperature control for the load pieces, not necessarily for the furnace. This is discussed in detail in chapter 6, but this box gives a generalized preview of *load* temperature control philosophy.

In earlier practice, if load pieces were loaded with their long dimension crosswise to the direction of load travel, T-sensors were located high in the zone and near the end of the zone (where the pieces were about to move into the next zone). Now, it is suggested that the T-sensors be positioned just above the level of the tops of the tallest loads. These sensors are now positioned about one-third of the load travel distance into each zone rather than near the exit from each zone. The rational for these decisions comes from experience with mill delays.

The so-called accordion effect upsets the supposedly steady pattern of temperature progression as load pieces move through the zones of multizone reheat furnaces, whether rotary, pusher, walking beam, or walking hearth. (See chap. 6.)

The charge zone was formerly unfired, hoping to recoup heat from the gases exiting as an endwise drift from the other (firing) zones (this attempt at heat recovery is now better accomplished by regenerative burners in the charging zone). The main reason for firing the charge zone is to help the newly charged cold pieces entering the furnace after a delay catch up with the pieces that have been heating in the furnace during the delay. Without charge zone firing, delay will build upon delay.

In average practice, the aforementioned rigid specifications do not apply. In conformity with varying requirements, the length of the soaking zone ranges between one-fifth and one-third of the furnace length.

***4.6.1.1. Flue Gas Exit Temperature.*** (See also sec. 2.4, 5.1, 5.2, and 5.6.1.) In any type furnace, calculating the firing rate requires determining the flue gas exit temperature, which is often underestimated. Its measurement is difficult, so "guestimates" may prevail, and the easiest number to guess is the measurable furnace wall temperature. That may work if a furnace has had poor care and suffers from considerable cold-air infiltration. In general, however, assuming that exit gas temperature equals furnace temperature is incorrect and leads to incorrect answers. Heat is a form of "potential flow," which always goes downhill—that is, to a point of less temperature (potential). If this were not so, how would the furnace wall get hot? This is as fundamental as the laws of thermodynamics. The temperature elevation of gases above furnace wall temperature is difficult to judge and measure! Obviously, heat transfer can be increased by raising the temperature differential ($\Delta T$), but then the $\Delta T$ becomes less as the better heat input accumulates in the form of higher furnace wall temperature. In steel heating, the rate of heating is limited by the strength of the refractory materials in only a few unusual designs.

When estimating the furnace temperature, the previous ideas must be used to properly design a furnace and estimate its fuel rate. Predicting the fuel rate if operating with delays is very questionable because the quantities of air infiltration with loss of furnace pressure can vary widely. Engineers must remember that the furnace heating capacity is determined by the actual furnace temperature, and not by the installed firing rate.

Developing a load heating curve (chap. 8) is the fundamental method for determining the following characteristics of a furnace: (1) zone firing rates, (2) waste gas temperature, (3) zone heat losses, and (4) temperature differences within the load throughout the heating cycle and at discharge. Some contend that heating curve work can be avoided by using rules of thumb (which invariably have limitations), but furnaces designed by rules of thumb are often poor performers with excessive firing rates in some zones and deficiencies in other zones.

***4.6.1.2. Rotary Hearth Furnaces*** Rotary hearth furnaces have no water-cooled skid pipes, so the soak zone can be less than one-fifth of the total furnace length. Very rapid heating results in a short heating zone, but requires a long soak zone for thick material. Rotary hearth furnaces have problems, such as:

1. Combustion gases move in two directions toward the flue.
2. Water seals reduce air infiltration around the outer periphery of the
3. hearth (and inner periphery for large "doughnut" rotary hearth furnaces. These seals limit, but do not completely prevent, air infiltration.
4. To reduce fuel rates, the first fired zone should be controlled by temperature measurement in the roof about 6 ft from the uptake flue in the direction of load

movement. Measurements at that point will adjust the firing rate of the first fired zone in accordance with the mill production rate.

5. Charge and discharge doors are usually very large, allowing large quantities of poc to escape, and making furnace pressure control difficult. This problem can be reduced by baffles on the right of the discharge door and on the left of the charge door (with the hearth rotating clockwise as viewed from above). Manually adjustable baffle heights should be used to further reduce the loss of poc. With larger load thicknesses, an air curtain must be added at the bottom of the baffle between charge vestibule and charge zone.
6. Indexing the positioning of shorter-than-design load pieces should place the loads as close to the sensors as possible, near the outer wall to take advantage of the greater hearth area there. This also allows wider spaces between the pieces for faster and more even heat transfer.
7. Rotary furnaces once had flues in each fired zone, which reduced thermal efficiencies to 30 to 35%. Most such furnaces have been rebuilt with one flue in the roof of the charge area, except where they supply a waste heat boiler, and all the steam generated is used in the operation.
8. The height of the baffle between the charge and discharge vestibules should be adjustable during operation. This allows operators to change the minimum clearance between the bottom of the baffle and the hearth to reduce hot gas flow from the high-temperature zones to the flue. With this baffle arrangement, nearly all furnace gasses will flow from the area of discharge toward the charge area, that is, around the full circle. (See also sec. 7.5.)

***4.6.1.3. Upgrading a Rotary Hearth Furnace.*** *Overcoming Problem 1.* The charge and discharge of a rotary (circular) furnace are connected; thus, the combustion gases can move in two directions to the flue and/or charge and discharge doors. As long as a door is open, large quantities of combustion gases can leave or much ambient air can enter, or both simultaneously. To remedy these effects, two baffles are necessary—one to separate the last zone from the discharge vestibule and one to separate the first zone from the charge vestibule.

With these two baffles, furnace pressure can be controlled, and practically all the hot combustion gases from the last zone would be forced to move to the first zone via all the other zones in the circle. In so doing, these gases would be forced to transfer more heat to the loads.

In addition to the previous two baffles, another baffle is necessary between the charge vestibule and the discharge vestibule to reduce the short circuiting of combustion gases from the last zone direct to the first zone. This baffle should be movable from a clearance between itself and the hearth of about 2" to 18" (51 to 457 mm).

*Overcoming Problem 2.* Furnace designers usually expect furnaces to operate in an equilibrium situation, in which case, the first zone could be unfired. However, delays are all too common with most operations, and must be considered. When a delay occurs, the products in a furnace will be heated above normal, especially in the first zone (many times to 1600 F to 1900 F). When the delay is completed, one, two, or three pieces are rolled to adjust product size off the mill; then the mill is ready to begin serious rolling. The new cold pieces charged into the first zone will be exposed to nothing but minor quantities of hot combustion gases (and minor radiation) from the other zones. As these pieces pass through succeeding zones, they may not encounter adequate gas flow and radiation because those zones' burners have been down or idling during the delay. The pieces that were left sitting in the furnace during the delay may be overheated or may not be up to satisfactory temperature for rolling. The differential temperatures in the loads are just too large to roll properly, and so the mill must close down due to lack of hot steel. Depending on the length of the delay, the new cold charges may not receive much hot gas convection or radiation until they are 50% through the furnace, so they may be inadequately heated, causing another delay.

Firing the first zone with main burners plus enhanced heating burners and controlling it by a T-sensors approximately 6 ft (1.8 m) into the first zone at the load level, the newly charged material will catch up to the material that had been held in the zone during the delay. That way, the productivity of the mill can be maintained even though there may have been "accordion effect" and "domino effect" delays during the heating of the product.

Admittedly, the total firing capability of the furnace as proposed previously will seem too high relative to conventional practice. Remember, however, that the full capacity of all the burners may never be used all at once. Flexibility to cope with delays will provide enough productivity capability and improved temperature uniformity (product quality) to balance any added fuel cost. The cost of delays cannot be ignored. Everyone must realize that even during delays, burners will be balancing heat losses, so fuel meters will be spinning.

Here are some numbers illustrating the need for built-in flexibility in a five-zone reheat furnace (rotary, end fired, side fired, or top fired). Main burners fire at very high rates in zone 1 (charge end) to heat the newly charged load pieces after a delay—because burners in zones 2, 3, and 4 stayed at low fire while the already-hot pieces in those zones were worked out. (Low-firing rates in zones 2, 3, and 4 reduced the quantities of hot gas normally available to assist in the heating of product in zone 1.) For example, normally zones 2, 3, and 4 will fire 20.8 kk gross Btu/hr providing 2.56 kk net Btu/hr of heat. After a delay, the firing rate would be on the order of 8.52 kk gross Btu/hr providing only 0.85 kk net Btu/hr. This net heat loss will require an increase in firing rate of zone 1 regenerating burners of 2.4 kk Btu/hr or 29% more fuel than a running rate of 8.4 kk gross Btu/hr. Because of this and other scenarios where additional firing rates are necessary, it is advisable to add a safety factor of at least 20% to cover unusual conditions.

To remedy the delay caused by delay situation so that the regular production rate can be maintained, it is wise to use enhanced heating to accelerate the heating. Enhanced heating provides more heat transfer to the cooler load surfaces in Zones 1 and 2. The temperature control measurement should be accomplished by using two sensors instead of one. The first sensor should be placed 6 ft (1.83 m) into the zone from the charge door and another sensor at about 90% through the zone. Both measurements must be controlled through a low select device to either the fuel or air valve. The first sensor is to measure the temperature of the cold material entering the zone for input control, and the second is to prevent overheating of the loads leaving the zone. The second sensor measurement's setpoint should be as high as any setpoint in the furnace. For example, if the zone 4 control temperature setpoint is 2300 F, the second (high limit) sensors of zones 1, 2, and 3 also should be set for no more than 2300 F. This control scheme should be reproduced in all zones, and enhanced heating used in the first two zones, to minimize delay problems. This control/heating scheme helps the newly charged loads to catch up to those that were in the furnace during any delay.

*Overcoming Problem 3.* In rotary hearth furnaces, load piece length and placement are very important. If the furnace is designed to heat 24 metric tons per hour (mtph) of 9 ft (2.74 m) long pieces but is used to heat 6 ft (1.83 m) long pieces, the capacity will be two-thirds of 24 or 16 mtph. Shorter pieces such as 5 ft (1.52 m) long will further reduce the furnace heating capacity and will heat only $(1.52/2.74) \times 24 = 13.3$ mtph.

The use of regenerative burners in Zone 1 will provide the input necessary without flue gases being part of a gas movement direction problem in the furnace. For example, firing Zone 1 with conventional burners would increase the flue gas flow moving toward the discharge vestibule. The reason for this is the division of gas flow in two directions as divided by the minimum cross-sectional area through which the gases must pass, as charge/discharge areas are generally built. If the firing rates are increased in the early zones, more flue gases must flow toward the discharge in ratio again to the two minimum areas in the directions of the two flows. However, with regenerative burners which have nearly all their gases move out of the furnace through their beds and their own flue system, the flue direction problems do not exist.

**Summary: Actions to Improve Heating Capacity of Rotary Hearth Furnaces**

1. Install a minimum of two fixed baffles and one movable baffle. Provide a furnace pressure control system if the present control is inadequate.
2. Provide main regenerative burners in zones 1 and 2, with enhanced heating in the form of small, high-velocity burners directed down at 10° to 25° to move the gases in the alleys between the pieces. The exposure increase will provide a remedy for delay problems, plus improved heat transfer in zone 1.

> Before regenerative burners, energy czars wanted to prevent the increasing of continuous furnace capacity by installation of added burners in unfired preheat zones because the poc of such burners could escape through a nearby charging entrance or flue without having delivered much of their heat to the loads. Regenerative burners, however, capture their own "waste heat" and send it back into the furnace; thus, they are a good way to increase furnace capacity without wasting fuel.

3. Install a new two-sensor control scheme in all zones to overcome delay difficulties.
4. Reduce the NOx generation by installing low-NOx regenerative burners.
5. Replace large burners in the center (doughnut hole) of large rotary hearth furnaces with high-velocity burners for better crosswise gas and temperature distribution.

*Overcoming Problem 4.* Another rotary furnace problem is the positioning of rounds on the hearth. Some operators index all the load pieces to one stop on the inlet roller table, which sets the pieces at a common point near the inner wall of the furnace. Others index the pieces to straddle the hearth centerline. In either case, short pieces may be 1 to 4 ft (0.3–1.3 m) from the outer wall of the furnace. One negative result of this is use of less hearth for heating loads. A second and critical problem is that the T-sensors will be farther away from the loads, causing the sensor to be less and less reflective of the pieces' temperature and more of a representation of furnace temperature. This problem is especially critical in the final zones where very responsive temperature control is needed.

For example, if the loads are 75°F (42°C) below the furnace roof temperature, and the outer wall temperature control sensor registers 25°F (14°C) below the roof, the control sensor will raise the firing rate promptly to perhaps 2 to 5% above its previous rate. That will increase heat transfer by about 4000 Btu/ft$^2$hr. If the T-sensor were more responsive to the actual load-piece temperature, it could raise the firing rate appreciably with a more prompt response. The effect would be that the hot zone would be two to three times as effective in heating the rounds because the roof temperature would have risen perhaps 100°F (56°C) above its former temperature to satisfy the more load-temperature-oriented control sensor. This increase in roof temperature would have increased heat transfer by 12000 to 15000 Btu/hr ft$^2$, or three times the previous scenario. If the loads had been 6 in. (0.15 m) from the sensor, a more beneficial response could have been achieved.

**Conclusion:** For maximum furnace productivity, multiple stops need to be available on the entry roll table to index the load pieces to an average of 9 in. (0.23 m) from the control sensor, or ideally 6 in. (0.15 m) from the sensor.

**Another Example:** Coauthor Shannon was controlling a 50 ft diameter rotary furnace, heating short rounds indexed near the inner wall of the furnace, when a $\frac{1}{2}$ hr mill delay occurred. When rolling resumed, several rounds were pierced until the tube size from the mill was considered satisfactory, and a rolling rate of 40 tph was begun. Zone 1 went to full fire in response to the control thermocouple located about 20 ft from the charge vestibule. At zone 2, the firing rate went up about 10% in response to a T-sensor located 15 ft inside zone 2 and 15" above the hearth. When the first cold round reached the T-sensor in the final zone, the firing rate went up in that zone about 10%. The final zone control sensor was about 15 ft before the discharge and 15" above the hearth. When the cold rounds reached the discharge, they were so cold they could not be pierced, requiring a heat delay of 15 min. Had the rounds been indexed to 6 in. (0.015 m) from the outer wall and the sensors 2 to 3 in. (0.051 to 0.076 m) above the hearth, no delay would have occurred because the zone 2 firing rate would have gone up 30 to 50% and the zone 3 firing rate would have risen to bring the rounds to piercing temperature.

### 4.6.2. Front-End-Fired Continuous Furnaces

Many believe that for greatest uniformity of temperature in top- and bottom-fired continuous furnaces, it is desirable to favor almost constant temperature from furnace end to end plus a soak zone for the ultimate heat flow rate per unit of time. This is not true if reflecting scale forms in the charge or preheat zone at temperatures above 2320 F (1270 C). Such scale will reduce heat transfer so that the product will be colder and productivity will be lower than if the charge zone had been limited to between 2250 F and 2300 F (1232 C and 1260 C). Reflecting scale develops when scale softens and becomes very smooth and the steel temperature under the scale has relatively low conductivity, preventing the steel from absorbing heat from the scale.

An example of this problem was in the operation of a large rotary furnace heating large rounds. All five fired zones were operated above 6.F. At the end of the first heating zone, the scale was soft and reflective while the bottom of the rounds were very cold black.

After the first piercer, the maximum surface temperature was 2100 F, and when the round was rolled down into the discharge conveyor, distinctive barber poling was seen. Maximum furnace production was 110 tph.

When charge Zones 2 and 3 were reduced to 2000 F and 2350 F, respectively, the temperature after the first piercer increased to 2200 F and the furnace averaged 125 tons/hr for several days. The scale was very thin and dull black without a reflective layer. (See discussions of scale formation and decarburization in chap. 8.)

Front-end-fired furnaces should have *soak zones* to allow equalization independently of the heating zones. Otherwise, (see fig. 4.18) the heating zones must be limited to maximum soak-zone temperatures when the heating zone temperature could be higher for maximum productivity.

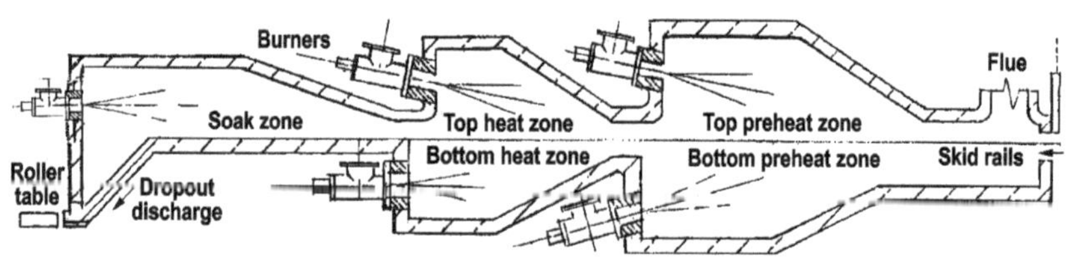


**Fig. 4.18.** Continuous steel reheat furnace, longitudinally fired in all five zones. Unless a recuperator will be above the furnace, flues at the *far right* bottom zone would be better than the up-flue shown (*a*) to minimize cold air inflow around the charge entrance and (*b*) for better circulation in the *bottom right* end of the furnace.

Soak zones with dropouts or extractors would best have screen burners through the roof to prevent air infiltration through the discharge opening. Such "screen burners" help build up a positive pressure to stop inleakage. DO NOT locate screen burners at the bottom of the furnace because they will create an eductor effect, pulling in more cold air and chilling the discharging pieces. (See more about soak zone and discharge in sec. 4.6.10.)

The soak zone should be divided into three zones across the furnace width to permit profiling of the temperature of the product. With small to medium sized bars in a straight ahead mill, the head ends should be approximately 50 F above the body temperature and the tail should be about 60 F above the body temperature. The reason for the higher temperatures for the head and tail is overfill and underfill of the roll passes when the head and tail of the billets are not being stretched between mill stands, which is a problem even with loopers between roll stands.

If firing only the outside zones does not suppress the body temperature enough, increase the minimum air flow on the center zone burners to actually cool the center of the billets.

### 4.6.3. Front-End Firing, Top and Bottom

Heating capacity of furnaces with top and bottom firing is less than twice that of furnace with top heating only because (l) the required water-cooled supports reduce the loads' exposed heat transfer area; and (2) the cold supports also act as heat sinks, stealing heat from the load and from the hot furnace gases, and (3) bottom-zone heat transfer also is reduced by movement of the hot furnace gases from the bottom zone to the top zone.

Minimization of problems 1 and 2 is difficult with conventional burners as their temperature profiles (that vary with input) limit temperature control setpoints in bottom zones because of excessive liquid scale in that zone. Problem 3 would be minimized with modern regenerative burners because 80% or more of the poc must flow to the off-cycle regenerative burner(s) in the bottom zone.

Water-cooled skid supports are a big factor in increasing bottom-zone firing rates. Coauthor Shannon has felt that an adjustable baffle just before the rabbit ears (uptakes or downtakes at the charge end of the furnace) would solve the problem by preventing movement of top or bottom gas to the other zone. The clearance under the baffle could be automatically or manually controlled to adjust flow patterns to nearly eliminate migration of furnace gases between bottom and top.

### 4.6.4. Side Firing Reheat Furnaces

Continuous furnaces with *rotating hearths* have no ends and thus cannot be end-fired, but must be side fired or roof fired through a sawtooth roof or with type E flat-flame burners. (See fig. 6.2.) Heating capacity of continuous *rectangular hearths (pusher, walking, or conveyorized)* is greatly increased by side firing for almost full furnace length, by increasing the number of temperature control zones, and by limiting the charge zone setpoints to 2250/2300 F for steel. (See figs. 4.19 and 4.20.)

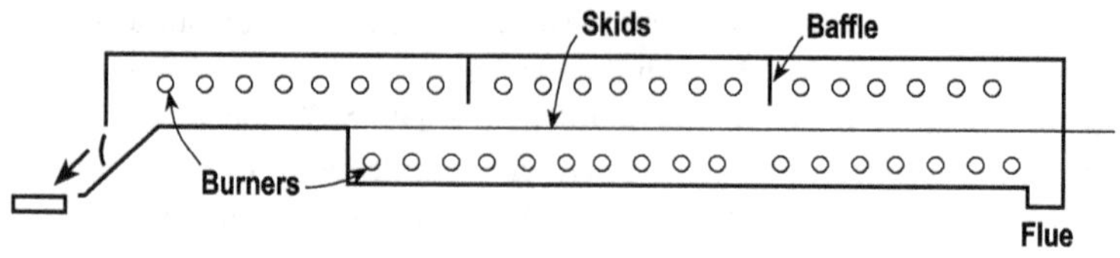


**Fig. 4.19.** Continuous steel reheat furnace, side fired from both sides, staggered, not opposed, in all top and bottom zones.

Emissivity and conductivity at low product temperatures can have major effects on heat transfer and therefore furnace capacity. Higher gas temperatures in the furnace can increase heat transfer, which is why recuperation, oxygen enrichment, or regenerative burners can increase furnace capacity by as much as 15% *and* reduce fuel rates from 20 to 45%.

Another problem that limits furnace capacity is *bowing* in top-fired-only furnaces wider than 25 ft (7.6 m). Excessive bowing in the charge zone is due to large temperature differentials between billet top and bottom. If the billet bows more than its thickness, pileups are sure to result. Pileups result in huge mill delays. Therefore, the furnace throughput must be reduced to a production rate that avoids serious bowing.

To increase furnace productivity in wide furnaces, underfired "enhanced heating" burners should be used at the charge end of the furnace to reduce top-to-bottom temperature differentials within the load pieces.

*Temperature differentials across the hearth* have caused engineers to avoid side firing. The first crosswise $\Delta T$ error was the installation of burners directly across from each other because the opposing flame streams stopped one another in the center of the furnace, sometimes causing completion of combustion at that point and resulting in a large temperature rise in the center of the furnace. The solution was to shut off every other burner on alternating sides of the furnace, reducing furnace capacity.

*A second crosswise $\Delta T$ error* is the variable temperature profile of the combustion gases across the furnace depending on the firing rate. With only one temperature measurement in a zone, the zone setpoint must be conservative to prevent rapid scale melting in any part of the zone; hence, productivity is sacrificed. Modern burners

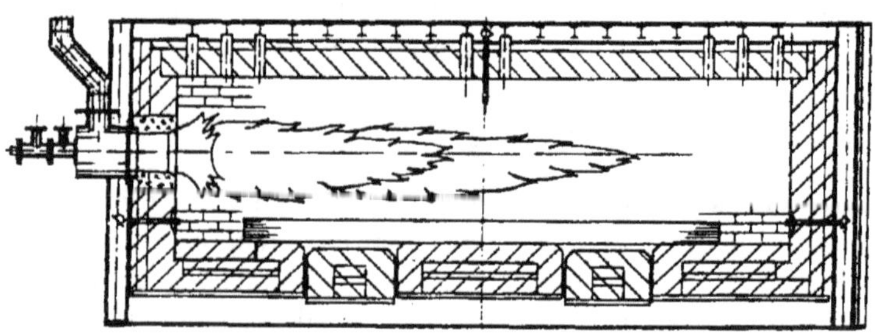

**Fig. 4.20.** Walking hearth furnace, cross-section detail.

can be controlled to avoid both problems by adjusting the energy to spin the poc to provide a level temperature profile to the poc (or a slope if desirable).

*A third crosswise* $\Delta T$ *error* can result from combining side firing with upstream longitudinal end firing. The flow lines of the longitudinally fired gases collide with the side-fire burner gases, causing the side-fired gases to turn toward the charge end of the furnace, raising the sidewall temperatures and lowering the temperature of the furnace center. The result is a reduced furnace heating capacity, high exit gas temperature, nonuniform heating of loads, and consequent high fuel rates. The solution to this problem is to install a baffle in the furnace between the longitudinally fired burners and the side-fired burners to interrupt the combustion gas flow from the longitudinal burners. After the baffle, the gases will then flow with a velocity close to that calculable using the whole furnace cross section downstream of the baffle. This will cause the longitudinal flows to have minimal effects on the gases from the side-fired burners. Another improvement may be air lances through the centers of the side-fired burners.

Generally, side-fired burner problems in continuous furnaces can be avoided by a baffle upstream of the side-fired burners, combined with automatically controlled ATP side-fired burners. Side firing in booster zones with pure oxygen or regenerative firing is ideal to raise productivity with minimal fuel problems. Long-term cost results favor regenerative firing, but with high capital cost. Oxygen firing has minimal capital requirements, but the oxygen costs remain an operating-cost problem forever.

### 4.6.5. Pusher Hearth Furnaces Are Limited by Buckling/Piling

Safe length of hearth is another factor that limits the capacity of pusher continuous furnaces (with regard to pounds heated per hour, but not with regard to pounds heated per square foot per hour). "Safe length" means a length that avoids upward buckling and piling. The safe length depends on the flatness of the hearth, the thickness of the stock being heated, and the shape of the contacting surfaces of the stock. Thin billets are seldom straight, and often have sheared ends that are irregular. Very cold bars rise in the middle when heated. A hearth length that is safe in one mill may cause buckling in another mill. Longer load pieces are more prone to thermal buckling.

If the hearth is horizontal, the pusher force is (weight of stock, W) multiplied by (friction coefficient, fr). The W is proportional to the length of the hearth. The pusher force for unit width of stock is proportional to Length of Hearth × Thickness of Stock. Although the equation for buckling of columns does not exactly apply in this case, it gives a general idea of the relation between thickness of stock and safe length of hearth. A rule of thumb to avoid pileups is to limit the ratio of furnace length to billet thickness (both in the same units) to 240/1.

Inclining the hearth increases the safe length. This is the principal reason why furnaces for heating thin stock have inclined hearths. Hearth inclination reduces pusher force in accordance with the equation

$$\text{Pusher force} = (W)(f)(\cos j) - (W)(\sin j) \tag{4.1}$$

where force and weight ($W$) can be in pounds or kilograms, but must be consistent; fr is the coefficient of friction (dimensionless), and j is the angle between the hearth and the horizontal. (If tan $j = fr$, the pusher force is reduced to zero).

Inclined hearth furnaces tend to create more natural draft, pulling in cold air at the low end of the incline. Excessive hearth inclination interferes with pressure conditions in the furnace. (See chap. 7.) An inclination of more than 8 degrees is rare. The safe length of hearth also depends upon the shape of the contacting surfaces of the billets. If the billets or slabs have round edges, climbing occurs easily. Crooked billets also tend to climb.

The as-built capacity of a bar mill often turns out to be a small fraction of the actual production capacity that mill operators finally attain. For example, a mill in coauthor Shannon's background was designed for 175 tph. Several years later, it rolled 268 tph for an 8-hr turn. Of course, everyone is pleased with such results, but furnaces generally cannot accomplish such production increases without major improvements. Furnaces may have been designed for the minimum heat transfer area to meet their original mill capacity. If a furnace is pushed beyond its capacity, bowing of the bars causes pileups that cause long delays. Such delays are so costly that the operators often become cautious and take a large step backward in their drive to greater productivity. Cutting slots in furnace hearths was tried for other reasons, but the slots filled up with scale. The scale could not be removed unless each end of every slot was open.

***4.6.5.1. A Solution to Bowing Problems in Reheat Furnaces.*** To move ahead to greater productivity without pileup concerns, the authors suggest that a major portion of the solid hearth in the furnace be dug out (down about one ft, 0.3 m) and replaced with rows of refractory blocks or skid pipes installed diagonally to allow added small, high-velocity burners to pump hot gases under the billets, between the blocks or skid pipes. Spaces ("tunnels") between the blocks or skids should be 6 to 8 in. (0.15 to 0.20 m) deep and about 4 ft (1.2 m) wide. A fairly large air lance should be installed beside each new underfiring burner to blow scale out the far end of each "tunnel" and up into the furnace, where it will be carried out with the billets. The top of the ends of the diagonal tunnels must be open so scale can be blown up into the furnace. Thus, enhanced heating can extend the furnace capacity by as much as 30% without danger of pileups.

***4.6.5.2. Round Billets.*** This type of billet cannot be pushed through a furnace, therefore, rotary furnaces or walking beam or walking hearth furnaces must be used. Rotary hearth furnaces need water seals, and walking beam furnaces need water seals on both sides of each walking beam. All have maintenance problems. The heat losses of these features may be very large due to both radiation and air infiltration through the seals. With enhanced heating, the capacities of rotary hearth and walking hearth furnaces can be increased 30%.

***4.6.5.3. Plate Heating.*** Generally, long, thin plates cannot be pushed through furnaces without buckling, so they are usually heated in roller-hearth furnaces. Plate

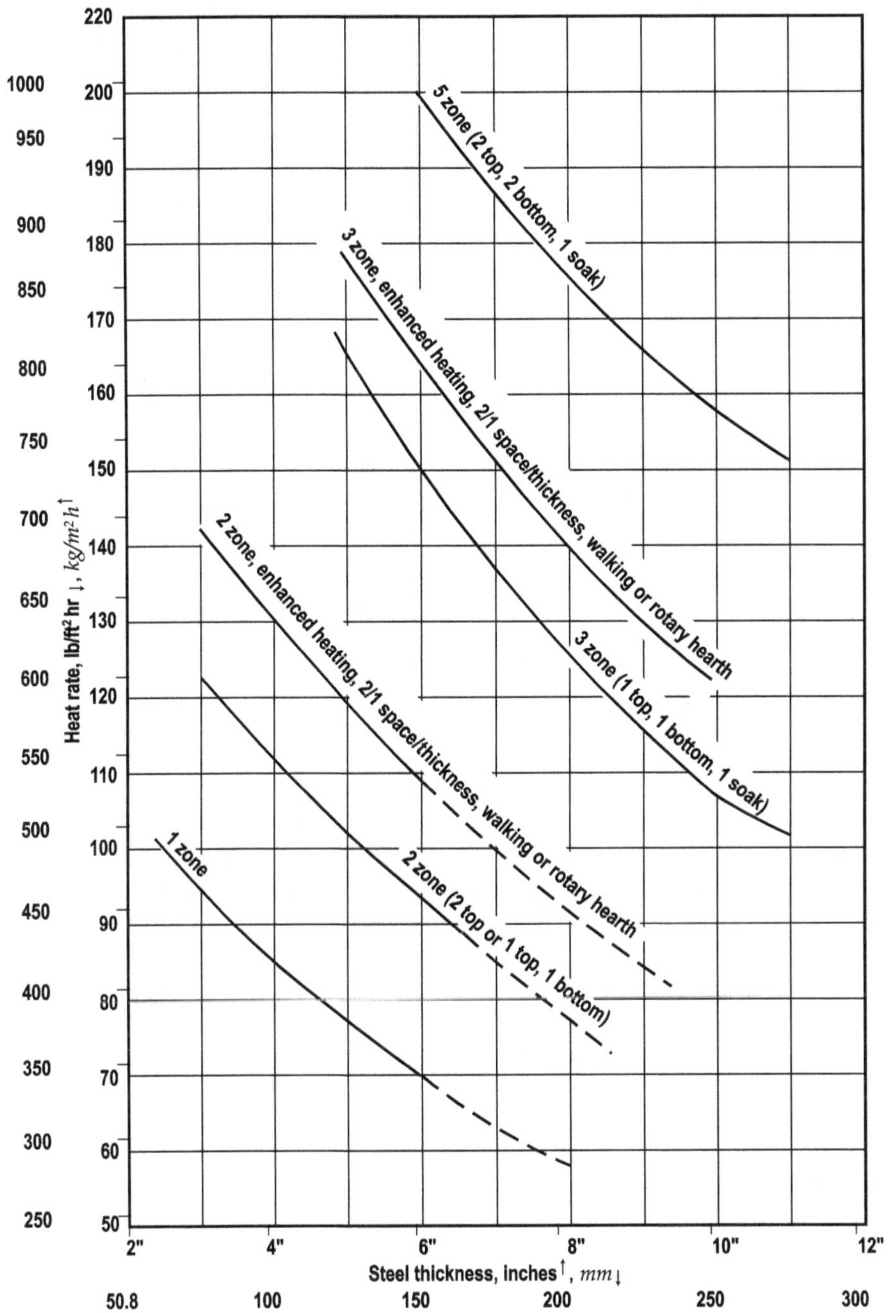


**Fig. 4.21.** Heating rates for various steel thicknesses. (See also fig. 3.12.)

heating is generally for annealing, bending, or preheating for welding. These are low-temperature operations, therefore, roller hearth furnaces can be safely used for these purposes.

Plates are usually annealed at low rates, such as 30 to 40 min per in. of thickness (12 to 16 min per cm of thickness). Where the gas blanket temperature above and below the plate can be held constant, 20 min/in. (or 8 min/cm) of plate thickness has been satisfactory. The graph of figure 4.21 suggests rates at which various load thicknesses and numbers of heating zones can be heated.

### 4.6.6. Walking Conveying Furnaces

***4.6.6.1. Walking* beam *reheat furnaces.*** This type of furnace uses a bell-crank mechanism to regularly lift longitudinal beams supporting all of the loads (billets, blooms, bars) a small clearance distance above water-cooled skid pipes, then advance them a step toward the discharge end of the furnace, and finally lower them back onto the skid pipes. Benefits of the walking process over a solid refractory hearth as in a pusher furnace are (1) underfiring forms an additional zone for heating the bottom sides of load pieces, (2) spaces between the load pieces for better exposure of their sides to radiation and convection, (3) prevention of pieces sticking together, (4) minimization of pileups when moving various sizes of billets through a furnace (whereas multiple sizes can be a problem in a pusher furnace), (5) the furnace can be emptied for repairs relatively quickly, (6) a possibility of a second (faster) set of walking beams for zones nearer the discharge end of the furnace (so that higher carbon steels can be protected from decarburization by varying the time at high temperature without changing charging rate, and (7) minimization of surface marks on the loads.

Disadvantages of walking beams relative to pushers are that walking beams have nearly twice as much skid-mark area and heat loss to water as pusher furnaces because of the walkers of the walking beams. However, these can be eliminated by a short soak zone at the discharge end of the furnace. (See *reference 3*.)

***4.6.6.2. Walking* hearth *reheat furnaces.*** These furnaces are mostly used for making bar and pipe products, and have many of the advantages of walking beam furnaces. The moving walking beams are replaced with moving refractory hearths.

**TABLE 4.1. Comparison of walking hearth heating curves with and without enhanced heating. (See figs. 6.26–6.29.)**

| Figure | Type Design | Time | Length | Capacity |
|---|---|---|---|---|
| 6.26 | Regenerative | 86 min. | 78 ft (23.8 m) | 100 tph |
| 6.27 | Recuperative | 110 min. | 100 ft (30.5 m) | 100 tph |
| 6.28 | Regenerative w/Enhanced Heating | 69 min. | 78 ft (23.8 m) | 125 tph |
| 6.29 | Recuperative w/Enhanced Heating | 86 min. | 78 ft (23.8 m) | 100 tph |

**An Honest Mistake—A Case Study**

Low capacity in a reheat furnace was blamed on ineffective heat transfer in the charging ("convection") zone, but that zone appeared to be hot.

**Problem 1**

In several places the height of the bottom of the entry zone below crossover support beams for the skid rails was less than 1 ft (0.3 m), but the top zone height was 3 ft (0.9 m). (*a*) A major portion of the bottom gases migrated to the top zone. (*b*) The crossovers inhibited flow in the bottom zone. Both (*a*) and (*b*) reduced the possible convection heat transfer to the load in the bottom zone.

To avoid these problems DO NOT reduce the height of the charge zone roof, and do not raise the floor level in the bottom of the charge zone.

**Problem 2**

Heat transfer by gas radiation was greatly reduced because the gas blanket was so thin—12" (0.3 m) versus a desirable 36" (0.9 m). From figure 2.13, the coefficient of gas radiation for 2200 F (1204 C) was only 10.6 instead of 22.5 Btu/ft$^2$hr°F (54 instead of 112 kcal/°cm$^2$), or about 50% less.

**Explanation**

With these reductions in both convection and gas radiation, the furnace capacity suffered terribly. In addition, the bottom zone refractory appeared very hot, causing the observer to believe that the bottom zone was indeed heating well. (This is similar to the conclusion that productivity is very high because the products are moving through a hot zone very quickly. In the formula, $q = hA\Delta T$, the $A$ and $\Delta T$ may be high, but the low h cuts the value of $q$.)

**Review**

Variables that regulate gaseous heat transfer radiation are: (1) blanket thickness, (2) average temperature of the complete blanket including flame, if any, and (3) concentration of triatomic molecules (principally $H_2O$ and $CO_2$).

*Disadvantages of Walking Hearths Relative to Walking Beams.* A bottom-firing zone cannot be made available for maximum heat transfer, so the capacity is less, or the furnace needs to be longer than with walking beams. Slabs are not heated on walking hearths because their width and thickness requires the extra bottom heat available with walking beams.

Combining the walking hearth system with enhanced heating results in the furnace length needing to be only about 26% longer than with a walking beam with all of its problems. Experimentation has shown that the exposure factor for a full walking beam furnace peaks at approximately 82% at about 2.6:1 space-to-thickness ratio whereas the walking hearth reaches 65% exposure when the space-to-thickness ratio is just slightly more than 2:1, thus making a best-of-all compromise. If it is possible to fire the enhanced heating slots alternating side to side, exposure can be practically that of a walking beam, avoiding a bottom heating zone.

### 4.6.7. Continuous Furnace Heating Capacity Practice

Capacities for steel heating furnaces are based on uninterrupted operation throughout the work week. (Delays in the mill or forge shop reduce the weight of steel heated in the furnace, but do not reduce the heating capacity of the furnace.)

Figure 4.21 gives approximations of the pounds of steel that can be heated per $ft^2$ of hearth with various steel thicknesses and numbers of heating zones. Heating curves (chap. 8) must be generated to verify whether a specific furnace can heat a certain product to the desired uniform temperature. From figure 4.21, it can be concluded that for reasonable temperature uniformity, loads more than 6" (150 mm) thick must be heated from both top and bottom, or separated on the hearth of a rotary or walking hearth furnace. The following example shows a simplified method for estimating the size of a steel reheat furnace. Plotting a heating curve (chapter 8) would be more precise, and assure adequate furnace size.

**Example 4.1**: Determine the size needed for a three-zone 1200 C, top-fired-only walking hearth furnace with half the furnace using enhanced heating for 100 tph of 127 mm × 127 mm × 6.71 m (5" × 5" × 22') steel billets.

**Solution 4.1**: Entering the bottom scale of figure $4.21_{SI}$ at 0.127 m (5") thickness, and moving up to the appropriate curve, read a guideline of 880 kg/h $m^2$ of hearth area as the heating capability. (100 tpr) (1000 kg/ton)/(880 kg/h $m^2$) = 113.6 $m^2$ of hearth required. If 100% coverage were used, the furnace length would need to be 113.6 $m^2$/6.71 m = 17 m. To allow for some future production growth, it would be wise to design an 8 m × 18 m furnace hearth area. Plotting a heating curve (Ch. 8) would assure adequate furnace size.

#### *4.6.7.1. Heat Transfer by Hot Gas Movement.*

(See also chap. 7.) An axiomatic thought that must be reviewed when calculating heat transfer in furnaces is: High-temperature areas must be provided with constant source of a high-temperature gas or 'solids' radiation from refractories for equilibrium conditions to be maintained. For example, for hot walls, roof, and hearth to sustain heat transfer between themselves and the load pieces, hot gases must provide a constant supply of gas radiation or convection to the hot refractory; otherwise, their temperature will fall to some lesser temperature and the heat transfer rate to the loads will be reduced.

Another case is the gas movement or lack of movement of hot gases between product. With the movement of hot gases between product (e.g., rounds on a rotary hearth on 1.6 to 2.0 space [centerline of product to the adjacent centerline of product at the average length of the center of the product diameter] to product thickness), the temperature of the gases in the space between can be a temperature of nearly product temperature with no hot gas flow (velocity), thus no additional heat transfer over and above solid radiation and furnace hot gas radiation from the furnace chamber above. The other extreme is to have very high hot gas flow between products providing furnace temperature between products. Even though the temperature is furnace temperature, heat transfer will not be as great as the top surface fully exposed to the furnace chamber because the hot gas blanket thickness in the between-piece space is generally less than one-fifth the thickness of the furnace chamber above the product. However, other variables that can improve the heat transfer to the load are:

1. The gases flowing between and around the product can be at much higher momentum than furnace chamber gases on the top furnace, thereby increasing convection transfer from 5 to 7% of the total heat transfer at that position in the furnace.
2. The refractory hearth, walls, piers, kiln furniture, and so on between the load pieces will be at much higher temperatures with the high gas momentum between the product supplying additional heat units. With the exposed hearth at high temperature, the hearth will supply its heat losses and provide heat to the hearth under the product and to the sides of the product.

With these two benefits, the effective use of the four long sides of the product for heat transfer can reach between 85 and 90% of two-side heating in a full walking beam furnace without the water losses and maintenance of the water-cooled support structure. Therefore, the need for two-side heating with a full walking beam furnace can be avoided, except for slab heating where spaces between product are not available.

Another phenomenon, which sometimes seems to defy logic, occurs when firing a "batch heating furnace"—we desire to maintain as uniform temperature as possible beneath the product supported on piers. What potential should the height of the piers be? Because there are two directions: (1) Do we want nearly the same transfer below and above the products, or (2) do we desire uniform temperature below the products across the hearth? We must study each option, as follows:

Let us say we expect to transfer nearly the same quantity of energy from below as above. To do this, the thickness of the gas blankets should be essentially the same above as below. For maximum heat transfer above and below, the gas blanket thickness should be at or above 36" because heat transfer rates reach near peak by 36" thickness. To get uniformity across the hearth, the pier height should be between 8" and 12" to hold transfer very low to have a minimum temperature drop across the furnace below the product. Alternating both top and bottom burners assists good results because the burners on each side partially compensate for their changing flux profile from low to high flow. As we have mentioned elsewhere, the maximum heat flux from the burner's poc moves away from the burner as the firing rate increases and vice versa.

Another problem with firing below the loads results from reducing the furnace crosssection in a continuous reheat furnace at about 50 to 60% of the furnace length from the discharge. This design spread across the furnace industry because fuel rates improved because solid radiation to the preheat zone from the heat zone was interrupted by the sloped roof, allowing a larger $\Delta T$ between the hot gases and load. However, the total heat transfer to the loads was less because the hot gas blanket was often only 1 ft (0.305 m), resulting in less production. Using a thin baffle instead of lowering the roof could have avoided the reduction in gas blanket thickness.

Designers made the distance between the roof and the top of the product the same as the bottom of the product to the bottom of the preheat area to hopefully divide the gas flow equally between the top flow area and the bottom flow area. However, a major error was committed because the crossover piping below the product was not considered, which reduced the bottom flow height by 1 ft and more, reducing the gas flow under the product to about one-half the top. This problem is compounded by scale dropping into the bottom gas flow area, further reducing the flow area. With this scenario, the top of the product heated much faster than the bottom, increasing the problem of the top of the product being hotter than the bottom due to the top heat input only in the soak zone.

***4.6.7.2. Gas Flow Directions.*** To provide the hot gas for heat transfer in furnaces, the burner or other sources of energy must be provided for the movement of these gases from the burners to the space between products for the heat transfer to take place. Just to supply the space will not necessarily mean that the gas will go there, so energy and direction must be provided. Sometimes designers have separated multilayered product loads with spacers, but failed to follow through by supplying the energy to move hot gases through the spaces. The result is only a minor improvement in cycle times. It also must be accepted that only a fuel meter can tell the operator when the heating cycle is complete. The cycle is complete when the fuel meter is at minimum flow, which indicates the product is no longer accepting energy. Even if the load is known to be nonuniform by peepholes or load thermocouples, additional time in the furnace with minimum fuel flow will probably not help improve uniformity of temperatures. Under these conditions, the product must be repositioned in the furnace to improve temperature uniformity. (See chap. 7.)

### 4.6.8. Eight Ways to Raise Capacity in High-Temperature Continuous Furnaces

Higher furnace capacity is necessary to keep pace with other mill improvements. Recommendations 1 to 8 below suggest ways to match the furnace capacity to the production line equipment "in series" with it. Furnace types such as rotary hearth, walking beam, walking hearth, pushers, and some other high-temperature continuous furnaces can benefit from one or more of these recommendations.

Before beginning to study the means to increase furnace heating capacity, everyone should review the fundamentals of heat exchange. First, there can be no heat exchange if there is no temperature difference. The simplified equation for heat transfer or heat flow rate is $Q = UA\Delta T$ wherein $U = h_r + h_c$ in units such as Btu/ft$^2$hr°F or kJ/m$^2$h°K. Both $Q$ and $U$ are functions of time, the variable we are attempting to reduce. To do this, we try to increase the coefficient of heat transfer "$U$," increase the effective area of heat transfer "$A$," and increase the temperature differential "$\Delta T$" that is the driving force of heat transfer. As we describe the means for increasing heat transfer, we will explain which variable or variables in the heat transfer equation we are attempting to increase.

*Recommendation 1.* Use enhanced heating, that is, small high-velocity burners between and over the load(s) to pump hot gases from above or below. Hot gases moving in this manner can raise the furnace heating capacity by 20 to 35% above what is possible by radiation alone. The hot gases are pumped from the space above the load to the spaces between the load pieces and along the tops (and sometimes bottoms) of the load pieces. The result is to replace the stagnant cool gases between the pieces. These hot gases moving between the load surfaces raise the rate of convective *and* radiative heat transfer to not only the sides of the load pieces but also to the hearth below, providing additional radiation and conduction heat transfer to the load, which previously had suffered heat loss to the colder hearth.

Enhanced heating not only raises $U$ by adding convection heating but also increases the effective area of heat transfer, $A$, by more exposure to higher $\Delta T$ from hotter gases and exposed refractory hearth, possibly raising productivity by another 5 to 7%. Pushers and other furnaces with no separation of load pieces can be improved by raising the temperature and velocity of gases in contact with the top and/or bottom of the loads. This capacity gain may be as much as 10% over radiation heating only.

*Recommendation 2.* Use regenerative air-preheating burners. They can raise productivity approximately 20% and maintain or improve fuel efficiency. They should be installed very near the charge doors to raise the furnace temperature in that area, for more capacity without increasing stack loss. (Regenerative burners have very low exit poc temperatures—usually about 500 F, 260 C.) If the flue system capacity is marginal, regenerative burners can be applied to the furnace because their exit gases are cooler than with traditional burners and because 80 to 90% of their exhaust gases are flued to the atmosphere through separate piping via exhaust fans.

Generally, regenerative burners will reduce the overall fuel rate and air rate of a furnace. Their available heat on steel mill continuous-reheat furnaces is often in the 70% bracket. If the whole furnace is converted to regenerative burners, the fuel rate will be reduced to about 1.0 kk Btu/ton. Many have feared that NOx generation would increase many fold, but this is not the case with modern regenerative burners because (*a*) many modern regenerative burners have low-NOx designs and (*b*) their reduced fuel and air rates result in fewer pounds of NOx generated per year, comparable to conventional burners. The latter has been called the "recuperator effect," but it now can be called the "regenerator effect." Summarizing, regenerative burners improve capacity by raising $\Delta T$.

*Recommendation 3.* Using oxy-fuel burners, usually added at the charge end, can increase furnace capacity by 25% because of (*a*) increased furnace temperature and (*b*) the higher concentration of triatomic molecules in the poc (almost no $N_2$) increases gas radiation. Theoretically, the triatomic concentration rises from 26 to 100%.

If the flue system capacity is marginal, oxy-fuel firing will help because it makes one-third the volume of poc as does air-fuel firing. To get quick productivity increases, installation of oxy-fuel firing is generally the best path. Summarizing, oxy-fuel firing improves capacity by raising the $\Delta T$ via higher flame temperature, and by raising $U$ by more intense gas radiation.

*Recommendation 4.* Install and use baffles effectively. Rotary furnaces have been poor performers over the years because engineers have treated them the same as rectangular furnaces joined at the charge and discharge vestibules, with one baffle between. Additional baffles are needed to separate the charge and discharge vestibules from the charge and discharge zones. Operators often leave charge and/or discharge doors open, resulting in uncontrolled furnace pressure with 30 to 40% of the combustion gases moving to the doors via the soak zone instead of the charge zone.

In many cases, the clearance beneath a baffle is as much as 20 in. (0.53 m), which is entirely too great, causing reduced productivity and increased fuel use. With laser devices to prevent baffle damage during loading and unloading, minimum clearance baffles should be used. Combining three properly sized baffles with the control system in Recommendation 5 below and with increased firing rate in the first heating zone (practical with a lower charge zone baffle) will permit 20 to 30% capacity increases.

One of the authors of this book increased productivity of a rotary furnace from 18 tph to 40 tph by using these techniques. In another case, a pipe mill rotary furnace, capacity was increased by 37% using these same techniques. A later rebuild by design engineers unfamiliar with operating practice lost these benefits. Summarizing, minimum clearance baffles prevent reverse flow of furnace gases, and thereby maintain much hotter gas blanket and refractory $\Delta T$ in the charge end.

*Recommendation 5.* Use dual-temperature control sensors, located as near the loads as possible and tied together by a low-select system, can help productivity. One sensor about 10% into the zone should control piece temperature, and a second sensor about 15% from the zone discharge should prevent overheating. Benefits will be greater if the loads are positioned to the side of the furnace where the sensors are located.

This novel control system can raise productivity by 10% or more, depending on the mill operation. Maximum benefits will be gained in a mill with many delays. After a delay, the early temperature sensor will detect the newly cold pieces much earlier, thereby promptly increasing firing rate to prevent further delay. The second sensor prevents the very hot load pieces in the furnace during the delay from being overheated.

In summary, this control improvement will result in increasing the time at optimum $\Delta T$ for each heating zone. Basically, control is shifted from refractory and gas temperatures being held constant while the load temperature varies to holding the load to a constant temperature by varying the refractory and gas temperatures. It is important to recognize that the sensors do not read the exact load temperature, but they are much closer than other temperature measurements.

*Recommendation 6.* Charge the loads hot where possible. This benefit depends on the melt shop location relative to the mill. When the load is charged very hot (over 1800 F or 982 C), the product will crack excessively during rolling. A high-temperature limit is needed for heating some products, especially alloy grades that tend to resist plastic flow at hot rolling temperatures, causing the steel to rupture along the columnar crystals during hot rolling. Coauthor Shannon has witnessed the use of a water quench on the product to break up the columnar crystals to avoid this problem.

*Recommendation 7.* Install firing capacity 1.4 times the expected rate to more quickly reestablish zone temperatures after delays, and during start-ups. Furnace designers generally limit firing capacity to only 1.15 times the expected running rate to save first cost and to hold fuel costs low. This is done at the expense of quality and productivity, which are more important than cost of fuel or equipment.

*Recommendation 8.* Use more short heating zones and side-fired burners to help maintain the burner wall temperature very high during maximum firing rates. Flat-flame roof burners also can help maintain nearly constant across-furnace temperatures throughout the maximum heat transfer period. The benefit will come from increased $\Delta T$ as needed to control load temperature in many small zones in stead of a few large zones.

When the cost of capital investment is high, some tend to reduce the number of control zones to lower first costs. However, for improved heating results (higher furnace capacity and better flexibility, plus lower fuel consumption), the number of firing zones should be increased. Zone lengths should vary between 12 and 20 ft (3.7 and 6.1 m), but should not exceed 30 ft (9.1 m).

With the many small zones controlled by the two-sensor approach (Recommendation 5), and with furnace heating curves supplying the needed zone setpoints through a computer program, a major improvement in quality, productivity, and fuel efficiency will result.

### 4.6.9. Slot Heat Losses from Rotary and Walking Hearth Furnaces (add this heat requirement to the available heat required in 2.1)

With moving hearths, there must be clearance (slot) between the movable and stationary parts. Water and sand seals have been used to control hot gas loss out and cold air loss in through such slots. The term "seal" implies complete stoppage of gas flow in or out of the furnace. Coauthor Shannon has worked with rotary furnaces in which seals held the leakage to near zero with a positive furnace pressure of 0.1" of water (2.54 mm), but that is rarely the case. To estimate the heat loss, multiply the slot area by the radiation per unit area at the zone temperature.

**Example 4.6.9:** Find the heat loss from the slots of a 20 ft long (6.1 m) furnace zone that has two walking beams with 1" (25 mm) wide slots on either side of each beam, when the average refractory temperature is 2300 F (1260 C). The heat loss area is 2 beams × 2 slots each × (1/12) ft × 20 ft = 6.67 ft$^2$. The black body radiation rate from 2300 F to 100 F is 99 200 Btu/hr ft$^2$. Assuming an effective emissivity of 0.85, the heat loss through the slots of one zone is 6.67 × 99 200 × 0.85 = 563 000 Btu/hr.

The heat loss illustrated by example 4.6.9 is not the only loss. When furnace pressure is high, there may be so much hot gas flow through the slot that it will raise the temperature of the adjacent parts far above their design temperature, resulting in tearing loose parts that will widen the gap and affect temperature uniformity of the loads in the furnace. If the furnace pressure should go negative, the slots will admit cold air, again affecting the product quality and costing more fuel to make up for the chilling effect of the cold air infiltration.

### 4.6.10. Soak Zone and Discharge (Dropout) Losses (see also sec. 4.6.2., add this heat requirement to the available heat required in 2.1)

Heat losses at the discharge of a reheat furnace are an almost universal problem, whether by dropout, extractor, roller, or pushbar. In all of these cases, there are additional radiation and air infiltration losses, which are often overlooked. Dropout losses are most difficult to correct because: (*a*) the irregular opening requires a large closure, (*b*) high furnace pressure will limit the life of the steelwork near the opening, (*c*) preventing infiltration is a nearly impossible task when considering the "chimney effect" of elevation change at the opening, and (*d*) they are unable to balance heat losses that cool the next load piece to be discharged.

The required available heat for the soak zone will be the sum of (*a*) the remaining heat needed into the loads to heat them to good quality; (*b*) heat losses to and from refractory, hearth materials, openings, and water-cooled devices; and (*c*) heat absorbed by infiltrated air in warming to zone temperature.

Figure 4.22 (top and bottom drawings) shows soak zone side-sectional views with T-sensor and burner locations (original and recommended). The two middle drawings show temperature profiles at three soak zone firing rates, plus heat consumption rates for losses, for cold air infiltration, and for heating the loads. The sum of these is the heat flux, which corresponds to available heat.

In both middle drawings of figure 4.22, the load piece at the discharge loses heat to the dropout, extractor, roller, or push bar. When the burner is at low input, such as 30%, the peak heat flux will be very near the burner wall; thus, the burner will then provide most of the discharge heat loss. When the burner firing rate is increased, the flame's heat flux moves away from the burner wall, providing less and less of the discharge heat loss; thus, the piece at the discharge will be heated less.

All three remedies for this situation involve forcing the flame's heat flux to remain strong near the burner wall at higher firing rates: (1) Spin the combustion gases as they enter the burner tile, (2) reform the tile into a more divergent angle, and (3) reduce the combustion gas momentum leaving the burner. However, these may raise the specific fuel consumption.*

*Specific fuel consumption, SFC = Btu or joules for each ton heated.

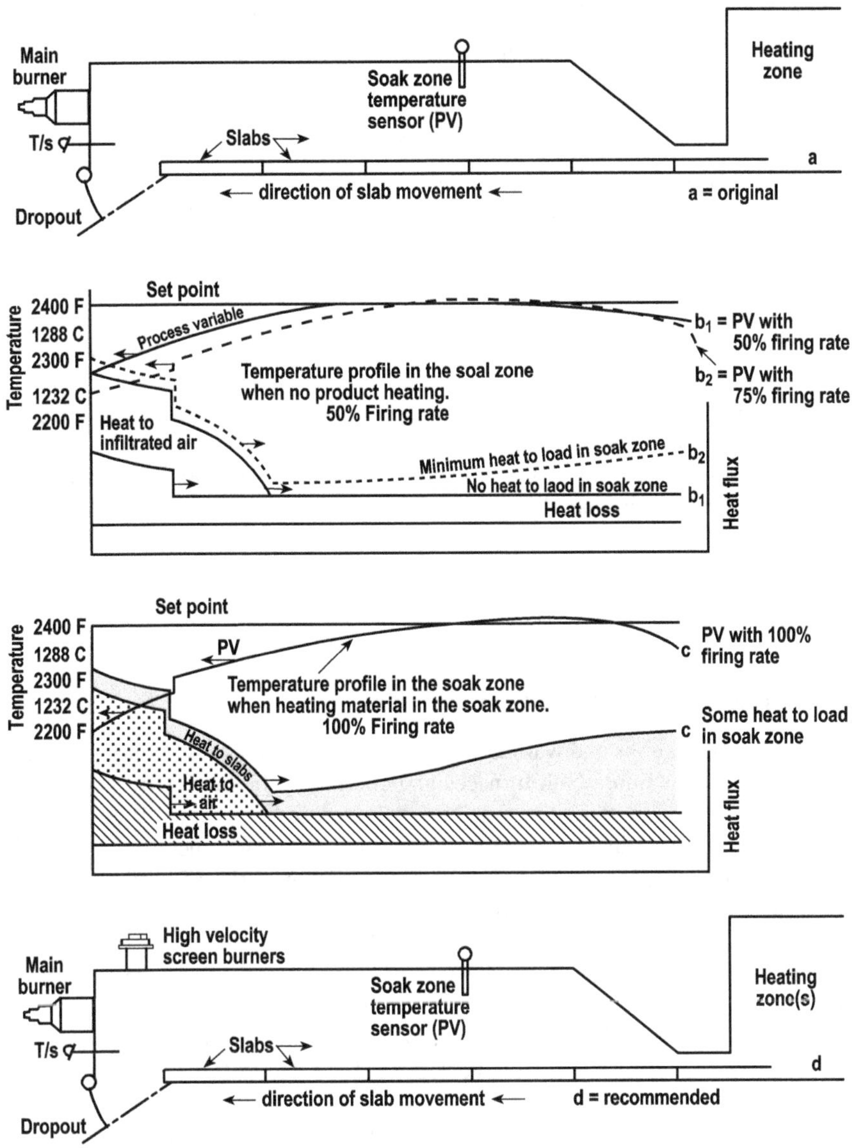


**Fig. 4.22.** Soak zone and dropout of a steel reheat furnace. *a*, original soak zone, side-sectional view; $b_1$, 50% firing rate; $SZT_{max}$ at 5% of *SZLfD*; 2280 F (1248 C) load discharge; $b_2$, 75% firing rate; $SZT_{max}$ at 53% of *SZLfD*; 2240 F (1227 C) load discharge; *c*, 100% firing rate; $SZT_{max}$ at 80% of *SZLfD*; 2200 F (1204 C) load discharge; *d*, recommended soak zone retrofit with high-velocity burners added at discharge. (*SZLfD* = soak zone length from discharge).

To prevent the resultant increase in fuel required per unit weight of load is to limit the volume of infiltrated air moving through the discharge opening

1. by holding the furnace pressure at the knuckle as high as reasonable, for example, 0.06 to 0.1" wc (0.149 to 0.249 kPa) so that all of the discharge slots have positive pressure for outleaking poc, not inleaking cold air
2. by lining the discharge doors and door seals with ceramic fiber or other pliable, high-temperature sealing material to minimize both inleakage and outleakage, and by maintaining these seals
3. by installing a row of down-firing high-velocity burners through the roof crosswise above the dropout doors, using their velocity pressure to exclude infiltration and their heat input to balance dropout heat losses. These burners should fire downward between the centerlines of the horizontally firing end-wall burners. They should be controlled separately from the soak zone, using a T-sensor low in the burner wall at the dropout. (See figures 6.24 and 6.25.) With these improvements, product delivery temperature to the mill can be more uniform, production higher, and fuel use lower.

## 4.7. CONTINUOUS LIQUID HEATING FURNACES

### 4.7.1. Continuous Liquid *Bath* Furnaces

Many of the suggestions and warnings given for batch liquid bath furnaces also may apply to continuous liquid bath furnaces and continuous liquid flow furnaces; thus, the reader is advised to review section 3.8.6 in the preceding chapter. Whereas batch liquid bath furnaces may be used for melting and alloying a metal as well as for coating solids by dipping into a molten bath, the great majority of continuous liquid bath furnaces are for the latter purpose. In many cases the liquid is not a metal, but glass, a salt, or a coating material (e.g., fig. 4.23.)

Glass melting furnaces range from batch-type "day tanks" to unit melters to large end-fired continuous melters (up to 1200 ft$^2$ bath area), and huge 3000 ft$^2$ side-fired melting furnaces. The continuous furnaces usually have integral regenerative checkerworks and are operated without stopping for a 0.5- to 15-year campaign. The ratio of tank area versus tons/day (tpd) melted ranges from 4 to 20 ft$^2$/tpd (0.41 to 2.04 m$^2$/tpd), depending on the type of glass. Fuel consumption in practice varies with the type of glass, ranging from 10 to 16 kk Btu/ton (11 600 to 18 560 mj/tonne).

The capacity of metal, glass, or salt baths for continuous operation differs from that of batch-type (dipping) baths because the coefficient of heat transfer is increased by the movement through the bath of the strip or pieces being coated. That movement also enhances temperature uniformity as well as finished product quality.

An empirical relation, developed by J. E. Keller, equation 4.2 is for the heat transfer coefficient between a moving molten liquid and a solid.

$$h_{US} = 80 + 540(V_{US}) \tag{4.2}$$

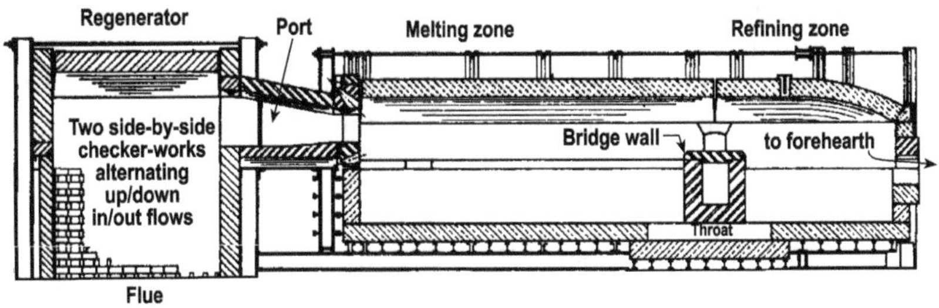


**Fig. 4.23.** Longitudinal section, end-fired glass melting tank. Far-side checkers feed preheated air to far firing ports (burners). Flames and poc take a U-path over raw batch and molten glass, returning to exit through near-side end ports (flues) to near-side checkers. After a designated number of minutes, or in response to automatic hot air temperature controls, flows reverse so that near-side ports act as burners and far-side ports act as flues.

where $h$ = heat transfer coefficient in Btu/hr°F ft$^2$ and $V$ = velocity in ft/sec, or

$$h_{SI} = 454 + 10\,050(V_{SI}) \tag{4.3}$$

where $h$ = heat transfer coefficient in $W/°\mathrm{C}m^2$ and $V$ = velocity in m/s.

The capacity of a bath also depends on the purpose for which the bath is to be used. The time required to heat wire for coating in a metal bath is considerably less

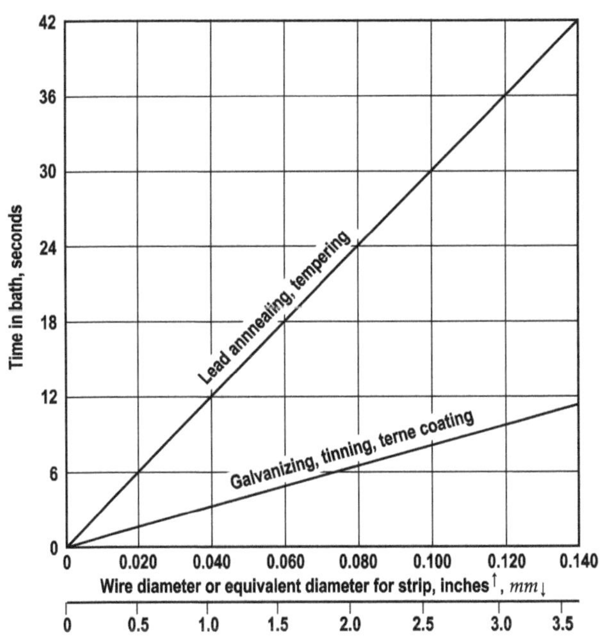


**Fig. 4.24.** Heating time required for steel wire or strip in molten lead, tin, or salt. Equivalent diameter for strip is twice its thickness. When heating for coating, the wire or strip may not need to be thoroughly heated to its center.

than the time needed to heat wire for metallurgical purposes, where the wire must usually be heated uniformly to its core. (See fig. 4.24.)

Burner input should be enough to maintain the bath temperature at least 100°F (55°C) of superheat above the liquid metal's melting point when operating at the maximum production rate.

## 4.7.2. Continuous Liquid *Flow* Furnaces

Continuous liquid flow furnaces include boiler furnaces, fluid heaters (such as 'Dowtherm' heaters), evaporators, cookers, and many liquid heaters used in the chemical process industries. (See figs. 1.12 and 4.25.) The tubing through which the liquid fluids flow is often built as an integral part of the furnace, for which many textbooks are readily available; therefore, they will not be discussed at length here.

The boiler and chemical process industries also have learned (1) that the flame and hottest poc should traverse a radiation section first, then flow through a convection

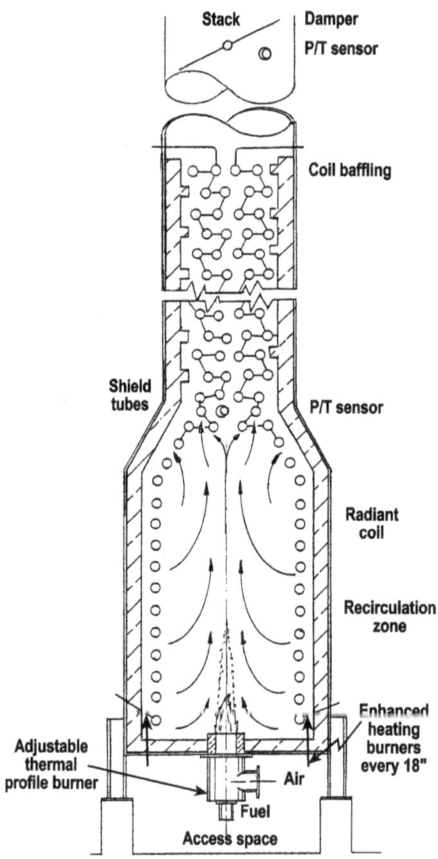


**Fig. 4.25.** Forced draft heater for petrochem processing—may be *cylindrical* with one burner as shown, or a circle of vertically up-fired, high-velocity type H burners (fig. 6.2) or rectangular (a "cabin heater") with rows of up-fired burners, or rows of side-fired type E flat-flame burners, shown in fig. 4.26 and 6.2.

Circulation by the burner gases helps convection, raises triatomic gas concentration (for more gas radiation to all sides of the tubes), and lowers NOx emissions. With large burners, use of adjustable thermal profile burners can optimize uniform heating to the coils.

Many small, high-velocity burners might improve heat transfer if installed to fire between the tubes and the refractory walls.

**Fig. 4.26.** Petrochem "cabin heater" process furnace for a vinyl chloride monomer process at 932 F (500 C) in Europe. This unit has a twin in Texas. Type E flat-flame burners (fig. 6.2) provide uniformly high-flux radiation transfer to the tubes without flame impingement.

section, and (2) that the radiation section should be a "room" shaped around the flame whereas the convection section needs more exposed surface area and enhanced velocities. In radiation sections, there is an advantage from wider tube spacing and from spacing the tubes out from the wall so that both convection and re-radiation can occur on the back sides of the tubes.

If the first bank of convection tubes can "see" the burner flames or hot refractory, its life may be shortened by the overdose of radiation. These are therefore called "shock tubes." The shock can be lessened by piping the coldest feed liquid into those tubes first. If hot combustion products are on one side of the heater (heat exchanger), and if the fluid "feed" on the other side of the heater tubes is a gas or vapor, the danger of tube burnout is greater because gases and vapors generally have poorer thermal conductivity than most liquids.

Most of the preceding discussions related to liquid flow heaters in which the liquid was inside tubes and the furnace gases outside the tubes. Figure 4.27 shows some "fire-tube boilers" wherein the opposite is the case; that is, furnace gases inside tubes that are surrounded by liquid water. These are mostly used in smaller boiler installations.

**Warning**: *In any job where equipment failure or downtime cannot be allowed (such as the school building boiler room shown in figure 4.27), designers must insist on multiple units, trusting that all units will not go down at once.* This is also good

**Fig. 4.27.** Fire-tube boilers with packaged automatic gas, oil, or dual-fuel burners having integral fans. These three-pass boilers have a large "Morrison tube" into which the burner fires as the first pass (radiation), and two banks of many small tubes (convection) for the second and third passes. Fire-tube boilers are more compact and less expensive than water-tube boilers, but they are limited in steam pressure and size, typically 150 psig (1030 kPa) maximum steam pressure and 33 kk Btu/hr (35 000 MJ/h) maximum input.

advice in situations having widely varying production demands (high turndown ratio). Multiple smaller furnaces (boilers, ovens, heaters, incinerators) may be able to save fuel and offer greater flexibility than one or two large units.

## 4.8. REVIEW QUESTIONS AND PROJECTS

**4.8Q1.** List all the ways you can think of to improve production capacity of high-temperature furnaces.

**4.8Q2.** Why is fuel economy so important to users of high-temperature furnaces?

**4.8A2.** Because fuel costs are much higher in high-temperature furnaces than in lower temperature furnaces as a result of the higher flue gas exit temperature causing higher stack loss.

**4.8Q3.** List advantages, then disadvantages, of continuous furnaces compared to batch furnaces.

**4.8Q4.** What is the driving force that causes each of these four forms of *potential flow*: fluid flow? electric current? heat transfer? drying (mass transfer)? Identify the resistance for each.

**4.8A4.** Fluid flow is driven by pressure difference. Fluid flow resistance can be a baffle, an orifice, a valve, a fitting, and so on. Electric current is driven by difference in potential (voltage). Electric resistances can be resistors, coils, or low-conductance materials. Heat transfer is driven by temperature differentials ($\Delta T$). Heating and cooling resistances can be insulators, poor conducting materials, air gaps, low-emissivity sources, or low velocity. Drying (mass transfer) is driven by difference in vapor pressure. Mass transfer resistances can be low velocity, imperviousness).

**4.8Q5.** How does convection by poc and air have an advantage over radiation from refractory or an electric element?

**4.8A5.** Convection can go around corners and reach long distances. Convection is not hindered by radiation's "shadow problem" because radiation must travel in straight lines. Convection also can provide mass transfer (drying).

**4.8Q6.** Why is it misleading to guess that a furnace zone's flue gas exit temperature is the same as the zone's inside refractory surface temperature?

**4.8A6.** Because the refractory at the exit could not have reached its temperature unless the passing furnace gases were hotter than the refractory itself. Those poc are the source for heat in the refractory walls, and there must be a difference in temperature to drive the heat from the gases to the walls.

**4.8. Problem 1.** Size a 3-zone, 2200 F top-fired-only walking hearth furnace with half the furnace using enhanced heating for 100 tph of 5" × 5" × 22' steel billets.

**4.8. Solution 1.** Entering the bottom scale of figure 4.21 at 5" thickness, and moving vertically up to the appropriate curve, read a guideline of 179 lb/hr $ft^2$ hearth for the heating capability. 100 tph × 2000 lb/ton = 200 000 lb/hr. Then, 200 000 lb/hr/179 lb/$ft^2$ = 1117 $ft^2$ of hearth required. If 100% coverage was used, the furnace length would need to be 1117 $ft^2$/22 ft = 50.8 ft. To allow for some future production growth, a 25 ft wide × 60 ft long furnace would be wise. Plotting a heating curve would assure adequate furnace size.

## 4.8. PROJECTS

### 4.8.Proj-1.

Refer to figure 4.10 of a catenary furnace. The inside length between hot refractory surfaces at left and at right is $L$, and the mean inside height between hot refractory faces at top and bottom is $H$. Use the mathematical formula for a catenary curve to write a formula for $P$, the percent of $H$ to specify end roll stand and slots height to attain equal areas under and above the catenary curve. This will provide equal average "beams" for gas radiation over and under the strip. Further refine the above to allow the user to specify desired other than equal average gas radiation beam lengths over and under the strip, biasing the average beam lengths tocompensate for the fact that the roof temperature may run hotter than the floor temperature.

### 4.8.Proj-2.

Design data are needed for enhanced heating, a mean for increasing heat transfer by moving stagnant cool gases from the surfaces of furnace loads and/or hearths by using high-velocity burner gases diluted with very hot furnace gases. Experimental work is needed to determine how the increase in heat transfer can be applied to the calculation of an *exposure factor*, which can be one of the variables involved in the calculation of a *heat transfer coefficient.*

The following heat transfer effects need to be analyzed individually, and a determination made whether they can all be added to each other:

1. Convection to the top and sides of the product
2. Gas radiation heat transfer from the furnace chamber
3. Gases radiation heat transfer from spaces between products
4. Solids radiation heat transfer from the hearth to the product sides
5. Solids radiation heat transfer from the furnace chamber to the loads
6. Conduction to/from the hearth from/to the bottoms of the load pieces

These effects also should be investigated for heating furnace loads to rolling/forging temperatures, quenching/hardening temperatures, tempering temperatures, and annealing temperatures.

This study and tests first should be made for bar heating. Then slab, strip, and plate heating also should be investigated to determine whether enhanced heating can be of value in those cases as well.

At this writing, coauthor Shannon is using a conservative exposure Improvement for bar heating of 25% with a belief that the actual improvement may be above 35%. Having the benefits quantified is very important to industry.

# 5

# SAVING ENERGY IN INDUSTRIAL FURNACE SYSTEMS

## 5.1. FURNACE EFFICIENCY, METHODS FOR SAVING HEAT

In some industrial heating processes, fuel represents only a very small fraction of the total cost of manufacturing. But in most industrial heating processes, fuel represents a considerable expense. Although fuel and electric energy generally cost less in the Americas, costs are continuously rising. Since about 1940, the rise in fuel cost has accelerated from its 4% rate of the previous 50 years. Since the last decade of the twentieth century, embargos, wars, regulations, and deregulations have caused the costs of oil and gas to go through unsettling fluctuations. Costs of electric energy also rise because of the increasing cost of fuels, wages, and equipment. The difference between fuel saving and fuel wasting often determines the difference between profit and loss; thus, heat saving is a must.

Side effects of fuel saving often include better product quality, improved safety, higher productivity, reduced pollution (including reduced noise), better employee and public relations, and long-range fuel supply extension.

Many furnace engineers, owners, and operators could benefit by the following *check list of ways to save heat*:

1. Better heat transfer by radiation exposure and convection circulation
2. Closer to stoichiometric air/fuel ratio control
3. Better furnace pressure control to minimize leaks and nonuniformities
4. More uniform heating for shorter soak times
5. Reduction of wall losses, wall heat storage, heat leaks, and poc gas leaks
6. Minimizing heat storage in, and loss through, conveyors, trays, rollers, kiln furniture, piers, spacers, packing boxes, and protective atmospheres
7. Losses to openings, cooling water, loads projecting out of a furnace, exposed liquid bath surfaces, terminals and electrodes, water seals, slots, dropouts, doors, movable baffles, and charging equipment
8. Avoiding use of high-temperature heat for low-temperature processes

9. Preheating furnace loads by using waste heat
10. Preheating air or fuel (or both if fuel has low heat value) by waste heat
11. Waste heat boilers
12. Reduction of flue gas exit temperatures by computer modeling
13. Rezoning of furnaces into more small zones (chap. 4 and 6)
14. Better location of zone temperature control sensors
15. Oxy-fuel firing
16. Enhanced heating (sec. 2.4.1 and 4.6.1.3)

The words "economy" and "efficiency," when used in their true sense in connection with industrial furnaces, refer to the *heating cost* per unit weight of finished, sellable product. 'Heating cost' includes not only the fuel cost but also the costs of operating and superintending, amortizing, maintaining, and repairing the furnace, plus the cost of generating a protective atmosphere and the costs of rejected pieces. The costs of rejected pieces (poor quality, poor temperature uniformity) include the costs of reworking pieces found defective because of improper heating and the costs of handling the material into and out of the furnace. With so many items entering into the total cost of heating, it is possible that in some cases the highest priced fuel or other heat energy source may be the cheapest.

Some engineering companies use the heat of oxidation of the load itself to reduce their estimate of required furnace fuel rate. Load oxidation heat is a very small fraction of the heat in most furnaces, except incinerators, and it is usually very expensive. For steel loads, heat from oxidizing steel costs more than 20 times that of heat from natural gas. One cannot measure the quantity of load oxidized or where it occurs in the furnace.

In many furnaces, fuel cost may be a major item of expense. Therefore, economy is worthy of constant watching for reasons discussed earlier and because of frequent vacillation of fuel prices and availability. In designing or selecting a new furnace, it is necessary to know its probable fuel consumption beforehand. This information also is necessary to select the correct size and number of burners, to figure sizes of ports, vents, and stack, and to select auxiliary equipment of proper size.

When some first observe furnaces, they are astonished by the low thermal efficiency of industrial furnaces. Whereas boiler efficiencies range from 70 to 90%, industrial furnace fuel efficiencies are often half as much. Electrically heated furnaces may appear to have higher efficiencies—if one forgets to consider the inefficiency of generation of electric energy, which includes the inefficiencies of converting fuel energy to steam energy, then to mechanical energy, and finally to electric energy. When crossing these many process boundaries, it is often wiser to make comparisons of total heating costs in dollars (or other currencies) per ton of material processed.

With good design and operation, fuel-fired furnace efficiencies of 60% or higher can be had, depending much on process temperature. "Efficiency" here is the ratio of heat input into the load/hr to the gross heat released by the fuel used/hr. The Glossary compares efficiency terms. When comparing costs, always ask for clarification as to what is meant by "efficiency."

The major reason for the difference in efficiencies between boiler furnaces and industrial furnaces is the final temperature of the material being heated.

Furnace gases can give up heat to the load only if they are hotter than the load. Therefore, the flue gases for high-temperature process heating must leave industrial furnaces at a very high temperature (except shortly after a cold start). By comparing (*a*) the available heat from figures 5.1 or 5.2 at the exit gas temperature of the poc leaving a 2400 F (1316 C) industrial furnace, with (*b*) the available heat (best possible efficiency) for poc of a 300 F (150 C) boiler, one can see that there can be a great difference between their efficiencies.

### 5.1.1. Flue Gas Exit Temperature

The *flue gas exit temperature* will always be higher than the furnace temperature at the flue because otherwise heat would not flow from the furnace gases to the walls and loads. Accurate measurement of flue gas exit temperature can be difficult. A high-velocity thermocouple with several radiation shields is essential. Figure 5.3 helps estimate the temperature elevation of the exiting gases above the furnace temperature. The sum of the furnace temperature and this elevation is the temperature that should be used to enter the bottom scale of available heat charts 5.1 and 5.2 to determine the %available heat.

A quicker approximate estimate of the temperature to use when entering the bottom scales on figures 5.1 and 5.2 is via fig. 5.4, from the empirical formula of equation 5.1.

Approximate flue gas exit temperature (fgt), in Fahrenheit =

$$740 + (0.758 \times \text{furnace temperature}) \tag{5.1}$$

For a furnace temperature of 1600 F, this equation says to use $740 + 0.758 \times 1600 = 740 + 1213 = 1950°F$ to enter figures 5.1 or 5.2. This agrees with Figure 5.3, but other conditions will be too low by equation 5.1 (especially with high velocity and low furnace temperature) and too high with low velocities. Use equation 5.1 only with careful judgment.

A higher temperature process must exhaust more heat to heat a load hotter. Similarly, there is a great difference between efficiencies of high-temperature industrial furnaces and lower temperature industrial ovens.

With regenerative burners, industrial furnaces can reach 70 to 80% efficiency because the regenerative bed determines the combustion efficiency, not the temperature of the load being heated. With regenerative burners, the average waste gas temperature can be as low as 600 F (317 C). With recuperators, vigilance is necessary or extensive damage can take place (1) if the flue gas temperature is too high, (2) if burning takes place in the flue or recuperator, or (3) if the air flow through a recuperator is reduced below 10% of maximum. In contrast, regenerative burners can reduce fuel rates to a minimum by returning a major portion of the sensible heat from the flue gas to the furnace. Therefore, the chances of these three recuperator problems occurring are much less with regenerators.

TABLE 5.1. Fuel saved by use of various degrees of air preheat with #6 fuel oil with 10% excess air. For other fuels, send higher heating value and fuel analysis (volumetric for gas, gravimetric with liquid or solid fuel) to North American Mfg. Co. (Cleveland, OH 44105). Reproduced with permission from *Ref. 49.*

| % Fuel saved | $t_2$, Combustion air temperature, F | | | | | | | | | | | | | | | |
|---|---|---|---|---|---|---|---|---|---|---|---|---|---|---|---|---|
| $t_3$, Furnace gas exit temperature, F | 400 | 600 | 700 | 800 | 900 | 1000 | 1100 | 1200 | 1300 | 1400 | 1500 | 1600 | 1700 | 1800 | 1900 | 2000 |
| 800 | 7.85 | 12.04 | 14.03 | – | – | – | – | – | – | – | – | – | – | – | – | – |
| 900 | 8.07 | 12.37 | 14.41 | 16.37 | – | – | – | – | – | – | – | – | – | – | – | – |
| 1000 | 8.32 | 12.73 | 14.81 | 16.82 | 18.75 | – | – | – | – | – | – | – | – | – | – | – |
| 1100 | 8.58 | 13.11 | 15.25 | 17.30 | 19.28 | 21.18 | – | – | – | – | – | – | – | – | – | – |
| 1200 | 8.87 | 13.52 | 15.71 | 17.82 | 19.84 | 21.78 | 23.65 | – | – | – | – | – | – | – | – | – |
| 1300 | 9.17 | 13.97 | 16.22 | 18.37 | 20.44 | 22.43 | 24.33 | 26.16 | – | – | – | – | – | – | – | – |
| 1400 | 9.51 | 14.45 | 16.76 | 18.97 | 21.09 | 23.12 | 25.06 | 26.93 | 28.72 | – | – | – | – | – | – | – |
| 1500 | 9.87 | 14.97 | 17.34 | 19.61 | 21.78 | 23.86 | 25.85 | 27.75 | 29.58 | 31.33 | – | – | – | – | – | – |
| 1600 | 10.26 | 15.53 | 17.98 | 20.31 | 22.54 | 24.66 | 26.69 | 28.64 | 30.49 | 32.27 | 33.98 | – | – | – | – | – |
| 1700 | 10.70 | 16.14 | 18.67 | 21.07 | 23.35 | 25.53 | 27.60 | 29.59 | 31.48 | 33.29 | 35.02 | 36.67 | – | – | – | – |
| 1800 | 11.17 | 16.82 | 19.42 | 21.89 | 24.24 | 26.47 | 28.59 | 30.61 | 32.54 | 34.38 | 36.14 | 37.82 | 39.42 | – | – | – |
| 1900 | 11.70 | 17.55 | 20.24 | 22.79 | 25.20 | 27.49 | 29.66 | 31.72 | 33.69 | 35.56 | 37.34 | 39.04 | 40.66 | 42.21 | – | – |
| 2000 | 12.28 | 18.37 | 21.15 | 23.77 | 26.25 | 28.60 | 30.82 | 32.93 | 34.93 | 36.83 | 38.64 | 40.36 | 42.00 | 43.57 | 45.06 | – |
| 2100 | 12.93 | 19.27 | 22.15 | 24.86 | 27.41 | 29.82 | 32.10 | 34.25 | 36.29 | 38.22 | 40.05 | 41.79 | 43.45 | 45.02 | 46.52 | 47.96 |
| 2200 | 13.66 | 20.28 | 23.26 | 26.07 | 28.70 | 31.17 | 33.50 | 35.69 | 37.77 | 39.73 | 41.59 | 43.34 | 45.01 | 46.60 | 48.11 | 49.54 |
| 2300 | 14.49 | 21.41 | 24.51 | 27.41 | 30.12 | 32.66 | 35.04 | 37.29 | 39.39 | 41.38 | 43.26 | 45.04 | 46.72 | 48.31 | 49.82 | 51.26 |
| 2400 | 15.44 | 22.69 | 25.92 | 28.92 | 31.72 | 34.33 | 36.76 | 39.05 | 41.19 | 43.21 | 45.11 | 46.90 | 48.58 | 50.18 | 51.69 | 53.13 |
| 2500 | 15.55 | 24.17 | 27.53 | 30.64 | 33.52 | 36.20 | 38.69 | 41.02 | 43.20 | 45.24 | 47.15 | 48.95 | 50.64 | 52.23 | 53.74 | 55.16 |
| 2600 | 17.84 | 25.88 | 29.39 | 32.61 | 35.59 | 38.33 | 40.88 | 43.25 | 45.45 | 47.50 | 49.43 | 51.23 | 52.92 | 54.50 | 56.00 | 57.41 |
| 2700 | 19.40 | 27.90 | 31.56 | 34.91 | 37.97 | 40.79 | 43.38 | 45.78 | 48.00 | 50.07 | 51.99 | 53.79 | 55.46 | 57.04 | 58.51 | 59.90 |
| 2800 | 21.30 | 30.32 | 34.15 | 37.62 | 40.78 | 43.65 | 46.29 | 48.71 | 50.94 | 53.00 | 54.91 | 56.69 | 58.34 | 59.89 | 61.33 | 62.69 |
| 2900 | 23.70 | 33.30 | 37.31 | 40.90 | 44.14 | 47.06 | 49.72 | 52.15 | 54.37 | 56.41 | 58.29 | 60.03 | 61.65 | 63.14 | 64.54 | 65.84 |
| 3000 | 26.83 | 37.08 | 41.27 | 44.97 | 48.26 | 51.21 | 53.86 | 56.26 | 58.45 | 60.44 | 62.26 | 63.94 | 65.49 | 66.92 | 68.24 | 69.47 |

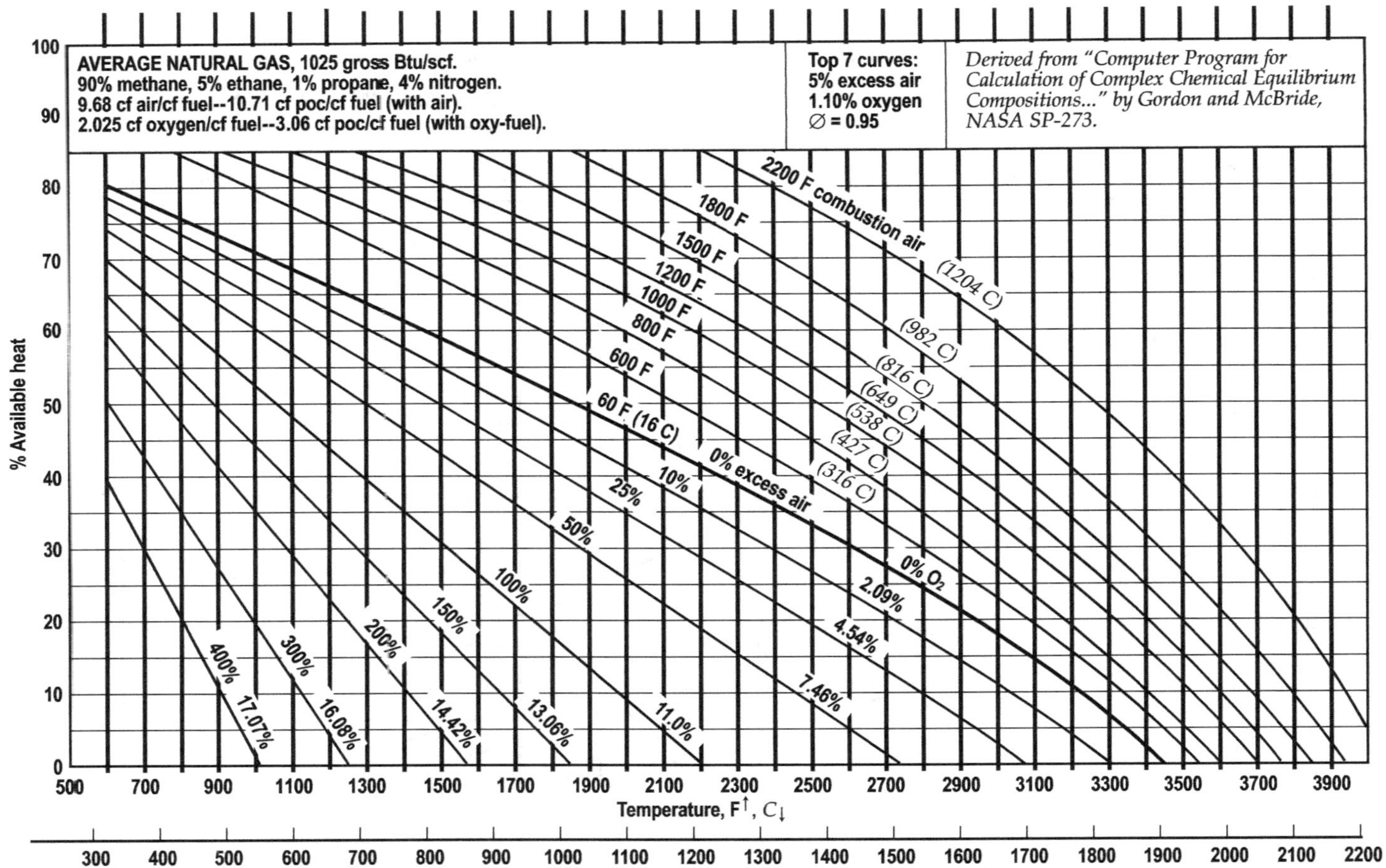


**Fig. 5.1.** Percents available heat for an average natural gas with cold air and with preheated air. (See fig. 5.3 for estimating flue gas exit temperature.) For other fuels, send fuel analysis and higher heating value to North American Mfg. Co., Cleveland, OH 44105–5600. Reprinted with permission from *reference 52*. (See also figs. 5.2, 5.3, and table 5.1.)

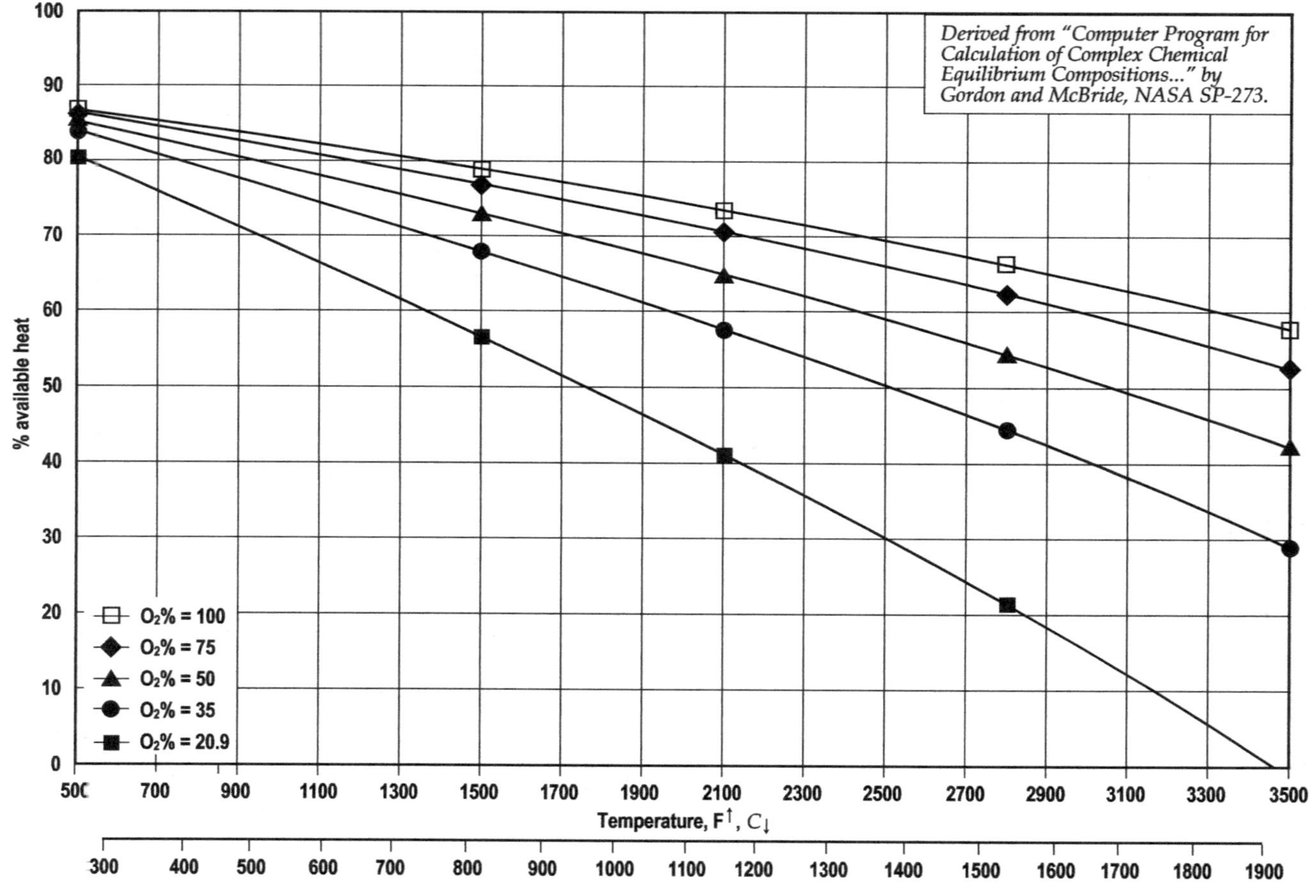


**Fig. 5.2.** Percents available heat for an average natural gas with oxygen enrichment or with oxy-fuel firing. (See fig. 5.3 for estimating flue gas exit temperature.) For other fuels, send fuel analysis and higher heating value to North American Mfg. Co., Cleveland, OH 44105-5600, developer of this chart.

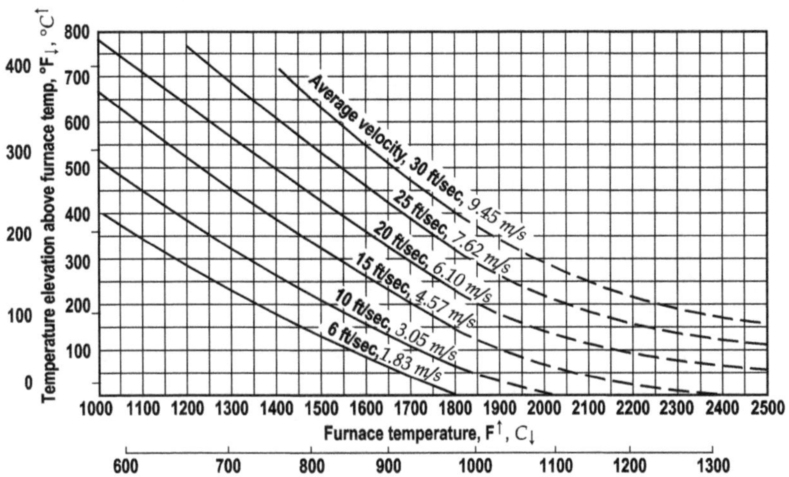


**Fig. 5.3.** Elevation of flue gas exit temperature above furnace temperature, for a variety of stp velocities (average across-the-*furnace* cross section where the poc approach the flue). The stp velocity = stp volume divided by the cross-sectional area of the flowing stream. (Same as fig. 2.2.) *NOTE: The convention used in this book is to omit the* degree mark (°) *with a temperature level (e.g., water boils at 212 F or 100 C) and to use the degree mark only with a temperature difference or change (e.g., the difference, ΔT, across an insulated oven wall was 100°F or 55.6°C, or the temperature changed 20°F or 11.1°C in an hour).*

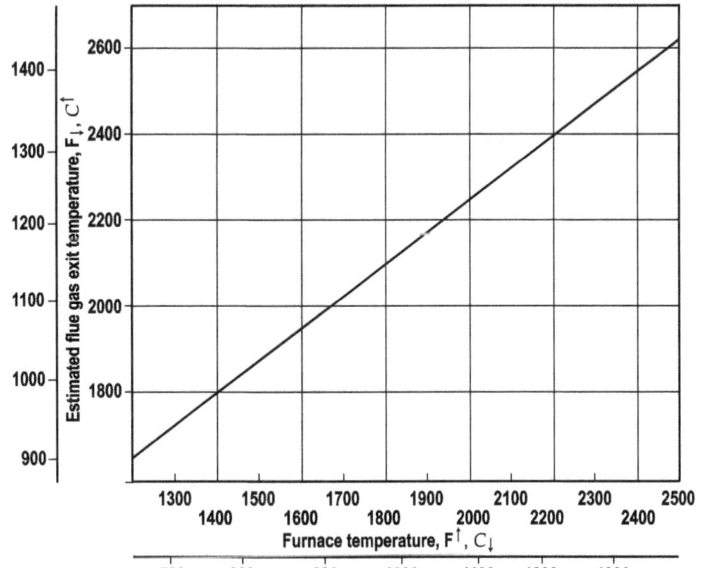


**Fig. 5.4.** Quick method for estimating flue gas exit temperature from the measured furnace temperature near the flue.

Regenerative burners have the following benefits:

1. The fuel efficiency has only a minor dependency on the furnace temperature. Their high efficiency results from the fact that their regenerative beds preheat the combustion air temperature within about 300°F to 400°F (167°C to 222°C) of the furnace exit gas temperature.
2. The air/fuel ratio is not as critical as with recuperators and cold air firing, provided that all of the fuel is burned completely. An increase of 50% excess air at 2400 F (1316 C) furnace temperature with air preheated to 2000 F (1093 C) reduces the efficiency only 2%.
3. During mill delays, efficiency remains very high, supplying heat losses and some heat to the product. Conventional burner systems lose efficiency as gas exit temperatures rise and infiltrated air increases.

## 5.2. HEAT DISTRIBUTION IN A FURNACE (see also chap. 7 and 8.1.2)

### 5.2.1. Concurrent Heat Release and Heat Transfer

*Phase 1.* A portion of the heat released in the combustion zone is transmitted by radiation (which 'travels' in straight lines) to the load(s), and to furnace inside surfaces (roof or 'crown', sidewalls, and floor or 'hearth').

*Phase 2.1.* As combustion gases (poc and excess air) flow from flames, they pass over load pieces, and may be directed across walls, roof, hearth, baffles, and piers in a circulation pattern, eventually finding their way to the flues. This flow phase delivers heat to loads and walls by convection and by gas radiation (largely from carbon dioxide and water vapor molecules).

*Concurrent Phase 2.2.* As all of the solid heat-receiving surfaces in the furnace begin to absorb heat, their surface temperatures rise. The refractory surfaces, being poorer conductors, experience a more rapid rise in their surface temperature, and therefore become good re-radiators, helping to transfer more heat to the loads. This secondary radiation (fig. 5.5) has always been considered to be a major portion of all the heat transferred to the loads in furnaces operating above about 1400 F (760 C). Many people have ignored gas radiation, but it is a big factor in furnace heat transfer.

*Phase 3.* The furnace gases may then be directed through some heat recovery device (covered later in this chapter), and maybe through some induced draft device, then finally to the stack.

If a long furnace is fired from one end, the cooling gases set up temperature differentials that affect the load heating rate. (See fig. 5.6.) Attaining a flat temperature profile along the length of a one-end-fired furnace requires burners with adjustable spin controlled by $\Delta T$ sensors. (See chap. 6.)

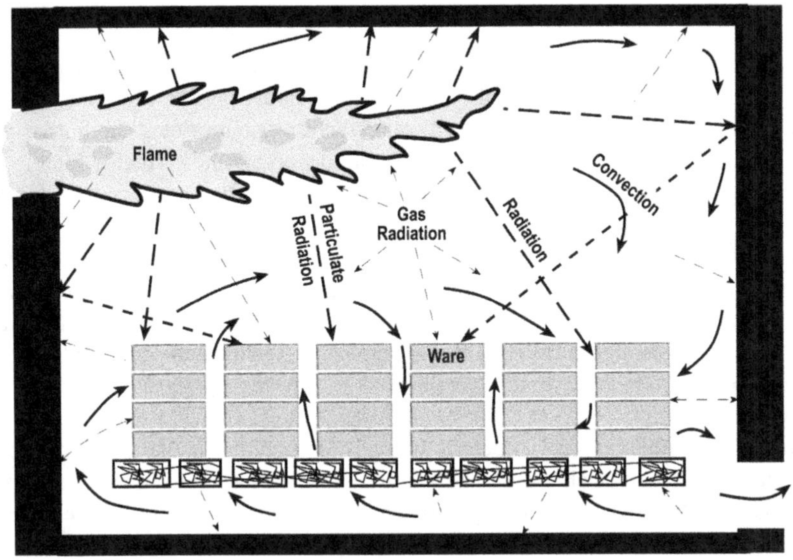


**Fig. 5.5.** Solids' and flames' radiant energy (*long-dashed arrows*) and convective energy (*curved arrows*) are absorbed by refractories, raising their temperature; then the walls re-radiate to the loads. *Triatomic gases in the flame and everywhere in the furnace* radiate everywhere (*light, short-dashed arrows*).

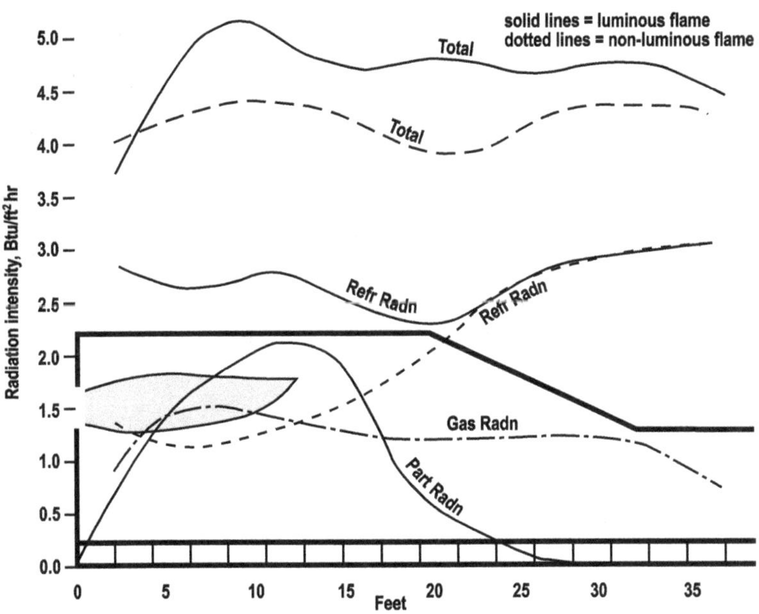


**Fig. 5.6.** Some relative values of refractory radiation, gas radiation, and particulate radiation intensities for a specific flame and furnace. Total radiation is 6.5% higher with a luminous flame than with a nonluminous flame. Multiply Btu/ft$^2$hr by 0.01136 to obtain MJ/m$^2$h. Multiply feet by 0.3048 to obtain meters. Adapted from a paper by Mr. K. Endo of Nippon Steel, presented at the International Flame Research Foundation, Ijmuiden, Netherlands, about 1980.

### 5.2.2. Poc Gas Temperature History Through a Furnace

To reduce fuel cost and improve productivity, an engineer must be able to adjust furnace gas temperatures to change the furnace temperature profile. In a longitudinally fired furnace, shortening the flame will raise the temperature near the burner wall. This can be accomplished by spinning the combustion air and/or fuel, which in turn spins the poc. The resultant increase in heat transfer near the burner wall will reduce the flue gas exit temperature, raising the % available heat.

In furnaces with top and bottom heat and preheat zones, there is greater resistance to poc gas flow below the loads and their conveyor. That resistance causes the bulk of the bottom gases to flow into the top zones, reducing the effective heat transfer exposure areas significantly. This movement of combustion gases into the top zones reduces productivity and lowers available heat, increasing fuel use per ton of product.

Another variable that can affect the flue gas temperature is the length of the gas flow path, which can be changed only by altering the furnace design configuration or size—not by changing an operating variable. This factor is sometimes referred to as "residence time," but that term is often misinterpreted because time in the furnace is not just a function of length of the gas flow path but also the velocity of the gases, which is a function of an operating variable, namely firing rate. (See the adjacent box.)

Flue gas exit temperature rises or falls with flame length, firing rate (furnace gas velocity), heat transfer to loads, and refractory. Longer flame length increases flue temperature. Longer flame length may result from increased inerts (as with fgr), less spin, lower combustion air presssure drop across the burner (poorer mixing), or changed combustion air temperature or excess air.

Lowering the firing rate will lower flue gas exit temperature because of lower poc temperature, thus raising %available heat. However, if the firing rate is so low that

---

*Residence time* was mentioned as a factor in cumulative heat transfer as gases flow through a furnace, but its function is often misunderstood.

Fossil fuel combustion transforms chemical energy into sensible heat, raising the temperature of the combustion gases. The resultant hot poc immediately transfer heat by convection and gas radiation to cooler solids and gasses, at rates proportional to their temperature differences.

If the burner firing rate is increased, the gas volume and temperature increases; thus, the gas flow velocity increases. The cumulative heat transfer from hot gases to loads (directly, and indirectly via refractory to loads) is a function of time. Higher velocity shortens the time for heat transfer to be accomplished within a given flow path length (furnace size); thus, the gases remain at higher temperature.

When the firing rate is lowered, the reverse phenomena take place: Gases take longer to traverse the same path, and so each molecule of poc has more 'residence time' during which to deposit its heat on the loads, but its coefficient of heat transfer is less (a function of velocity to only the 0.52 to 0.80 power).

---

it fails to provide adequate circulation to all loads and all their surfaces, the result will be poor temperature uniformity and the need to soak longer, or do the job over (doubling the fuel bill). As the firing rate is lowered with conventional forward-fired burners in longitudinally fired furnaces, the burner wall temperature rises whereas the gas temperature farther away from the burner drops.

**Generalizations**

Lower flue gas exit temperature saves fuel

Better heat transfer rate lowers gas exit temperature

Lower firing rate lowers gas exit temperature

Excess air can absorb heat intended for the load

Long flames or added burners near the flue raise flue temperature, and thus waste fuel

Inerts in flames reduce NOx formation

**Exceptions**

Low firing rate may reduce circulation and create nonuniformities that cost more fuel

Limited amounts of excess air may enhance circulation or complete mixing at low firing rates

Regenerative burners save fuel with very low exit gas temperatures

Inerts in flue gas recirculation endanger flame stability and steal heat

## 5.3. FURNACE, KILN, AND OVEN HEAT LOSSES

Predicting losses is difficult, particularly losses through and around doors, jamb, sills, tramp air, cooling losses, and losses through conveyor equipment and gaps around it. Assigning safety factors or security factors to cover these matters requires experience and careful judgment.

### 5.3.1. Losses with Exiting Furnace Gases

(*a*) via gases intentionally exhausted through the flues and (*b*) via outleaking gases. (See also sec. 5.3.5.) Both carry away valuable energy that could have been delivered to the loads in the furnace. Both (*a*) and (*b*) involve convection (flow losses) and radiation losses. All of these losses tend to worsen as furnaces age. If the leaking gases include unburned fuel, the loss is more than doubled. To remedy such a problem, check for poor mixing and consider changing to better burners. For the purpose of evaluating these losses, with properly mixed air and fuel and with complete combustion, both the poc exiting via flues, those exiting through leaks can all be considered "flue gas loss" and evaluated as the difference between the fuel's net heating value and its "available heat."

Total "flue gas loss," with excess air loss = (5.2)

$$(\text{Fuel used/hr})(NHV^*)(1 - \frac{\%\text{ available heat from figs. 5.1 or 5.2}}{100\%}).$$

Evaluation of radiation loss through furnace cracks and other leaks is very difficult. The best policy is to deal with them by constant surveillance combined with immediate repair. Operators and maintenance persons must understand that they can only get worse, and will do so at accelerating rates.

Sensible heat carried out of the furnace by the furnace gases (poc) is often the largest loss from high-temperature furnaces and kilns. It is evaluated by the available heat charts mentioned in section 5.1: 100% − %available heat = %heat carried out through the flue. It can be reduced by careful air/fuel ratio control, use of oxy-fuel firing, and good furnace pressure control.

***5.3.1.1. Air/Fuel Ratio Control.*** Careful air/fuel control avoids excessive rich burning, which results in incomplete combustion with partially burned or unburned fuel escaping from the furnace without releasing heat where it can be used effectively. This is rarely a problem with modern burners, with excellent mixing of fuel and air, resulting in very low ppm of CO emissions. Hydrogen emissions (another evidence of incomplete combustion) are typically close to the same low ppm level. Measuring the *flue gas analysis* (usually for oxygen or CO) must be done with a probe carefully located to get a true sample of the flue gas mixture. At least two traverses of the flue duct should be taken at each of several different firing rates. Do not allow amateurs to do this. Use a refractory probe.

Air/fuel ratio control also prevents excessive lean burning, which results in extra unused air passing through the furnace, absorbing heat, and carrying that heat out the flue, unabsorbed by the loads. Chapter 7 of *reference 52* describes how a variety of air/fuel ratio control systems work and how to evaluate the savings from their use.

***5.3.1.2. Oxy-Fuel Firing.*** The use of oxy-fuel firing (pure oxygen, no nitrogen as with air-fuel firing) eliminates about 80% of the heat-stealing capacity of hot flue gases. (See pt 13 of *reference 52*.)

***5.3.1.3. Furnace Pressure Control.*** This type of control prevents excessive outleakage of unburned air, unburned fuel, poc, and pic (products of incomplete combustion) before they have had time to transfer heat to the loads. Chapter 7 of *reference 51* describes how a variety of furnace pressure control systems work and how to evaluate the savings from their use. Furnace pressure control also prevents unnecessary infiltration (inleakage) of unwanted 'tramp air,' which is excessive excess air.

Heat also is lost if air *leaks into a furnace* because (*a*) that air absorbs heat directly from the load pieces, chilling them, requiring longer soak time for good product temperature uniformity, and (*b*) it also picks up heat from flame, refractory, and piers or kiln furniture, and carries that heat out the flue (greater mass of hot waste gas up the stack). Imperative solutions to this problem are: (1) Constant vigilance for, and

*Net heating value. (See glossary.)

immediate repair of, leaks, and (2) control of furnace pressure at a slightly positive pressure (at least +0.02"wc, or +0.51 mm $H_2O$) at all elevations down to the lowest possible leak. (See also sec. 6.6, 7.2, and 7.3.)

### 5.3.2. Partial-Load Heating

Long load pieces may have to protrude out the furnace door. This poor practice allows heat to escape by conduction out along the piece from the part in the furnace to the part outside, dissipating heat to the surroundings. This practice should be avoided because of (*a*) high heat losses, (*b*) poor control of temperature of the load piece(s), and (*c*) poor control of the furnace atmosphere. A similar loss occurs by conduction through the *terminals or electrodes* of electric furnaces. In tall electric furnaces, the loss of heat due to outflow of hot air through the annular spaces between the terminals and the sleeves in the walls through which they pass may be considerable. Tight sealing is difficult because of electrical insulating requirements.

***5.3.2.1. Exposed Hot Liquid Surfaces.*** Other partial-load heating losses may occur by radiation and convection from exposed liquid surfaces, as salt and lead baths (chap. 4), or from water baths (table 4.23 of *reference 51*).

### 5.3.3. Losses from Water Cooling

***Water cooling*** (to protect skid pipes, conveyor rollers, and door frames from overheating) absorbs much heat, lowering thermal efficiency. It is rarely practical to recover the low-level heat from cooling water (except possibly for locker room showers with a generously sized mixing tank and good automatic temperature control). Water-cooled door frames cause so many accidents when they spring leaks that they are being replaced with hoselike door seals of braided ceramic fiber (some, air inflatable). (See sec. 8.1.4.)

***5.3.3.1. Water Seals.*** In many modern furnaces—rotary, walking hearth, walking beam, car hearth, and pellet hearth—there are sizeable losses through the clearances that allow facilities to move the load pieces in and out of the furnace. Mechanical closures, to allow loading and unloading, can be maintained in most batch heating operations. However, in furnaces where movement is almost constant, the use of small clearances and water sealing is practically universal.

**TABLE 5.2. Door leak losses with slight positive furnace pressure control**

| | Complete Combustion | Incomplete Combustion |
|---|---|---|
| Batch furnaces | (1) | (3) |
| Continuous furnaces | (2) | (4) |

*Note.* All losses are much greater with negative furnace pressure.
(1) = least loss; (4) = worst loss.

When new, water seals provide a complete (100%) seal, but after years of operation they may no longer be gas tight. Unfortunately, many seals become overheated at times as a result of a cooling water loss or perhaps because a piece of refractory falls into the seal and causes a mechanical wreck. Furnace pressure then becomes uncontrollable, breaking through the water seal, and exacerbating overheating and warping.

When any one of these problems happens, the seal usually drops to about 50% effectiveness, and no one has any idea as to the magnitude of hot gas movement through the seal. Some designers use a rule of thumb of 600 Btu/hr for each linear foot of seal. Others try to estimate the clearance area and multiply it by the difference in radiation from each zone's average temperature to furnace room temperature. Some managers rationalize that they can save on furnace capital costs by downsizing the furnace input, which turns out to be inadequate to balance seal heat losses after their deterioration.

Coauthor Shannon has equipped furnaces with inputs 30 to 40% greater than the calculated need when new. He has found that they have used all the fuel capacity at some occasion in the first three years, and that after ten years all the furnaces have used all the available fuel input rate, quite often to make up for aging losses or because of a need (by the process) to extend the heating capacity of the furnace.

***5.3.3.2. Sand Seals.*** The sand seals on rotary- and car-hearth furnaces minimize heat loss, but require frequent refilling and attention. A miniature metal plough near the leading edge of an "insertion blade" attached to the car(s) of rotary- or car-hearth furnaces can push the sand against the blade for a sure seal. A large piece of scale, refractory, or tramp metal may fall into the sand trough and spill sand or possibly damage the blade and/or trough.

### 5.3.4. Losses to Containers, Conveyors, Trays, Rollers, Kiln Furniture, Piers, Supports, Spacers, Boxes, Packing for Atmosphere Protection, and Charging Equipment, Including Hand Tongs and Charging Machine Tongs

If loads are heated using these items, they themselves may absorb much heat and carry that heat out into the cool room as they return for emptying and reloading. This not only wastes energy but the cyclic heating and cooling causes oxidation loss and change of grain structure, thus shortening the useful life of the containers and conveyors. Wise designs of continuous furnaces and ovens incorporate conveyor return within the hot furnace or in an insulated tunnel. In batch furnace operations, charging and removal equipment may absorb considerable heat from the furnace.

### 5.3.5. Losses Through Open Doors, Cracks, Slots, and Dropouts, plus Gap Losses from Walking Hearth, Walking Beam, Rotary, and Car-Hearth Furnaces (see also sec. 4.6.9)

***5.3.5.1. Flow (Convection) Heat Losses.*** These losses occur when furnace gases exit around doors and through cracks or dropout load discharge chutes, sometimes burning as they go but always carrying away heat. Major heat loss occurs whenever a door is opened. Every operator must understand this horrendous energy waste, and *make a habit of closing doors and peepholes* ***promptly.***

Flow heat losses may involve cold air leaking into a furnace as well as hot gases leaking out. The losses from cold air inleakage are usually larger than those from hot gas outleakage. Cold air inleakage occurs if the opening is at a level where the pressure inside the furnace is less than the pressure outside at the same elevation, thus sucking 'tramp air' (excess air) into the furnace through any cracks or openings. This cold air inleakage may chill some of the load pieces, turning them into rejects, or else requiring a longer heating cycle to achieve good temperature uniformity, and therefore using more fuel. (See figs. 5.7, 5.8, and 5.9.)

The tramp excess air also will absorb some heat from the load or furnace, and carry that heat out the flue. The cold excess air tends to creep across the hearth and up the flue without helping to burn fuel or circulate heat. For this reason, industrial furnace engineers advocate holding a slightly positive furnace pressure (+0.02"wc, +0.51 mm $H_2O$) at the level of the lowest possible leak. (See "Furnace Pressure Control" in pt 7 of *reference 52*.)

**5.3.5.2. Losses from Exposed Bath Surfaces.** (See also sections 3.8.3 and 3.8.9 relative to galvanizing tanks and pp. 125 to 126 of *reference 51* for water

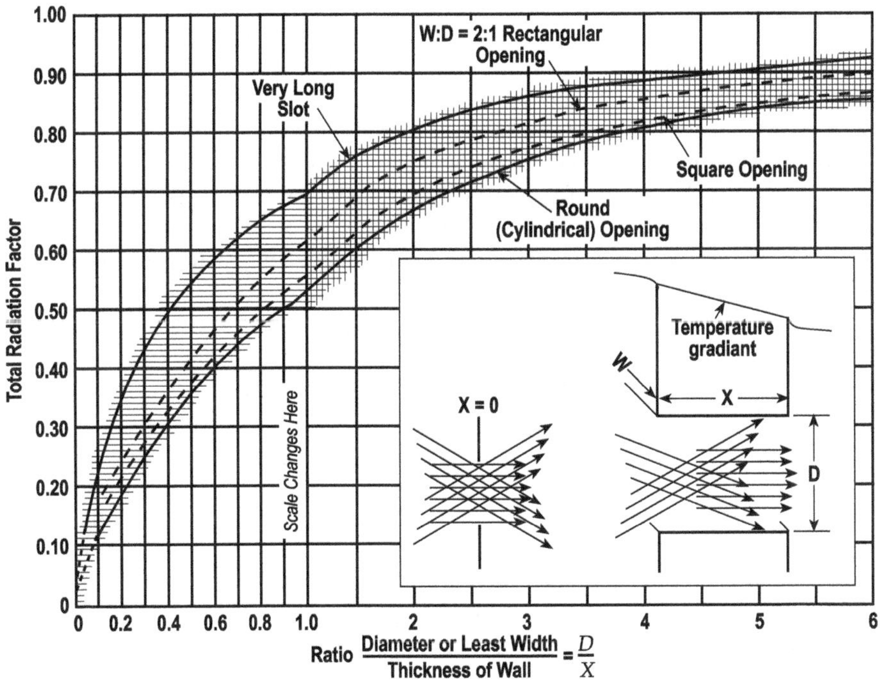


**Fig. 5.7.** Radiation through openings of various shapes as a fraction of the radiation from an exposed surface of the same cross-sectional area.

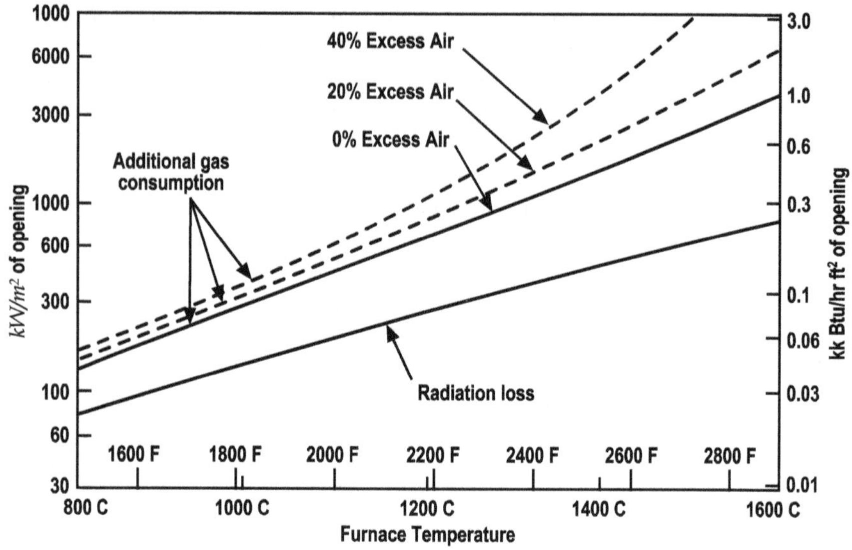


**Fig. 5.8.** Radiation loss and additional fuel consumption of openings. (Based on British Gas R&D Report MRS E 478 by N. Fricker.)

(immersion) tanks.) In exposed molten metal baths, the loss from an exposed surface may far exceed the sum of wall losses and useful heat. Data on radiation constants for molten metals are scarce, but for a bright surface of molten lead, the emissivity is apparently about 0.35. If the surface is covered with scum formed by oxidation, the emissivity increases to 0.63. In *wire patenting baths*, the surface loss is decreased by covering it with a layer of crushed or powdered charcoal to a depth of about 1 in. (.025 m). That covering also reduces metal loss by oxidation. The third edition of Trinks' *Industrial Furnaces, Vol. II,* shows the following radiation heat losses for uncovered salt baths:

| | | | | |
|---|---|---|---|---|
| Bath temperature, F | 1000 | 1500 | 2000 | 2350 |
| Bath temperature, C | 538 | 816 | 1093 | 1288 |
| Heat loss, kW/ft$^2$ | 2.3 | 7.7 | 19.2 | 31.9 |
| Heat loss, kw/m$^2$ | 24.7 | 82.6 | 206 | 343 |

***5.3.5.3. Radiation Heat Losses.*** through all small furnace openings follow the Stefan-Boltzmann law as discussed in section 2.3.3. An emissivity of 1.0 may be used because the radiating source surface is most of the furnace interior surface, giving a pinhole camera effect with the radiation coming from a surface that approaches infinite area relative to the actual area of the opening. Furthermore, the thickness of the furnace wall often results in a considerable portion of the radiation (that enters the opening) striking the sidewalls of the opening, thus, it is not completely lost from the furnace. Figure 5.7, from Trinks and Mawhinney's fifth edition,

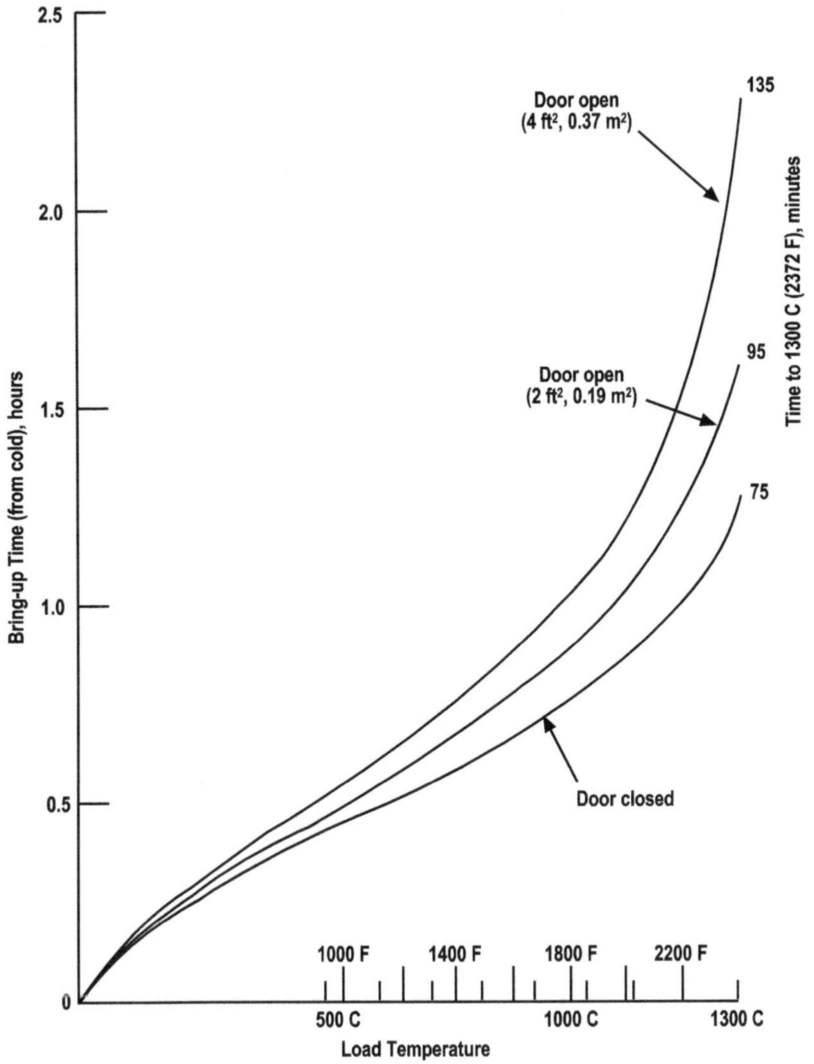


**Fig. 5.9.** Bring-up time increases because of loss through openings. (Based on British Gas R&D Report MRS E 478 by N. Fricker.)

gives correction factors for this beam-narrowing effect with four different shapes of openings—very long slot, 2:1 rectangle, square, and circular. The insets show why the full cross-sectional area of an opening in a thick wall (*right sketch*) does not radiate like a pinhole (*left sketch*). It is not clear whether the original data took into account the effect of temperature gradient through a thick wall (*top of right sketch*) on the variable intensity of re-radiation from the interior surfaces of the thick wall opening.

Figures 5.8 and 5.9 emphasize another aspect of most furnace heat losses, namely, that these losses should be labeled "added available heat requirements." *Example:* Loss through an opening has been evaluated at 100000 Btu/hr. The 2300 F furnace has a flue gas exit gas temperature of 2450 F. From figure 5.1, available heat is 28%, so the cost of the opening loss is 100000/0.28 = 357000 Btu/hr. This should convince everyone that the rewards of minimizing furnace losses can be large fuel savings.

### 5.3.6. Wall Losses During Steady Operation (see chap. 4 of *reference 51*)

Many modern furnaces are well insulated, but the heat lost by conduction through the furnace walls and then by radiation and convection from the outside furnace surfaces may have a significant effect on furnace economy. Furnace walls built of insulating refractories and encased in a steel shell reduce flow of heat to the surroundings. The loss is further reduced by the insertion of fiber block between insulating refractory and the steel casing. (See sec. 5.3.5 and 8.2.1.4 regarding doors and sealing.)

Furnace walls built of successive layers of hard refractories, insulating refractories, and fiber block, encased in a steel shell, reduce heat loss to the surroundings. No form of insulation should be outside the metal shell because (*a*) trapped furnace gas condensed during downtimes will corrode the metal shell, and/or a leak of hot furnace gas through the hard refractory may melt the casing (shell).

The walls of tall furnaces are often built of strong, dense refractories ("hard refractories"), which have greater strength but higher heat storage and wall loss. A question then arises: "How much can the heat loss be reduced by the application of insulation?" The answer depends on thicknesses and types of refractories and insulations as well as on continuity of furnace operation. The manner in which the heat saving varies with three of these variables can be seen in table 5.3, which refers to wall losses only and not total heat consumption of the furnace.

Recommended maximum insulation thickness in combination with thickness of hard refractory is given in *reference 51*. Saving of heat does not necessarily mean saving money because the fixed charges on the cost of insulation may exceed the cost

**Preparation for Wall Loss Study**

Before proceeding with any study of wall losses, the engineer should determine the make-up of the refractories, insulations, and casing of the furnace walls, roof, and hearth. This requires going back to the furnace drawings and material specifications of the most recent rebuild or relining. When the engineer is certain that he or she has all the details of materials and their thicknesses, he or she can (*a*) ask a refractory supplier to plug the wall information into their wall-loss computer program or (*b*) use the method of pp. 107 to 111 of *reference 51*. (See also wall loss information in chap. 8 and 9 of this (Trinks 6th).)

**TABLE 5.3. Percent reduction of wall loss during *continuous* operation, by adding insulation**

| Heavy Refractory Wall Thickness | 2.5" (6.3 cm) Insulation | 5" (12.5 cm) Insulation |
|---|---|---|
| 4.5" (11.4 cm) | 62% | 76% |
| 9" (22.8 cm) | 46% | 65% |
| 13.5" (34.2 cm) | 38% | 57% |
| 18" (45.6 cm) | 35% | 53% |

of the fuel that is saved. Although this is seldom the case, it must be taken into consideration. Another factor that reduces the profitability of insulation is its application to walls that are subject to frequent repairs. Examples are furnaces near steam hammers and furnaces that are heated up too quickly after a prolonged shutdown. In such furnaces, spalling may occur. The original insulation usually cannot be salvaged after extensive repairs.

### 5.3.7. Wall Losses During Intermittent Operation (see also chap. 4 of *reference 51*)

The relative rates of heat conduction and temperature leveling when burners are intermittently off, as in batch furnaces, can change the justification for added insulation. This depends on the thicknesses of heavy refractory and insulation, on the types of each, and on the continuity of furnace operation. The way in which the %heat saving changes with three of these variables can be seen in table 5.4, an extension of table 5.3, which was for steady operation only. Both tables refer to wall losses only and not to the total heat consumption of the furnace. "One-week cycle" means continuous operation for 6 days, 24 hr per day. For 5-day, 24 hr per day operation, the savings would be reduced by about 10%. "One-day cycle" means 8 to 10 hr per day. The tabular values must be reduced somewhat if the wall is thick relative to the interior dimensions of the furnace. The tabular values apply only to those furnaces entirely covered with insulation.

**TABLE 5.4. Percent reduction of wall loss, during *intermittent* operation, by adding insulation**

| | Continuous Operation | | Intermittent Operation | |
|---|---|---|---|---|
| | (Repeated from table 5.3) | | 1-week cycle | 1-day cycle |
| Heavy Refractory Wall Thickness | 2.5" (6.3 cm) Insulation | 5" (12.5 cm) Insulation | 2.5" (6.3 cm) Insulation | 5" (12.5 cm) Insulation |
| 4.5" (11.4 cm) | 62% | 76% | 58 | 25 |
| 9" (22.8 cm) | 46% | 65% | 36 | 18 |
| 13.5" (34.2 cm) | 38% | 57% | 20 | 14 |
| 18" (45.6 cm) | 35% | 53% | 15 | 12 |

## 5.4. HEAT SAVING IN DIRECT-FIRED* LOW-TEMPERATURE OVENS

In all but intentionally designed flame-impinging† operations, the poc should be cooled below flame temperature before they contact the loads. This is not difficult in high-temperature furnaces, but if the stock is to be heated to temperatures between 800 F (427 C) and 1300 F (704 C), finding a good solution is more difficult. The poc temperature is often "tempered" by mixing with excess air or with flue gas recirculation. The cost of excess air can be analyzed by use of an available heat chart (sec. 5.1) for the specific fuel involved. Further waste may occur if the mixing results in incomplete combustion from either quenching by the cooler air or poc steams or by dilution with inert gases. A warning signal of the latter is less than about 16% oxygen in the furnace or oven atmosphere. The cost of flue gas recirculation for reducing NOx emissions is analyzed in section 5.12.

In low-temperature furnaces, fuel is saved, if the poc transfer part of their heat to the charge by radiation before physically contacting the loads. This principle has been successfully applied in refining petroleum and in the radiant (water wall) section of large water-tube boilers. A flame located in the center of a large furnace radiates to pipes that almost cover the surrounding walls. After the poc gases are partially cooled, they then contact other heat transfer surfaces for convection heat transfer. (See sec. 4.7.2.) *The radiation section should always precede the convection section* (usually a tube bundle), that is, radiation upstream along the poc flow path and convection farther downstream along that path. The reasoning is that radiation heat transfer from solids varies as the fourth power of the absolute temperature of the radiation source and thus is most powerful while the poc are hottest. In contrast, convection is only proportional to the first power of its $\Delta T$.

Pulse-controlled firing, where burners are cycled on and off systematically, has attracted many adherents. Stepped pulse firing (an alternative to excess air firing) saves fuel while maintaining maximum circulation (to assure temperature uniformity) and high convection heat transfer.

Ovens operating in the 400 F to 1200 F (204 C to 649 C) range, including some dryers, are often direct-fired recirculating ovens, wherein in-duct burners fire into a stream of oven gases being recirculated by a large fan pulling exhaust gases from the bottom of the oven, past the burner flame, and returning to the oven/dryer space through a multitude of specially directed inlets with louvers for direction and flow control. Loads are usually stacked on racks or in trays, largely filling the oven space. Mixing the hot poc with the cooler recirculated gases that have already passed over the loads may be accomplished by the jet action of the flame, and/or by a circulating fan capable of withstanding the temperature of the stream between the burner

*Unless otherwise specified in this book, "furnaces" and "ovens" are assumed to be direct fired. Indirect-fired units use radiant tubes or muffles to protect the load from contact with the poc.

†Impingement heating machines are not very common, being custom designed for long runs of identical loads. Even for these, "flame impingement" is a misnomer, as the combustion should be completed before the stream of pic and poc contacts the load. Otherwise, the pic may be chilled to the point where combustion can never go to completion or maintain maximum gas blanket temperature uniformity, or achieve maximum triatomic gas concentration or high gas radiation heat transfer.

and the oven. Those "cooler recirculated gases" produce a cooler "hot-mix temperature" in a manner similar to (but less effective than) that of using excess air. (See fig. 3.18.)

If combustible volatiles are evaporating from the load, NFPA standards require that the atmosphere in the oven never exceed one-fourth or one-half (depending on the control system) of the lower explosive limit of the volatile gas. For noncombustable volatiles, the required volume for circulation is less severe, but based upon the ability of the circulating stream to absorb the vapor. If the vapor is water, humidity sensors should be used to automatically adjust burner input, circulated volume, and/or exhaust damper. If humidity is not a sensitive factor, simple temperature controls will suffice.

## 5.5. SAVING FUEL IN BATCH FURNACES

The fuel economy of furnaces is commonly expressed in units of fuel or electrical energy expended to heat a unit weight of load. A generalized way to compare furnaces is furnace efficiency, or %thermal efficiency = 100% × (heat absorbed in the load)/(heat in fuel consumed for the load).

From the preceding study of heat losses, one can conclude that the heat efficiency of a furnace depends not only on its design but also, to a large extent, on its operation and on the requirements for uniformity of heating. For example, if a few small pieces are heated in a large furnace, the fuel consumed per unit of material heated will be extremely high—whether the furnace was heated up especially for those pieces, or whether it had been kept hot all the time.

If the furnace was heated up just for a specific load, a large part of the heat would have to be used to raise the temperature of the walls, hearth, and roof of the furnace. If the furnace had been kept hot and empty, the continued heat losses through its walls and the continued flue gas losses would depress the heating efficiency to a very low value. Furnace builders are aware of these problems and are careful to make their efficiency guarantees quite specific regarding operation (e.g., *not* with partly opened or broken or leaky doors; high excess air or fuel, or poor mixing; or poorly controlled, stuck, or otherwise inoperable stack damper). In most modern furnaces, the effects of the human element have been minimized by automatic control of furnace temperature, air/fuel ratio, and furnace pressure; but those controls themselves need watchful and knowledgeable attention.

*Location of T-sensors* in continuous furnaces requuires much more important consideration than logic would indicate. In many furnaces, for example, the furnace exit temperature is higher at 50% furnace capacity than at 100% of furnace capacity, which will result in very high flue gas losses and high fuel rates. To avoid this problem, the first fired entry zone should be controlled by a T-sensor approximately 6' (1.8 m) from the flue opening and in the hot gas stream, and in a position to "see"* the loads. With this arrangement, if no adjustment is made to the control setpoint, at least the flue gas temperature will not exceed that of high furnace capacity during any lower capacity operation.

*i.e., to receive (straight line) radiation from . . . or emit radiation to . . .

The general method for calculating the energy consumption of a furnace heating a given amount of material is:

$$\text{Energy input to furnace} = \frac{\text{`Heat needs' for load + furnace}}{\text{\%available heat}/100\%} \qquad (5.3)\ \text{(same as 2.1, 5.4)}$$

**Step 1.** Add together all amounts of heat going to different areas in the Sankey diagram (fig. 5.11)—load and furnace, including walls, hearth, roof, cooling water, conveyors, and openings (except for heat carried out by gases exiting via flue and leak openings, covered by step 2).

**Step 2.** Predict the "%available heat" (which is 100% − %flue losses) by reading it from an available heat chart (figs. 5.1 or 5.2). Section 5.1 explains how to determine flue gas exit temperature.

**Step 3.** Divide the total required heat for load and furnace (from step 1) by the %available heat divided by 100% (step 2 as a decimal).

## 5.6. SAVING FUEL IN CONTINUOUS FURNACES

Continuous furnaces should be more fuel efficient than batch furnaces because they do not cool down during and after every load is removed, throwing away the heat stored in their walls. In addition, they are usually longer furnaces, and if fired only from one end, they give their hot gases more time and more surface contact with which to transfer heat to their loads, reducing the flue gas exit temperature.

When managers seek more productivity, they often add input along more of the furnace length, and in so doing, lose the fuel economy advantage mentioned in the previous paragraph. If the input were added with regenerative burners, they would achieve the best of both fuel economy and productivity because each regenerative burner lowers the throw-away flue gas temperature to the 400 to 600 F (200 to 316 C) range, regardless of furnace temperature and burner positioning

### 5.6.1. Factors Affecting Flue Gas Exit Temperature

To reduce fuel costs and/or improve productivity, it is important to be able to change the furnace temperature profile, which may lower or raise the furnace gas exit temperature. In a longitudinally fired continuous furnaces, and those fired only from one end, shortening the flame will be effective in raising the temperature near the burner. This can be accomplished by faster mixing (usually by spinning the combustion air and/or fuel and poc.* The resultant increase in heat transfer near the burner will reduce the ultimate flue gas exit temperature, thus raising the %available heat.

In furnaces with bottom-fired heat or preheat zones (firing below the work load), there is often greater resistance to poc gas flow in the bottom zones than in the top zones because the bottom zones usually contain conveying equipment, support

*poc = products of combustion = furnace gases.

rails, and cooling water crossovers that tend to block the gas flow passages. These cause the bulk of the bottom gases to flow up into the top zone, reducing the bottom zone's effective heat transfer exposure areas significantly. Increasing the depth of the bottom zones might help the bottom side heat transfer, thus improving the temperature uniformity between bottoms and tops of the load pieces and reducing the necessary length of soak zone, correspondingly reducing fuel consumption.

Flue gas exit temperature is affected by (*a*) flame length, (*b*) firing rate (furnace gas velocity), and (*c*) heat transfer from the furnace gases to the loads, and from furnace gases to the refractory and then to the loads.

Longer *flame length*, higher combustion air temperature, use of oxygen, or change in excess air may affect flue temperature. Longer flame length can be the result of increased inerts (as with flue gas recirculation for NOx reduction), poor mixing, fuel and air pressure drops across the burner, reduced burner tile (quarl) diameter, or direction of the flame.

*Firing rate* affects flue gas exit temperature because it affects flame and poc temperature. For example, in conventional straightforward firing, as the firing rate is increased, the burner wall temperature drops and the poc temperatures rise farther away from the burner. Higher firing rates raise flue gas exit temperatures; lower firing rates lower flue gas exit temperature. Higher combustion air temperature, use of oxygen, or change in excess air also may affect flue temperature.

*Heat transfer* lowers flue gas exit temperatures. Heat transfer rises if

1. the thickness of the gas cloud (blanket) increases,
2. the concentration of triatomic molecules increases, or
3. the average gas blanket temperature increases.

Increasing flue gas recirculation (FGR) to reduce NOx emissions raises the concentration of inerts in a flame, thereby increasing the flame length. The longer flame raises the flue gas exit temperature and also lowers the reaction (flame) temperature, thereby raising the fuel rate. Using FGR to lower NOx can raise fuel costs considerably. (See sec. 5.12.)

## 5.7. EFFECT OF LOAD THICKNESS ON FUEL ECONOMY

When heating material of low absorptivity (and emissivity) and high conductivity (such as aluminum), the stock thickness does not affect fuel economy. However, for a material such as steel (high absorptivity, but low thermal conductivity), the load thickness has a major effect on fuel economy because (*a*) the surface will be hotter than the interior, and (*b*) the poc must leave with a higher temperature. Of course, if the loads were left in the furnace longer in hopes of lowering the gas throwaway temperature, the production rate would drop.

If the load material is easily oxidized, other factors enter. Scale has a higher absorptivity than bright metal; thus, in the initial stages of heating, it promotes heat absorption. However, thick scale can act as an insulator, requiring a longer heating time. If the operator attempts to increase the heat input, the scale will be softened and become shiny, reflecting the heat.

Fuel economy calculations are more complex for multizone furnaces, including rotary furnaces—side fired, roof fired, or longitudinally fired—with or without baffles between zones. (See sec. 2.6, 3.4, 3.5.) With thick loads, load placement is more critical. (See sec. 3.5, 6.9, 6.10.)

## 5.8. SAVING FUEL IN REHEAT FURNACES

### 5.8.1. Side-Fired Reheat Furnaces

Side-fired reheat furnaces can be troublesome in two ways: (1) When conventional burners are installed directly opposite one another, the center of the furnace becomes very hot because the velocity pressures of the poc from the opposing burners negate each other and because the completion of the fuel burning is concentrated in the furnace center; and (2) with staggered long-flame burners, a wide furnace's center gets hotter than the sides when on high fire, but at low fuel inputs the sidewalls get hotter than the centers. Both troubles can be prevented with controlled temperature profile burners and added T-sensors/controls. (See chap. 6.)

In addition to the usual factors affecting fuel saving (e.g., rate of heating, final stock temperature, type and thickness of refractories), other fuel economy factors are heat flux distribution lengthwise and crosswise of the furnace, and location of the flue(s). With heavy firing at the entering end, the poc leave a side-fired furnace at a higher temperature than they do with discharge-end-firing, thus higher fuel consumption is the price paid for increased heating capacity coupled with good temperature uniformity. With the advent of regenerative burners, operating with high temperatures all the way to the charge entrance does not significantly lower the furnace fuel rates, because the regenerators are themselves a heat recovery zone. (See fig. 5.10, for which a control discussion is included at the end of Section 6.11.) However, charge zone temperatures are limited in many furnaces by scale softening with the resultant reflective (non-heat-absorbing) surfaces mentioned earlier.

### 5.8.2. Rotary Hearth Reheat Furnaces

Little difference exists in the fuel economy of end-fired, side-fired, and rotary* continuous furnaces operated above 2200 F (1204 C) and properly designed and operated, and using a fuel of high calorific value (not blast furnace gas or producer gas).

For metallurgical reasons, some rotary hearth furnaces are divided into sections by radial baffles. Rotary furnaces designed to heat rounds for seamless tube mills have some very special problems: (1) furnace pressure control, (2) air/fuel ratio

*Rotary furnaces cannot be end fired, but they can be roof fired with type E flat flame burners or with a sawtooth roof. They may be side fired on the outside only, or inside and outside with a donut design.

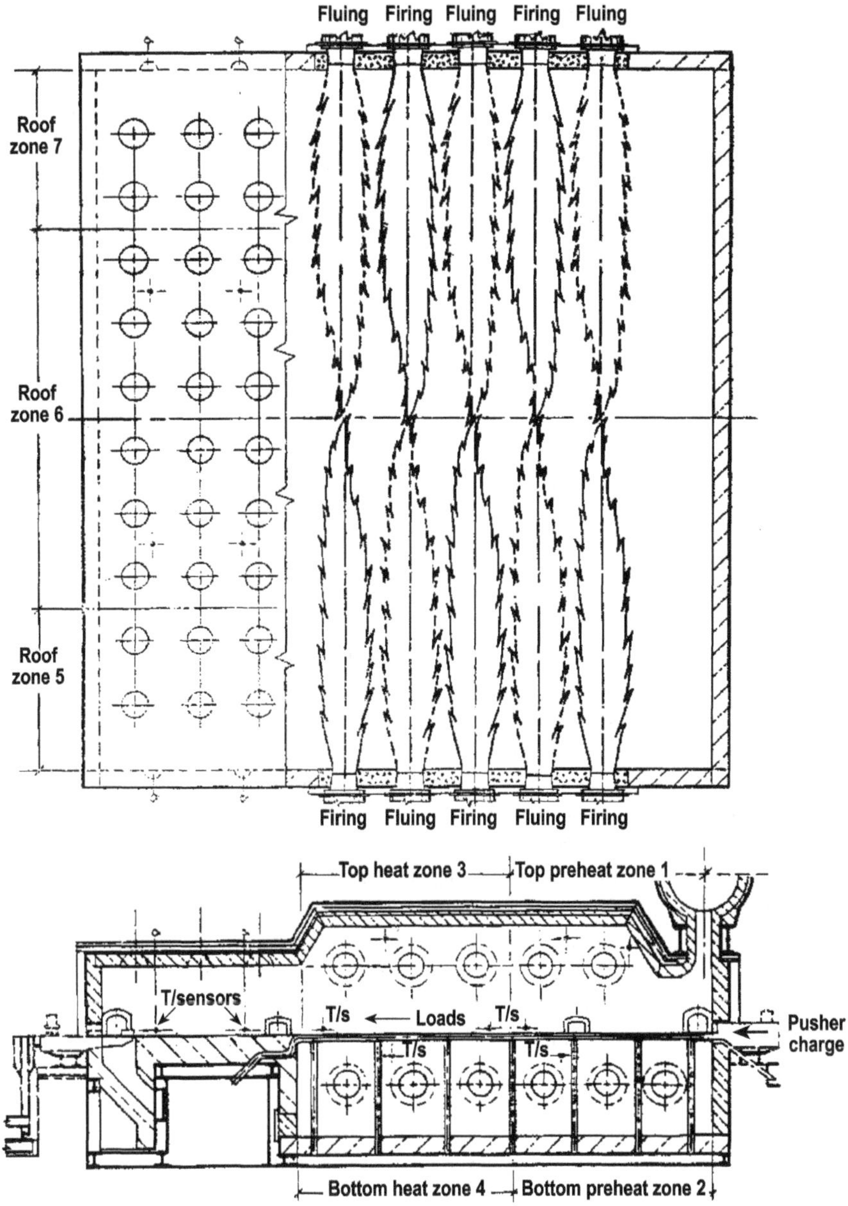


**Fig. 5.10.** Continuous steel pusher reheat furnace side fired with regenerative burners in the top and bottom heat and preheat zones, and roof fired in the soak zone. Preheat zones often have been designed as unfired preheat zones, which are good for fuel economy. However, also firing the preheat zones with regenerative burners would add capacity while retaining high fuel efficiency. (For a discussion of controls for this furnace, see sec. 6.11.1.)

control, (3) gas flow direction control, and (4) burner placement. (Problems 3 and 4 are discussed in detail in sec. 7.5.3.)

***5.8.2.1. Furnace Pressure Control.*** Extraction of load pieces may be as frequent as one to four pieces per minute; therefore, door maintenance is difficult, with the result that discharge doors are often left open. These doors may be very large to accommodate a peel bar mechanism, so leaving a door open permits a large quantity of furnace gas to escape and results in loss of heat and furnace pressure. This problem, combined with the two-way combustion gas flow of a rotary hearth furnace, necessitates three baffles. This solution is described in the following paragraph.

*Three Baffle Solution.* One baffle separates the charge vestibule from the first heat zone, a second (center) baffle is between the charge and discharge vestibules, and a third baffle is between the discharge vestibule and the soak zone (final heat zone). The center baffle, between charge and discharge vestibules, is to limit heat and gas flow between the vestibules. The other two baffles are to limit gas movement out the doors to maintain furnace pressure with the doors open. In theory, this is excellent, but these three baffles must have clearance above the hearth for the largest product thickness, plus a minimum of 3 in. (76 mm). Thus, the total in many cases may be 18 in. (460 mm).

With the previous arrangement, furnace pressure can be controlled with the doors open and no product under one of the baffles, but the reverse furnace gas flow from the soak zone to the zone 1 and flue will be very large, often more than 20% of the total poc. To minimize this part of the problem, an air curtain is recommended on the bottom of the baffle separating the charge vestibule from the first heating zone to limit the reversed gas flow to perhaps 5% of the total poc. The air curtain should be aimed 20 to 40 degrees from the vertical toward the charge vestibule. This replaces an earlier idea of providing adjustable height for the center baffle.

Another problem to be resolved required limiting the poc gas flow from the soak zone to the discharge vestibule and out the discharge door. The solution to this is installing high-velocity burners, one above the other in the inner and outer walls immediately below the baffle between the soak zone and the discharge vestibule. These burners firing at one another will build positive pressure in the furnace center and negative pressure near each burner wall, causing circulation that will practically stop hot gas flow from the soak zone to the discharge vestibule.

These suggested modifications will minimize the problems of controlling furnace pressure and limiting poc flow toward the discharge, without limiting operator functions such as backing up the hearth during delays.

***5.8.2.2. Air/Fuel Ratio Control.*** Air flows may differ to burners in parallel in the same zone on the inside and outside of a rotary hearth furnace donut because of the long runs of air duct and the large number of tees and elbows. High design air velocity creates very different air flows to burners in a zone. One such furnace was designed for an air flow of 70 ft/sec (21 m/s) with three elbows and four tees to each burner. The fan's discharge pressure was 14"wc (3.5 kPa), but the pressure delivered to one burner air connection with the air control valve wide open was only 1.75"wc (0.43 kPa)! The air pressures from one burner to another differed widely. With only one air/fuel ratio control for the whole zone, only one burner had the desired air/fuel ratio.

The two possible solutions are to increase the size of the piping and install cross-connected regulators on each burner, or raise the discharge pressure of the combustion air blower and add a cross-connected regulator to each burner, accepting different firing rates from the individual burners.

If the combustion air is preheated, repiping with mass flow air/fuel ratio for the zone is a must. To reduce burner-to-burner differences in air/fuel ratio, design the air velocities in the piping to a maximum of 40 ft/sec (12.2 m/s) actual velocity, and add air and gas flow meters and a limiting orifice valve in each burner's gas line for setting the air/fuel ratio at each burner.

## 5.9. FUEL CONSUMPTION CALCULATION

Use the graphs and diagrams from section 5.1, repeating the three steps from section 5.5, with equation 2.1 = equation 5.3 = equation 5.4.

$$\text{Energy input to furnace} = \frac{\text{`Heat needs' for load \& furnace}}{\text{\%available heat/100\%}} \qquad \begin{array}{r}(5.4)\\ (\text{same as 2.1, 5.5})\end{array}$$

**Step 1.** Add together all of the amounts of 'heat needs' going to all areas and heat sinks within the load and furnace as shown in the Sankey diagram (fig. 5.11)—including walls, hearth, roof, openings, cooling water, conveyors, radiation losses through openings, and for batch furnaces, heat storage in the furnace enclosure, conveyors, piers, and containers.

**Step 2.** Predict the "%available heat" (which is 100% − %flue losses) by reading it from an available heat chart (figs. 5.1 or 5.2). Section 5.1 explains how to determine flue gas exit temperature.

**Step 3.** Divide the total 'heat need' for load and furnace (from Step 1) by the %available heat divided by 100% (from step 2, as a decimal).

**Example 5.1:** *Given data* for a CPI cabin heater for monomer process:
*Loading:* Cracking vinyl chloride at a rate requiring 40 kk Btu/hr
*Outside dimensions:* 72' × 10' × 23' high.
Wall, roof, and hearth heat loss when operating with an inside refractory face temperature of 2000 F has been calculated to be 2.3 kk Btu/hr.
To be equipped with 220 type E burners using natural gas with air at 400 F.
**Solution**: Find gross fuel input required.

**Step 1.** This is a modern steel-encased furnace with steady flow through its pipelike retorts; thus, its 'heat needs' are only heat losses through its insulated walls and heat to the product load = 2.3 + 40 = 42.3 kk Btu/hr.

**Step 2.** The type E flames already selected are primarily radiation burners, so the flow of poc across the retort surfaces will be quite low, estimated at 15 fps. From figure 5.3, at 2000 F furnace temperature, read 60°F elevation of the flue gas exit temperature (fget) above furnace temperature, or fget = 2000 + 60 = 2060 F. If the furnace will have sophisticated automatic air/fuel ratio control, and is constructed with a steel outer shell so that tramp air will be minimal—say 5% excess air, then extrapolating at 5% XS air from figure 5.1 at 2060 F flue gas exit temperature and 400 F preheated air, read 49% available heat.

**Step 3.** Dividing the total 'heat need' by the decimal %available gives required gross heat input = (42.3/0.49) = 86.3 kk Btu/hr. Adding a security factor to counteract leak development in the future, a wise design input rate might be 100 kk Btu/hr. For natural gas, typically 1000 Btu/ft$^3$, the predicted fuel consumption would be 100 kk Btu/hr/1000 Btu/ft$^3$ = 10 000 ft$^3$ of natural gas per hour. The burners should be selected for (100 kk Btu/hr)/220 burners = 455 000 Btu/hr through each burner, or (455 000 Btu/hr × 10.5 ft$^3$air*/ft$^3$fuel)/1000 Btu/ft$^3$ fuel = 4780 ft$^3$ air through each burner. QED†

## 5.10. FUEL CONSUMPTION DATA FOR VARIOUS FURNACE TYPES

The heat energy consumption by furnaces varies widely with the design, fuel, controls, operation, need for tight temperature control, and use of heat recovery. Tables 5.5 and 5.6 list some specific and average values. The reader must understand that the actual fuel consumption of a given furnace may depart considerably from the figures in this table. The lowest fuel consumption will seldom go below 60% of the average values; the highest may exceed the average values by 100%. Readers should modify the experience data of tables 5.5 and 5.6 to compare with any specific job. If large pieces are placed tight to sidewalls or tight together (reducing sides exposed to heat transfer and limiting passage for hot gases), lag time may increase by 200%.

In one soaking pit, installation of adjustable heat-release burners controlled by T-sensors behind the ingots reduced the cutback period from 3+ hr to 40 min even with 10 hot ingots (23.6 in., 0.6 m, square) charged at the wall opposite the burner and six cold ingots charged at the burner wall. Larger ingots require longer "cutback periods" (see glossary), proportional to the ratio of squares of thicknesses. For 30 in. (0.76 m) ingots, cutback time would be [40 min. × (30"/23.6")$^2$] = 65 min.

For hot charged ingots, fuel rates will be at least 10% less because of shorter heating time to the 'cutback point' (beginning of cutback or soak period). The time at high fire (up to the cutback point) can be as much as 8 hr with cold steel, but 1.5 hr when charged with hot ingots. However, the actual fuel use depends on the length of the cutback period, which in some instances can be 7 hr or more. Generally, long cutback periods are caused by poor charging practice (pieces too close together) or

*10 ft$^3$air/ft$^3$ of natural gas (typical) + 5% excess air. (Useful numbers for natural gases are 1000 gross Btu/ft$^3$ of natural gas, 100 gross Btu/ft$^3$ of air, 10/1 stoichiometric air/gas ratio). (See pp. 16, 17, 34–36 of *reference 51*.

†See glossary for abbreviations and definitions.

**TABLE 5.5. Typical gross heat inputs, steel/iron processing furnaces**

| Heating Process | Approximate Temperature | Furnace Description | Gross Heat Input, kk Btu/ton~MJ/tonne average | minimum |
|---|---|---|---|---|
| Anneal, shorts | 1650 F, 900 C | B, car | 3.0+ | 1.2 |
| | 1290 F, 690 C | B, in & out | 2.0 | 0.8 |
| Anneal, strip stl | max 1290 F, 690 C | C, catenary | 2.0 | 1.6 |
| 300 stainless | 2000 F ±50°F | C, catenary | 3.0 | 1.2 |
| 400 stainless | 1400–1750 F | C, catenary | 3.0 | 1.2 |
| Direct reduce, ore | 1550–1850 F, 843–1010 C | B, DRI | 12.0 | 8.4 |
| Forge, ingots | 2100–2350 F, 1150–1290 C | B, in & out | 2.0+ | |
| | | B, car or box | 5.0+ | 2.5 |
| Forge, misc. | 2100–2350 F, 1150–1290 C | C | 2.8 | 2.5 |
| | | C, Rec | 2.5 | 2.0 |
| | | C, Reg | 1.8 | 1.3 |
| Pelletize | 2300–2450 F, 1260–1343 C | C, arch over bed | 0.8 | 0.45 |
| Roll, longs | 2000–2250 F, 1090–1230 C | C, Hr | 2.5 | 1.5 |
| Roll, longs | 2000–2250 F, 1090–1230 C | C, Hr, Hc | 2.0 | 0.9 |
| Roll, longs | 2000–2250 F, 1090–1230 C | C, Rec | 1.7 | 1.3 |
| | | C, Reg | 1.5 | 1.15 |
| Roll, longs | 2000–2250 F, 1090–1230 C | B | 3.5 | 2.5 |
| Roll, longs | 2000–2250 F, 1090–1230 C | B, Rec | 2.0 | 1.3 |
| | | B, Reg | 1.5 | 1.2 |
| Roll, longs | 2000–2250 F, 1090–1230 C | C, axial barrel | 4.0 | 3.5 |
| Roll, rounds | 2000–2250 F, 1090–1230 C | C, rotary hearth | 3.0 | 2.0 |
| | | C, Rec, rotary hearth | 2.5 | 1.5 |
| | | C, Reg, rotary hearth | 1.5 | 1.2 |
| Roll, ingots | 2100–2400 F, 1150–1320 C | B, pit* | 2.0 | 1.5 |
| | | B, pit,* Hc | 1.1 | 0.5 |
| Roll, ingots | 2100–2400 F, 1150–1320 C | B, pit,* Rec | 1.7 | 1.5 |
| | | B, pit,* Rec, Hc | 0.9 | 0.4 |
| Roll, slabs | 2250–2350 F, 1230–1290 C | C, Rec | 1.4 | 1.1 |
| | | C, Reg | 1.2 | 1.0 |
| Sinter | 2200–2400 F, 1205–1314 C | C, arch over bed | 1.5 | 2.5 |
| Smelt | 2500–2700 F, 1370–1480 C | C, blast (shaft) | 11.0 | 7.0 |
| Weld, skelp | 2500 F, 1370 C | C, axial | 4.0 | 3.5 |
| | | C, Rec | 3.0 | 2.5 |

*Regenerative burners and oxy-fuel firing lack mass flow to load bottoms in pits, therefore increasing top-to-bottom temperature differentials from 40°F to 100°F (22°F to 56°C). (See sec. 7.4.6.) B = batch. C = continuous. Hc = hot charge. Hr = heat recovery. Rec = recuperative. Reg = regenerative. "longs" = billets, blooms, pipe, rails, and structurals (but not rounds or short pieces).

by a large $\Delta T$ between the burner wall and its opposite wall, as when the burner's peak heat release is far from the burner.

Using a burner with variable poc spin and with T-sensors at each end of a sidewall about 3 ft (0.95 m) above the ingot bottoms to control the heat pattern will reduce the cutback period to about 1 hr with 30" (0.76 m) square ingots. If an ingot is charged into a pit at 1800 F (982 C), it already contains 80% of the heat required to get to rolling temperature. If charged cold, 100% must be added by burner input. For each 20-ton ingot, that would be 14.4 kk Btu (15.2 GJ) divided by (%available heat/100).

## 5.11. ENERGY CONSERVATION BY HEAT RECOVERY FROM FLUE GASES

Sankey diagrams (visual heat balances) assist overseeing the Btu checkbook, that is, to analyze where heat is being wasted and how to divert wasted heat to optimum use. Figures 5.11 and 5.12 are Sankey diagrams before and after addition of heat recovery equipment to a furnace.

$$\%\text{furnace efficiency} = 100\% \times (\text{useful output})/(\text{gross input}) \quad (5.5)$$

$$\therefore \text{gross input} = 100\% \times (\text{useful output})/\text{furnace efficiency}$$

$$\%\text{available heat} = \text{best possible efficiency after flue loss, that is,}$$

$$\%\text{ of gross input used to heat the load and any losses other than flue losses}^*$$

$$= 100\% \times (\text{required available heat input}^*/\text{gross heat input}) \quad (5.6)$$

$$\therefore \text{gross input} = 100\% \times (\text{required available heat})/\%\text{available heat}$$

The loss caused by sensible heat in the flue gases (stack loss) can be evaluated as the %net heating value (90% for natural gas) minus the %available heat at the flue gas exit temperature, from Figure 5.1. At high temperature, the loss becomes excessive, especially with high excess air; thus, such cases give payback by using heat recovery. (See figs. 5.13 to 5.16.)

The need to reduce stack loss should lead furnace engineers to *first* seek faster and more uniform heat transfer to the loads in a furnace, as discussed in chapters 1 to 7, and *second* to use heat salvaging methods, discussed later. All heat salvaging or heat recovery methods have a potential problem if they carry the reduction of exit gas temperature too far and lower the gas below its dew-point temperature. Steam-generating engineers encountered "rain in the stack" which rusted out the breaching. $H_2O$ condensation is not as harmful as acids formed from gaseous oxides in the poc—sulfuric, carbonic, nitric. Condensing moisture combines with acid-generating combustion gases to damage recuperators, waste heat boilers, ducts, and preheated furnace loads. Natural gas may have sulfur-based mercaptan added as an odorant for leak detection. $SO_3$ has a catalystlike effect in raising acid dew point. (See fig. 5.13; pp. 118–119 of *reference 52.*)

### 5.11.1. Preheating Cold Loads

Preheating cold loads with flue gases can be accomplished in preheating chambers, in a preheat zone of a continuous furnace, or in the first part of the time cycle of a batch or shuttle furnace. (See sec. 4.3.)

*heat to load + losses other than flue losses = required available heat = heat needs.

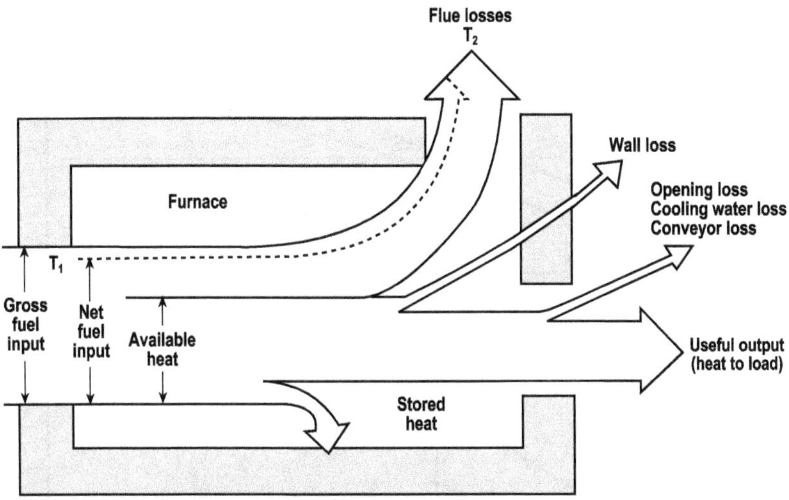


**Fig. 5.11.** Sankey diagram before addition of heat recovery. This is the origin of the ditty: "Lower the T2, for less waste up the flue." (See fig. 5.12.)

For *batch furnaces*, preheating the load is often done as the first segment of a timed program, but that can lengthen the time in the furnace. Another approach is to build a preheat oven immediately adjacent to the furnace and feed the furnace's exit gases through the preheat oven, but that increases the load handling and heat loss during transit. *Continuous furnaces* usually offer a better opportunity for load preheating.

*Unfired preheat vestibules* take many different forms, such as (1) an elongated conveyor though a furnace extension, (2) loading cold charges down the stack of a

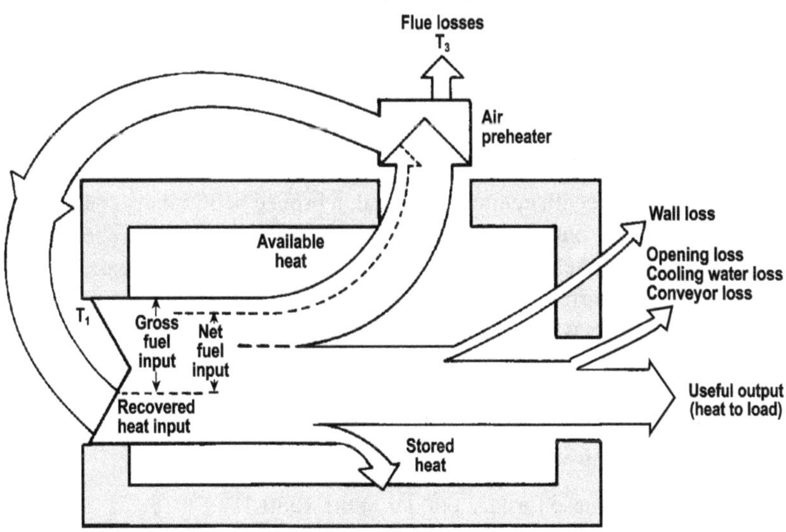


**Fig. 5.12.** Sankey diagram after addition of a heat-recovering air preheater.

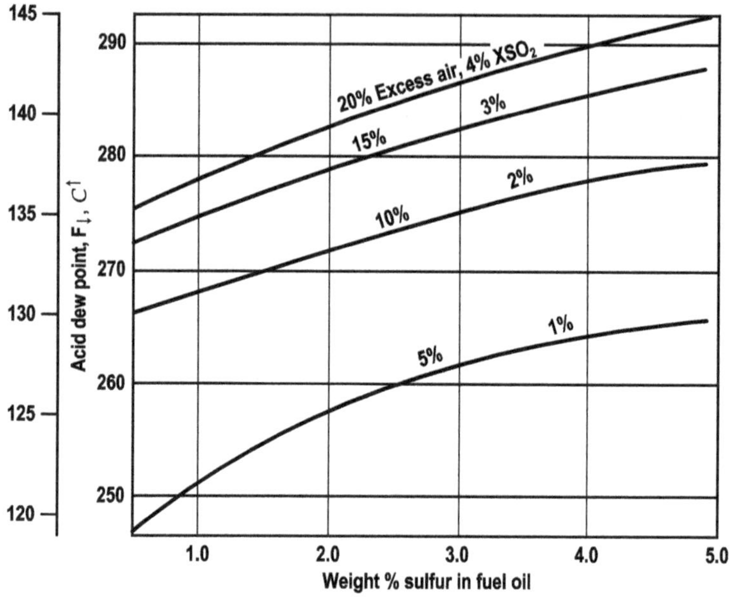


**Fig. 5.13.** Effect of oxygen concentration in poc on acid dew point. Shown for 10 to 12° API crude oil. Courtesy of *reference 58.*

melting furnace, or (3) a pair of adjacent furnaces that alternate preheating and final heating, each receiving waste gas heat from the other when in the preheat mode. These are just a few of many possibile schemes. The sizes, shapes, and properties of the variety of furnace loads in the world should encourage furnace engineers to apply their imagination and ingenuity to their own particular situations. Few industrial furnaces are duplicates. Most are custom-made; thus, their designs present many unique and enjoyable challenges to engineers, of which adding unfired preheating is not the least.

At the site of a thirteenth century cathedral, a bronze bell foundry loaded their melting furnace by putting raw pig metal down the stack for preheating* to save time and fuel each morning while the women of the town carried wood from diminishing surrounding forests.

**Preheating loads with waste gases** has been widely practiced in the forging and hardening of tools . . . from the village blacksmith to slot forge furnaces where extra loads were placed in the slot for preheating. Their fuel efficiency may not have been so crude after all. Fuel was often scarce or dear. Necessity was the mother of invention.

*Patented by a Japanese furnace builder in the 1980s!

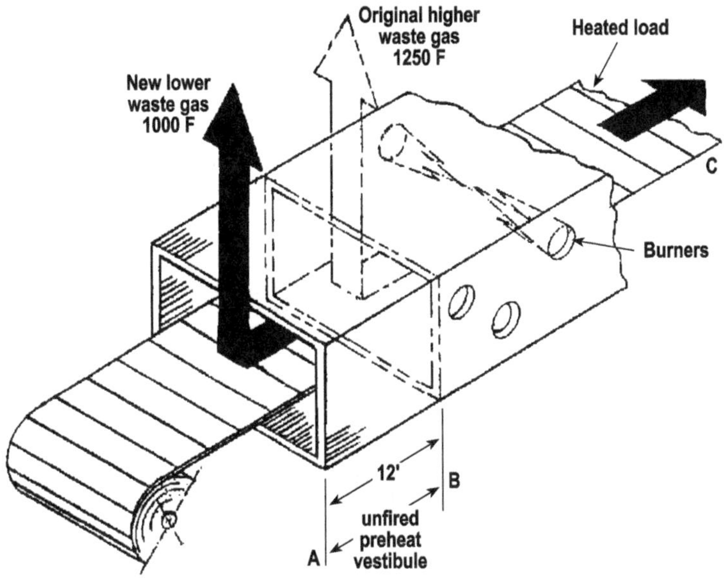


**Fig. 5.14.** An unfired preheat vestibule is an inexpensive way to practice heat recovery. The only extra expenses are an insulated extension of the furnace (no burners), extension of the conveyor, and some floor space.

Figure 5.14 shows how an unfired preheat vestibule works as a heat recovery device—for heating either strip material or load pieces on a belt conveyor. The cold load enters the vestibule at A and is preheated in the vestibule by absorbing heat from the furnace gases exiting through the vestibule at B. The load then enters the original furnace at B preheated to a higher temperature, thereby allowing the burners to be throttled to a lower input, saving fuel. The load exits the original furnace at the same controlled temperature as before.

Figure 5.15 shows a common practice in ceramic tunnel kilns, where the more gradual warm-up of the preheat vestibule has the added bonus effect of less sudden expansion damage to the raw ware.

**Warning**: In all heat recovery schemes, it is very important to minimize transport losses: keep ducts and pipes (for hot flue gas, hot air, and steam) short and very well insulated. Similarly, when preheating loads, if they must be transported hot, keep the distances short and cover them with insulation while being transported.

The unfired charging zones of most continuous furnaces serve as preheating zones. As demand for more production has increased, however, many of those furnaces have been fired harder, which does increase furnace productivity—but at the expense of higher exit gas temperatures and resultant higher fuel use. Some cases even have had burners added in the charge zone, which can greatly reduce the fuel efficiency. An exception to this is the addition of regenerative burners in the charging zone, which gives the best of both worlds—efficiency and productivity—because the exit

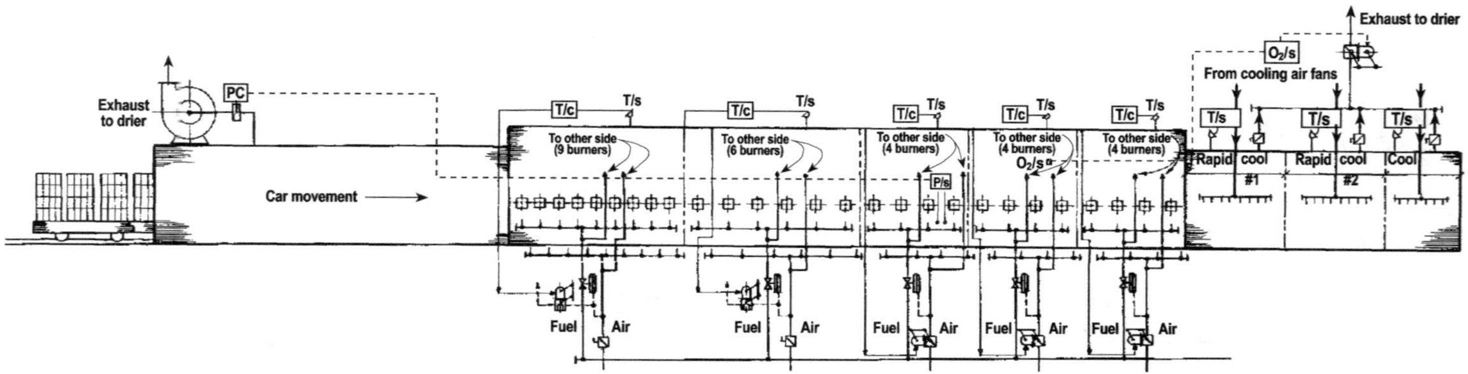


**Fig. 5.15.** Ceramic tunnel kiln (not to scale) with unfired preheat vestibule for heat recovery. Long, narrow kiln or furnace geometry minimizes the proportion of heat loss at the conveyor entrance and exit. Air-lock chambers are even better.

gas temperature is still held very low by virtue of the heat recovery by the regenerative bed. In fact, with regenerative burners, simple preheating of the loads to save fuel may no longer be justified because the thermal efficiency of the regenerative burners can be as high as 75%.

If an unfired preheat vestibule is selected as the vehicle for heat recovery, there may be a great temptation later to add burners to the preheat section for higher capacity. With any preheat section, unfired or fired, careful attention must be paid to gas flow patterns.

### 5.11.2. Steam Generation in Waste Heat Boilers

If there can be good load-related scheduling between hot flue gas generation by the process furnace and the need for steam nearby, waste heat boilers can convert much waste heat to useful free steam, allowing the boiler to use less fuel. Figure 5.16 shows a special fire-tube boiler (with no burner) located close to forging furnaces. A steam header pressure signal controls the induced draft fan's "pull" of hot flue gases through the boiler from the stack. Precaution is necessary so that the pressure in the furnaces is not upset by demand for more free steam.

When waste heat recovery boilers are used with process heating furnaces, they fail to get prime attention from their owners and operators. It may be that the plant managers have no training in boiler operation or hazards, and they try to operate the waste heat boiler with no licensed fireman or engineers. That can lead to a catastrophic steam explosion.

When waste heat boilers are used with steel reheat furnaces, they are often fed gases that are far above the boiler design temperature. Depending on the tightness of the furnace, 2300 F gases may reach the boiler every time there is more than a 15-min delay in mill operation.

The major boiler safety concern is maintaining proper water level. Some sections of fire-tube boiler's plate or tube sheet may sometimes not be protected with water backing—when water level is below the gauge glass.

It is imperative that this compartment, which provides a passage of gases to the very highest fire tubes, have water above it all times. If not, the plate will overheat, its strength will decrease, and the boiler will fail with explosive violence. Water-tube boilers have all heat-exposed surfaces water backed, but control of their water level is more difficult because the water-tube boiler has much less water in its system per unit area of heat transfer surface. Hence, fire-tube waste heat boilers are more widely used for waste heat boilers. Petrochem plants have had good success with water-tube waste heat boilers.

The feed water supply is most important to protect against boiler failure. Complete dual systems to the de-aerator are essential. When the water level falls to near the bottom of the water level gauge glass, the source of heat to the boiler must be removed immediately! Unlike fuel-fired boilers, where removal of the heat sources is generally not complicated, removal of the heat source from a waste heat boiler applied to a steel reheating furnace may involve large dampers that move slowly and do not shut tightly.

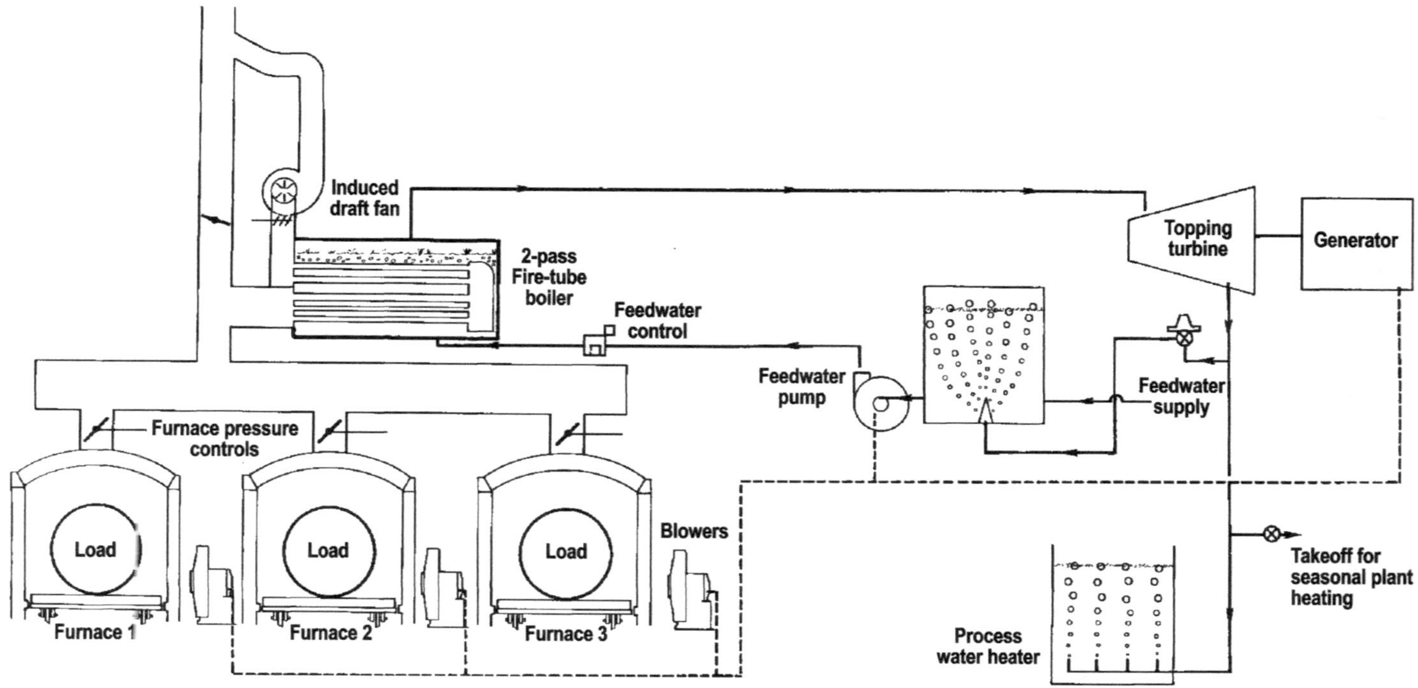


**Fig. 5.16.** A waste heat boiler can save much fuel if there is need for steam concurrent with availability of hot flue gases. The need for steam must never be allowed to reduce the positive pressure in the process furnaces supplying the waste heat for making steam. Courtesy of North American Mfg. Co.

With these waste heat boiler problems—managers with no boiler training, water systems and hot gas shutoff systems inadequately designed, and no operators in attendance—it might be advisable to select an alternate heat recovery system to reduce fuel consumption.

If the plant may grow to depend on the output of a waste heat boiler to make up for inadequate capacity in the main boiler house, consideration should be given to equipping a waste heat boiler with an emergency burner system to keep steam available when waste flue gas is not available.

In countries with high fuel cost and low labor cost, even the heat in the water that flows through skid pipes is utilized in waste heat boilers. To prevent scale deposits in the skid pipes, the circulating water must be treated with an oxygen scavenger and scale treatment. The water is under pressure and may be heated to a high temperature, depending on the steam pressure of the boiler. With the high pressure of a modern boiler, say 500 psi (3448 kPa), steam bubbles that happen to form in the skid pipes are very small and are less likely to cause overheating damage to the skid pipes, but coordination between furnace operators and power plant operators is always wise.

Installations using a waste heat boiler with a single furnace are unusual, but in small forge plants, a waste heat boiler connected to two or more furnaces is not uncommon. An emergency flue-relief valve from furnace to stack (required by law in some European countries) can be opened if the boiler has to be shut down, allowing continued furnace operation (without saving fuel). The emergency flue-relief valve also can be opened if there is danger of overheating any part of the boiler that could cause an explosion.

If a waste heat boiler is the best choice of heat recovery system, the following check list should be observed: (*a*) a licensed engineer in charge of all boilers, (*b*) a complete duplicate water supply system, and (*c*) automatic means for removing the heat source (venting the hot waste gas) using an air-cooled or water-cooled upstream shutoff valve designed to handle 2400 F gases.

The reader should be aware of the differences between the usual boiler installation and a waste heat boiler installation. In the former, the greater part of the heat transfer is effected by radiation from the flame or fuel bed. In the latter, all the heat transfer is effected by convection and by radiation from clear gases. Therefore, in waste heat boilers, not only is the heat transfer coefficient lower but also the average temperature difference is considerably less, requiring a larger amount of heating surface for a required output. Additional "pumping power" (induced draft fan) is recommended to pull the flue gases through the additional resistance of a waste heat boiler in the exhaust system, as shown in figure 5.16.

For the extraction of waste heat, the single-pass horizontal fire-tube boiler having a very large number of small tubes is now widely used in the United States. For a given available draft, a higher heat transfer rate can be obtained in a fire-tube boiler than in the water-tube type. In fire-tube boilers, there is less danger of a gas explosion if the waste gases contain unconsumed combustible, and less chance of air infiltration. Scale must be minimized by thorough water treatment before and during each use cycle. Water-tube boilers and fire-tube boilers have been found to have about the same efficiency of heat recovery when the gases are above 1800 F (982 C), but at lower temperatures the water-tube type falls behind, partially because of air infiltration. Despite its name, do not waste 'waste heat'!

Flue gas temperatures of waste heat boilers are only 100 to 150 F lower than from regenerative systems; thus, fuel savings may be marginal. Waste heat boilers have proved effective with stainless-steel annealing catenary furnaces. They have adjacent steam requirements all year for cleaning their product after annealing. Their firing rates, flue gas temperatures, and heat stored in refractory are moderate, so water problem shutdowns are fewer.

### 5.11.3. Saving Fuel by Preheating Combustion Air

To determine how much fuel can be saved by preheating air, read %available heat from figure 5.1 with and without preheated air, and use equation 5.7. In rare cases, fuel also can be preheated, but not if the fuel contains hydrocarbons that may crack when heated and deposit on the heat transfer surfaces. Preheating fuel usually is not justifiable if the fuel has a heating value greater than about 350 Btu/ft$^3$ (13 MJ/m$^3$).

$$\%\text{Fuel saved} = 100\% \times [1 - (\%Av\ Ht_c/\%AvHt_h)] \qquad (5.7)$$

where subscripts $c$ and $h$ are for cold air and hot air, respectively.

**Example 5.2**: A furnace is needed to melt 25 000 pounds of aluminum per hour from cold to 1450 F, which requires 505 Btu/pound, or 25 000 × 505 = 12 625 000 Btu/hr heat to the load. It is estimated that the wall, storage, opening, and water-cooling losses are estimated as 1 000 000 Btu/hr. Thus, the "heat need" or "required available heat" = 12 625 000 + 1 000 000 = 13 625 000 Btu/hr.

To heat the aluminum to 1450 F, it is estimated that the furnace temperature will be 2200 F and the flue gas exit velocity about 23 fps. Therefore, from Figure 5.3, the flue gas exit temperature will be about 2200 F + 200°F = 2400 F. From figure 5.1, at 2400 F, read 30% available heat with 60 F air and 10% excess air, or read 48% available heat with 800 F preheated air and 10% excess air. Using equation 5.7, the %fuel saved with 800 F air instead of 60 F air will be $100\% \times [1 - (\%Av\ Ht_c/\%AvHt_h)] = 100\% \times [1 - (30/48)] = 100\% \times [1 - 0.625] = 37.5\%$.

If it is then decided to add an air preheater to accomplish heat recovery, the *required gross heat input* to the furnace will equal required available heat or heat need ÷ (%available heat/100) = 13 625 000 ÷ (48%/10) = 28 400 000 gross Btu/hr. A security factor* of at least 25% should be used; therefore, the design input should be (28.5 kk Btu/hr) (1.25) = 35.6 gross kk Btu/hr.

Added benefits from preheating combustion air are faster burning, resulting in a hotter burner wall, and lower flue gas exit temperature. The desired prompt heat release is difficult to evaluate. An interesting facet of the available heat charts (figs. 5.1 and 5.2) is that the curves' $x$-intercepts (where available heat is zero) are 'theoretical adiabatic flame temperatures,' or 'hot-mix temperatures' mentioned earlier. For the

*See glossary.

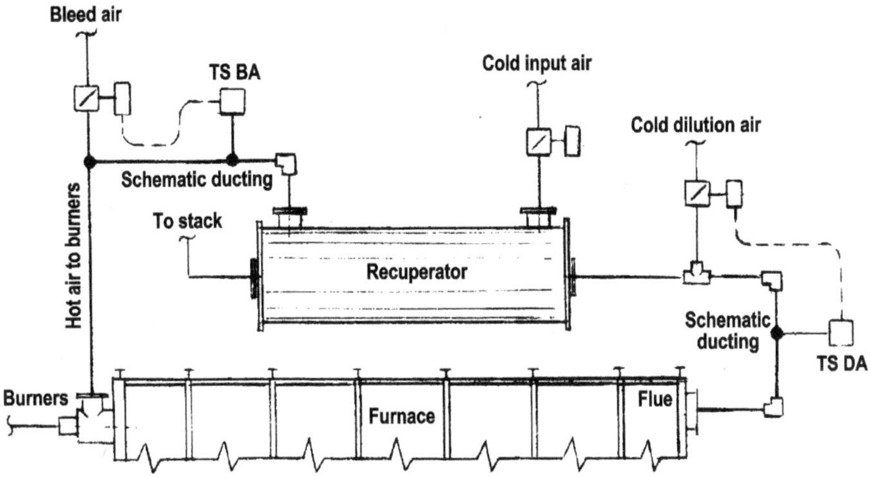


**Fig. 5.17.** Schematic piping for dilution air for a recuperator. TSBA = temperature sensor for control of bleed-off air, TSDA = temperature sensor for control of dilution air. Both elbows at the right function as in fig. 5.21 to prevent radiation between recuperator and the furnace load from damaging either. Both elbows also assure good mixing between the furnace poc and dilution air, and both elbows prevent the TSDA from being "fooled" by "seeing" hotter or colder surfaces in the furnace or recuperator. If a velocity thermocouple at or near the same location, or a wall-mounted sensor, is found to be reading, say, 50° low, the setpoint should be adjusted 50° lower to protect the recuperator.

previous example, the hot-mix temperature is 3300 F with 60 F air and 10% excess air; or 3600 F with 800 F preheated air and 10% excess air.

***5.11.3.1. Recuperators*** Recuperators are heat exchangers that use the energy in hot waste flue gases to preheat combustion air. The poc gases and air are in adjacent passageways separated by a conducting wall. Heat flows steadily through the wall from the heat source (hot flue gas) to the heat receiver (cold combustion air). Recuperators are available in as many configurations as there are heat exchangers. Common forms are double pipe (pipe in a pipe), shell and tube, and plate types. All may use counterflow, parallel (co-current) flow, and/or cross flow. (See figs. 5.18, 5.19 and 5.20.)

Counterflow types deliver the highest air preheat temperature, but parallel flow types protect the recuperator walls from overheating. Therefore, the hot flue gases are often fed first to a parallel flow section and then to a counterflow section to benefit from both advantages.

If the heat transfer coefficients, $h$, were constant, the curves in figure 5.18 would be logarithmic. As was shown in chapter 2, however, there is considerable variation in the value of the coefficient, depending on the temperature of gas and air, density and velocity of gas and air, after-burning, radiation, leakage, and the character of the heat exchanging surface. In view of these many variables, the necessity for approximation is no drawback.

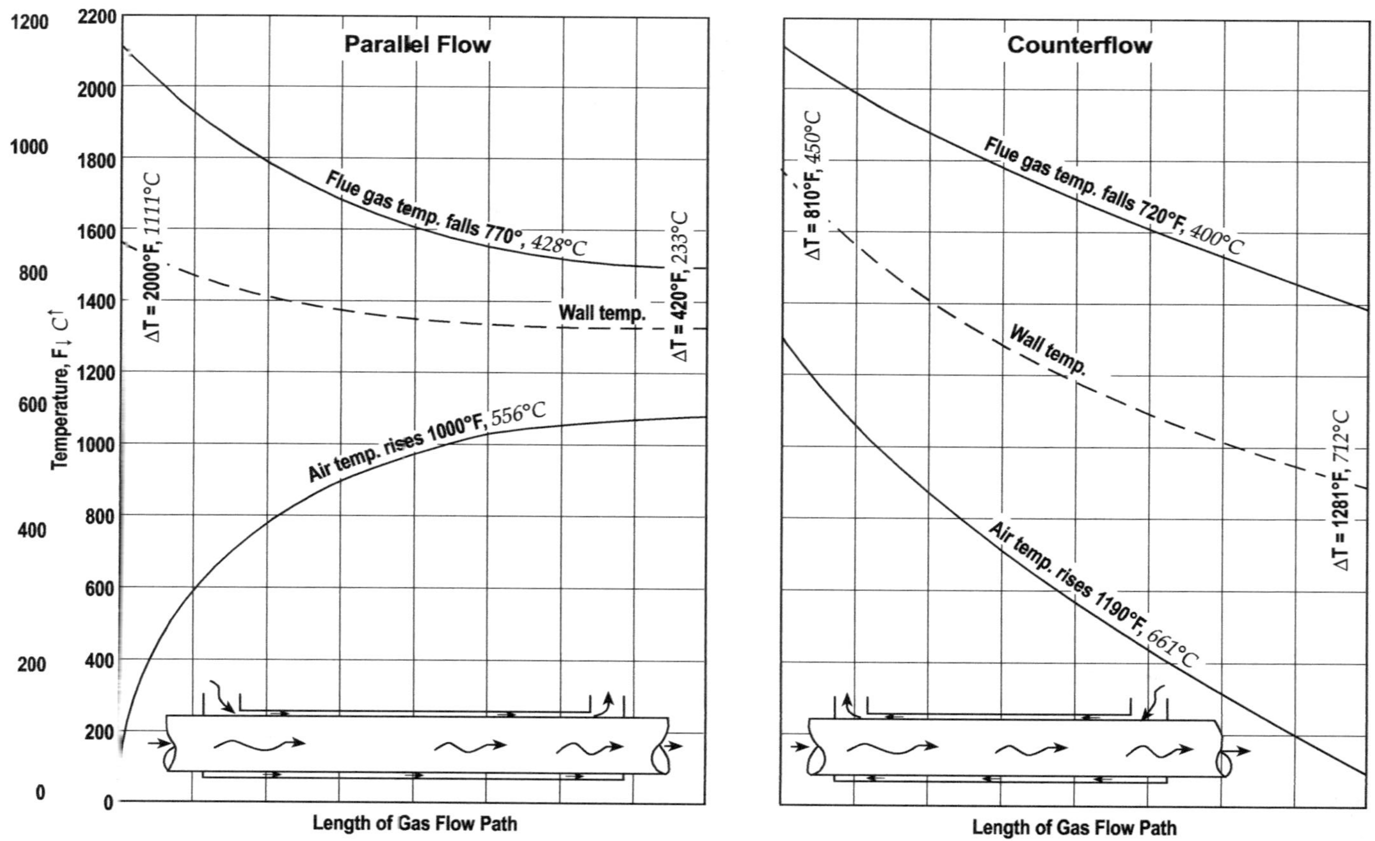


**Fig. 5.18.** Comparison of temperature patterns in parallel flow and counterflow recuperators—applicable to types other than the double pipe shown. Calculate heat transfer using LMTD, pp. 127–128 of *reference 51*. There may be a burnout danger at the flue gas entry with counterflow.

A *heat balance* for a recuperator should be: heat input from flue gas, $q$ = heat output in preheated air or

$$Wt_g(c_p)(T_{g1} - T_{g2}) = Wt_a(c_p)(T_{a2} - T_{a1}) \tag{5.8}$$

where

$Wt$ = weight flow rate, in lb/hr or kg/hr,
$c_p$ = specific heat at constant pressure, in Btu/lb°F or cal/g°C,
$T$ = temperature, in Fahrenheit or Celsius,
$_g$ = flue gas, $_a$ = air to be preheated,
$_1$ = incoming, $_2$ = outflowing.

This can be equated to the total rate of heat transfer, $q$, in the recuperator:

$$q = U \times A \times LMTD \tag{5.9}$$

where

$q$ = heat flow rate in Btu/hr or Kcal/hr,
$A$ = heat transfer surface area = (total length) (π) (OD + ID)/2
$U$ = overall coefficient of heat transfer = $1/h_g + x/k + 1/h_a$ as described in chapter 2. (See $h$ values in figure 5.19.)
(LMTD = log mean temperature difference. See glossary and pp. 126–128 of *reference 51.*)

In a cross-flow recuperator, $T_{g2}$ is the temperature of that portion of the flue gases leaving the tubes in the center of the tube bank, and $T_{a2}$ is the temperature of the preheated air beyond the middle of the last tube.

The heat exchanging surface needed with a cross-flow recuperator is greater than that required by a counterflow recuperator of equal heat transfer. When applied to existing recuperators, the preceding equations 5.8 and 5.9 are used to find values of the overall heat transfer coefficient, $U$. For new recuperators, the equations are used to determine the needed heating surface, if there are no gas, air, or heat leaks.

On the air side of recuperators, heat transfer from the separating wall to the air takes place almost entirely by convection. The radiation absorbing capacity of the small amount of water vapor in the air is practically zero. The coefficient of heat transfer by convection increases rapidly with the mass velocity (i.e., the product of Velocity × Density) of the air or gases.

Figure 5.19 gives convection heat transfer coefficients for flow along flat surfaces, through the inside of tubes, and across tube banks. For flat surfaces, the air coefficient can be approximated by the following equation.

$$h_a = 1.0 + 2.71\,\rho\,\upsilon \tag{5.10}$$

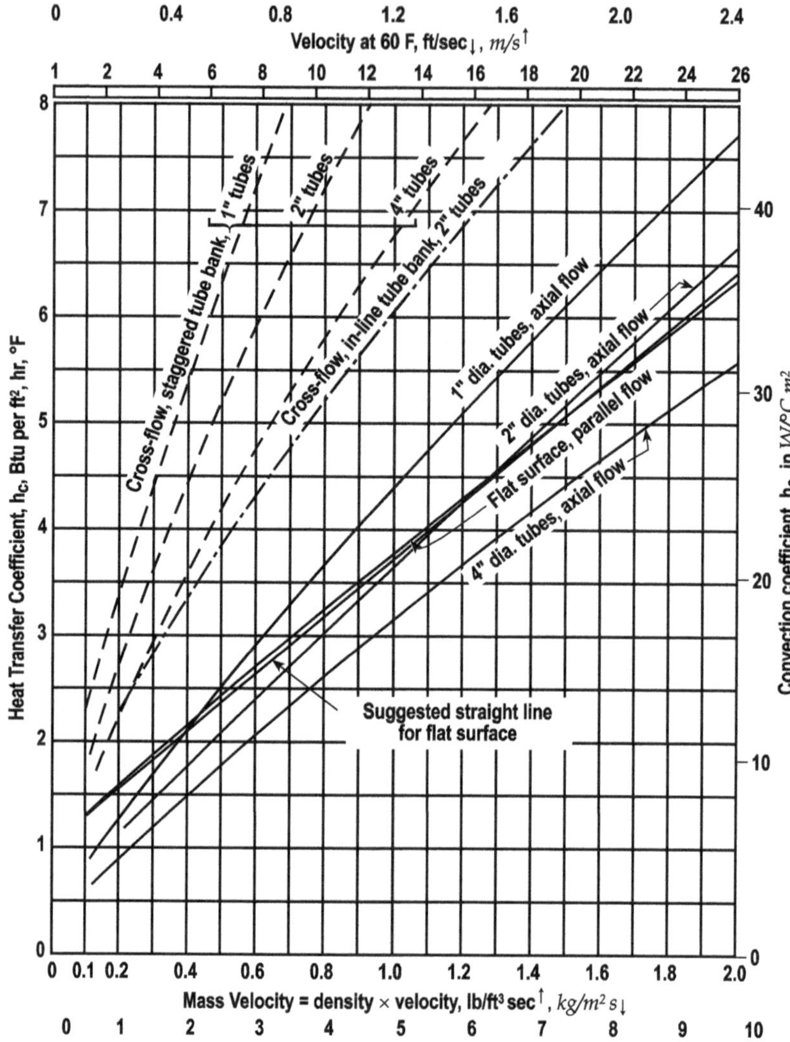


**Fig. 5.19.** Convection heat transfer coefficients for gases.

where

$h_a =$ convection film heat transfer coefficient flat surface to air, Btu/ft²hr°F;
$\rho =$ density of air in pounds per cubic foot; and
$v =$ velocity, feet per second.

Figure 5.19 also provides convection heat transfer coefficients from tube walls to air. The convection heat transfer coefficient in a 1-in. tube is approximately 1.4 times

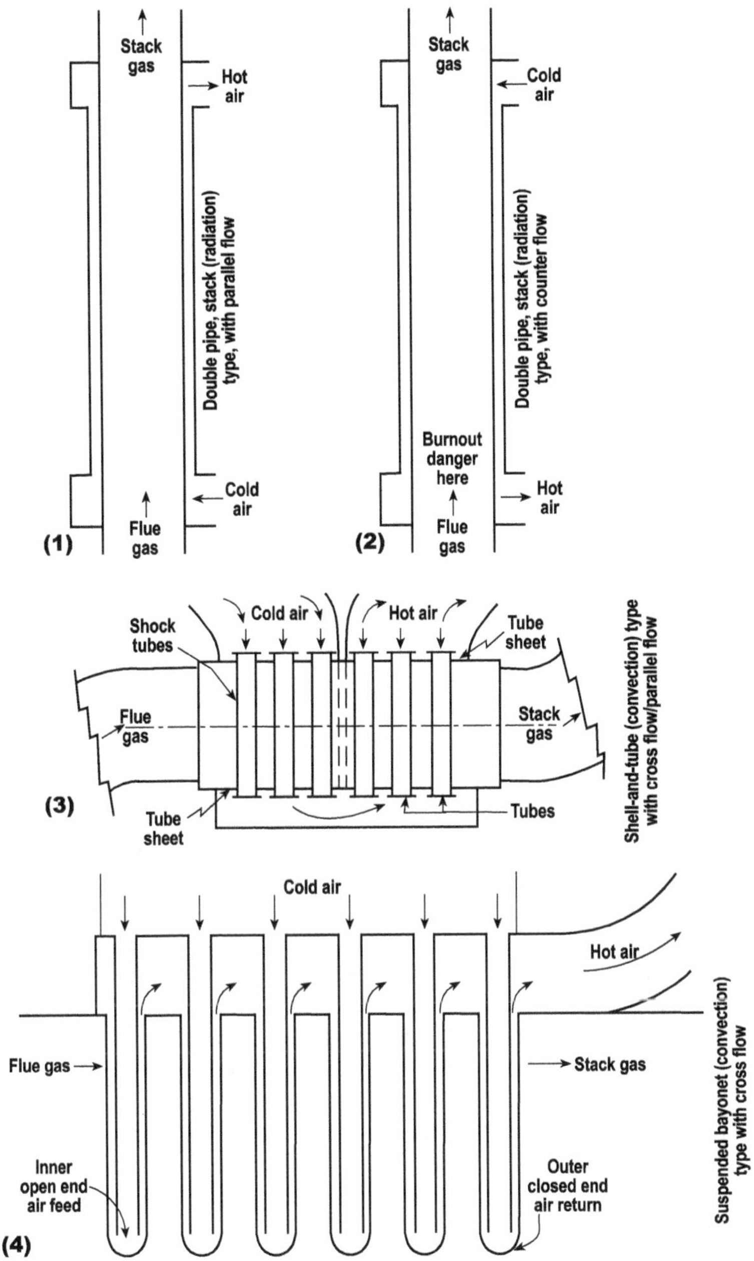


**Fig. 5.20.** Recuperator flow types, shown schematically. All but types 1 and 2 have many, many tubes. Cross-flow recuperators (types 3, 4) often have the configuration of a square shell-and-tube heat exchanger. For the same heat exchanging area, temperature levels, and type, the average heat flux rates (see glossary) of parallel flow, cross-flow, and counterflow are about proportional to 1.00 to 1.40 to 1.55, respectively.

as great as it is in a 4-in. tube, with the same velocity. The same relations hold for convective heat transfer from the poc to the separating wall of the recuperator.

Heat also is transferred by gas radiation, which may outweigh the effect of convection, especially in a straight duct feeding poc to a recuperator, which provides a large radiating beam length. The coefficient of heat transfer by gas radiation is independent of the velocity of flow, but varies with the temperature of the gases, their composition, and the thickness of the gas layer. Values from figures 2.13 and 2.14 are averages for the poc, without excess air, of high-calorific fuels such as natural gas, coke oven gas, and oils or tar. The values must be multiplied by the radiation absorptivity* of the receiving surface. For typical gas layer thicknesses in recuperators and regenerators, an increase (or decrease) of 1% in the $CO_2$ content from 12% raises (or lowers) the gas radiation about 1% whereas an increase (or decrease) of 1% in the $H_2O$ content raises (or lowers) the gas radiation about 1.75%.

Example 5.1 illustrates calculation of the overall coefficient of heat transfer. *Convection/conduction* heat transfer from hot flue gases through a separating wall, with conductivity k and thickness x, to cold air on the other side of that wall is like three resistances in series, totaling to $R_t$. From that, equation 5.11 solves for $U$, the overall (total) heat transfer coefficient.

$$U = 1/R_t = 1/(R_g + R_w + R_a) = 1/\left[1/h_g + 1/(x/k) + 1/h_a\right]. \quad (5.11)$$

The $h_g$ involves convection *and* gas radiation to or from a surface, and it is like two resistances in parallel, thus $h_g = h_c + h_r$. Similar to Ohm's Law, ($I = E/R_t$), heat flux, $q = Q/A = \Delta T/R_t$, or $Q = UA\Delta T$, which is the basic equation of heat transfer. Example 5.1 illustrates the method for calculating $U$, the overall coefficient of heat transfer.

**Example 5.3:** Flue gases at an average 1600 F flow in a 2" wide passage along one side of a flat recuperator wall at a velocity of 20 fps while air at an average of 300 F flows along the other side of the same wall at a velocity of 30 fps. Calculate the resulting overall heat transfer coefficient.

If the wall is metal, its resistance, Rw, is probably so small that it can be neglected. Use figure 5.19 to determine the air-side convection coefficient, $h_a$. Calculate the air-mass velocity (for the bottom scale), getting air density at 300 F from any standard tables, such as p. 247 of *reference 52*, as 0.0523 lb/ft$^3$; then $\rho V = 0.0523 \times 300$ fps $= 15.7$, and on the flat surface, parallel flow curve, read $h_a = 5.2$ Btu/ft$^2$hr°F. *(An alternate way is to figure that the air at 300 F and 30 fps has the same mass velocity as 60 F air moving with a speed of 30 × [(60 + 460)/(300 + 460)] or 20.5 fps. Then use the top scale of fig. 5.19 to drop down to the same flat surface parallel flow curve and read $h_a = 5.2$).*

Use figure 5.19 again to determine the $h_{gc}$ of the flue gases. The flue gases at 1600 F and 20 fps have a mass velocity the same as gases at 60 F moving at 20 × [(460 + 60)/(460 + 1600)] or 5.05 ft/sec. From figure 5.19, the corresponding

*The value of *absorptivity* is usually very close to the same value as the *emissivity* of a material. (See both terms in the glossary.)

convection coefficient is 2.12 Btu/ft$^2$hr°F. The gas radiation coefficient. $h_{gr}$, for a 2-inch thickness of gas layer at 1600 F, from figures 2.13 is 3.0, which must be multiplied by an absorption coefficient of 91% for the rough metal wall, giving 2.73 Btu/ft$^2$hr°F. Then,

$h_g = h_c + h_r. = 2.12 + 2.73 = 4.85$ Btu/ft$^2$hr°F, and

$U = 1/\left[1/h_g + 1/(x/k) + 1/h_a\right] = 1/[1/4.85 + 0 + 1/5.2] = 2.50$ Btu/ft$^2$hr°F.

On the air side, the heat transfer coefficient grows with the air flow velocity. It is therefore desirable to pass the air through at high velocities, which also helps to reduce the size of the recuperator. This becomes impractical when the increased power cost for moving the air against the increased back pressure exceeds the reduction in cost of system.

On the flue gas side, however, this rule does not apply. Although an increase in waste gas velocity increases the convective heat transfer, it requires that the gas passages be reduced in cross-sectional area (for a given quantity of gases), and thereby decreases gas radiation from the $CO_2$ and $H_2O$ vapor in the poc. The net result may actually decrease the total heat transfer on the gas side of a recuperator.

From a heat transfer standpoint, the best recuperator design is usually one in which the flue gas is pulled though relatively large passages while the air is pushed through smaller passages at high velocity. This also assures that any leaks (and there will eventually be some leaks) will not dilute the combustion air and upset control of the combustion process.

If leaks should happen to occur from air side to gas side, they will (1) reduce the quantity of preheated air (lowering overall combustion efficiency) and (2) cool the flue gases, lowering the $\Delta T$ that is the driving force for heat flow from flue gases to combustion air.

*Recuperator concerns* stem mostly from fouling of the heat transfer surfaces, overheating damage, and leaks. Flame, pic, direct furnace radiation, or condensation should never be allowed to enter any heat recovery equipment. *The air flow through any recuperator must never drop below 10%* of its maximum design flow until the furnace has cooled several hours after the time when none of its refractory showed even a dull red color.

Ducting between a recuperator and a furnace must follow the dictates of figures 5.21 and 5.22. The top views of figure 5.21 are concerned about damage to the recuperator; the lower two views are concerned about damage to the furnace load. The solutions for both are the same, and apply to most types of recuperators.

*Thermal expansion* is the bane of a recuperators' existence. With conventional shell-and-tube heat exchanger configuration (two tube sheets), tube expansion tears a tube sheet; therefore, a single tube sheet is sometimes used with suspended open-end hot gas feed tubes inside concentric closed-end suspended outside tubes. The thermal expansion problem is exacerbated by the much higher heat transfer to the front row of tubes (shock tubes) because of (*a*) highest convection $\Delta T$ from the hottest (entering) flue gases, (*b*) gas radiation from the long 'beam' of triatomic gases in the duct

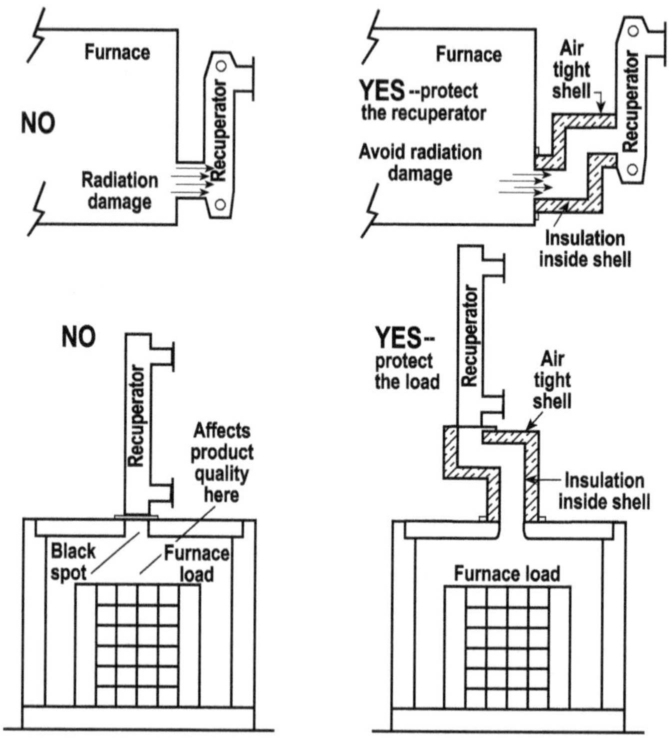


**Fig. 5.21.** Correct recuperator installation prolongs recuperator life and avoids temperature nonuniformity in the heated loads. An air-tight connector should be used between the furnace and the recuperator, with elbows and with inside insulation throughout its length.

approach, and (*c*) 'solid' radiation from the hot walls of the approach duct. Never locate a recuperator or damper where it can receive radiation direct from the furnace.

Recuperator damage happens with changing temperatures, especially when the furnace goes offline and then back online. Tube-sheet breakage and tube buckling result from heat transfer surfaces changing length because of changing temperatures. This problem can be reduced by use of expansion bellows or packing glands on each tube, if temperatures permit. If the bellows or expansion joints become work hardened, however, the tube sheet may still be torn.

Direct furnace radiation (direct lines of sight from hot furnace interior surfaces into a recuperator) often causes overheating damage, usually thermal stress damage, within recuperators. The *top left view* of figure 5.21 illustrates this, and the *top right view* shows a solution. Metalpipes and ducts conveying hot gases *always must be insulated on the inside to protect the air-tight metal pipe or duct from heat damage and corrosion.*

Anything that affects the exhaust loop will result in higher than desired furnce pressure, tending to force final zone flames to exit through the discharge, and/or it may affect mixing or air/fuel ratio at the burners. Damaged or missing recuperator

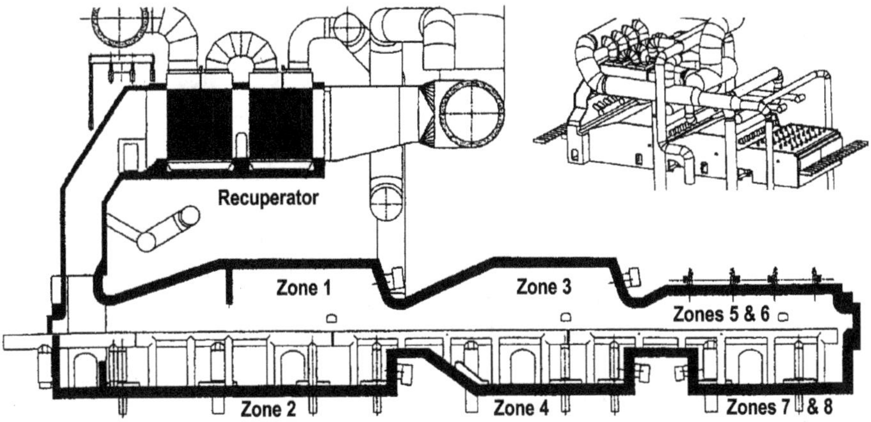


**Fig. 5.22.** Eight-zone reheat furnace, side sectional view with an aerial perspective view inset at *top right.* This furnace has longitudinal firing in all but zones 5 and 6, which are roof fired. Billets or slabs move from left to right, and poc move from right to left. An unfired preheat zone is *left* of zones 1 and 2.

tubes may harm operation in two ways: (1) air leaks from the cold air side to exhaust side may load up the exhaust fans with cold air or (2) air pressure will drop after the recuperator during high firing, thereby causing a deficiency of air and incorrect furnace atmosphere.

Bottom fluing is preferred, that is, from furnace bottom into a recuperator, (*a*) to avoid hot furnace gases from fluing through the recuperator after the air has been shut off (which could overheat the recuperator when it has no air cooling) and (b) to give better poc gas circulation through the furnace loads, avoiding accelerating up-channeling of hot gases.

Recuperators are usually designed with very low pressure drop on the flue gas side. In a shell-and-tube recuperator, the flue gas is generally on the shell side, with the air in the tubes, requiring more $\Delta P$. In a vertical pipe-in-pipe recuperator such as a "stack" or "radiation" recuperator, the flue gas goes up the middle pipe (*a*) to take advantage of the additional stack or natural convection draft, (*b*) to allow a wider gas

> A recuperator has a 10"wc pressure drop on the air side (2.5 kPa drop) at design capacity. By the square root law, from Bernoulli's equation, at 10% capacity it will have only a 0.1"wc (0.025 kPa) pressure drop. Below that, much of the heat transfer surface will "feel" no cooling because of poor air distribution with the low flow rate. For good recuperator life, (1) waste gas temperature should not exceed 1600 F (870 C), and the high-limit sensor must not "see" cold recuperator tubes, (2) flue products must never contain reducing (unburned) gases, and (3) air flow must never drop below 10% of design flow.

radiating beam for the flue gases, and (*c*) to avoid the high surface-to-sectional area ratio of the annulus. The radiation recuperator can act as the stack for the furnace.

Recuperators usually have more pressure drop on the air side. Forced draft is preferred because of the higher cost of handling hot air or gases with induced draft fans or blowers for hot gas or hot air. In addition, forced draft keeps the furnace under a positive pressure, causing any leaks to be outward rather than inward on the furnace, piping, and recuperators.

Any attempt to increase a recuperator's effectiveness or capacity without increasing its size will necessitate a higher blower pressure rating as well as a higher blower capacity rating because pressure drops through recuperators and everything else in the system increases as the square of the flow throughput. This markedly increases the first cost of the blowers.

After careful heat exchanger calculations are completed, the authors advise specifying a size 25% greater than calculated to cover loss of effectiveness with aging, due to fouling of surfaces and leaks, and because needs invariably arise for temporary or permanent increases in throughput. This foresight will diminish future drops in fuel efficiency; thus, the increased capital investment will be rewarded with lower operating costs.

The term "heat exchanger effectiveness" called 'pickup' as applied to recuperators, means the actual air temperature rise expressed as a percent of the maximum possible air temperature rise. Commercial recuperators are usually designed for a 60% to 75% range. Higher pickup ratios result in larger and more expensive recuperators. Regenerators (discussed in sec. 5.11.3.) have higher heat exchanger effectiveness than recuperators, and they avoid some of the difficulties inherent in recuperators.

***Dilution air*** is sometimes purposely added to the furnace's waste gas stream to protect the materials of heat exchange and air handling equipment from overheating by exposure to excessive poc temperature. The design of dilution air systems would seem simple enough, but unfortunately many furnace dilution air systems are undersized by 30 to 50%, perhaps because (1) a low bidder gets the contract, (2) waste gas temperature and/or firing rates were underestimted, and/or (3) a faulty waste gas temperature measurement for control.

1. The low-bidder problem results from designing all parts of the furnace to just do the theoretical heating required at a most efficient time where the firing rate will be minimal with a minimum of excess air and no infiltrated air. Under those conditions, a minimum amount of dilution air will be required. The sizing of the dilution air system should be based on the maximum firing rate of the *whole furnace* to be able to dilute all the possible combustion gases. Some assume that all burners will probably never fire at maximum rate simultaneously, *but they will when coming off a mill delay*. Operating with all burners at 100% is a life-threatening situation for a recuperator without adequate dilution air!
2. Designers tend to assume perfect mixing of the dilution air and flue gases without regard for real-world mixing situations. In addition, some designers fail to realize that with a single nozzle, the energy available at high flow due to the acceleration effect will decrease as the square of the flow. In actuality,

with a turndown to $\frac{1}{5}$ of maximum flow, only $\frac{1}{25}$ of the maximum energy is available in the dilution air for mixing the two fluids. In a properly designed system, the maximum energy (pressure drop) must include mixing energy for both fluid streams in addition to energy to overcome flow resistances in the system.

Coauthor Shannon has redesigned numerous systems with an experience factor of maximum dilution air velocity of *160 fps* entering the flue at elbows. This has produced good resultant mixing even at low flow rates. Failure to use this much velocity (price buying) neglects the need for mixing energy at turndown conditions. Engineers writing furnace specifications should make certain that the 160 fps mixing velocity is spelled out, and that all bidders conform to it.

3. Faulty waste gas temperature measurement for control. If the recuperator tubes can 'see' (i.e., interchange radiation with) the control T-sensor, the control temperature reading may be low by 100°F to 250°F (55°C to 140°C).

A typical control thermocouple may read 100°F below a high-velocity thermocouple measurement. The ideal system has two elbows as shown in figure 5.21. When it is not practical to install a second elbow, a hemispherical depression in the flue wall (8" in diameter and 4" deep) can hide the thermocouple hot junction from the recuperator tubes and will provide a reasonable measurement. Check it with a high-velocity T/sensor.

A dilution air system designed for fuel-oil firing requires about 5% less dilution air than for natural gas firing; therefore, a natural gas system design will perform satisfactorily while burning fuel oil.

**Example 5.4: Sample Capacity and Head Calculation for a Dilution Air Fan**
*Given*: Cool the waste gas of a 180 kk Btu/hr gross input with natural gas and 20% excess air from 2000 F to 1600 F. This means that 180 000 000 Btu/hr / 1000 Btu/cf = 180 000 cf/hr of natural gas is being fired. That would require 1 800 000 cf air/hr for stroichiometric firing, or

$$1.2 \times 1\,800\,000 \text{ cf air/hr} = 2\,160\,000 \text{ cf air/hr with the chosen 20\% excess air.}$$

$$1CH_4 + 2O_2 + 8N_2 \rightarrow 1CO_2 + 2H_2O + 0.4O_2 + 8N_2 \text{ with 0\% excess air.}$$

$$1CH_4 + 2.4O_2 + 9.6N_2 \rightarrow 1CO_2 + 2H_2O + 0.4O_2 + 9.6N_2 \text{ with 10\% excess air.}$$

From table 3.7a of *reference 51,* the heat in the flue gas at 2000 F will be:

$$\begin{aligned}
1CO_2 \times 61.9 \text{ Btu/cf} &= 61.9 \text{ Btu/cfh fuel}\\
2H_2O \times 48.0 &= 96.0\\
0.4O_2 \times 40.8 &= 16.3\\
9.6N_2 \times 38.8 &= \underline{372.5}\\
&= 546.7 \text{ Btu/cfh fuel,}\\
&\quad \text{which} \times 180{,}000 \text{ cfh fuel} - 98\,400\,000 \text{ Btu/hr.}
\end{aligned}$$

Similarly, the heat in the flue gas at 1600 F will be:

$(1 \times 47.4$ Btu/cf$) + (2 \times 36.9) + (0.4 \times 31.8) + (9.6 \times 30.2) = 423.8$ Btu/cfh fuel, which $\times$ 180 000 cfh fuel $=$ 76 300 000 Btu/hr.

The quantity of heat that must be absorbed in heating the dilution air = 99 400 000 − 76 300 000 = 22 100 000 Btu/hr.

From Table A.2a of *reference 51*, raising the dilution air temperature from 100 F to 1600 F requires 30.4 − 0.74 = 29.66 Btu/ft$^3$ of air. Therefore, the ft$^3$ of dilution air needed = 22 100 000/29.66 = 745 000 ft$^3$/hr or 745 000/3600 = 207 scfs *minimum required dilution air fan capacity*. It should be increased for inlet temperatures above 60 F (above 16 C).

For proper mixing, the experience factor mentioned previously says that the dilution air velocity at maximum firing rate should be no less than 160 fps. The *pressure head required* with air at 100 F (from equation 5/6, p. 132, *reference 51*, where $G =$ air density relative to stp air $= 1 \times (60 + 460)/(100 + 460) = 0.929$) is $\Delta P$, osi $= 0.000132 \times G \times (V_{fps})^2 = 0.000132(0.929)(160)^2 = 3.14$ osi, or 3.14 osi $\times$ 1.732 in. wc/osi = 5.45 in. wc.

The *nozzle size* to pass the calculated 207 scfs of air into the waste gas for mixing (with a 1.2 safety factor) and corrected for temperature $= [(100+460)/(60+460)] \times$ $1.2 \times 248$ scfs/160 fps = 2.00 ft$^2$ which would be a 20" OD schedule 20 round pipe nozzle, or a 17" inside square nozzle.

From the pipe velocity guidelines on pages 175 to 176 of *reference 51*, the air piping should have an stp velocity of 40 ft/sec. Therefore the *"cold" air feed pipe* from the blower to the air preheater should have an inside pipe area of (248 cfs/40 fps) $\times$ (460 + 100)/(460 + 60) = 6.68 ft$^2$. The *hot air feed pipe* from the air preheater to the hot air burner manifold should have an inside pipe area of (248 cfs/40 fps) $\times$ (460 + 1600)/(460 + 60) = 24.6 ft$^2$. For square ducts, the cold air feed duct should be the square root of 6.68 ft$^2$ = 2.6 ft $\times$ 2.6 ft, and the hot air feed duct should be the square root of 24.6 ft$^2$ = 4.6 ft $\times$ 4.6 ft.

***Hot air bleed*** is an alternate way to protect a recuperator from heat damage by hot flue gas when burners are at low fire and air flow through the recuperator is too low. (High air flow through a recuperator is its only coolant to prevent burnout.) Both hot air bleed and dilution air protect a recuperator from burnout, but also waste energy. Care must be used in design and piping of the air/fuel ratio control system so that it does not count bleed air as combustion air. The primary control sensor actuating a bleed (dump) valve in the hot air exit line from a recuperator should be a high-velocity (aspirated) sensor.

***5.11.3.2. Regenerators.*** The first major use of regenerators in industrial heating was by Sir William Siemens in England in the 1860s. His purpose (rather than to save fuel) was to preheat air to achieve higher flame temperature from the only gaseous fuel then available (made from coal). His regenerative air preheater used a refractory checkerwork. Figure 5.23 shows the principle of a type of regenerative melting furnace.

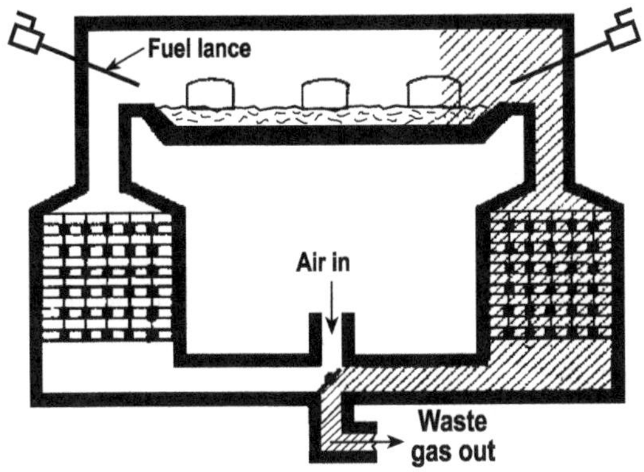


**Fig. 5.23.** Refractory checkerwork regenerator, widely used with steel open-hearth furnaces, and still used with large glass-tank melting furnaces. Positions of the bottom valves and fuel lance valves are reversed about every 20 min.

The same principle applies to blast furnace stoves and to the multiple-tower heat recovery units positioned around the periphery of vertical cylindrical incinerators for waste gases or liquids. For furnaces with lower temperature waste gases, such as boilers or steam generators, a Ljungstrom all-metal recuperator, rotating on a vertical shaft, is used.

Horizontal flows in regenerators are usually unstable and not self-regulating, so vertical stacking in towers is usually the configuration of choice to avoid *"channeling,"* the same problem as with bottom firing and top flueing in ceramic kilns and in heat treating furnaces filled with stacked loads. Here is how channeling occurs: If one piece should happen to get hotter than surrounding pieces, it will create more natural convection (stack effect), causing a faster flowing up-channel for adjacent gases. That pulls even more gases to that vertical channel. Meanwhile, flow is reduced in other

> Particulates are a pain in many heat recovery devices, but especially in checkerworks and other packed tower type recovery equipment. Dust deposits cause difficulties in furnace operation by choking flow passages, necessitating higher pressure drops to maintain flows of air and poc. The necessary higher pressures can cause leaks of air, poc, and heat through walls and by dampers.
>
> Particulate accumulations can cause a negative pressure, resulting in cold air being sucked in and diluting the preheated air.
>
> On the flue side, the dust deposits create high pressures, causing hot flue products to escape before they can transmit their heat content to cold air.
>
> Over time, these pressure difficulties become so great that the furnace productivity decreases enough to warrant an end to the "campaign", initiating a furnace rebuild.

vertical paths for gas flow; therefore, the load pieces in those areas are heated less, leading to "snowballing," a compounding acceleration of differences in temperatures and flows.

Modern compact regenerators are arranged in pairs, close coupled to burners, which alternately serve as burners or flues. They use small refractory nuggets or balls (with high surface-to-weight ratio) that have short heat-up and cool-down cycle times, using the benefit of a "pebble heater" without the problems of a moving pebble heater. Figure 5.24 is a schematic diagram showing how they are applied to batch furnaces, such as steel-forging and aluminum-melting furnaces. Regenerative burners also have been used very successfully for ladle dryout/preheat stations.

Figure 5.25 compares the heat recovery effectiveness of typical recuperators with a modern compact regenerator. With the latter, thermal efficiencies have reached 75% to 85%, with air preheat temperatures within 600 to 900 F (330–500 C) of furnace temperature. Exhaust gas temperatures overall average 600 to 700 F (315–371 C) regardless of furnace temperature. Figure 5.26 shows integral regenerator-burners in use on a batch-type furnace, such as used for melting aluminum or glass.

*Continuous Steel Reheat Furnaces* can benefit from the use of compact regenerative burners as shown in figure 5.27. For this arrangement with cross firing and longitudinal firing (side and end burners), it is important that the end burners have low input or momentum so that their jet streams do not interfere with thorough coverage of the full hearth width by the side burners. The graph in figure 5.27 shows the experienced variation of fuel consumption versus throughput rate for this furnace rated at 89 tph, which has reached input rates as low as 0.94 kk Btu/USton (1.09 GJ/tonne).

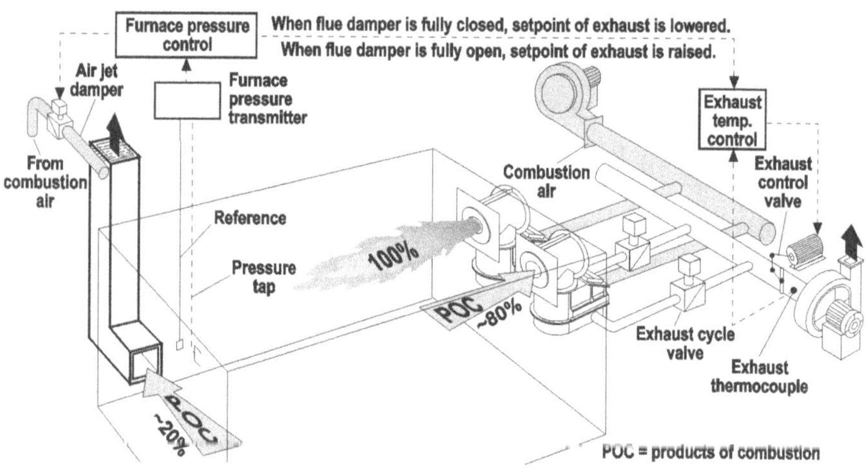


**Fig. 5.24.** Batch furnace with one pair of regenerative burners. Recovery is so good that not all poc need to be sent through the air heater, leaving some to help control furnace pressure. For faster bring-up from cold (when waste gas temperature is low and efficiency high), both burners can be fired simultaneously. After about 20 sec of firing as shown, the system automatically interchanges the left and right burner functions. (See also fig. 5.26.)

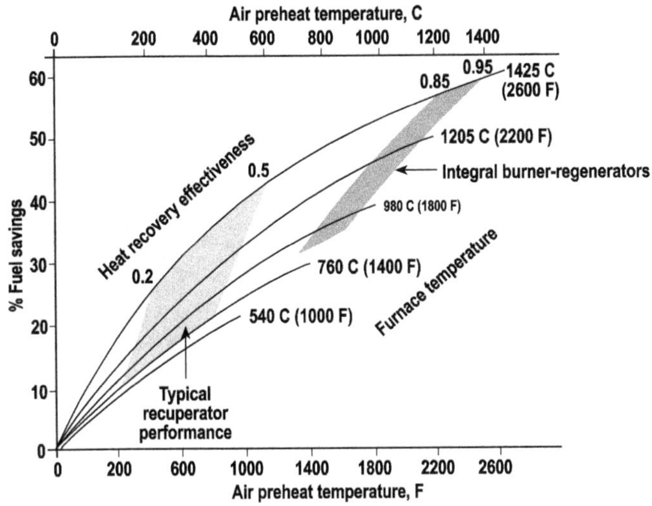


**Fig. 5.25.** Heat transfer effectiveness of a compact integral burner-regenerator compared to a typical recuperator. From *reference 52*.

Preheat zones of steel reheat furnaces were formerly unfired, in line with the "unfired preheat vestibule" philosophy (advocated earlier in this chapter) for recovering heat from the gases exiting the soak and heat zones. However, the regenerative burners are so effective at recovering heat that their final throwaway temperature is just as low with, or lower than, an unfired preheat zone. *And* the furnace now has much additional

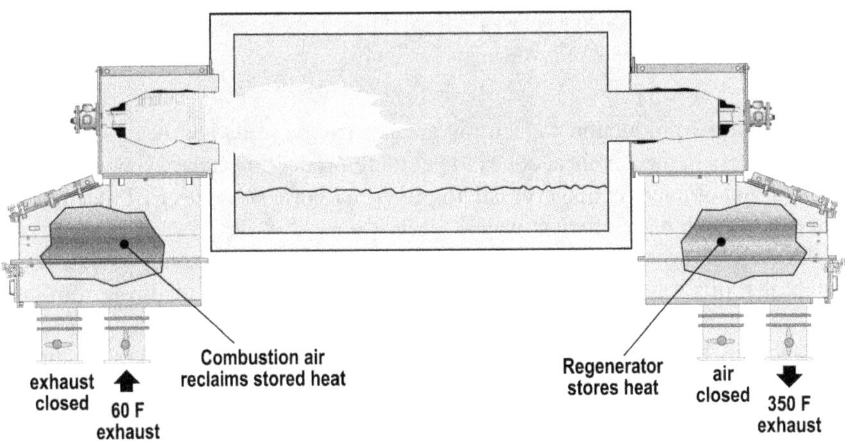


**Fig. 5.26.** Melting furnace with a pair of compact regenertive burners. After about 20 sec of firing as shown, the system automatically switches to firing the left burner and exhausting through the right burner by closing the right air and fuel valves plus left exhaust valve, and (not shown) opening the left air and fuel valves plus right exhaust valve. Then, the regenerator on the right will be storing waste heat, and the burner on the right will be receiving reclaimed stored heat in the form of preheated combustion air.

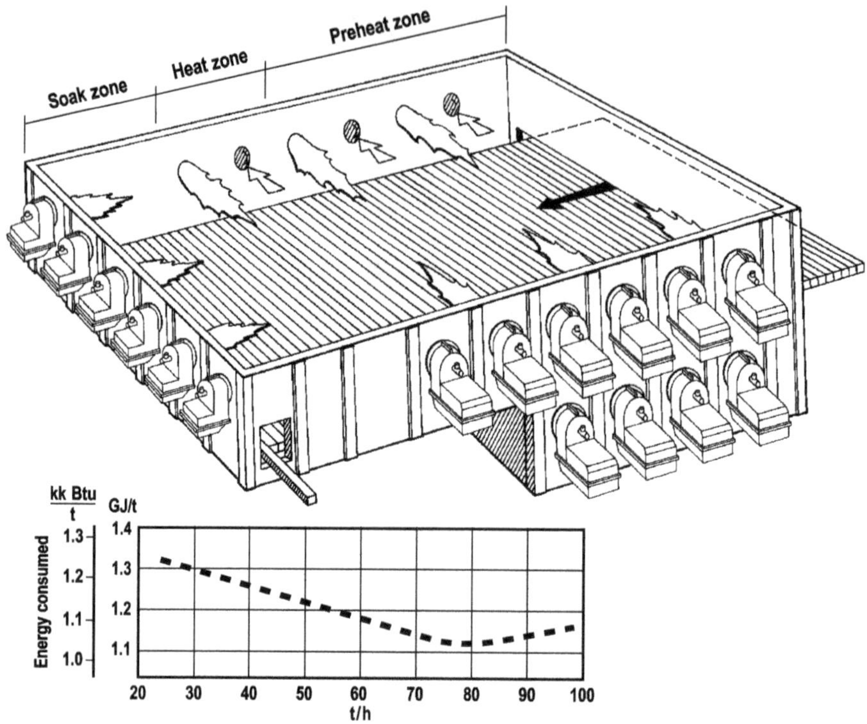


**Fig. 5.27.** Continuous steel reheat furnace with nine pairs of regenerative burners in three top control zones and four pairs in a bottom zone. The sweep of hot poc from side burners can alternately proceed all the way across the furnace width, avoiding the former uneven heating when opposed burners created a hot spot "pileup" of heat in the center when on high fire, and a cool stripe down the middle on low fire.

input, so that its production capacity is greater. (Some mills had been adding roof or side burners in their preheat zones to get more production capacity, while foregoing good fuel efficiency; however, adding oxy-fuel burners or compact regenerative burners is a much more efficient way.)

Older reheat furnaces often had lowered roofs in their preheat zones because it was thought that this was an all-convection zone (no radiation), and the lower roof gave less cross section for gas flow, so velocity would be higher, enhancing convection. This was true, but the convection gain was small compared to the gas radiation loss because of less triatomic gas beam height. The power of gas radiation has only very recently been recognized by furnace engineers. (See the review problem at the end of this chapter.)

To hold low fuel rates with cold air firing or recuperative air firing, a furnace capacity must be moderate and the load entry zone unfired so that the furnace exit gas temperature will be very low. With regenerative firing, on the other hand, this need not be the case because regenerative heating beds perform both functions—air heating as well as final exit gas cooling. With recuperative air heating or with cold air firing, the furnace and loads must lower the exit gas temperature to 1000 F (538 C) or lower to compete with regenerative air heating fuel rates. Charge zone temperatures can vary by more than 500°F (278°C) between regenerative, recuperative, and cold air systems, so the furnace heating capacities can be very different. At least one of the several regenerative burners on the market gives a throwaway gas temperature of about 350 F (177 C) immediately after the regenerative bed, regardless of furnace temperature.

Fuel consumption rates are profoundly different with recuperative and regenerative air preheating. During a delay on a furnace with recuperation, the furnace exit gases may rise to 2000 F (1093 C), then be diluted to 1500 F ± 250°F (816 C ± 140°C) by infiltrated air from many causes, resulting in very low air preheat. Regenerative air preheating depends only on the regenerative bed; thus, as the furnace gas temperature rises, the air preheat rises. The result is that the available heat falls during a delay with a recuperator, but may even rise with a regenerator during a delay.

**Aluminum-melting furnaces** are often fired with regenerative burners (fig. 5.26), but care is necessary to prevent fouling the regenerative beds with carry-over from the melting process such as flux, oxides, and aluminum droplets (an operational mistake).

Flux is used only for drossing off and for cleaning in some aluminum melters. Others use no flux. Some use flux only with dirty scrap.* When drossing off or furnace cleaning, it is safer to operate integral regenerator-burners either on "stop cycle" or in direct-fire mode so that none of the furnace fumes are pulled through the regenerative beds. With flux feed into a sidewell-charged furnace, the flux feed rate must be even, making certain that all pieces are immersed immediately.

Oxides can be a problem with thin aluminum sections melted at too high a rate. In direct-charged melters, charges of thin sections should be charged at the bottom of the furnace, with heavy-section material above. An alternative is to charge thin-section material by submerging it in a molten pool. In any event, never allow any thin shredded material to be charged on top of a molten bath because it will float, burn, waste metal, and create oxides.

Well-charged melters rarely have problems with oxides. Continuous flux fed into sidewall furnaces causes trouble. Use an even feed rate, and make sure that no one uses excessive flux. Good flux immersion practice permits no large clumps (which may float to the surface and vaporize immediately). Excessive amounts of flux must be avoided. Metal can recyclers must take care to feed flux continuously with a shredded used beverage containers (UBC) charge. With a liquid-metal recirculating pump, the vortex at the liquid surface is a place to feed a stream of chopped UBC.

Flying metal droplets may be a problem with charges of thin section, such as extrusion scrap. If a load is piled high before firing up, it is best to operate the burners in nonregenerative mode until a "tunnel" is melted into the charge pile by ablative melting. This prevents molten droplets from 'raining down' and being entrained in the exhaust stream entering a regenerative bed.

*Typical cleaning cycles for direct-charged melters may be 3 to 6 months; for well-charged melters, as often as every 5 to 7 days.

Fig. 5.28. Tilting batch aluminum melting furnace with a pair of integral regenerator-burners for heat recovery. Courtesy of Deguisa S.A.

Fig. 5.29. Sixty-four pairs of regenerative radiant tube burners annealing steel strip in a galvanizing line.

In **radiant tube furnaces**, each radiant tube can be fired from both ends with a pair of smaller regenerative burners. This achieves longer tube life by leveling the average temperature profile along the tube length. This same principle can be applied to pot or crucible furnaces by firing tangentially around the pot alternately in opposite directions to assure longer pot life by more even heating.

Figure 5.29 shows the boxes containing the regenerative beds on both ends of radiant U-tubes. Evidence of the lower final exhaust temperature with regenerative burners was shown by the fact that it was no longer necessary to pay double time to persons working around the regenerative radiant tubes because of lower ambient temperature.

### 5.11.4. Oxy-Fuel Firing Saves Fuel, Improves Heat Transfer, and Lowers NOx

Although oxy-fuel firing is not exactly what is normally considered a method of heat recovery, it does save energy by reducing the mass of hot waste gas thrown away through the flue. Therefore, the authors have chosen to treat it here as an alternate form of heat recovery.

"Oxy-fuel firing" means substituting "commercially pure oxygen" for air in a combustion system. For 1 volume of methane (the principal constituent of natural gas), the combustion reaction with air,

$$CH_4 + 2O_2 + 7.57^*N_2 \rightarrow CO_2 + 2H_2O + 7.57N_2 \text{ (10.56 volumes flue gas)},$$

is replaced with the reaction for oxy-fuel firing,

$$CH_4 + O_2 \rightarrow CO_2 + 2H_2O \text{ (only 3 volumes of flue gas} = 28.4\% \text{ of w/air)}.$$

The convection heat transfer will be lower because lower volume means lower velocity. But convection is a minor fraction of the total heat transfer in furnaces above about 1200 F (650 C). Because about the same amount of chemical energy is released with oxy-fuel firing as with air-fuel firing, the adiabatic flame temperature as well as the triatomic gas radiation intensity from the poc[†] of oxy-fuel firing will be higher.

When the last two sentences are related to heat transfer within heat recovery devices (instead of within furnaces), the low volume and velocity do present concerns with oxy-fuel firing. Heat recovery equipment with larger flow passage cross sections can benefit more from the triatomic gas radiation with oxy-fuel firing. A good example of this is the double-pipe "stack" or "radiation" type recuperator. However, they must have parallel flow at the recuperator's waste gas entrance to prevent overheating there.

With oxy-fuel firing, the existence of almost no nitrogen in the poc helps keep NOx formation to a minimum—if no air can leak into the furnace and if the oxygen is close

[*]The ratio of volumes of nitrogen to oxygen in air = $(100\% - 20.9)/20.9\% + 3.78$.

[†]poc = products of complete combustion, pic = products of incomplete combustion.

to pure (oxygen enrichment, wherein the air is enriched with some oxygen, can create much NOx because the atmosphere then contains considerable concentrations of both nitrogen and oxygen—the essential ingredients for making NOx.)

When contemplating oxy-fuel firing, one must be concerned about mass flow reduction, much higher flame temperatures, and very much higher gas radiation heat transfer in short, longitudinal paths. Batch processes that depend on high mass flow to provide uniform product temperatures—(in-and-out furnaces, car-bottom furnaces, box furnaces, soaking pits)—will suffer from the use of oxy-fuel firing because of its lower mass flow and lower volume for circulation.

**Example (a):** In a one-way, top-fired soaking pit without spin, control of its poc will have an end-to end temperature difference of about 175°F (97°C) at the time when the load is expected to be rollable, but with oxy-fuel firing and its lower mass circulation, the corresponding end-to-end temperature difference might be 400°F (222°C) or more.

**Example (b):** In a pit with bottom control of temperature opposite the burner wall, the top-to-bottom temperature difference will be 20°F (11°C) with cold-air firing, 40°F (22°C) with hot-air firing, and over 75°F (42°C) with oxy-fuel firing.

If someone wants to reduce fuel consumption or raise productivity for a heating process, oxy-fuel firing may be a short-term, minimum-investment option. There are times when additional thermal head is limited in increasing productivity because of quality control (poor temperature uniformity) problems. Oxy-fuel firing may be able to help increase heat transfer without raising furnace temperature by virtue of its higher percentages of triatomic gases.

Clauses in some mills' oxygen contracts have caused them to pay for oxygen not used. Unfortunately, they have gone to oxy-fuel firing to take advantage of paid-for-but-not-used oxygen without being certain that oxy-fuel firing was appropriate for their process for the long term.

For long-range reduction of fuel rates, a better alternative to oxy-fuel firing may be regeneration with compact integral burner-regenerators. (See sec. 5.11.3.) These can meet oxy-fuel efficiencies if the regenerative bed materials have a high surface-to-mass ratio, that is, small refractory balls or nuggets averaging less than $\frac{3}{8}$" (0.01 m) diameter. Use of thin bed material with irregular surfaces can raise thermal efficiencies to 78% or higher, lowering fuel rates by 16 to 20%. Reversal cycles should be timed to a practical minimum without causing the dead time between firing cycles to cause the furnace temperature to fall. Long cycle times severely affect the available heat.

> The principles of the preceding two paragraphs were found years ago by fuel experts assisting regenerative open-hearth operators. After World War II, open-hearth cycle times were near 40 min, and the fuel-off times were about 2 min. By the early 1950s, the cycle times were down to 20 min. By the end of the open-hearth era, cycle times were 4 to 6 min, with fuel-off times down to 13 to 20 sec.

Combining oxygen and air preheat may sound risky, but may be a way to higher efficiencies if carefully monitored by modern controls, and provided NOx generation in not increased.

## 5.12. ENERGY COSTS OF POLLUTION CONTROL (see also sec. 6.3)

Early days of pollution control aimed principally at "smoke abatement," that is, particulate emission control. For installations using solid fuels, it was often necessary to change to more expensive gaseous or liquid fuels, which later were less expensive. As better designs evolved to reduce particulates, users benefited because more complete combustion was achieved.

When pollution control people turned their attention to NOx emissions, it became clear that fast mixing and high flame temperatures aggravated this form of pollution. At first, it seemed that any way to lower NOx had to result in poorer heat transfer and poorer fuel efficiency. Other possibilities required longer, slower mixing flames which required larger furnaces or some form of steam or water-spray cooling, which were very fuel wasteful. Modern burner technology has found ways to lower NOx without these first-feared, unwanted consequences.

The formation of NO (which later becomes $NO_2$, both of which are collectively known as NOx) is a chemical process with a reaction rate that is a function of temperature. The NO formation rate doubles for every 16°F (9°C) of reaction temperature rise *if* sufficient nitrogen and oxygen ions are available. Therefore, prime goals of combustion engineers should be to (*a*) reduce reaction (flame) temperature as much as possible and (*b*) use mixing configurations that minimize concurrent availability of N and O.

Excess air can add oxygen which contributes to NO generation, the precursor for $NO_2$, but better burner designs then allowed reduction of excess air to 5% or 10% with complete combustion and was therefore encouraged as both a fuel saver and a NOx reducer. Type E (flat) flames (fig. 6.2) have such thin flame envelopes, often rapidly cooled by their "scrubbing" of burner and furnace walls, that they never achieved the high flame temperatures of large, intense flames; thus, they were rightfully touted as NOx-reducing flames. Similarly, type H (high-velocity) flames (fig. 6.2) have a natural Venturi effect, inducing flue gas recirculation (fgr) within the furnace. This type of "internal fgr" was highly desirable as an NOx-reducing method, unlike the "external fgr" method discussed later (which required extra gas-pumping power, extra piping, and special burner designs with less available heat). (See fig. 5.30.) Where emissions regulations have low allowable NOx levels, the fgr retrofit may not suffice.

Modern methods utilize the limiting of oxygen availability* in the hottest part of the flame. The aforementioned in-furnace fgr utilizes this as well as its natural flame cooling. Many modern low-NOx burners have special internal or external air, fuel, or oxygen-mixing configurations that are capable of reducing NOx to levels below current, most strict regulations.

*Oxygen enrichment (25–80% oxygen) in the "air stream" increases the O-ion availability and therefore worsens the NOx pollution, but oxy-fuel firing (96–100% oxygen as the air stream) practically eliminates the N-ions; therefore, it is a good method of NOx control.

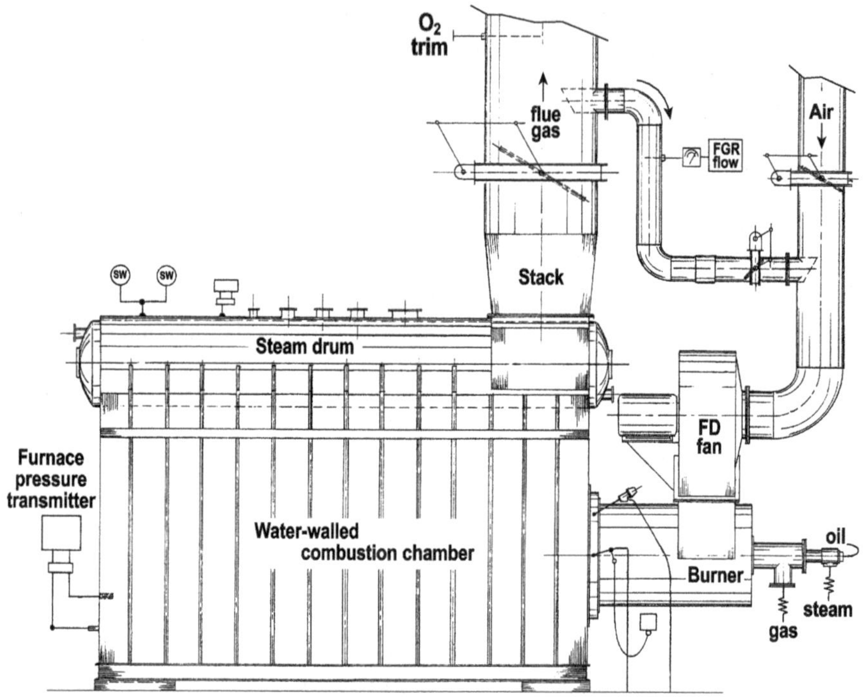


**Fig. 5.30.** Water tube boiler with flue gas recirculation to lower NOx emissions. Steam capacity rating is 88 000 lb/hr (4000 kg/h).

1 scf $CO_2$/scf fuel × 54.62 Btu/cf $CO_2$ + 2 scf $H_2O$/scf fuel × 42.37 Btu/scf $H_2O$ + 0.2 scf XS $O_2$/scf fuel × 36.3 Btu/scf $O_2$ + 8.27 scf $N_2$/scf fuel × 34.45 Btu/scf $N_2$ + 100 Btu latent heat/cf fuel = 54.62 + 84.74 + 7.26 + 284.9 + 100 = 531.5 Btu/scf fuel.

$$\begin{array}{c}\%\text{Available heat}\\ \text{with cold air}\end{array} = \frac{(100\%)\,(\text{gross hv} - \text{flue gas heat})}{\text{gross hv}}$$

$$= \frac{(100\%)(1000 - 531.5)}{1000} = 46.8\%.$$

Water or stream spraying are considered only emergency measures. "External fgr" is more effective than in-furnace recirculation of combustion chamber gases because its gases are usually much cooler, but it actually has to have a higher cost than most people realize, as shown in the following example 5.3 and its summary tabulation.

**Example 5.3 (Cost of fgr):** A furnace burning natural gas has 1800 F (1255 C) flue gas exit temperature with 10% excess air. Use %available heat calculations to compare fuel costs for Cases *a* to *e* discussed next.

**TABLE 5.6. Heat contents of gases[a]. Courtesy of North American Mfg. Co.**

| Gas temperature, F (Btu/scf) | Air[b] | CO | $CO_2$ | $H_2$ | $H_2O$[c] | $N_2$ | $O_2$ | $SO_2$ |
|---|---|---|---|---|---|---|---|---|
| 60 | 0 | 0 | 0 | 0 | 0 | 0 | 0 | 0 |
| 100 | 0.74 | 0.73 | 0.96 | 0.73 | 0.84 | 0.73 | 0.74 | 1.02 |
| 200 | 2.60 | 2.56 | 3.44 | 2.57 | 2.96 | 2.56 | 2.61 | 3.63 |
| 300 | 4.47 | 4.42 | 6.04 | 4.41 | 5.13 | 4.41 | 4.52 | 6.36 |
| 400 | 6.1 | 6.30 | 8.74 | 6.25 | 7.33 | 6.28 | 6.46 | 9.18 |
| 500 | 8.1 | 8.20 | 11.54 | 8.09 | 9.58 | 8.17 | 8.43 | 12.10 |
| 600 | 9.9 | 10.13 | 14.44 | 9.94 | 11.86 | 10.08 | 10.44 | 15.11 |
| 800 | 13.7 | 14.05 | 20.49 | 13.65 | 16.55 | 13.95 | 14.53 | 21.36 |
| 1000 | 17.8 | 18.05 | 26.87 | 17.39 | 21.39 | 17.9 | 18.73 | 27.89 |
| 1200 | 22.0 | 22.14 | 33.51 | 21.17 | 26.40 | 21.93 | 23.02 | 34.64 |
| 1400 | 26.2 | 26.30 | 40.38 | 24.98 | 31.56 | 26.03 | 27.39 | 41.58 |
| 1600 | 30.4 | 30.53 | 47.44 | 28.84 | 36.89 | 30.20 | 31.82 | 48.63 |
| 1800 | 34.6 | 34.84 | 54.62 | 32.76 | 42.37 | 34.45 | 36.30 | 55.77 |
| 2000 | 39.0 | 39.21 | 61.91 | 36.73 | 48.02 | 38.77 | 40.82 | 62.94 |
| 2200 | 43.5 | 43.65 | 69.23 | 40.77 | 53.83 | 43.15 | 45.36 | 70.08 |
| 2400 | 47.9 | 48.14 | 76.56 | 44.88 | 55.81 | 47.61 | 49.91 | 77.16 |
| 2600 | 52.4 | 52.70 | 83.85 | 49.06 | 65.94 | 52.14 | 54.47 | 84.11 |
| 2800 | 56.8 | 57.30 | 91.06 | 53.33 | 72.25 | 56.73 | 59.01 | 90.89 |
| 3000 | 61.3 | 61.96 | 98.13 | 57.69 | 78.71 | 61.39 | 63.52 | 97.46 |

[a] Computer-calculated data based on formulae from *Reference 28*, except b.
[b] Standard air at stp. (See Glossary.) From *Reference 77*.
[c] Superheat of the water vapor, not including latent heat.

**TABLE 5.7. Heat contents of gases[a]. Courtesy of North American Mfg. Co.**

| Gas temperature, C (kcal/$m^3$) | Air[b] | CO | $CO_2$ | $H_2$ | $H_2O$[c] | $N_2$ | $O_2$ | $SO_2$ |
|---|---|---|---|---|---|---|---|---|
| 15.6 | 0 | 0 | 0 | 0 | 0 | 0 | 0 | 0 |
| 100 | 25.0 | 24.7 | 33.2 | 24.7 | 28.6 | 24.6 | 25.2 | 35.1 |
| 200 | 52.9 | 54.5 | 75.5 | 54.1 | 63.5 | 54.3 | 55.9 | 79.4 |
| 300 | 83.2 | 85.0 | 120.7 | 83.5 | 99.5 | 84.6 | 87.6 | 126.5 |
| 400 | 113 | 116.2 | 168.6 | 113.1 | 136.6 | 115.4 | 120.1 | 175.9 |
| 500 | 145 | 147.9 | 218.8 | 142.9 | 174.9 | 146.8 | 153.3 | 227.4 |
| 600 | 178 | 180.3 | 271.0 | 172.9 | 214.4 | 178.7 | 187.3 | 280.6 |
| 700 | 212 | 213.2 | 325.0 | 203.2 | 255.0 | 211.1 | 221.9 | 335.3 |
| 800 | 245 | 246.7 | 380.5 | 233.8 | 296.7 | 244.1 | 256.9 | 391.1 |
| 900 | 280 | 280.6 | 437.1 | 264.8 | 339.6 | 277.6 | 292.4 | 447.7 |
| 1000 | 313 | 315.1 | 494.7 | 296.1 | 383.7 | 311.6 | 328.3 | 504.7 |
| 1100 | 348 | 350.1 | 552.9 | 327.8 | 428.9 | 346.1 | 364.4 | 561.9 |
| 1200 | 383 | 385.5 | 611.4 | 360.1 | 475.3 | 381.1 | 400.6 | 619.0 |
| 1300 | 419 | 412.3 | 669.9 | 392.8 | 522.8 | 416.6 | 437.0 | 675.5 |
| 1400 | 456 | 457.6 | 728.2 | 426.1 | 571.6 | 452.7 | 473.3 | 731.2 |
| 1500 | 490 | 494.2 | 785.9 | 460.0 | 621.5 | 489.2 | 509.6 | 785.8 |
| 1600 | 526 | 531.2 | 842.8 | 494.5 | 672.6 | 526.2 | 545.7 | 838.9 |
| 1700 | 562 | 568.6 | 898.5 | 529.7 | 724.9 | 563.6 | 581.6 | 890.3 |
| 1800 | 597 | 606.2 | 952.9 | 565.5 | 778.4 | 601.6 | 617.1 | 939.5 |

[a] Computer-calculated data based on formulae from *Reference 28*, except b.
[b] Standard air at stp. (See Glossary.) From *Reference 77*.
[c] Superheat of the water vapor, not including latent heat.

*An accurate method* would use available heat charts corrected for dissociation such as from *reference 52*, figure 9.7 and 13.4a, or figures 5.1 and 5.2 in this book, which give the following answers for a natural gas analysis of 90% methane ($CH_4$), 5% ethane ($C_2H_6$), 1% propane ($C_3H_8$), and 4% nitrogen ($N_2$), with 1800 F exit gas:

With 60 F air, 9.68 scf air/cf fuel, 10.71 scf poc/scf fuel, 48% available heat.

With 800 F air, 9.68 scf air/cf fuel, 10.71 scf poc/scf fuel, 62% available heat.

With 60 F $O_2$, 2.03 scf $O_2$/scf fuel, 3.06 scf poc/scf fuel, 76% available heat.

*A simplified method* is used here to show the reader an alternate calculation that gives an easy understanding of available heats. This simple method assumes the natural gas to be methane, which is about 90% of most natural gases. It assumes that the difference between gross and net heating values is 100 Btu/cf of fuel, typical for natural gases. (This is latent heat of water from burning hydrogen.)

For each cubic foot (cf) of fuel, assumed to be methane ($CH_4$),

$$CH_4 + 2.2^a\,O_2 + 8.27^b\,N_2 \rightarrow CO_2 + 2H_2O + 0.2O_2 + 8.27N_2,$$

(1 scf fuel) + (10.47 scf air/cf fuel w/10% XS air) → (11.47 scf poc).

a: 2.2 = (2 mols $O_2$/mol $CH_4$) (1.1) for 10% excess air.
b: 8.27 = (2.2) (3.76 mols $N_2$/mol $O_2$ in air).

*(a) Calculate %available heat using cold air and no fgr:* First determine the total heat lost in all the flue gases by adding the heat in each of the flue gases leaving the furnace, using heat contents of the exit gases, at 1800 F (1255 C) from tables 5.6 or 5.7 + 100 Btu/cf for the latent heat of vaporization of water formed from combustion of hydrogen:

| Constituent | (1) poc | (4) = heat in 1 scf of (1) at 1800 F[a] | (5) = (1) (4) = heat of (1) at 1800 F |
|---|---|---|---|
| $CO_2$ | 1 scf[b] | 54.6 Btu/scf | 54.6 Btu |
| $H_2O$ | 2 scf | 42.4 Btu/scf[c] | 84.8 Btu |
| $O_2$ | 0.2 scf | 36.3 Btu/scf | 7.3 Btu |
| $N_2$ | 8.27 scf | 34.5 Btu/scf | 285.3 Btu |
| Total | 11.5 scf | | 432 Btu (Dry flue loss) |

% available heat, without heat recovery = (100%)

$$\frac{\text{(gross hv} - \text{dry flue gas loss} - \text{latent flue loss)}}{\text{gross hv}} = 100(1000 - 432 - 100)/1000 = 46.8\%.$$

[a]per scf constituent From table 5.6 at 1800 F.
[b]per scf of fuel, e.g., 1 scf $CO_2$/scf of fuel.
[c]superheat only, no latent heat.

*(b) Calculate %available heat using 800 F combustion air (including 800 F excess air) and no fgr;* then compare it with the previous %available heat using cold air and no fgr. From table 5.6, heat (recovered from the exhaust poc by recuperator or regenerator) is (13.7 Btu/cf air) (2.2 $O_2$ + 8.27 $N_2$ or 10.47 cf air/cf fuel) = 143.4 Btu/cf fuel.

$$\%\text{available heat, w/heat recovery as 800 F air} =$$

$$(100\%)\,\frac{(\text{gross hv} - \text{dry flue gas loss} - \text{latent flue loss} + \text{ht recovered})}{\text{gross hv}} =$$

$$100(1000 - 432 - 100 + 143)/1000 = 61.1\%, \text{ an increase of}$$

$$61.1 - 46.8 = 14.3\% \text{ from } (a).$$

*(c) Calculate the available heat with cold air and 20% fgr* (fgr volume equal to 20% of the stp volume of the flue gas before installing fgr).[α] The following tabulation determines the heat content of the poc + fcg:

| Constituent | (1) poc | (2) = 0.2 (1) fgr | (3) = (1) + (2) poc + fgr, | (4) = heat content in (1) at 1800 F | (5) = (3) (4) = ht content in (3) at 1800 F |
|---|---|---|---|---|---|
| $CO_2$ | 1 scf* | 0.2 scf* | 1.2 scf* | 54.6 Btu/scf† | 65.5 Btu/scf* |
| $H_2O$ | 2 scf | 0.4 scf | 2.4 scf | 42.4 Btu/scf | 101.7 Btu/scf |
| $O_2$ | 0.2 scf | 0.04 scf | 0.24 scf | 36.3 Btu/scf | 8.7 Btu/scf |
| $N_2$ | 8.27 scf | 1.65 scf | 9.92 scf | 34.5 Btu/scf | 341.7 Btu/scf |
| | | | | Total dry stack loss = | 517.6 Btu/scf |

*per scf of fuel, e.g., 1 scf $CO_2$/scf of fuel.
†per scf of constituent, from table 5.6 at 1800 F.
[h]Superheat only, not including latent heat of vaporization

Total stack loss = dry + latent = 517.6 + 100 $H_2O$ stack loss = 617.6 Btu/scf fuel. %available heat with cold air + 20% fgr = (100%) (1000 − 617.6)/1000 = 38.2%, a decrease from (*a*). This assumes the fgr had been cooled all the way to 60 F (16 C) before it was returned to the combustion chamber. If the fgr were not cooled to 60 F (16 C), more fgr would be required to achieve the NOx reduction.

*(d) Calculate the %available heat with 800 F combustion air, 10% excess air, and 20% fgr.*

From table 5.6, heat (recovered from the exhaust poc by recuperator or regenerator) is now heat recovered from air + fgr. Heat recovered by preheating the air is 13.7 Btu/scf of fuel, the same as in Part (*b*) of this example, which when multiplied by 10.47 scf air/scf fuel, = 143.4 Btu/scf fuel. The heat recovered from fgr is determined in the following table.

[α]There are many ways to express the extent of flue gas recirculation. Note carefully the one used in this example.

| Constituent | (1) poc | (2) = 0.2 (1) fgr | (3) = (1) + (2) poc + fgr, | (6) heat content in (2) at 800 F | (7) = (2) (6) = ht content in (2) at 800 F |
|---|---|---|---|---|---|
| $CO_2$ | 1 scf* | 0.2 scf* | 1.2 scf* | 20.49 Btu/scf** | 4.10 Btu/scf* |
| $H_2O$ | 2 scf | 0.4 scf | 2.4 scf | 16.55 Btu/scf | 6.62 Btu/scf |
| $O_2$ | 0.2 scf | 0.04 scf | 0.24 scf | 14.53 Btu/scf | 0.58 Btu/scf |
| $N_2$ | 8.27 scf | 1.65 scf | 9.92 scf | 13.95 Btu/scf | 23.02 Btu/scf |
| | | | Total heat recovered from the dry fg = | | 34.32 Btu/scf |

*per scf of fuel, e.g., 1 scf $CO_2$/scf of fuel.
**per scf of constituent, from table 5.6 at 800 F.

The %available heat, with fgr and heat recovery = 100% × (gross hv − flue gas heat + ht recovered from air & fgr)/(gross hv) = 100% × [1000 − 617.6 (from *c*) +143.3 + 34.32]/1000 = 56.0%. Thus, the loss in %available heat due to fgr with 800 F air is 61.2% − 56.0% = 5.2%.

*(e) Further Considerations.* A larger recuperator will be needed to handle the larger volume and hotter exit gas. An additional blower and piping will be required with fgr. The inerts in the fgr stream may reduce the stability of the burner.

Higher flow through the furnace with fgr will raise exit flue gas temperature from 1800 F to about 1870 F for case calculated, necessitating another iteration of the preceding calculations (not shown here), resulting in 53.7% available heat.

*(f) Summary tabulation.* The findings for the previous furnace are compared in the following tabulation. Lines *(a), (b), (c), (d)* are for 1800 F (982 C) flue gas exit temperature, but Line *(e)* is for the 1870 F (982 C) flue gas exit temperature that ultimately results with fgr in (*d*).

| Combustion air temperature F/C | W/or w/o fgr | %available heat | Gross fuel input required for 100 kk Btu/hr available for loads and losses other than stack loss | %fuel used[X] |
|---|---|---|---|---|
| *(a)* 60 F/16 C | w/o fgr | 46.8% | 100 kk/0.468 = 213.7 kk Btu/hr | 100 |
| *(b)* 800 F/427 C | w/o fgr | 61.1% | 100 kk/0.611 = 163.4 kk Btu/hr | 76 |
| *(c)* 60 F/16 C | w/fgr | 38.2% | 100 kk/0.382 = 261.8 kk Btu/hr | 122 |
| *(d)* 800 F/427 C | w/fgr | 56.0% | 100 kk/0.560 = 178.6 kk Btu/hr | 84 |
| *(e)* 800 F/427 C | w/fgr[β] | 53.7% | 100 kk/0.537 = 186.2 kk Btu/hr | 87 |

[β]Corrected for fg temperature rise from 1800 F to 1870 F (982 C to 1021 C) as a result of higher volume flow through the furnace with fgr.
[X]%fuel used = 100% (original %available heat/new %available heat).

## 5.13. REVIEW QUESTIONS, PROBLEMS, PROJECT

**5.13Q1.** List the ways in which it may be possible to increase efficiency (reduce fuel consumption) of an industrial furnace.

**A1.** a. By excluding infiltrated air (tramp air).

b. By reducing excess air.

c. By recovering heat from the exiting flue gases by preheating air in a recuperator or in a regenerator.

d. By recovering heat from the exiting flue gases by generating "free" steam in a waste heat boiler.

e. By recovering heat from the exiting flue gases by preheating the cold loads entering the furnace.

f. By insulating the furnace better.

g. By closing furnace doors and peepholes promptly after use.

h. By installing an insulated ell (elbow) at every flue so that the hot interior walls or loads cannot radiate to cold outside surfaces.

i. By minimizing water cooling of furnace components by keeping abreast of modern furnace construction and operating techniques.

j. By controlling the first fired zone with a T-sensor 6' to 10' before the flue exit, high in a sidewall, and making sure the sensor "feels" the hot furnace gases and "sees" the loads. This way, the first fired zone will quickly follow production rate changes, especially after a delay.

k. By following heating curve when adjusting control setpoints, particularly in the first fired zones, both top and bottom. If curves are not available, set up a plan to weekly reduce the first fired zone setpoint by 50°F (28°C). When the plan has gone too far, raise the setpoint by 50°F (28°C).

l. By shortening the firing length of the first fired zone as much as possible to increase the slope of the thermal profile of that zone.

m. By shortening the heating cycle time of batch furnaces by using direct hot gases to heat all surfaces as nearly alike as possible.

n. By increasing firing rates in batch furnaces to reduce firing time to zone setpoints, reducing the overall cycle time.

o. By locating T-sensors as near to the loads as possible to assure that they are sensing load temperatures, not furnace temperatures.

p. By attempting to heat the product in continuous furnace as late in the furnace as possible—to keep the thermal slope as steep as possible, for high productivity combined with low fuel use.

q. By using burners with controllable thermal profile—to keep heat as late in the zone as late as possible, for maximum thermal slope in the zone.

**5.13Q2.** A five-zone slab heating furnace had a very high fuel rate because the operators believed it was necessary to maintain the top and bottom preheat zone temperature setpoints (with temperature measurements about 60% through the zone) the same at all production rates. What can be done to reduce fuel rates of such a furnace?

**A2.** The answer revolves around reducing the flue gas temperature as follows:

a. A very expensive solution is to purchase a computer model to adjust temperature setpoints using heating curves.

b. Change the location of the control measurement in the top preheat zone from the roof near the flue to 6 to 10 feet toward the furnace discharge. There, it can "feel" the gas temperature and "see" the product.

c. To control the bottom zone, use the present top preheat temperature measurement as a remote setpoint for the bottom zone's control. That will assure that the bottom zone's thermal profile will be nearly identical to that of the top preheat zone.

d. Use experimental evidence to adjust the top preheat zone setpoints for different products and productivity rates. The key point is to avoid flue gas and furnace flue temperatures being higher at low productivity than at high productivity. In one large rotary furnace that coauthor Shannon followed, the fuel rate dropped from 3.0 kk Btu/ton (0.83 kk kcal/mton) to 1.5 kk Btu/ton (0.417 kk kcal/mton) when the control temperature measurement was moved and the setpoint adjusted for product thickness.

**5.13Q3.** Why are steel reheat furnaces without waste heat recovery so thermally inefficient in compared to boilers?

**A3.** If the furnace were used to near its heating capabilities, the entry furnace temperature could be 1600 F (871 C). The flue gas temperature would be about 1950 F (1066 C). If the furnace air/fuel ratio were held to 10% excess air, the available heat would be 42%.

In addition, heat losses could be held to 10% of the heat required for the load. In general, boilers would have a waste gas temperature of 300 F (150 C), resulting in about 86% available heat, if using natural gas. Heat losses would be less than half as much as with a reheat furnace.

**5.13Q4.** Why is the flue gas exit temperature always higher than the furnace temperature?

**A4.** For heat to be transferred from the furnace (walls, flame, gas) to the loads, there must be always a higher temperature in the heat source than in the heat receiver. Heat flows "downhill," temperature-wise.

**5.13Q5.** If furnace temperature at the furnace entry (flue gas exit) is 1800 F (982 C), what will the flue gas exit temperature be?

**A5.** A quick approximate estimate, via equation 5.1, would say 740 F + (0.758) (1800 F) = 2104 F, but from figure 5.3, using a typical gas velocity of 20 fps, the flue gas exit temperature will be 240 F + 1800 F = 2040 F.

**5.13Q6.** Why is it advantageous to have a positive furnace pressure at the point where the temperature control sensor is located?

**A6.** When a T-sensor is located in an area of negative pressure, air inleakage may cool the sensor, so that it will call for more input, raising the flue gas temperature, reducing fuel efficiency, and perhaps endangering product quality.

**5.13Q7.** Why should multiple flues be avoided?

**A7.** Multiple flues should be avoided because it is very difficult to balance and to predict circulation with them, often raising flue gas temperatures. In addition, in a batch furnace, having gases from one zone flowing through other zones can prevent proper temperature control in the downstream zone(s), increasing flue gas exit temperature, raising fuel rate, and causing nonuniformities in product temperature.

**5.13Q8.** Why are adjustable thermal profile burners generally more efficient in continuous longitudinally fired reheat furnaces?

**A8.** For maximum heat transfer at minimum fuel cost, short flame burners are ideal. However, if higher production with reasonable efficiency is needed, flame lengthening is often necessary. This change can be made manually or automatically with adjustable thermal profile burners. Most other burners cannot be adjusted without part changes.

**5.13Q9.** Why is it advisable to analyze furnace gas flow patterns before building or modifying a furnace?

**A9.** Temperature uniformity cannot be achieved without first knowing combustion gas flow patterns at various fuel inputs. Assuring uniformity requires longer cycle times and soak times.

**5.13Q10.** Why do pulse firing and step firing reduce fuel rates?

**A10.** Conventionally, excess air has been used to reduce temperature differences along the gas flow paths, but that approach costs more fuel. With pulsed flows, high mass flows accomplish the same more-level temperature profile as excess air but without the fuel cost and without the necessary added soak time. Stepped pulse firing allows soak times between its pulses.

## 5.13. PROBLEMS

### 5.13.Prob-1.

*This problem relates to figure 5.1, "Percents available heat for an average natural gas with cold air and with preheated combustion air."* All excess air curves are based on 60 F (16 C) combustion air. All hot air curves are based on 10% excess air. Computer printouts of available heat data for other fuels are available from North American Mfg. Co.

*Given:* $T_3 = 2300$ F = 1260C, $t_2 = 1000$ F = 538 C.
Required fuel with cold air = 10 000 000 Btu/hr = 10 550 MJ/h.
*Find:* The required fuel input with hot air, and the %fuel saved.

*Solution:* Interpolating with a millimeter scale on Figure 5.1, %available heat at $t_2 = 60$ F = 16 C with 10% excess air = 33%; %available heat at $t_2 = 1000$ F = 538 C with 10% excess air = 54%.

Required input with 100 F air =

= 10 000 000 Btu/hr × (33/54) = 6 110 000 Btu/hr,

or = 10 500 MJ/h × (33/54) = 6 417 MJ/h.

%fuel saved = 100 × (1 − 33/54) = 38.9%.

### 5.13.Prob-2.

*This question relates to table 5.1, Percents available heat for a typical #6 residual fuel oil with cold air and with preheated combustion air.* All excess air curves are based on 60 F (16 C) combustion air. All hot air curves are based on 10% excess air. Printouts for plotting available heat data for other fuels are available from North American Mfg. Co. Permission was granted by North American Mfg. Co to reproduce this copyrighted info.

*Given:* A heat-treat furnace has a flue gas exit temperature of 1800 F (982 C) and is running with 10% excess air while burning #6 fuel oil.

*Find:* The %fuel saved by preheating the air to 900 F (427 C) (using an air temperature compensator in the air/fuel ratio controller to continue to hold only 10% excess air at all firing rates).

*Solution:* Interpolating on table 5.1, with 1800 F (982 C) flue gas exit, available heat with 900 F (427 C) combustion air and 10% excess air = 70%. For 1800 F (982 C) flue gas, but with 60 F (15.6 C) air, the available heat is only 53%. The additional savings from use of preheated air will be 100% × [1 − (53/70) = 24.3% fuel saved.

### 5.13.Prob-3.

The procedure of section 5.9 and the exercise of example 5.1 need a lot of practice. Design a parallel problem based on a furnace with which you are familiar. Search out the needed given data for your furnace, solve the problem again for your case, write up your solution, and submit it to your group's instructor for use by others not familiar with your kind of furnace.

## 5.13. PROJECT

*This project relates to section 5.11.3.* Compare (*a*) the gain from more gas radiation via a raised preheat section roof with (*b*) the loss from reduced convection.

# 6

# OPERATION AND CONTROL OF INDUSTRIAL FURNACES

## 6.1. BURNER AND FLAME TYPES, LOCATION

### 6.1.1. Side-Fired Box and Car-Bottom Furnaces

Side-fired box and car-bottom furnaces are ideally fired with main burners on 2.5-ft to 4.5-ft (0.6 m to 1.4 m) centers along the top on one side, and small "pumping" high-velocity burners on the opposite bottom side. (See fig. 6.1.) The main burners should have ATP technology so that the temperature can be controlled to a flat profile with the T-sensors located at the level of the top of the load through each of the two long sidewalls.

The loads should be on piers so that small, high-velocity burners can be fired underneath. For practically constant temperature under the loads, the base pier height should be 5" to 9" (0.13 to 0.23 m) and the burners fired with constant air. Uniform temperature will result from the fact that the thin gas blanket will transfer only about one-third as much heat as above the load, so the blanket temperature will fall very slowly as it moves under the load. Therefore, load temperature profile across the furnace and below the load as well as above will be practically flat, leading to less than $\pm 10°F$ ($\pm 5°C$) temperature differential throughout the load.

When conventional burners are used to side fire a furnace, they produce larger differentials across the furnace. These larger temperature differences stem from the changeable thermal profile of the burner at different firing rates. At high-firing rates,

SAFETY SHOULD BE THE UTMOST PRIORITY of all furnace engineers . . . above quality, before productivity, preceding pollution control, outprioritizing labor minimization, and overshadowing fuel economy!

Thorough study of section 6.6.2, plus "Combustion Supervising Controls" in pt 7 of *reference 52,* is imperative for *your own personal safety, for your job, and for the whole organization in which you work.*

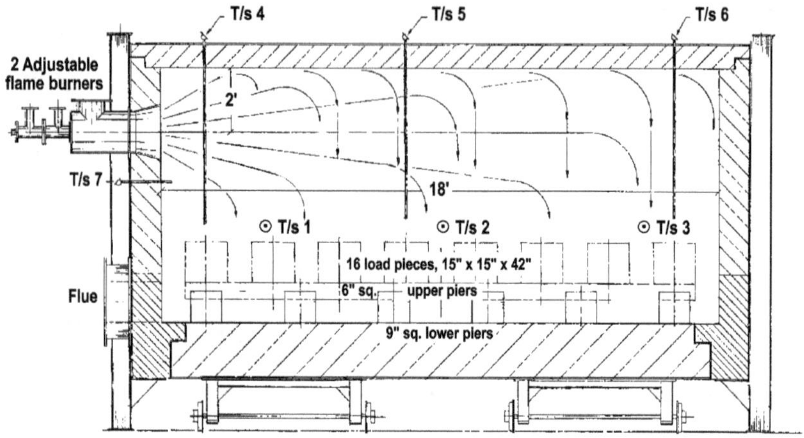


**Fig. 6.1.** Side-fired in-and-out furnace (with car-hearth), 18' wide × 12' deep × 8' high ID. Adjustable flame burners give uniform heating width-wise/depth-wise; double-stacked piers help bottom uniformity. (See also figs. 3.26 and 6.23.)

the thermal profile has the peak temperature far from the burner wall, with the burner wall temperature very low relative to the setpoint temperature. At low-firing rates, the thermal profile peaks near the burner wall and is very low at points far from the burner wall. With the ATP burners, automatic control can hold the whole profile flat at all firing rates.

If using conventional burners to side fire thin stock where only ±25°F (±14°C) is satisfactory, ATP burners are not necessary. Use of high-velocity burners high in both long walls (top firing only) alternating on 8-ft (2.44 m) centers will produce a good-quality product; however, to reduce temperature differences in the product, bottom flues are recommended in both sidewalls. (With no bottom burners, flues are needed to pull hot gases to all areas for reasonable temperature uniformity.)

With thick loads, the pieces should be on piers with high-velocity burners located in rows near the bottoms of both sidewalls, alternating on 4-ft (1.22 m) centers. With this arrangement, flues can be in the roof. *One important point:* In batch operations, do not pass the poc gases of any zone through another zone because that will result in loss of temperature control for the second zone.

Burners should have capacity for 60 000 to 125 000 Btu/ft$^2$hr hearth, preferably about 75 000 Btu/ft$^2$hr. A heating curve is preferred to select a firing rate accurately.

### 6.1.2. Side Firing In-and-Out Furnaces

Side firing in-and-out furnaces is more difficult because generally one long wall is a door or row of doors, which makes it difficult to measure temperature, increases heat losses, and prevents use of burners on the door wall. However, if the temperature uniformity requirements for the product are not stringent, the burners can be located in the back wall firing toward the doors with control thermocouples inserted through the roof.

### 6.1.3. Side Firing Reheat Furnaces

Side firing reheat furnaces with low NOx requirements is a problem because it is difficult to hold a flat thermal profile across the furnace with current low NOx techniques. The result may be a hot furnace center with cold sidewalls or vice versa, depending on whether the firing rate is high or low and whether the burners are alternated side to side or opposite. At firing rates above about 50%, opposite burners produce a hot furnace center. At firing rates below 30%, they produce hot burner walls. Alternating burners firing above 50% will give a cool furnace center and hot furnace walls. It is hoped that soon a low NOx burner will be developed with the ability to control a flat temperature profile across a wide furnace.

### 6.1.4. Longitudinal Firing of Steel Reheat Furnaces

Longitudinal firing of steel reheat furnaces in top and bottom heat and soak zones, including sawtooth-roof rotary furnaces, is used to reduce the number of burners and to develop a uniform temperature across the hearth. Otherwise, most of these furnaces would be side fired to hold the heat transfer temperature higher and longer (many times for as long as 40 ft, perhaps 25 ft, for longitudinally fired zones).

Determining firing rates (burner sizes) for top and or bottom zones of reheat furnaces is difficult without first developing heating curves. (See chap. 8.)

An effective and practical control is described next for a three-zone walking hearth furnace. The preheat zone should have a control T-sensor about 6 feet from the zone, with entry either through the roof or preferably high in the sidewall, in the exhaust gas flow. At that location, the T-sensor will be very sensitive to productivity and will prevent the waste gas temperature at low production from being hotter than it is during high production.

The heat zone should have a thermocouple in the sidewall about 6" (0.15 m) above the hearth and about 5 feet (1.52 m) into the zone, plus a thermocouple 6" (0.15 m) above the hearth and 2 or 3 ft (0.6 or 0.9 m) from the zone end. These two controllers should operate through a low select device to the energy input control. The inlet thermocouple should be set for several hundred degrees below final temperature—for example, 1600 F to 2000 F (870 C to 1090 C). The discharge T-sensor should have a setpoint of 2450 F to 2490 F (1340 C to 1365 C) to prevent damage to the product or the melting of scale. This system was devised to reduce the heating problems caused by delays.

### 6.1.5. Roof Firing

Roof firing can provide uniform temperature across a hearth, especially in soaking zones. An almost-standard practice for soaking zones has been to use roof burners in three zones across the width of the furnace. Attempts to cut costs with only two zones have given very poor results.

Roof firing can be accomplished either with type E ("flat" flames) in a *flat roof* or with conventional (type A) flames or long, luminous (type F or type G) flames in a *sawtooth roof.* (See fig. 6.2.)

## 6.2. FLAME FITTING

Table 6.1 provides a guide for burner selection—a list of industrial heating processes preferably heated by convection heat transfer, and another list of processes usually better done by radiation heat transfer. Many jobs end up being done by a combination of convection and radiation. A simplistic, three-step order for decisions might say:

*First*, if mass transfer (such as drying) is involved, choose convection because it simultaneously provides heat delivery *and* mass transfer (movement of whatever was vaporized).

*Next*, choose radiation, often more powerful than convection.

*Finally*, fill in with convection where radiation cannot go because of its straight-line delivery limitation.

*Radiation* is usually more intense at temperature levels above 1400 F (760 C). It is best used for well-exposed surfaces such as thin flat loads, thin rotatable loads, and thin cylindrical or spherical loads, loads encased in valuable containers, and ablative melting (see footnote in Table 6.1), plus holding of stirred liquids.

*Convection* is usually preferred below the 1400 F (760 C) level. The big problem with radiation is its "shadow problem" because radiation travels in straight lines, making it difficult to heat stacked or loosely piled loads, granular materials such as fluidized beds, or to get to 'reach' or 'wraparound' configurations. Thus, in those cases, convection has to be the prime (or at least a fill-in) heat-delivery mechanism. Convection (sometimes combined with gas radiation, as in "enhanced heating"), is often the best vehicle for improving productivity through better temperature uniformity.

### 6.2.1. Luminous Flames Versus Nonluminous Flames

Luminosity is generated by the cracking of fossil fuels into micron-sized solids and gaseous hydrocarbon compounds. The heaviest of those compounds, perhaps with some solid carbon, is called "soot." When the soot particles become very hot and begin to burn, they radiate like other solids. Since solids radiate in all wavelengths and follow the rules of heat transfer between solids, luminous flames transfer more heat

**TABLE 6.1 Suggested primary heating modes for industrial loads**

| Radiation | Convection |
|---|---|
| Thin flat loads | Mass-transfer processes |
| Thin rotatable loads | Recirculating ovens <1200 F (<650 C) |
| Thin hollow loads | Granular or loosely piled loads |
| Liquid holding | Reach or wraparound configurations |
| Ablative melting* (dry-hearth) | Impingement heating |
| Loads in valuable containers | Fluidized bed heating |

*ablative melting, as opposed to submerged and un-stirred melting, allows the newly melted liquid to flow away (by gravity) so as to expose more solid surface to all forms of heat transfer for further melting.

A *candle flame* is a miniature example of a type F long, luminous, laminar flame. Author Reed has often demonstrated some of the features of type F flames with a candle—polymerization soot formation, flame quenching, flame holders, starved air incineration, natural convection, particulate emission, streams in laminar, transition, and turbulent flows, aeration (by exhaling through a tiny straw across the blue base of the candle flame) changes it to a compact, all-blue flame that demonstrates combustion roar. Some of these demonstrations were recently found to have been alluded to in Professor Michael Faraday's famous candle lectures of the 1850s (*reference 19*).

than nonluminous flames. The "skin" of a luminous flame is the locus of points where the soot combines with oxygen to self-incinerate to carbon dioxide and water vapor.

Luminous flames can transfer about 7% more heat than nonluminous flames. However, modern nonluminous flame and heat transfer techniques, together, can be more effective overall than luminous flames.

Until recently, all long flames were luminous, but that is not true of several modern burners. Flame lengths are important to deliver heat flux as needed by the product and fit into the space available. For example, high-velocity burners were added to a 15 ft (4.6 m) wide car furnace between the piers, which were about 12" (0.3 m) high, with much scale accumulated on the hearth. The scale displaced all but 10" (0.25 m) of the gas blanket; thus, the heat transfer coefficient was only 10 Btu/ft$^2$hr°F (57 W/°Cm$^2$) versus 25 Btu/ft$^2$hr°F (142 W/°Cm$^2$) for a 36" blanket. Therefore, the gas $\Delta T$ drop across the 15 ft (4.6 m) wide car was low. The wall opposite the burner took a beating, its thickness halved in a few months. Reduced flame length was needed, by spreading the gases or reducing the firing rate.

### 6.2.2. Flame Types (see fig. 6.2)

In many cases, space limits the firing rate and the type of flame; so it is necessary to use type E burners, which have very short flames with large diameters. For larger firing rates, ATP burners can vary the flame length from short to very long for the needed temperature profile across the length of the space.

### 6.2.3. Flame Profiles (see figs. 4.22 and 6.3)

## 6.3. UNWANTED NOx FORMATION (see pt 11 of *reference 52*)

Low NOx injection (LNI) of fuel and air into the furnace chamber provides the highest potential efficiency and lowest NOx. The LNI system takes advantage of the furnace itself, which is the largest source of "free" flue gas recirculation (FGR) to produce uniquely low NOx emissions from high-temperature systems.

**Flame type A.** Conventional forward, feather-shaped flame. No swirl. Little recirculation. Axial jet. Moderate to fast mixing. Used in all-purpose combustion chambers.

**Flame type C.** Ball. Considerable swirl, >0.6*. Hot reverse flow into center; cold forward flow at sides. Secondary jet velocity >primary jet velocity. Intense mixing. Used for combustion chambers that are more or less cubical, such as stoker-fired boilers, fluidized beds, and Dutch ovens.

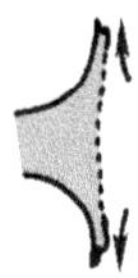

**Flame type E.** Flat, coanda. Very high swirl ~2*. Some recirculation. Fast mixing. Swirled flow contained by a refractory quarl. The convex version (shown) is used to avoid flame impingement on furnace loads, and provides a wider heat spread. A concave version (not shown) focuses radiation in large "hot spots". Both versions enhance and direct refractory radiation.

**Flame type F.** Long, luminous, lazy. No swirl, nor circulation. Low fuel and air jet velocity. Laminar jet, buoyancy-controlled. Delayed, slow, diffusion mixing. Used for coverage in long chambers, and to add luminous radiation.

**Flame type G.** Long, luminous, "fire hose". No swirl. Fuel and/or atomizing medium jet velocity much greater than air jet velocity. Thrust controlled. Fuel directed. Stretched-out mixing. Used for uniform coverage in long chambers, and to add luminous radiation.

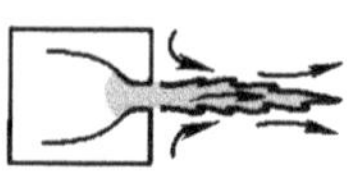

**Flame type H.** High velocity. Low swirl. High furnace recirculation. Flow contained by refractory tile, burning inside and outside tile. Fast mixing. Used to drive heat into loosely-piled loads; to force flow around backs of loads, to enhance convection, to reach long distances; and to fire between piers.

* Swirl number. See Reference 52.

**Fig. 6.2.** Typical industrial flame types. Arrows show furnace gas flows induced by the flames. With natural gas, dark gray = blue flame, light gray = yellow flame. With fuel oil, all flames would be yellow. Adapted with permission from *reference 52.*

The principal variable in NOx generation is the temperature at which the combustion reaction takes place. Anything that can be done to reduce the actual combustion reaction temperature will reduce NOx, and anything that results in a higher combustion reaction temperature will increase NOx. LNI increases the inerts in the combustion reaction. They absorb heat, lowering the reaction temperature, thereby lowering the NOx.

NOx formation is a chemical reaction that is part of the combustion reaction of fuels. As in all chemical reactions, the rate of the reaction increases with temperature,

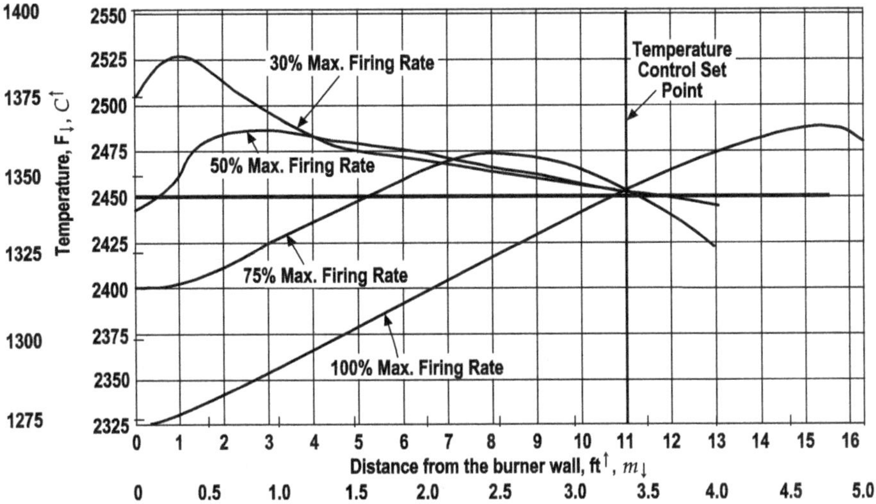


**Fig. 6.3.** Flame profile of a conventional type A flame (fig. 6.2) on a steel reheat furnace. The vertical (temperature) scale reflects the heat flux profile. ATP burners can operate at a constant high input while switching temperature profiles, for example, from 30% to 100%.

as long as the reagents are available to sustain it. Very little NOx is generated below 2800 F (1 C to 93 C), but above that temperature level the rate doubles, about every 16°F (8.9°C) as with most reactions; thus, lowering the reaction temperature can be a primary way to forestall $NO_x$ generation. Therefore, the principal routes to low NOx are:

1. Add materials to the fluid stream that must be heated to the reaction temperature, but do not contribute additional energy. In this way, the reaction temperature is lowered.
2. Expose the actual combustion reaction to inert furnace gases, furnace walls, and products so that some of the reaction heat is transferred while the reaction is taking place.

A technology often used delays the burning so that most of it occurs out in the furnace rather than inside the burner tile (or quarl), then it is possible to inspirate inert furnace gases into the combustion air and/or fuel being supplied to the combustion reaction.

With this LNI technology, essentially all combustion takes place in the furnace chamber where refractory, furnace gases, and product all receive radiation from the combustion reaction, lowering the flame temperature. In addition, the combustion air and the fuel are supplied at high velocity and separated from each other to inspirate furnace gases into their individual streams without purposely discharging the streams into each other. The reasons for so doing are:

1. To inspirate as much inert furnace gas as possible into both the air and fuel streams before burning takes place so that the reaction must heat those inerts to the lowered reaction temperature
2. To have the reaction take place where it can transfer heat to furnace gases and solids, thereby further reducing the reaction temperature

Coauthor Shannon encountered an opposite effect in a large pelletizing plant in Mexico that was a very large producer of NOx. It used a regenerative system to preheat air to about 1750 F, but with conventional burners. The very high flame temperature sometimes melted the burner tile ports. A large reduction in NOx could be accomplished with injectors directed into the furnace with very high velocity, perhaps at 350 ft/sec (107 m/s). This gas velocity would entrain large volumes of furnace gases with large percentages of $O_2$, perhaps as high as 18%. Some might fear that this high percentage of $O_2$ would raise NOx. This is true to perhaps 5%, but beyond that the oxygen acts as an inert because it would not be involved in the reaction. It would act as $N_2$ or $CO_2$, absorbing heat. This uncommon combustion air would then produce a lower combustion reaction temperature in the tile, lowering NOx emission.

Injectors should be developed to raise reentrainment to the highest possible level, perhaps using a closed-end tube with four jets at 90 degrees, as in existing low NOx roof burners. When the proportion of inerts is very large, the reaction temperature is lowered to a level at which the flame is barely visible. However, this is not simply a temperature effect, but due to a depletion of hydrocarbon cracking in the presence of $H_2O$ and $CO_2$.

In a conventional burner, the tile (quarl) shields the flame reaction from gaseous radiation and severely limits reentrainment of furnace gases, resulting in much higher reaction temperatures, hence higher NOx.

With preheated air, NOx generation increases as burning begins in the tile. However, if the combustion takes place outside the tile (in the furnace) with large quantities of inerts in the reaction, little effect is noted on NOx generation with preheated combustion air. If air preheat is used to raise the process temperature, NOx will again rise because the reentrained inerts will be at higher temperatures, thus raising the combustion reaction temperature.

When the oxygen concentration is only moderately above stoichiometric, the combustion reaction will speed up, raising the temperature, which in turn will raise NOx. As the oxygen quantities increase above 4 to 6%, depending on the specific burner, the combustion reaction will cool, lowering NOx. The local oxygen concentration at which this phenomenon occurs depends on the completeness of the mixing of reactants in the particular burner.

Some engineers are concerned about residence time as a significant factor in chemical reactions at high furnace temperatures. This is rarely the case because reaction rates are extremely fast. They double every 16°F (8.9°C) rise in reaction temperature; thus, equilibrium is attained extremely quickly at 1800 F and above, assuming excellent mixing. It has been said that NOx generation at equilibrium is 8,000 ppm. This is true, but only at a high temperature such as a theoretical adiabatic flame temperature at 3500 F. When there is gaseous heat transfer, plus large quantities of furnace gas reentrainment into the reaction, the actual temperature of the reaction may be 3000 F or less, where the equilibrium NOx would be lower.

Whether or not the inerts entering the combustion reaction are recirculated, they are at a temperature that is several hundred degrees higher than the furnace temperature. The inerts will require energy to reach the combustion reaction temperature, which must be at an even higher temperature, resulting in an overall lowering of the reaction temperature, hence generating lower NOx. In summary, NOx generation in the combustion reaction is mainly a function of the *actual* reaction temperature. (This discussion assumes no fuel-bound nitrogen, which increases NOx.) (See sec. 5.12.)

## 6.4. CONTROLS AND SENSORS: CARE, LOCATION, ZONES

Temperature control can be no better than the sensors upon which it relies. Although operators and engineers are inclined to trust the measurement of temperature to those who specialize in that field, the operating engineers must be aware that they cannot expect greater accuracy from a control than is put into it by the sensors. (This applies to pressure and other sensors as well.). While T-sensors are usually very good at replicating, they need to be calibrated. And it is the duty of everyone involved around a furnace to be alert to conditions that may cause sensors to deteriorate.

If T-sensors, including thermocouples, are covered by a protective tube, that builds in an error and a time delay. Cooling air jets or water-cooled surfaces anywhere near sensors can be misleading. Try to locate T-sensors close to the load pieces that are to be heated—not the walls, hearth, or roof. Of course, they also must be somewhere where they are never subject to damage during loading or unloading—and watching out for them must be stressed over and over to operators.

Cold junction temperatures should be uniform for all sensors. Check regularly for causes of either hot or cold junction degradation. Avoid exposure to high temperature, oxygen, moisture (condensation), or corrosive atmospheres or liquids.

Unless it is physically impossible to place T-sensors in tight physical contact with load pieces, one must expect delays in temperature reaction. Controlling gas or wall temperature is a poor substitute for controlling load temperature. If thick, heavy pieces have to be heated all the way through, time delays in conducting heat to their centers can result in a hysteresislike roller-coaster ride for the temperature controls. This same sort of time delay versus control setpoint can apply to furnace pressure control when repressurizing a large furnace volume. Make changes slowly, with a lot of patience.

Remember that many control measurements are implied or indirect or have a time delay, and need study to improve operations.

Control of input, flow, or pressure is generally more gradual and more precise with *variable frequency drives* (VFD; see glossary) on pumps, blowers, and fans than with control motors and valves, or (worse yet) with dampers. If many zones are supplied from one blower, VFD is not practical; therefore, careful linearization of both actuators and valves is necessary.

*Moisture control in drying processes* has conventionally been done inferentially by humidity sensors in the discharge air stream, but moisture content sensors at the discharge end of the dryer are preferred. Both amount to feedback control, which responds more slowly than feedforward control. For thick load pieces, the mass transfer time to their surfaces may dictate use of feedforward control by locating sensors within the loads (usually difficult) or earlier in the traverse time within continuous dryers. In view of the dead time of some moisture sensors, locating the control moisture sensor(s) at or nearer the entrance will help improve production, product quality, and energy conservation.

Many *reheat-furnace* managers have spent their limited capital budget on new controls, hoping to reduce fuel costs and improve product quality, but results have been disappointing. The real cause of the imperfect results has been the length of the heating zones.

To understand this zone length problem, the reader should envision a 100 ft (30.5 m) long furnace, top and bottom fired for heating 8.5" to 10" (0.216 m to 0.254 m) thick load pieces.

| Zone | Past Practice Zone Lengths |
|---|---|
| Unfired charge zone | 15 ft (4.57 m) |
| Preheat zone | 30 ft (9.14 m) |
| Heating zone | 30 ft (9.14 m) |
| Soak zone | 25 ft (7.62 m) |

Except for the soaking zones, these zones are far too long to adequately control the furnace, especially after productivity adjustments. For example, after a delay, the newly charged product must move through the unfired zone and 50 to 60% of the preheat zone before the control temperature measurement senses the newly charged, much colder material. This happens in both the top and bottom preheat zones and again in each of the heat zones, resulting in the new material discharged too cold to roll.

This "accordion" or control wave problem is caused by greatly extended heating time for all material in the furnace during the delay. All material will be more uniformly heated, top to core and bottom to core, and to higher temperatures than intended. After the end of a delay, several pieces would be discharged to check the gauge. When the gauge is found satisfactory, rolling begins at a rate of, say, 80% of maximum.

The load pieces charged at the time of gauge checking usually can be rolled without difficulty. However, after the 80% mill speed is in effect, the new cold material entering the furnace will be heated at very low rates in the unfired zones and in the first 50 to 60% of the preheat and heat zones. If the temperature measurements in the preheat and heat zones are sensitive, the firing rates of the preheat and heat zones, top and bottom, will be driven to 100% for the balance of the time the new material is in those zones. With these higher firing rates, the material now entering the furnace will be heated above the uniform conditions desired. After this instability begins, it is difficult—if not impossible—to achieve uniform heating, regardless of the control program.

If the heating zones from the charge door to the soak zone were shorter and more numerous, for example, seven instead of three top and bottom zones (and if firing were added in the charge zone), the furnace program would enter the correct action at the second or third piece extracted. Instability of the firing rates would be avoided, fuel rates reduced, and product quality improved.

Some might say that this solution would be too costly, but they have not experienced actual heating problems that operators have after delays or considered the cost of all the scrap made while waiting for the "accordian effect" to settle out. It is unfortunate that new equipment installers and mill managers who make new equipment decisions do not stay around long enough to suffer the day-to-day heat/control problems of the operators.

With the seven heating zones (four top and three bottom), the temperature measurement would control each small zone as the heating curve predicts, and would not get out of step as was the case with larger zones. To build a furnace with many zones, as indicated, it would probably be roof or side fired. If a furnace is to be side fired, it would need control of the product length temperature, using ATP technology.

A side effect of the "accordion problem" with reheat furnaces having too few and too large zones (that could be avoided by many heating zones), would be charge zones hotter during low productivity than during high productivity. For example, if the program calls for the product leaving the heat zone at 2200 F (1200 C) but, as a result of a mill productivity upset (delay), after which cold loads have moved into zones that had throttled to low firing rate, 2100 F (1150 C), the control cranks its way up and up to perhaps 100% input because it lacks the wall temperature to transfer the heat needed for the new cold load. Under this scenario, the waste gas temperature leaving the heat and preheat zones will be very high, contributing to high fuel consumption. With shorter zones, only the few small zones needing to raise firing rates would fire harder, not the balance of the furnace, so the flue gas temperature will rise slightly but not to the point that high-productivity flue gas exit temperature will be lower than it will be with low productivity.

The authors hope that these ideas will help managers and operators understand the real control problem after delays and figure out how it can be corrected to reduce fuel rates, reduce rejects, and improve product quality.

### 6.4.1. Rotary Hearth Furnaces

The reader is urged to review sections 1.2, 4.3.2, and 4.6.1.2 for descriptions of rotary *hearth* furnaces—not to be confused with rotary *drum* furnaces described in section 4.2.3.

**Example 6.1:** This is a case study of a 45 ft (13.7 m) diameter donut (see glossary) rotary hearth furnace, similar to figure 1.8, that was having problems with low production capacity. The inside cross-section dimensions of the donut-shaped, circular gas and load passageway (a circular tunnel furnace) are 4.5 ft (1.37 m) high × 12 ft (2.66 m) wide. Most of the furnace gas flow is counter to the load movement.

The gases from the burners in zones 5, 4, 3, and 2 may exit through the flue, some via the space under the present single baffle to the flue, or through the discharge and charge doors. About 20% of the total gas flow is in the same direction as the product movement. If the baffle clearance were reduced, the hot gas moving in the same direction as the loads would be reduced to 5.8%. The flue and a short stack are sometimes put at the base of the outside wall to minimize short-circuiting of furnace gases along the ceiling and inner wall.

Furnace problems uncovered were:

a. a need for two more baffles
b. lack of burners in zone 1
c. instability of temperature control necessitates optimizing the PID loop and linkage settings, plus relocation of temperature control sensors
d. needed repositioning of the load pieces relative to the outer wall and
e. advisability of enhanced heating for crosswise uniformity, and more hot air capacity

**Add baffles**, and make the existing baffle adjustable. Install two additional baffles (one between the final zone and the discharge vestibule, and the other between Zone 1 and the charge vestibule). These will allow control of furnace pressure by greatly reducing furnace gas loss through the charge and discharge doors. (See also sec. 1.2.2, 4.6.1, 4.6.7–4.6.9, and 5.8.2.)

Reducing hot gas leakage by adding two baffles will reduce the aforementioned difficulty. One of the two additional baffles should be between the final heat or soak zone and the discharge vestibule, and the other between the preheat zone and the charge vestibule. These baffles should have only 2" to 3" (50 to 75 mm) clearance above the maximum load height. This reduction of gas escape area results in a proportional reduction of furnace gas loss through the discharge vestibule (typically reduced to one-fourth of the flow without the baffle addition). This forces most of the poc to flow with the load piece movement and exit via the flue adjacent to the baffle by the charge door. (See fig. 6.4.)

If three baffles had been used, with a moveable baffle between the charge and discharge vestibules, the sawtooth roof rotary furnace would have delivered at least

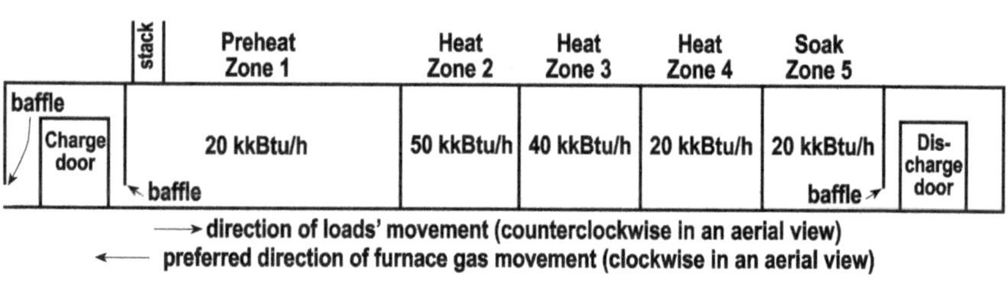


**Fig. 6.4.** Unrolled side view from outside a side-fired donut rotary hearth furnace. The baffle (at *left*) between the charge and discharge doors is moveable and/or has an air curtain. (See also fig. 1.8.)

*Evolution of firing methods for large rotary furnaces.* Round furnaces had limited capacity and poor control of gas flow pattern. The first donut rotaries had burners through the sides of both inner and outer walls, but the inner circle of burners were difficult to get to and to work on.

The next method was called the *sawtooth* roof system, wherein each fired zone had one tooth of the sawtooth roof with burners firing through the vertical wall of the tooth toward the charge door, firing counter to the direction of product movement. This system was less expensive for larger diameter products and furnaces because it required fewer burners and less piping, especially if preheated combustion air was used.

The sawtoothed roof furnaces sometimes had several zones practically unfired, but they at least had some firing even with reversed gas flow. Furnaces side fired, or roof fired with flat-flame (type E) burners had burners all along the walls or roof. Sawtoothed roof furnaces may have cost less, but with large loads and one fixed baffle, control was difficult. Regardless, a move to sawtooth roofs proceeded because of less cost.

acceptable tons per hour. With large-diameter products, the moveable baffle can be closed during operation, and only opened during a delay to allow the hearth to be backed up so that a load or loads that had been discharged or were about to be discharged could be returned to the soak zone to keep them hot. At the same time, newly charged pieces would be backed temporarily into the discharge vestibule.

In the arrangement before this recommended improvement (i.e., with only one baffle), a 12" diameter round load would require a clearance to 16" in normal practice. When no piece was under the baffle, up to 25% of the poc was allowed to move in the direction of the product (parallel gas and load movement instead of the preferred counterflow). In one instance, this leaking caused nearly half of the furnace zones to be underfired, and with little, if any, hot gas flow in the entry part of the zone where the gas turned around. Each zone downstream from this gas-turnaround point all the way to the discharge would be controlled by the thermocouple at the discharge of the preceding zone. The result was that calculated furnace capacity could not be met! This may have caused the removal of burners from zone 1.

Furnaces heating product pieces of 8" diameter and less can be corrected for the previous problem by the addition of two baffles with 2" clearance as discussed earlier. For furnaces that must heat larger diameter products, the problem can be solved by installation of a moveable baffle between the charge and discharge vestibules, and holding a 2" clearance while operating, raising the baffle when product must move past it.

With the suggested change, the quantities of furnace gases escaping through the charge and discharge doors would be so small that the furnace pressure would be controllable, reducing infiltrated air, and would allow effective heat transfer from reburnering zone 1, increasing furnace capacity and reducing fuel rates. Hot gas leakage from zone 5 to zone 1 would be minimized. The two additional baffles also limit loss of combustion gases through the doors.

Because of operator resistance, a moveable baffle has never been accepted. Co-author Shannon therefore suggests an air curtain at the bottom of the baffle separating the charge vestibule and zone 1. The air curtain (a row of small air jets issuing from drilled holes in an air manifold on the bottom of the damper) should be aimed downward, but at a 20- to 40-degree angle from the vertical toward the charge vestibule. This curtain builds a barrier, preventing escape of hot gas from the discharge vestibule or entry of cold tramp air from the open charge door. In the event of a delay, the recently charged pieces can be backed temporarily through the air curtain's jets into the discharge vestibule.

To prevent gas flow under the baffle between the soak zone and the discharge vestibule, a pair of high-velocity burners are suggested, firing opposed to one another under that baffle—creating a 2500 F (1370 C) hot mix baffle. This not only stops poc or cold air flow under the baffle but also balances some of the heat losses from the discharge vestibule. With these arrangements, sawtooth-roof-fired furnaces (firing to the charge baffle) would finally reach the productivity expected of them.

**Add burners in Zone 1.** Originally, rotary-hearth-type furnaces had burners in zone 1, but hot gas leakage from the last zone toward zone 1 caused increased fuel rates. When firing in zone 1 rose from, for example, 0 to 20 million Btu/hr, it caused an additional 5 million Btu/hr of zone 6 gases to move toward the flue. As these hot gases moved past the (generally) open doors, some of the gases moved out through the tops of the doors while cold outside air moved into the hot gas stream, passing closer to the hearth. The result was less hot gas moved toward the flue at much lower temperatures, causing higher fuel consumption.

If any of the major heating zones experienced more of its poc moving toward the discharge zones, that could reduce the heat transfer to the loads in the entry end of that zone. In addition, the temperature of the gases passing the T-sensor increased because they did not have as much opportunity to transfer their heat, thus causing the temperature control to reduce the zone's firing rate. As the gases of smaller volume moved into the next zone (toward the discharge door), less heat was transferred into the entry space of the next zone than could have been transferred if the gases had been moving countercurrent to the loads. This difficulty repeated in each zone all the way to the discharge door, producing an "accordion effect" or control wave problem. (See glossary.)

Perhaps the operators did not realize that the difficulty was happening, but they found that if zone 1 was unfired, the fuel rate dropped and furnace capacity did not suffer (except when the number of delays was very high, causing a large loss in furnace capacity). Pleased with the fuel benefit, apparently operators did not worry about the capacity problem then, and so the first zone burners were removed. This unwise action removed heat input from 105 degrees of rotation, of a possible 340 degrees, or nearly one-third of the effective heating area of the furnace.

From furnace heating curves, assuming using cold air, zone 1 should be fired with 20 million gross Btu/hr to reach a capacity of 24 mtph. For zone 2 to reach 24 mtph, assuming 800 F (427 C) preheated combustion air, would require a firing rate increase from 10.8 to 23.17 kk Btu/hr.

**Stabilize temperature control** by (1) *optimizing the PID loop and/or linkage settings* to minimize cycling of energy inputs to the zones, and (2) *relocation of temperature control sensors*. A control system, patented by North American Mfg. Co., with two sensors per zone provides excellent heating in every zone under normal conditions and largely remedies problems from delays. This method of control requires that all T-sensors (except the zone 1 entry sensor) be inserted through the outside wall 2" to 3" (25 to 76 mm) above the hearth. This low location provides a measurement closer to the true product temperature. The material on the hearth must be indexed to about 6" from the furnace wall. All thermocouples should be placed in depressions in the wall for mechanical protection.

The charge zone (zone 1) entry thermocouple should be placed high in the furnace outer wall in a position where it can "see" the load material and "feel" the hot gases moving though the zone. The position of this "early" thermocouple should be about 6 feet into the zone. The zone 1 discharge thermocouple should be near the hearth about 4 to 6 feet from the end of the zone to protect the product from overheating. (Depending on the process, if there is no likelihood of material damage at the end of the zone, the discharge thermocouple and control may be omitted.) Normally, the entry and discharge thermocouples should be within 6 feet of their respective ends of any particular zone.

Present temperatures in zone 1 are very difficult to understand because there are two gas paths that supply zone 1, even though the primary measurement senses only gases from zone 5. The two paths are gases from zone 2 and gases from zone 5. After two additional baffles and a nearly closed middle baffle are in place, gas from zone 5 will be of no significance while gases from zone 2 will generally be all the furnace gases. zone 1 gases will be fired to hold the waste gas temperature constant. With a constant temperature at the flue, heat input to zone 1 will stabilize heating needs in the balance of the furnace, without the present cycling of load temperatures. In addition, zone 2 will add more stability with the rounds indexed to 6" from the outer wall and with T-sensors 2" above the hearth controlling temperatures of the loads. The rounds will be heated more effectively and steadily. With these improvements and with enhanced heating, rotary furnaces will be equal; rectangular furnaces in productivity per unit of hearth area.

In each zone, a controlling sensor should be positioned early in the zone so that it can react quickly to temperature changes. A second T-sensor, also with a controller, should be placed near the discharge of the zone with a setpoint just below the temperature at which damage to the product could occur. The control signals from these two sensors (inlet and outlet of each zone) would pass through a low-select device so that the control with the lowest output signal would have that signal sent to the control drive.

The two controllers should operate through a low-select device to gain heat head without damage to the product, yet providing automatic heat head adjustment to maintain constant product temperature.

The benefits of such a control method are that mill production changes will be "felt" quickly and a near constant load temperature will be accomplished by varying the zone temperature. Conventional systems hold zone temperatures constant while allowing the product temperature to vary whereas constant product temperatures are desired. This system is very effective when the furnace is starting up after a mill *delay*. The benefit is accomplished because the entry thermocouple very quickly senses the change in product temperature and actively pursues heating that load.

Capacity reduction due to a production delay results from cold product following much hotter-than-normal product after each delay. Once the mill has been readjusted for size after a delay, and has moved to perhaps 70 to 100% of maximum production, the next load piece entering the furnace moves nearly to the zone 2 T-sensor before that zone's firing rate control increases its input. With that measurement perhaps 80% through the zone, there was insufficient time to make up for lost heating time. This same difficulty will often be reenacted in each succeeding zone, frequently reducing heating capacity by 50% or more. This is the series of phenomena that coauthor Reed has termed the "accordion effect" or "control wave effect." (See glossary.)

*Heat head* (temperature) should be automatically added or subtracted as needed to hold product surface temperatures as desired. Heat heads to 100°F above normal furnace setpoints may be desirable. Holding the product at a near-constant distance from the thermocouple is necessary for the control to hold the product temperature near constant; therefore, the product should be charged at a fixed distance from the outside wall of the furnace chamber.

**Position loads relative to the outer wall:** Because of possible cooling of the ends of pieces if they are too close to either the inside or the outside wall of the donut, the maximum practical load piece length should be about 1 ft (0.3 m) less than the hearth width. If the lengths of the load pieces are less than the maximum usable inside width of the rotary hearth furnace chamber, it is usually preferable to locate them within about 6 in. (0.15 m) of the inside surface of the outer wall, permitting the greatest load in a circular furnace, with maximum space between pieces for good heat transfer exposure. (See fig. 6.5.) This leads to maximum furnace production with best possible temperature uniformity, minimizing "barber-poling" (see glossary) in seamless pipe and tube.

If the furnace is fired only with conventional (type A) burners or with long-flame (type F or G) burners (fig. 6.2), in its outer wall, the recommended positioning usually puts loads where they can benefit most from the radiation and convection characteristics of those flames. This combination plus two more baffles (to control gas movement and allow effective furnace pressure control, *and* reinstating the firing of zone 1 almost to the charge door) raised the furnace capacity (figure 6.7).

**Add enhanced heating,** with more input. Enhanced heating high-velocity type H burners (fig. 6.2) add effective heat-transfer area. The increased firing rate in Zone 2 will help provide extra heating capacity that the heating curves predict would be necessary to obtain a full 24 mtph furnace capacity. Figure 6.6 shows the existing furnace temperature curves at a production rate of 12 mtph.

More input will be necessary to raise the furnace output to a full 24 mtph capacity. (See fig. 6.7.) This will require more fuel and *additional combustion air supply* capacity in both zones 1 and 2, preferably via regenerative firing or with larger recuperators.

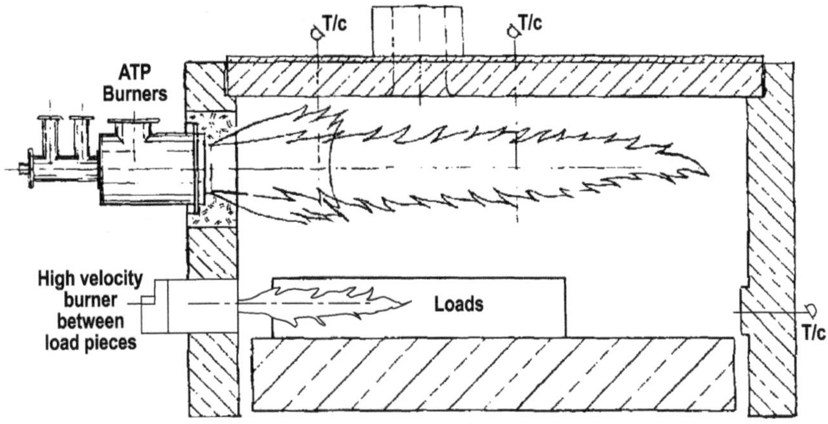


**Fig. 6.5.** Sectional view of a rotary hearth furnace (such as fig. 1.8) with enhanced heating. This also could be a car-hearth batch furnace or in-and-out batch-box furnace. In many cases, the higher velocity burners would be smaller (relative to the main burners above) than they appear in this drawing. In other than rotary hearth furnaces, the high-velocity burners should fire between piers and opposite the main burners—to further enhance circulation.

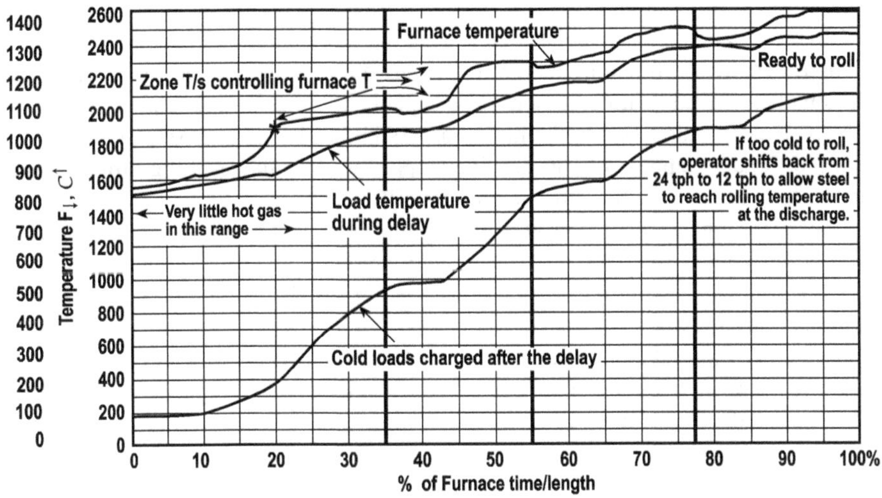


**Fig. 6.6.** Calculated time–temperature heating curves for a rotary hearth donut furnace showing the effects of delays *before* addition of enhanced heating burners. (Directions for calculating time–temperature curves are given in chap. 8.) The top two curves show what happens upon restart at normal tph after a delay. The bottom curve shows that loads charged after resumption will be too cold to roll, forcing a fall back to half the normal tph.

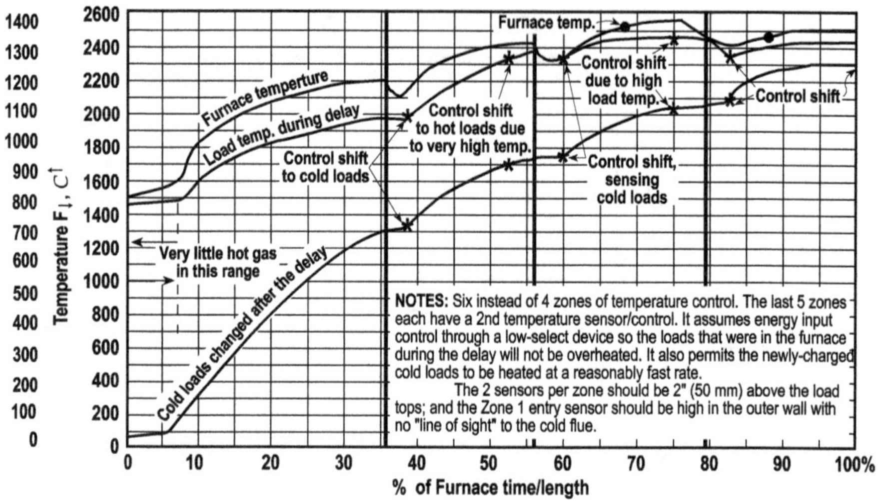


**Fig. 6.7.** Predicted time–temperature steel reheat curves showing better results *after* adding enhanced heating burners for the furnace of fig. 6.6 at a 24 tph production rate. Control T-sensors were added in positions nearer the charge end of the furnace. (See NOTES on the graph.)

If capital money is not available for either of these more efficient improvements and if production demands take priority over reducing fuel consumption, then more cold combustion air is an option.

Obviously, adding more fuel and air is necessary for doubling production capacity. A bonus benefit was found in the lower fuel rate during holding (for line stoppages). The small enhanced-heating burners were capable holding furnace temperature with only 10% excess air whereas the main burners had to be set to 100% excess air to hold the furnace temperature during line stoppages. This makes a big difference in the %available heat and therefore in the fuel bill.

The preceding improvements will provide more efficient heat transfer and reduced reject loss. When a product fails to meet quality requirements, the following must be reinvested all over again: *fuel, labor, power, materials that cannot be recycled, and prorated cost of capital investment.*

Figure 6.7 shows the proposed furnace temperature curves at 24 mtph production rate. Each zone now has a second T-sensor/control with energy input control through a low-select device so that the loads that were in the furnace during a delay will not be overheated. This also permits the newly charged cold loads to be heated at a reasonably fast rate. These curves show how a better understanding of the heat transfer phenomena can improve operation and control.

Each zone now has a second T-sensor/control with energy input control through a low-select device so that the loads that were in the furnace during the delay will not be overheated. This permits the newly charged cold loads to be heated at a reasonably fast rate. The improvements allow prompt input to the cold loads entering immediately after a delay, continuing the 24 mtph production rate.

In summary, the preceding discussions explain how furnace temperatures are produced from the present control temperature measurements (fig. 6.6) and the changes that must be made in the furnace to produce the furnace temperature curves of figure 6.7, raising furnace capacity from 12 to 24 mtph. Changes are:

*a.* Add two baffles plus a moveable section at the bottom of the center baffle to practically eliminate reverse poc flow in the furnace. This will redirect the gas flows so that the last 90% of furnace gases move countercurrent to the load movement. Furnace pressure then will be controllable even with charge and discharge doors open.

*b.* Install burners in zone 1.

*c.* Stabilize temperature control (1) by optimizing the PID loop and/or linkage settings to minimize cycling of energy inputs to the zones and (2) by relocation of control sensors.

*d.* Index the load piece positions to within 6" (0.152 m) of the outer wall hot face.

*e.* Install enhanced heating (high-velocity, type H) burners in zones 1 and 2 to provide additional effective heat transfer area. The increased firing rate in zone 2 helps provide the extra heating capacity that the heating curves predict would be necessary to utilize the full 24 mtph furnace capacity.

### 6.4.2. Zone Temperature in Car Furnaces

Car-hearth (batch) furnaces, commonly used for heat treating and in heating for forging, should be divided into zones in two ways, if a ±15°F (±8°C) temperature range must be certified on grid of T-sensors strung across the furnace. The floor plan of the furnace should be divided lengthwise into a minimum of three zones, and top to bottom in each of the longitudinal zones, for a minimum of six zones.

The lengthwise division of the furnace into three top and three bottom zones is necessary because of the differences in heat loss and in heat transfer between the center and the ends. Similarly, because of the difference between the two ends, usually only one end has a door (high loss) whereas the other end does not (low loss).

The reason for dividing the longitudinal zones into top and bottom zones is because there are usually considerable differences in the losses and the heat transfer rates at different levels. Door seals may leak more outward at top than inward at bottom. Car seals may leak more at front than at back, and more at front and back than at the sides. In some cases, the flow pattern of the flames' poc may completely upset the predictions of the previous two statements because of different impacts or suctions caused by the jet effects and heat transfer patterns of the many flames. Another reason for separate top and bottom zones is that cost and practical reasons often result in as much as 25% less clearance space below the loads than above them.

In furnaces loaded with pieces of very different front-to-back dimensions, three or more lengthwise zones are necessary for uniform heating. In furnaces loaded with pieces having very different thicknesses (vertically), two or more vertical zones should be used to achieve uniform heating.

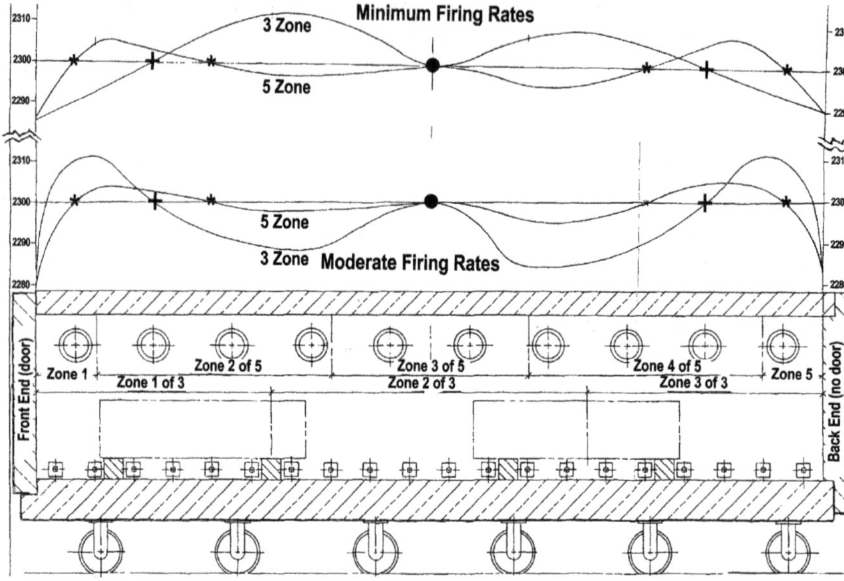


**Fig. 6.8.** Temperature patterns in a car-hearth furnace with three versus five zones, and modulated versus minimum firing rates. ⁺3-zone T/s *5-zone T/s

All variations of the previous paragraph are reasons for careful attention to (*a*) zoning for temperature uniformity control (this chapter) and (*b*) burner locations, burner flame types, and furnace flow patterns (chap. 7). (See fig. 6.8 showing soak temperature variations between three and five lengthwise zones at minimum firing rates (top set of curves) and at moderate firing rates ([bottom set of curves]).

***Constant and careful attention to load placements by those loading the furnaces is crucial in avoiding rejects and preventing customer dissatisfaction.*** Above all, the many factors affecting temperature uniformity make it extremely important that those placing the loads in the furnace have superior training and an understanding of temperature distribution of each of their furnaces at all firing rates and conditions.

When heating stock of thin cross section, it is often practical to reduce pier height to less than 1 ft (0.3 m) because the saving from reducing lag time does not justify the cost of higher piers. With large-diameter ingots, however, the reduction of lag time definitely justifies taller slots below the loads. For example, with a 78" (2 m) ingot, the lag time can be reduced from $(78/10)^2 \times 1.45 = 882$ min to $(78/10)^2 \times 1.05 = 638$ min, or a saving of 243 min = 4 hr. This results in a reduction in cycle time.

To limit temperature differences to ±15°F (±8.3°C), the top and bottom end zones (door and backwall) should be as short as possible. The minimum practical number of burners in these four end zones is one burner each. To limit the length of the temperature slope in each of these zones to the end zone itself, the temperature control sensors in each of these end zones should be located at the junction between the door or back-end zone and the adjacent zones, top and bottom.

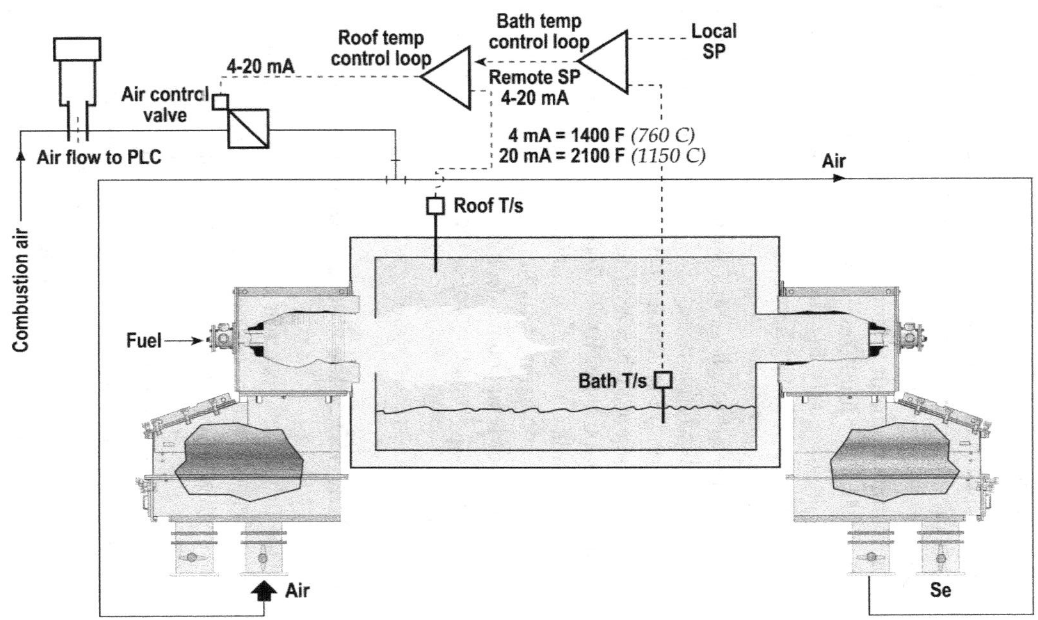


**Fig. 6.9.** Direct-charged aluminum melting furnace with cascaded temperature control and regenerative burners. On the next 20-sec cycle, two air valves, two exhaust valves, and two fuel shutoff valves will reverse positions. Ma = milliamps. Se = suction exhaust. SP = setpoint. T/s = temperature sensor. Courtesy of North American Mfg. Co.

If furnaces are expected to heat a wide variety of load shapes and sizes, the operator will need more zones between the two end zones if quality products and minimum cycle times are to be expected. If in doubt about the future loading, the furnace designer should err in the direction of more zones for future versatility.

### 6.4.3. Melting Furnace Control

A very carefully thought-out temperature control system is necessary on large metal melting furnaces if acceptably high production rates are to be attained without excess dross formation. Figure 6.9 shows only a suggested temperature control portion of a control system for an aluminum melting furnace fired with a pair of alternately fired, low-NOx regenerative burners. It utilizes a cascaded temperature control loop. Additional control systems are necessary for air/fuel ratio, furnace pressure, flame monitoring, high-limit temperatures, and perhaps pollution high limits.

In the aluminum melter of figure 6.9, the temperature in the furnace is automatically controlled by adjusting flow through the burner air control valve in response to a signal from the T-sensor in the furnace roof. The setpoint of that roof T-sensor is cascaded from the bath T-sensor. If the bath temperature is low, the roof temperature setpoint will be high, providing more heat transfer to the liquid metal surface. A typical setpoint range might be 1400 F to 2100 F (760 C to 1150 C). When the bath temperature approaches its setpoint, the output of the bath temperature control loop will decrease, lowering the roof temperature setpoint. As the roof refractory tranfers its stored heat to the bath, the roof temperature decreases. Thus, this system allows optimum melting rate without overheating the roof or the liquid metal surface (which would increase dross formation).

## 6.5. AIR/FUEL RATIO CONTROL (see also pt 7 of *reference 52*)

The chain of command for air/fuel ratio controls is usually as follows: The burner or zone input control responds to a T-sensor (or steam pressure sensor in the case of a boiler). The burner input control (also termed furnace input control, kiln input control, etc.) may actuate a burner or zone *air* valve ("*air* primary air/fuel ratio control") or a burner or zone *fuel* valve ("*fuel* primary air/fuel ratio control"). Air primary air/fuel ratio control is more common with smaller burners. Many problems are avoided if each burner is equipped with its own ratio control. Where multiple burners are "ganged" in parallel downstream from a single air/fuel ratio control, if one burner has a problem with its ratio, all parallel burners of that zone will have the opposite difficulty, the intensity of which will be divided by the number of burners in the zone.

### 6.5.1. Air/Fuel Ratio Control Must Be Understood

Furnace engineers and operators must understand the many aspects of air/fuel ratio control for safety and for equality. *Mass flow control* is essential if the combustion air is preheated. Changing air temperature affects the weight of air passing through a control valve, affecting input rate and air/fuel ratio. Control valves are volumetric devices, but temperature changes density, which changes the weight of air delivered. The air volume delivered to a furnace should be corrected for temperature changes because the chemistry of combustion really requires a constant weight (or mass) ratio of air to fuel. The magnitude of the correction will vary as the square root of the absolute temperature. Most larger modern air/fuel ratio controllers have an input port for a signal from an air T-sensor. This type of air/fuel ratio control is called "mass flow control."

*Individual ratio controls at every burner* make it easy to modify the input profile pattern up and down or across a furnace without having to reset the ratio of each burner afterward.

*Small burners without preheated air* are generally controlled by cross-connected air/fuel ratio regulators (one for each burner). This arrangement is ideal because it saves the operator from constantly having to adjust the ratio—until the paint is worn off the hand dial—because of changing maldistributions of flows in either air or fuel manifold.

*Air and Fuel Manifolds.* It is difficult to correct bad manifold designs; therefore, it is important to be generous in initial air and fuel manifold sizing, and get it right the first time. (See fig. 6.10.) Designers should think of manifolds as plenums that should be sized for low velocities. A nonuniform air or fuel distribution often changes its maldistribution as burners are turned up and down. An easy, safe design has the manifold cross-sectional area equal to the sum of the cross-sectional areas of all of its offtake pipes. (See *references 54 and 60.*)

**Benefits of Good Air/Fuel Ratio Control** (see also sec. 6.5.2 and 6.5.3)

1. Safety from explosions and fuel-fed fires by minimizing the chance of accumulating a rich mixture in the confined space of a furnace or duct.
2. Lower fuel consumption because "ff-ratio" operation leaves fuel unburned if too rich but sends too much hot gas out the stack if too lean.
3. Better product quality, because the load surface is less likely to be oxidized when air/fuel ratio is too lean, and less likely to be carburized or have hydrogen absorption if too rich.
4. Rolled-in sticky scale is avoided by controlling air/fuel ratio to prevent a reducing atmosphere in the furnace. (Rolled-in scale causes pits which generally cannot be ground out.)
5. Less metal loss because less scale is formed.
6. Reduced scrap because poor air/fuel ratio control can result in the load being scrapped for fear of customer penalties.

### 6.5.2. Air/Fuel Ratio Is Crucial to Safety

Air primary control is generally preferred over fuel primary control for safety reasons. Burners are generally more stable if they should happen to go lean than if they happen

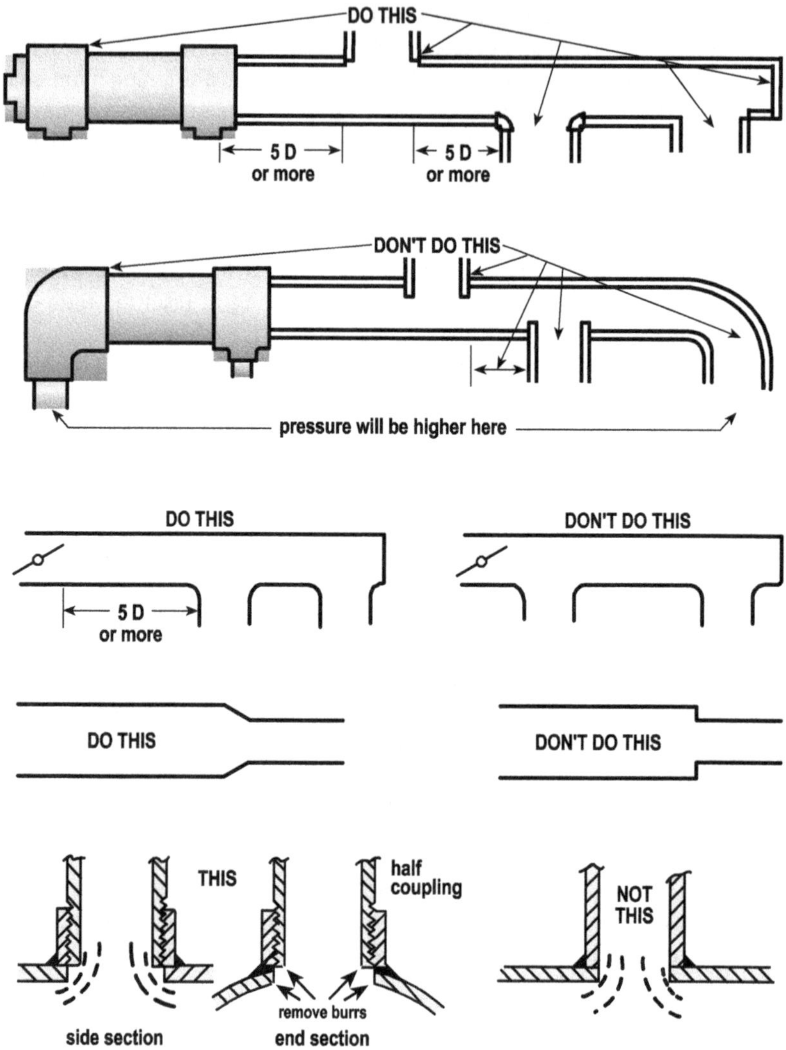


**Fig. 6.10.** Conservatively designed manifolds and headers assure uniform and easily adjusted distribution to all offtake pipes to individual burners. Streamlined computer-designed manifolds are for mass-produced internal combustion engines—not for a one-of-a-kind industrial furnace. (See *References 54* and *60*.)

to go rich. Having air lead the fuel (air primary) may avoid a dangerous flame-out when input is rising. If burners go rich, do not try a "soft shutdown" with a flame-out hazard impending. Do a *FULL shutdown* because otherwise unburned fuel may work its way back upstream into feed pipes and ducts, followed by hot furnace gases, followed by an in-duct explosion. "Soft shutdowns" that leave the air on low and do not trip the fuel safety shutoff valve (to avoid a time-consuming total restart) are very

**How to Burn Bunker Oil**

*Set the burners open wide.*
*Do not touch the valves at side.*
*Keep the pressure on the pump,*
*and up the bally steam will jump.*

*If the smoke is black and thick,*
*open up the fans a bit.*
*If the smoke is thick and white,*
*to slow the fans will be quite right.*

*For when sufficient air is given,*
*no smoke ascendeth up to heaven.*
*If the jets refuse to squirt,*
*assume the cause is due to dirt.*

*If the flame is short and white,*
*your combustion's complete, bright.*
*If the flame is sooty-orange and long,*
*your combustion is entirely wrong.*

*A wise man to his heater sees,*
*and keeps it at the right degrees.*
*To have it more is not quite wise,*
*because the oil may carbonize.*

*If you keep the filters clean,*
*no drop in pressure will be seen.*
*Should the pump kick up a ruction,*
*there's likely air within the suction.*

*There's more than what's said here.*
*To the rules you must adhere.*
*Junior engineers should know them,*
*or explosions may cause mayhem!*

*AUTHOR UNKNOWN.*

*Contributed by Gary L. Cline.*

likely to move the fans or blowers into the low end of their pressure curve, where surging may happen. Surging can pull unburned fuel into air-filled pipes or ducts, forming combustible mixtures, and then suck in hot furnace gas, providing a source of ignition, resulting in an explosion. *An explosion will be much more time consuming than a proper shutdown* (including fuel shutoff) than a restart.

If the fuel is not shut off *immediately* to prevent any unburned fuel accumulation or if the rich atmosphere has already accumulated considerably after loss of ignition, these situations are potential bombs. Do not open any furnace doors or other openings. Turn off air to any pilots or other sources of ignition that may still be burning, but do not change main gas or air flow. Let the furnace self-cool even though smoking. "Flood" the furnace with steam or other nonreactive gas such as argon, $CO_2$, or $N_2$, which are better coolants than a too-rich-to-burn fuel–air mixture.

Figure 6.11 cites two potential hazards leading to explosions and fuel-fed fires from using *constant* pilots instead of *interrupted* pilots when a single flame monitor is used to check both pilot flame and main flame. (See **pilot** in the glossary.)

The upper time-line diagram of figure 6.11 shows a burner startup situation where the air/fuel ratio control has erroneously been set too rich. The burner may have lighted as it entered the flammable zone (about 5% gas in a gas–air mixture, for natural gas), but its mixture soon became too rich to burn, exceeding the upper limit of flammability (about 15% gas in a natural gas–air mixture), exiting the flammable zone, with the flame going out. The pilot has its own controlled air and fuel supply, set at an air/fuel ratio between the flammability limits; thus, it stays lighted even

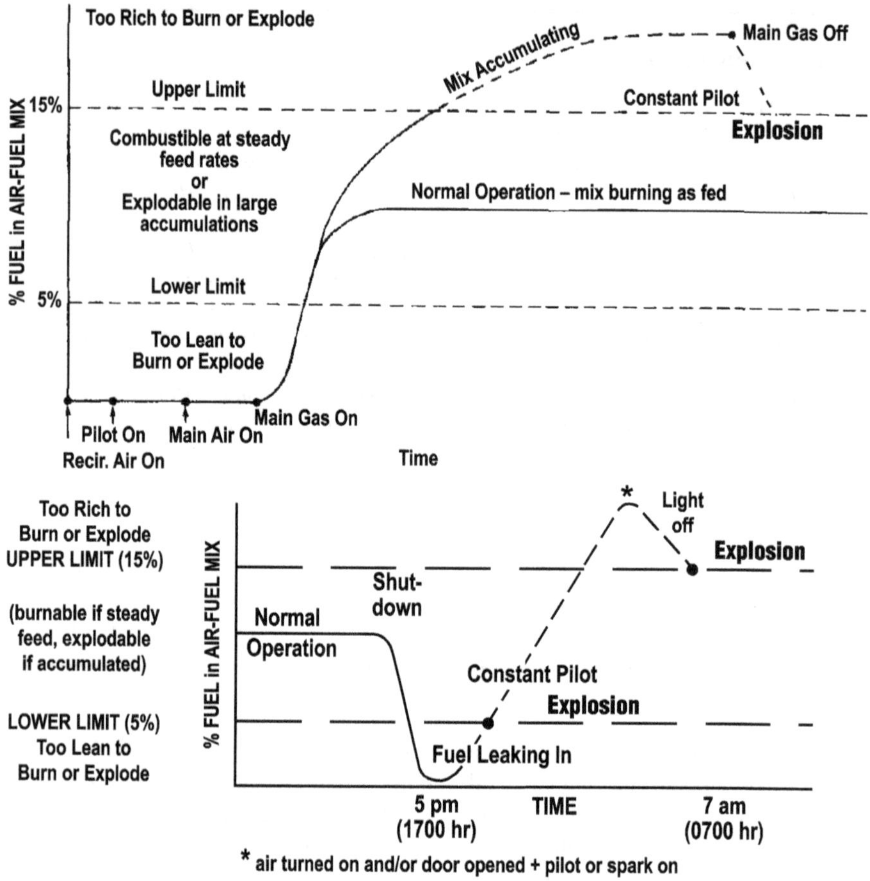


**Fig. 6.11.** Two time-line diagrams showing potential explosion situations. Use interrupted pilots—not constant pilots. (See glossary.) Courtesy of North American Mfg. Co.

though it is surrounded by a nonflammable atmosphere. The accumulated too-rich-to-burn fuel–air mixture will be ignited as an explosion when someone wonders why the burner went out after an assumed-to-be-normal startup and (*a*) opens the furnace door, letting in air, or (*b*) turns off the fuel to the main burners, allowing the continuing air supply to bring the accumulated rich mixture back to a combustible (explosive) mixture.

The lower diagram of figure 6.11 shows a situation where a burner fuel shutoff valve was not closed tightly or fuel somehow leaked into a furnace or oven overnight. If a pilot had been left running overnight, an explosion would occur as soon as sufficient fuel accumulated in the furnace to bring the fuel percentage up to the lower limit of flammability (about 5% gas in a gas–air mix, for natural gas). If there was no constant pilot or other source of ignition in the furnace while shut down, the air/fuel ratio could pass through the flammable (explosible) zone and rise above the upper limit of flammability (about 15% gas in a natural gas–air mix). The asterisk marks the

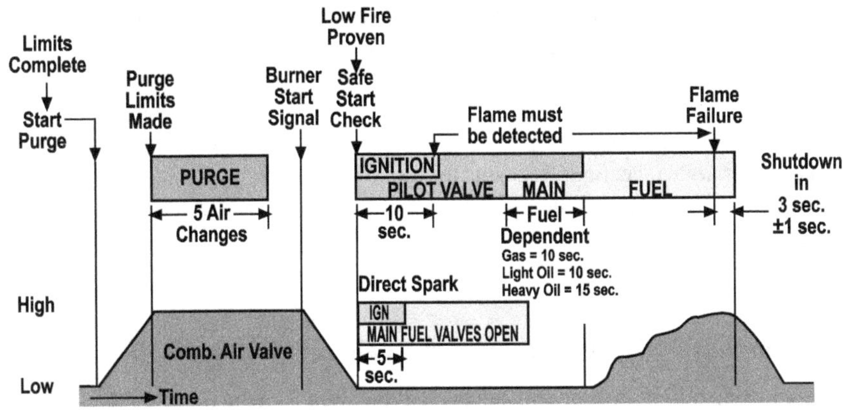


**Fig. 6.12.** Typical lighting/shutdown programs for a one-burner furnace. Some cases need more than five air changes. Courtesy of North American Mfg. Co.

point at which someone trying to light a burner the next morning (*a*) opens the furnace door, letting in air, or (*b*) turns on the main air, or (*c*) turns off the leaking gas valve.

Figure 6.12 shows a time line for a lighting and shutting down program for a one-burner furnace. The block diagram across the top shows the programmed functions designed to prevent accumulation of rich or combustible air–fuel mixtures. The bottom plot shows air flow during the programmed lightup and shutdown. This is for a system with interrupted pilot or direct spark ignition with a flame monitor that checks for presence of either pilot or main flame. All such programs should be designed, installed, and operated in compliance with insuring underwriter's requirements, those of government authorities, and recommendations of the U.S. National Fire Protection Association.

#### *6.5.2.1. Fan or Blower Surging Can Cause Explosions.*

There have been many explosions in air supply ducts that have not been adequately explained. A cause of explosions is surging of the air supply fan or blower as follows:

1. In an air-flow system that has been operating normally, the system resistances gradually increase, and as the air flow drops the fan discharge pressure rises, eventually reaching its maximum.
2. The fan surges, causing reverse flow in the whole air system including a burner.* That air flow reversal into a burner causes the fuel flow inside the burner to move into the air supply connections, followed by hot furnace gas.
3. The resultant air–fuel mixture in the air ducts is ignited by the hot furnace gases that flowed back through the burner.

*Fan surge also can occur if a fan's pressure versus flow curve has a hump as the flow demand moves back and forth across that hump, momentarily creating higher pressure downstream than upstream at the fan outlet, causing reverse flow and cycling.

4. The flame front is pushed faster than flame speed—up to sonic speed—by the expanding hot gases behind it. That is an explosion!

Small burners suffer little damage, but air control valves and dampers, the fan itself, fan inlet equipment, and people generally suffer damage. Coauthor Shannon was part of separate investigating teams for four different air supply/fan explosions. In each case, the teams were without solutions until the surge possibility was explained. In one of those cases, the team would not agree until after the second fan was destroyed.

### 6.5.3. Air/Fuel Ratio Affects Product Quality (see also sec. 8.3.1)

Oxides of iron, aluminum, copper, zinc, and glass often form on their molten surfaces, becoming inclusions in the final casting, probably causing it to be a reject. It is therefore desirable to minimize excess oxygen in contact with a molten metal bath; thus, a quality air/fuel ratio controller can be a major help in controlling product quality.

In heating the solid state of castings, forgings, or rolled products, there also is a danger of oxide formation on the product surface. This danger is less than in the molten state because the temperature level is less, reducing the probability of oxidation of the surface. Because of the higher temperature level of *steel* forging and rolling than of other materials mentioned earlier, however, the risk of unacceptable product quality from oxides (scale) is a great concern.

***6.5.3.1. Steel Quality Problems.*** Scale on steel is many different oxides of iron combined with sulfur, silicon, and alloying elements in the steel. The melting point of such mixtures varies from 1650 F to 2500 F (900 C to 1370 C), with a normal softening temperature of about 2300 F (1260 C). With large quantities of sulfur in the mixture or furnace atmosphere, the softening temperature may be as low as 1600 F to 1700 F (871 C to 927 C). Steel with high-silicon content may have a softening temperature as low as 2150 F (1177 C).

If the sulfur and silicon contents of a steel are not above normal, its scale melting temperature will be 2500 F (1371 C). If that temperature is reached on the steel surface, molten scale will run off the steel like water, a phenomenon termed "washing." If the melted scale is permitted to drop into a bottom zone, it will solidify and begin to fill the heating space, requiring jackhammers for its removal.

If scale softening occurs, the scale will have a highly reflective surface on its hot face, backed by a very porous dull material. If the reflective scale condition develops in the charge area of a reheat furnace, heat transfer to the steel in the remainder of the furnace will be significantly reduced. This "mirror effect" occurs above 2300 F (1260 C); therefore, charge zones should be limited to 2300 F (1260 C). Of course, tight control of oxygen in the furnace atmosphere (less than 2% $O_2$, with a quality air/fuel ratio control system) also helps minimize scale formation and therefore improves the heating efficiency in the charge zone.

If large percentages of *sulfur* are in either the furnace atmosphere or the steel, scale formation can easily be twice normal. If large quantities of *silicon* are in the steel, scale formation can be 30% larger than with normal silicon levels.

**Normal causes of scale formation are:**

1. *Atmosphere.* A slight deficiency of air forms about 20% of the scale formed with a slight excess of air. With only 50% of the air necessary to burn the fuel, almost no scale is formed. If the combustion air is increased to slightly above the minimum needed to burn all the fuel, the scale formed per hour increases by about five times. As the combustion air is further increased, very little additional scale is formed. Scale formed at higher levels of oxygen is usually from other causes.
2. *Temperature.* The most important factor in scale production is temperature of the steel surface. From 1900 F to 2000 F (1038 C to 1093 C), the rate of scale formation increases by 30%; from 2300 F to 2400 F (1260 C to 1316 C), 100%. At 2500 F, scale "washing" occurs.
3. *Time.* If time at temperature is doubled, scale formed increases by 40%.
4. *Velocity.* As the velocity of furnace gases flowing over a product surface is increased, the inert gas at the surface of the steel is stirred and enriched with more $O_2$, $CO_2$ and $H_2O$ (oxidizing agents), increasing scale formation. If the furnace gas velocity over the surface of the steel were doubled from 40 to 80 fps (12.2 to 24.4 mps), the scale formed would increase from 5#/hr to 8.1#/hr (2.27 kg/h to 3.69 kg/h), a greater than 62% increase.

### 6.5.4. Minimizing Scale

When excessive scale build-up occurs, it is often because of a problem with temperature measurement. Scale is oxide on the load surfaces. To melt scale, the temperature must exceed 2490 F (1365 C). If the control thermocouple is reading below this melting point, but scale is a problem, it becomes necessary to check the temperature measurement. Problems that may cause a T-sensor reading lower than the true furnace temperature are:

1. Using an "S" thermocouple (Pt vs. Pt-10% Rh), when an "R" thermocouple (Pt vs. Pt-13% Rh) should be used. Check whether the instrument that controls the temperature is calibrated for an "R" or "S." If an "S" thermocouple is calibrated for an "R," it may read 2292 F (1256 C), when the actual temperature is 2497 F (1370 C). If so, it is suggested that the setpoint be lowered by 50°F (28°C). If that only reduces the scale melting but does not stop scale formation, the setpoint should be lowered another 50°F (28°C).
2. T-sensor is reading low because of cool air entering the furnace through a T-sensor insertion hole in the furnace wall that is not properly sealed. Check this by visual sighting into the furnace. Is it blacker around the T-sensor?
3. T-sensor is not reaching the end of its protection tube.
4. T-sensor contaminated by furnace gases via a cracked protection tube.
5. T-sensor buried in scale.

Another condition that has caused numerous control problems (with both temperature and furnace pressure) is combustion gases and air leakage through cracks in the burner and/or the burner's refractory tile. These cracks may allow gases to flow laterally through the furnace insulation and/or refractories through a T-sensor opening, causing a misleading reading depending on the leakage path and whether the leaking stream is hot combustion gas or cold air. This may cause the actual furnace temperature to differ from the control temperature by as much as 100°F (56°C).

## 6.6. FURNACE PRESSURE CONTROL (see also sec. 5.3.1.3 and 7.2)

Controlling infiltration of air into a furnace is a major concern in maintaining high product quality and low fuel consumption. Any air *in*leakage, from negative furnace pressure,* (1) may chill part of the load causing inferior quality and (2) increase stack loss because of heat absorption by "tramp air."* Furnace gas *out*leakage will fail to heat the load as intended, (3) somewhat reducing production, and (4) raising fuel consumption. See a case history of benefits, table 6.3, page 278.

### 6.6.1. Visualizing Furnace Pressure

Visualizing furnace pressure requires measuring it by an inclined manometer with one leg connected to a tap through the wall to the furnace interior and the other manometer tap simply receiving pressure from the atmosphere just outside the furnace. To control the effects of furnace pressure, one must determine the elevation within the furnace of the zero pressure level (i.e., zero $\Delta P$ inside to outside the furnace) and understand how it affects interior furnace gas flows. (See pp. 58–69 of *reference 52*.)

The hottest gas within a furnace (or any enclosed chamber) rises to the top, creating a higher pressure at the furnace's higher elevations and a lower pressure at the furnace's lower elevations. (This is "stack effect"* within the furnace.) The zero gauge-pressure plane or "neutral pressure plane"* is the locus of points where the pressure inside the furnace is the same as the atmospheric pressure outside the furnace at the same elevation. The neutral or zero plane is the boundary between + and – pressures within the furnace. If there are leaks through the furnace walls, *furnace gases will leak outward from the space above the neutral plane* and *air will leak inward to the space below the neutral plane.* (See fig. 6.13.)

In most industrial heat-processing furnaces, it is desirable to have the entire furnace chamber at a positive pressure with an automatic furnace control system having a setpoint of 0.02 in. wc (0.5 mm) at the elevation of the lowest part of the load(s); or better yet, at an elevation just below the lowest leak. To keep out tramp air inleakage, raise the furnace pressure enough to drive the neutral pressure plane below the furnace bottom, in a liquid bath furnace, below the liquid surface level.

Furnace pressure or "draft"* is normally controlled by a damper in the stack, thus choking off the outflow of gases and pressurizing the furnace. (See sec. 6.6.3.) If negative furnace pressure is needed, use a speed control on an induced draft fan, a pressure (volume) control on an eductor jet, or a barometric damper.* (See sec. 6.7.1 on Turndown Devices.)

*See glossary for definitions, description, and discussion.

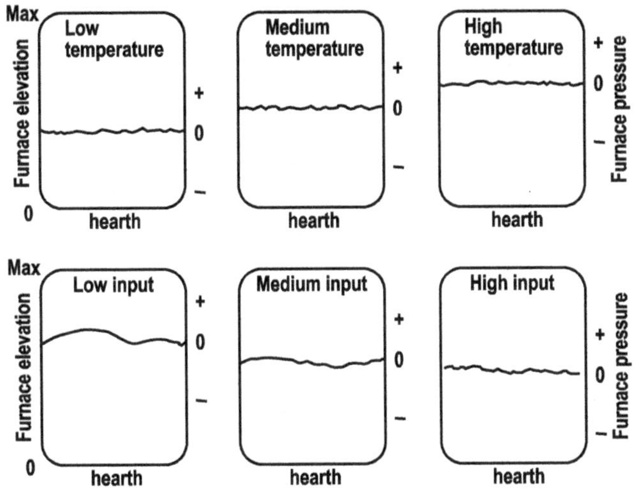


**Fig. 6.13.** Effects of furnace temperature and input on the level of the neutral pressure plane elevation shown on six sectional elevation views of a furnace with no furnace pressure control. If there were any gas flow in the furnace, the neutral pressure 'plane' would be more like a *wrinkled sheet* than a *plane*. The *top* three show the effect of temperature with no change in input. The *bottom* three show the effect of input rate with no change in furnace temperature.

For example, in a three-zone steel reheat furnace (soak zone, top heat zone, and bottom heat zone) with the zero line at the hearth level, any opening above the hearth will have furnace gases moving out of the furnace. Any opening in the bottom zone will have outside air moving into the furnace diverting hot gas flows from their normal paths. This infiltrated air will cause temperature nonuniformity; therefore, the working quality of the load will be affected adversely. If the furnace pressure was raised (by increasing the furnace pressure setpoint), the zero or neutral pressure plane would be lowered, less air infiltration would mean less oxidation of the product surface, and lower fuel consumption for unnecessary heating of tramp air.

### 6.6.2. Control and Compensating Pressure Tap Locations

Sensing taps for furnace pressure controllers are crucial in their design and location—not pluggable or oversensitive to transient vibrations and pressure blips. (See figs. 6.14 and 6.15) *references 55* and *56* show details of tap construction. Taps must be rugged, pressure tight, easily cleaned, and not damageable by heat. Pressure-sensing taps should not be opposite burners, beside burners, or anywhere they would be subject to the impact velocity from burner fuel, air, or flame jets. They should not be close beside fast-moving jets or streams where a suction effect would send a false signal. For these reasons, locating furnace pressure taps in the backs or sides of flues will lead to a lot of trouble because they will give obviously incorrect signals at some firing rates and not at other rates. (See fig. 6.13.)

The pressure-sensing tap must go all the way through the wall—metal skin and refractory. Flare the refractory opening into a cone so that crumbs of refractory and

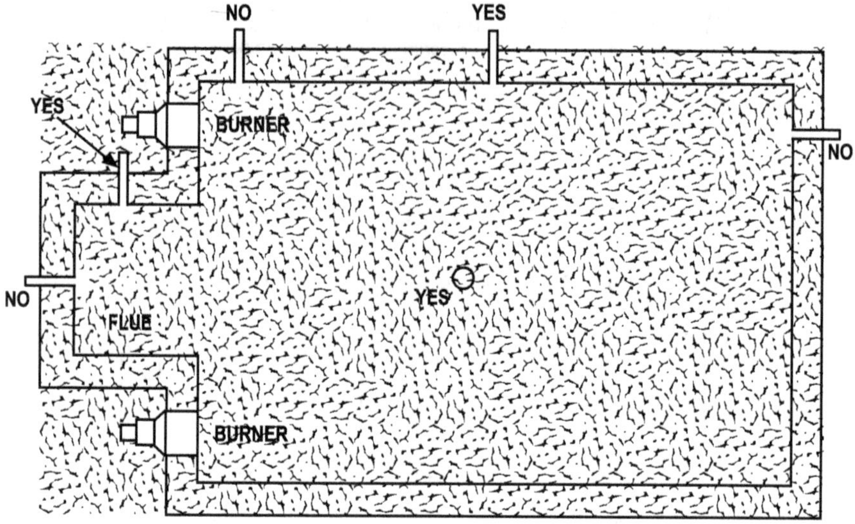


**Fig. 6.14.** Plan view of a melter furnace showing suggested furnace pressure tap locations selected to avoid both impulse and suction effects of burner jets or flue.

splashed metal can roll back to the furnace Hot, moist gases may get into pressure-sensing taps and condense there. All lines from taps to instruments should slope *up*hill away from the furnace and *down*hill away from the sensor so that condensate can flow back to the furnace by gravity—not into the instrument. If low spots (*U*s) in the signal tubing cannot be avoided, they should be fitted with reservoirs and drain taps.

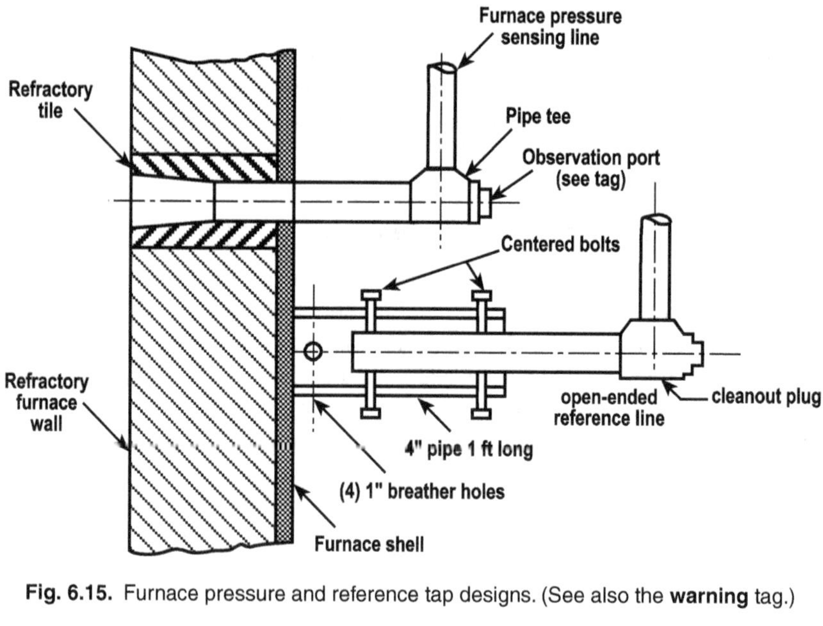


**Fig. 6.15.** Furnace pressure and reference tap designs. (See also the **warning** tag.)

| **Big tag** |
|---|
| **WHEN FURNACE IS NOT IS USE,** |
| **remove this observation port and tie it to this tag.** |
| ○ **CLEAN OUT hole through wall very well.** |
| **Clean glass (both sides), leave tag attached, and** |
| **REPLACE OBSERVATION PORT, hand tight.** |

The reference tap (measuring atmospheric pressure) should be on the outside of the furnace (*a*) at the same elevation as and close to the furnace pressure tap, and (*b*) protected from drafts, (*c*) where cleanout will be easy, and (*d*) not in a control room. The control room is sometimes thought by some to be a clean, cool place for the furnace pressure transmitter, but it is definitely bad because the control room air conditioner pressurizes the room, giving a faulty compensating reading, because opening and closing the control room door changes the sensed $\Delta P$ of the control, and the different elevation and long lines may cause error and longer reaction time.

A crossover with shutoff valve should be installed between the pressure tap and the compensating (atmosphere) tap immediately below the instrument, for "zeroing." Both the pressure tap and the compensating tap should have tightly piped lines all the way to the instrument. A pipe tee should be installed on the outside end of every tap—pressure and compensating—with a heat-resistant, glass observation port in the tee to allow operators to see that the measuring tap has not been plugged. Keep the pressure transmitter away from heat.

The elevation of the pressure-sensing tap does not necessarily have to be at the elevation desired for the neutral pressure plane. The most desirable height for the zero pressure plane may be at a point that turns out to be bad for good measurement, for example, below the hearth, at a level where scale might plug the pressure tap, or in a place where liquid metal may splash into the tap. In such cases, a very workable solution is to locate the sensor tap at a convenient higher position and then adjust the controller's setpoint in accordance with the correction for the rise in pressure for the chosen higher elevation from table 6.2. (See example 6.2.)

**TABLE 6.2 Draft or chimney effect at various furnace levels and temperatures**

| Temperature | 400 F | 800 F | 1200 F | 1600 F | 2000 F | 2400 F | 2800 F |
|---|---|---|---|---|---|---|---|
| Draft, $\frac{\text{"wc}}{\text{ft of height}}$ | 0.0058 | 0.0086 | 0.0101 | 0.0110 | 0.0116 | 0.0120 | 0.0123 |
| Temperature | 200 C | 400 C | 600 C | 800 C | 1000 C | 1200 C | 1400 C |
| Draft, $\frac{\text{mm water}}{\text{m of height}}$ | 0.484 | 0.718 | 0.840 | 0.915 | 0.946 | 0.975 | 1.012 |

**Example 6.2:** The proposed pressure control tap location on a 2200 F car furnace happens to be at hearth level and right in the line of fire of a low-level enhanced heating burner. The first choice would be to locate the tap on the opposite wall, between the burners, if space permits.

The next choice would be to locate the tap in the wall opposite the burners, but equally spaced between the burner centerlines and elevated 2 feet above the hearth. The setpoint of the furnace pressure control will have to be biased to correct for the difference in elevation between the pressure tap and the desired level of the neutral pressure plane (at the hearth). Interpolating from table 6.2, the setpoint bias should be 0.0118 in. × 2 feet of elevation = 0.0236, or say 0.025 or 0.03 in. wc to allow for expected wear on the car seals.

### 6.6.3. Dampers for Furnace Pressure Control

Many ingenious damper designs have been used for controlling positive furnace pressures in high-temperature furnaces. (See pp. 64–69 of *reference 52*, plus *references 53* and *54*.) Butterfly-type valve/dampers and sliding gate dampers in high-temperature flues or stacks are prone to having problems with thermal expansion, metal oxidation, wear, and lack of lubrication. Much effort has been devoted to locating the moving parts out of the hot furnace gas stream, as with clapper dampers, bell-crank mechanisms, and refractory-faced, cable-operated guillotine dampers. Smooth, sensitive motion is important to assure bumpless opening and closing, especially at the low-fire (high-turndown) end of the control range.

*Throttled air jet dampers* have often been found to be a welcome answer in avoiding or overcoming many of the aforementioned damper design problems. *Reference 56* gives suggested design criteria. A "sheet" of blower air is blown across the open end of a flue, choking off the effective exit area and thereby building up a back pressure in the flue and furnace. The sheet of air comes from a drilled-pipe manifold located slightly back from the edge of the flue exit. If there is concern about cold air being blown down into the furnace, an automatic control system can be put in place to automatically shut off an air-jet damper whenever the burners go off.

The manifold is out of the hot exit gas stream, but its choking jets can effectively cover an 18" (045 m) wide flue opening with 1 psi (6.9 kPa) air. If there is a problem with the 18" throw limitation of an air damper, the designer should consider changing the shape of the flue opening from square or round to an oblong rectangle with air jets on one of its longer sides (blowing across its shorter dimension).

The air control valve and its drive motor, controller, and transmitter can be located in any cool (but not freezing) environment away from the flue and not on top of the furnace.

Air damper jets (fig. 6.16) should be aimed slightly into the oncoming hot exit gases. If the flue flows vertically up, there may be a danger of backfeeding cold air down into the combustion chamber, possibly cooling the load(s). One solution to this problem is to corbel a shelf protruding into the flue passage from its wall opposite the air jets. A better solution is to build a 90-degree turn into the flue's exit as it emerges from the top of the furnace. This can usually be built with a ceramic-fiber-lined duct

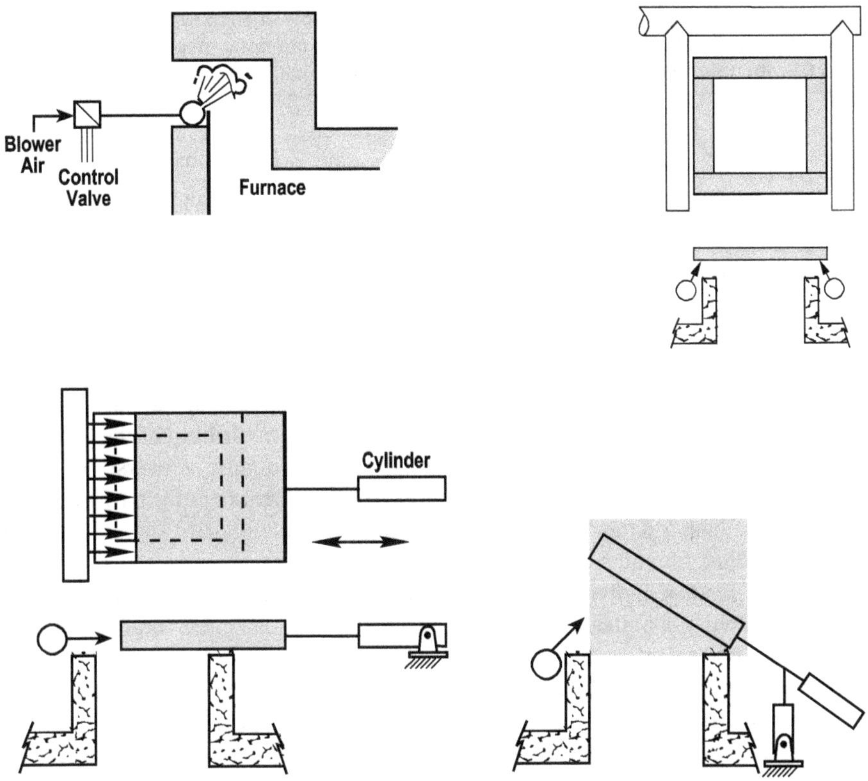


**Fig. 6.16.** Air-jet dampers (*top left* and *right*) can use throttled air (high pressure at low burner input, low pressure at high burner input). Constant air-jet-assisted mechanical dampers (*bottom left* and *right*) have a jet assist to provide better control sensitivity at low-firing rates (high-turndown). Another way to improve sliding damper sensitivity is with a v-notch (a right triangle with its hypoptenuse about one-third of the width of the damper's leading edge). Courtesy of *reference 56.*

fitting onto the furnace roof. Then, the throttled air-jet manifold can be positioned to blow across and slightly up into the exit of the duct extension, where backfeeding is much less likely to happen. Such a refractory-lined duct has an added advantage in that it prevents the precious load in the furnace from "seeing" a "cold hole" in the furnace ceiling, through which it might radiate heat, affecting load quality and/or requiring more fuel input.

All dampers and control valves have their most difficult sensitivity problems at low-firing rates (high-turndown), where they tend to "bump, hump, and overjump." For better sensitivity, a constant-pressure air-jet damper can be combined with a sliding-guillotine refractory damper, or a hinged clapper damper. (See fig. 6.16.) Dampers tend to lose usefulness with wear and lack of maintenance.

Multiple flues were once popular as a means of distributing the gas flows along the furnace length. That idea works only if there is a near-equal number of burners

**TABLE 6.3 Benefits of automatic furnace pressure control—A case history.[a] Batch forging furnace heating 5200 lb (2364 kg) of 3.9 in. (0.1 m) diameter steel parts to 2400 F (1316 C) with natural gas. Ceramic fiber walls 8" (0.2 m) thick.**

| Control | Cycle time to 2400 F | Natural Gas/Cycle | | Specific Fuel Use | |
|---|---|---|---|---|---|
| | | scf | $sm^3$ | Btu/lb | MJ/kg |
| off | 13.0 hr | 20 736 | 590 | 3981 | 92.6 |
| automatic | 11.5 hr | 16 612 | 475 | 3187 | 74.1 |

[a]Abstract from Gas Research Institute Report 5011–342–0120.

similarly positioned along the furnace length. It is difficult to damper such multiple flues because tiny inequalities in dimensions can cause uneven distribution. If a series of air dampers is used, great care must be taken for uniform drilling of the hole size and angle all along the manifold lengths, and the manifold must be oversized, like a plenum, to assure equal pressure at every hole. Another treatment for a row of flues is a series of clapper dampers on arms projecting from a long drive shaft. These are difficult to adjust for equal effect at every flue.

With any kind of individual vertical flue controls, a flue that happens to carry more hot gas will get hotter and natural convection will create more "draft" or "pull," causing that flue to get even hotter—a true "snowball in hell." If scale or refractory crumbs accumulate unevenly on the floor near multiple bottom flues, this same sort of acceleration will happen in the least-plugged flue. These sorts of problems have led many engineers to favor one flue per zone, or per furnace, and to use wise engineering in burner placement, and best control of furnace circulation. (See chap. 7.) This is more easily accomplished in continuous furnaces where the pieces "march" through several zones and past a number of burners.

In-the-wall flues or tall flue systems are not generally recommended unless barometric dampers or "air breaks" (see Glossary) are used to counteract the resultant changeable draft.

## 6.7. TURNDOWN RATIO

This ratio, often simply termed "turndown" or "t/d," is the quotient of (high-fire rate)/(low-fire rate). Typical values for industrial heating operations are in the range of 3:1 to 6:1. If higher ratios are needed, the cost of the control valve and burner will increase. Because of the square root law relating pressure drop to flow, a 10:1 flow turndown ratio requires a 100:1 pressure turndown ratio; a 40:1 turndown ratio requires a 1600:1 pressure turndown ratio. (See table 6.4.)

A higher than normal "effective" turndown ratio can appear to be accomplished by use of excess air, particularly at low-firing rates. The excess air lowers the available heat. (See fig. 5.1.) This literally throws away otherwise useful available heat, running up the fuel bill. Some pressure-balanced regulators are built with an extra-long spring that permits biasing the regulator to go lean (excess air) at low-firing rates.

Turndown may be limited by (*a*) burner stability range, flammability limits, mixing quality, (*b*) valve leak or process low-flow limit, either of which raises the denominator in the t/d equation. (*c*) flow controller range limit, (*d*) low-pressure air atomizer for liquid fuel, (*e*) flame detector range, and (*f*) transmitter turndown (4 to 20 ma ~ 5:1 t/d).

### 6.7.1. Turndown Devices

Turndown devices are most often control valves (not shutoff valves) or dampers. The best valve turndown characteristic is usually accomplished with adjustable port valves or with characterized globe-type valves. Butterfly valves usually have very poor characteristics (not straight-line), but their characteristic curves can sometimes be improved by undersizing or selecting reduced port models.

Speed controls on blowers (VFDs: variable frequency drives) are becoming more acceptably priced so that they can now accomplish a net saving over the old energy-wasteful method of controlling input by throttling flows with valves.

**Example 6.3:** If a 30-hp blower is operated at an average of 70% of its rated volume for 50 weeks per year, how much energy could be saved by using VFD?

From the fan laws, p. 200 of *reference 51*, flow is proportional to rpm, but power required is proportional to rpm$^3$, so when $hp_1 = 30$ hp rating,

$$hp_2 = hp_1\ (Q_2/Q_1)^3 = 30\text{ hp }(70/100)^3 = 10.3\text{ hp consumed with VFD.}$$

$$\text{hp saved} = hp_1 - hp_2 = 30\text{ hp} - 10.3\text{ hp} = 19.7\text{ hp saved.}$$

$$\text{kW saved} = 19.7\text{ hp} \times 0.746\text{ kw/hp} = 14.7\text{ kW.}$$

If the cost of power to drive the blower is \$0.05/kwh, the saving will be 14.7 kW × 24 hr/day × 7 days/week × 50 weeks/yr × \$.05/kWh = \$6,174.

A blower with VFD can take care of modulating the air flow, but the flow of fuel must still be reduced by a throttling valve in the fuel line, sometimes by a regulator, which is a form of globe-type control valve. This leads to a brief review of air/fuel ratio control systems.

*Area control of air/fuel ratio,* that is, "linked valve control," uses one common contol motor to drive a linkage to both air and fuel valves. The air and fuel valves must have very similar characteristic curves. VFD is not appropriate with this area control system, but can be used effectively with either pressure control or flow control, discussed next.

*Pressure control of air/fuel ratio* is usually an 'air primary' system, and VFD can be used with it. (See fig. 6.17.) The input signal (usually furnace temperature or boiler pressure) operates an air flow control. A "cross-connection" impulse, an air pressure signal, moves a regulator's valve until its output pressure sensor stops the fuel valve movement to "balance" the fuel pressure to match or follow the controlled air pressure.

Flow control of air/fuel ratio can be either air primary or fuel primary, and VFD can be used with either. This system actually measures the primary fluid flow and

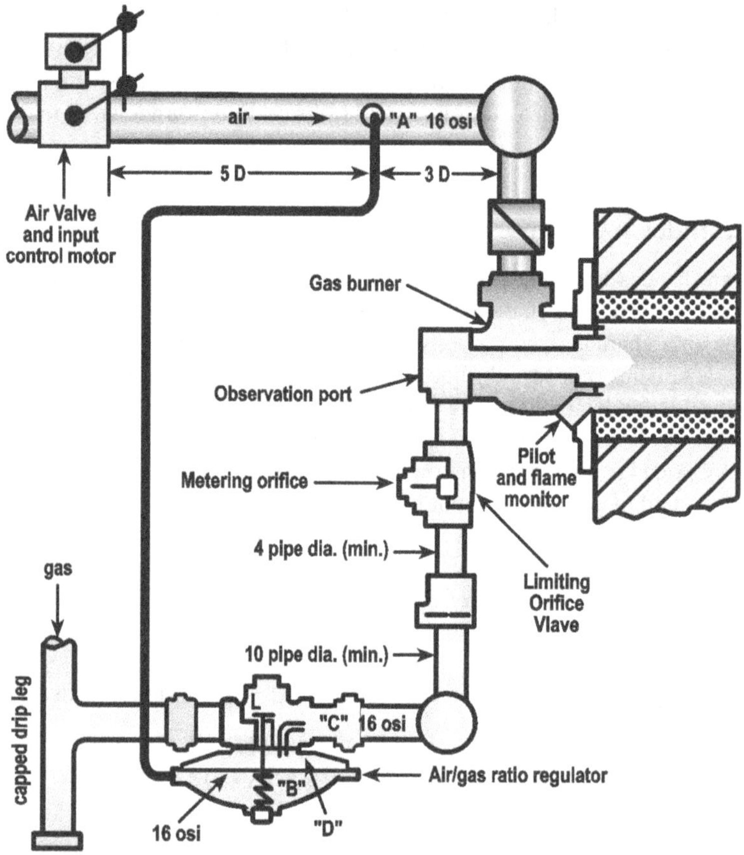


**Fig. 6.17.** Pressure-balanced air/fuel ratio control, usually limited to control zones with a fuel gas line smaller than 4" (0.1 m) pipe size. Sample pressures at A, B, C, D are 16 osi = 1 psi = 6.9 kPa= 27.7"wc = 0.70 m $H_2O$. A VFD blower could replace a constant speed blower and the air control valve (*top left*).

adjusts the secondary flow to the proper air/fuel ratio—typically with natural gas, one-tenth with air primary or ten times with fuel primary. (See fig. 6.18.)

## 6.7.2. Turndown Ranges

Some process designers start out saying they do not require any turndown because the process is so designed that it can always run flat out at 100% of design rate. As they start to get the kinks out of their system, and realize that neither they nor those who will run it are perfect, the designers will want a high-turndown ratio that would be beyond reason, costwise. Table 6.4 gives approximate turndown ratios possible with a variety of turndown control systems.

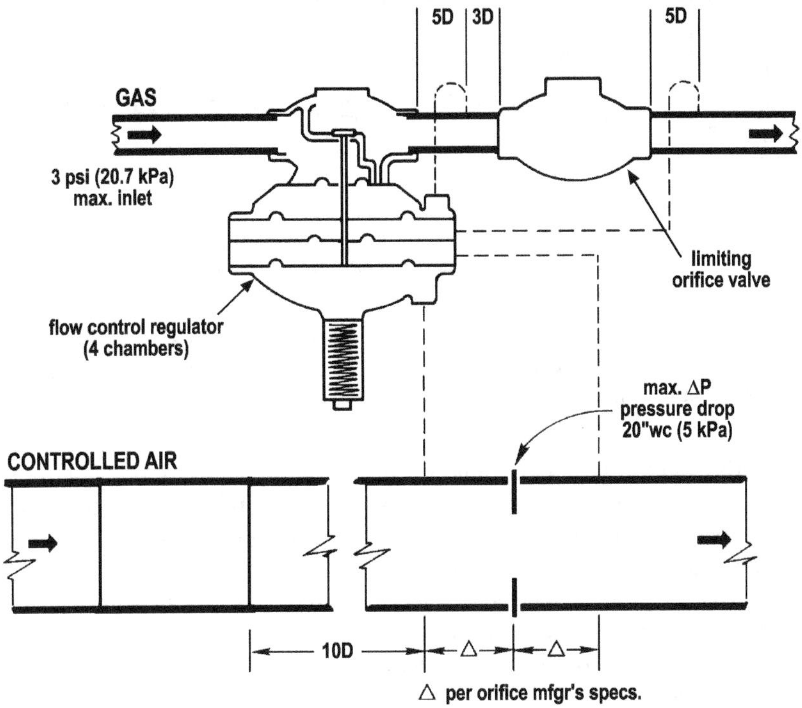


**Fig. 6.18.** Flow-balanced air/gas ratio control system, air primary. Air at lower left could come from a VFD blower or from an input-control-driven valve.

## 6.8. FURNACE CONTROL DATA NEEDS

The ideal way to get information on rate of heating and temperature uniformity (for avoiding undue stresses and for quality assurance) is to bury T-sensors within the piece(s) being heated. This may damage the piece; therefore, an expendable sample may be necessary, which hopefully can be placed where it receives exactly the same heat treatment as the real loads.

**TABLE 6.4 Some typical turndown ranges (for listed pressures only).**

| System | Description/Comment | Turndown Ratio |
|---|---|---|
| Inspirator | Cheap—no blower/with 25 psi gas | 2.5:1 |
| Aspirator | Zero gas pressure/with 16 osi air | 4:1 |
| Linked valves | Poor tracking unless with special linkage & valves | 4:1 |
| Pressure balanced | Cold air only/with 16 osi cold air | 5:1 |
| (Can be biased for gradually higher excess air at lower inputs.) | | |
| Flow balanced | Cold air only/with 10"wc max orifice ΔP | 7:1 |
| Electronic flow balanced | Accommodates $O_2$ trim, mass flow control, oxy-fuel firing | 7:1 |

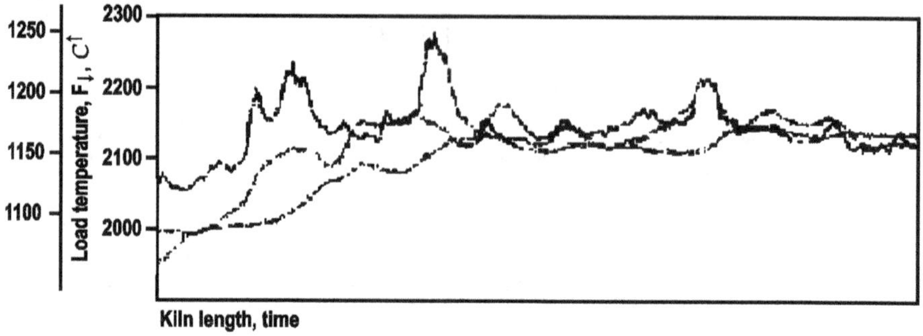


**Fig. 6.19.** Load temperature versus time (or furnace length) in a continuous furnace *before* use of data acquisition to modify the design, control, and operation. From Ruark, Ralph, "Making the Connection," *Ceramic Industry,* Vol. 150, No. 1, Jan. 2000, p. 14. Reproduced with permission.

Measuring only surface temperatures is much easier than measuring interior temperatures of the pieces being heated, but it gives only implied results relative to interior heat patterns within the load pieces. Batch heating processes are less difficult than continuous furnaces, where the measuring sensors need to "ride" along with the loads, necessitating long, protected lead wires or radio transmission of the data—both of which are difficult at high temperatures.

Figure 6.19 from *reference 75* shows temperature measurements of load pieces as they were moved through a continuous ceramic kiln. This data helped the operators and engineers to work together in deciding how to modify the furnace, burners, and controls, resulting in the temperature pattern shown in figure 6.20 (from *reference 73*). The result has been less product distortion and more consistent properties within each piece and throughout the year.

The ceramic industries are leading the way in kiln and furnace data-acquisition technology. Fixed noncontact thermocouples give only a general idea about the true thermal history of the molecules within a load. It behooves leaders within the industrial heating field to encourage cooperation with instrument and control experts by

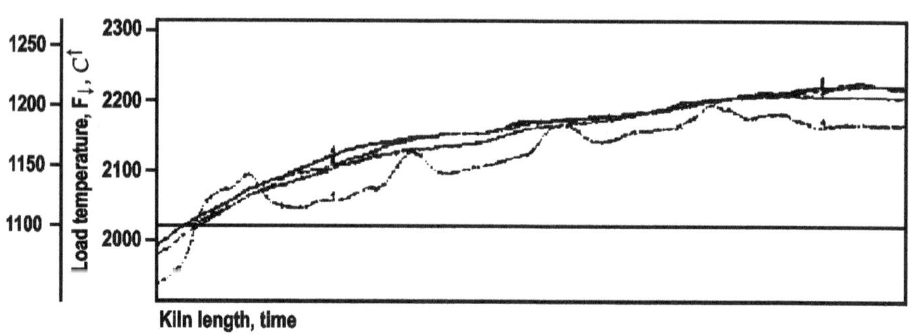


**Fig. 6.20.** Load temperature versus time (or furnace length) in a continuous furnace *after* use of data acquisition to modify the design, control, and operation. From Ruark, Ralph, "Making the Connection," *Ceramic Industry,* Vol. 150, No. 1, Jan. 2000, p. 14. Reproduced with permission.

their organizations and industry associations. Those who take the lead in new developments in data acquisition and application will be able to surpass their competition with precise quality-controlled products.

## 6.9. SOAKING PIT HEATING CONTROL

### 6.9.1. Heat-Soaking Ingots—Evolution of One-Way-Fired Pits

The steel industry has been using soaking pits for at least 125 years. Originally, they were simply refractory boxes in the earth with no combustion systems. From these simple units, the industry graduated to regenerative pits which had no instrumentation to the bottom-fired pits with ceramic recuperators to one-way top-fired pits with or without metallic recuperators. With the one-way top-fired pits, more pit area is under the crane per unit of real estate, so they became the universally accepted standard. Typical size: 22 ft (6.7 m) long, 8.5 to 10 ft (2.6 to 3.0 m) wide, and 10 to 17 ft (3.0 to 5.2 m) deep. The combustion system has one or two burners located high on one end of the pit with the flue directly beneath them.

These one-way-fired pits were fired with blast furnace gas, coke oven gas, natural gas, or heavy oil. With the number of these pits in operation, it is a wonder that more data are not available concerning their deficiencies. They were built to supply primary mills which rolled ingots into slabs, rounds, and bars, all to be reheated and rolled into finished products, but they had temperature differences longitudinally and top to bottom.

For example, when a pit would arrive at setpoint temperature (see glossary), the temperature difference between the burner wall and the opposite wall might have been 140°F to 300°F (60°C to 149°C), as measured by the control T-sensors in each end wall. The temperature differences longitudinally, near the bottom of the pits, was even greater. The temperature differences from the top to the bottom of the ingots at soak conditions was at least 40°F (22°C). After hours of soaking conditions, the bottom temperature difference burner wall to the opposite was 170°F or more. These temperature differences were caused by all the hot combustion gases flowing from the burner to the opposite wall in the combustion chamber above the ingots splashing against the far wall, then turning downward to the pit bottom, again splashing and turning toward the flue below the burner or burners. As the gases pass the ingots, they give up some of their heat, reducing their temperature.

***6.9.1.1. Attempts to Improve Temperature Uniformity.*** For the most part, heat transfer is by gaseous radiation. There is some (but not much) solid radiation from the combustion chamber walls. After one-way-fired pits were in operation for about 25 years, a burner with fixed spin was adapted to these pits to reduce the longitudinal differentials at the control thermocouple locations (generally near the top of the ingots in the wall opposite the burner(s). This fixed-spin burner rarely had the right spin. More often than not, it was not enough, but sometimes it was too much because of the type of fuel used. The result was washed ingots at the burner walls, burned-out recuperators, and ingots at the wall opposite the burner which were so cold they could not be rolled. Those fixed-spin burners were followed by 'variable heat pattern burners,' which had a movable spinner in the air passage. The spinner position was controlled to keep the longitudinal temperatures at the control T-sensor locations nearly the same. Maintenance of the variable spin vanes was a problem.

Many operators felt that this improvement was all that would ever be needed, but they were not aware that the bottom longitudinal temperatures, when the ingots were judged rollable, were 150°F to 200°F (83°C to 111°C) colder at the burner wall than the ingots at the opposite wall, and the top-to-bottom temperature difference at the burner wall was 40°F to 100°F (22°C to 56°C). A few individuals knew of these problems, but there were no solutions at that time except to raise the control temperatures until product quality was tolerable.

In the late 1970s, a burner became available that could change the spin by adjusting the gas flow between axial and tangential nozzles to control the spin necessary to hold two measurement locations at the same temperature. The ATP burner had no moving parts within. This burner made it possible to hold the temperatures at two longitudinal locations near the pit bottom to the same temperature. This technology was applied in France, where pits still had a top-to-bottom temperature difference of 40°F (22°C). The real difference is that now ingots are heated from top to bottom rather than end to end, which changes the fuel curve. High-fire time was much longer and cutback time much shorter, reducing the whole heating cycle by about two hours.

The aforementioned 40°F (22°C) difference was the result of the sensible heat of the combustion gas mass at minimum gas flows. With cold air combustion, the gas volume is approximately double that with hot air firing, and the top-to-bottom temperature differential is reduced to 20°F (11°C). With oxygen firing instead of hot air, the temperature difference (from ingot top to bottom) will likely be 80°F to 100°F (44°C to 56°C) because the gaseous heat transfer is so much greater, along with the gas mass being just one-third the mass of cold air firing.

The industry is still trying to reduce soak-pit fuel rates by regenerative air heating and/or oxygen firing, either of which can double the temperature differences from top to bottom of a pit. The real problem is a lack of understanding the problem; thus, product quality is the loser. It is the hope of the authors that this explanation will be spread to more operators and cause a better understanding of what is really happening in soaking pits. With either oxygen or hot combustion air, the lower mass flow of combustion gases will result in greater top-to-bottom temperature differentials. This will require changes in both oxy-fuel and regenerative air preheating burners to include the ATP feature. If it is necessary to make a choice between product quality and fuel economy, the authors favor product quality. The only factor that has a higher priority than product quality is safety. Both safety and product quality save money.

In summary, the major slab (instead of ingot) soak-pit problems are:

(*a*) The need to control the burner combustion gas movement to move down the long sidewalls behind the slabs leaning on the wall piers so that the slabs will be heated uniformly top to bottom. This can be accomplished by using a

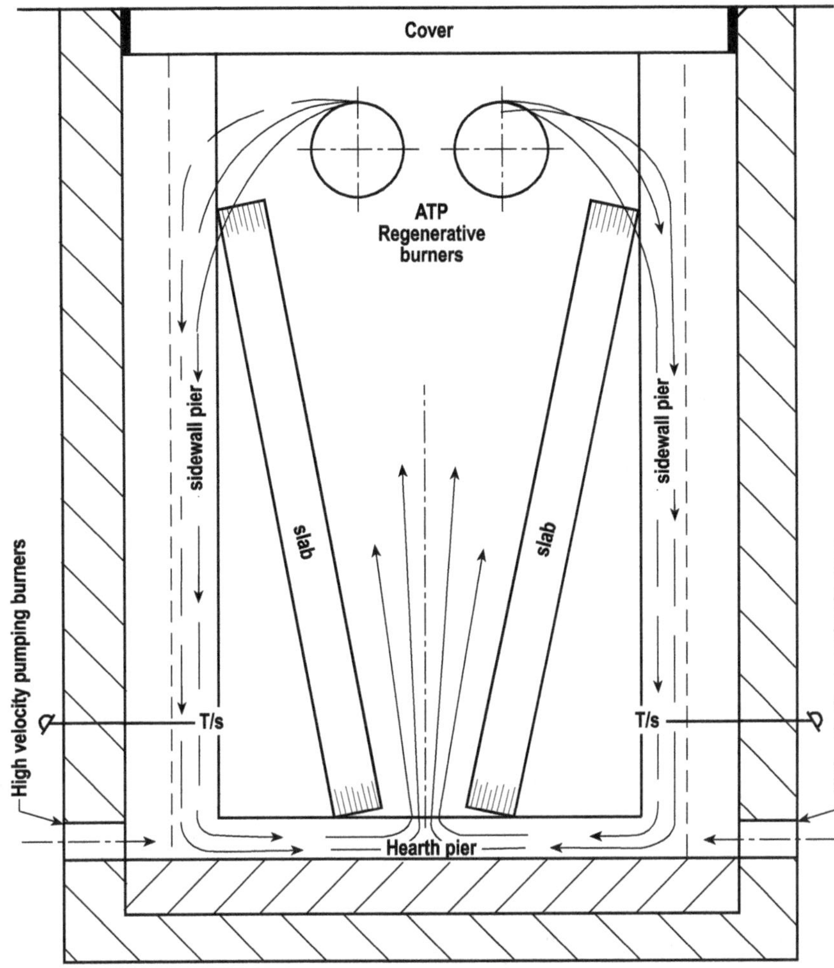


**Fig. 6.21.** Slab soaking furnace, end sectional view, example 6.7. Two ATP burners are end fired at the top, and flue at the hearth under the burners. The slabs stand on piers on the hearth, and lean against vertical piers in the sidewalls. Piers allow poc circulation behind and under the slabs.

minimum of two controlled, high-velocity air jets tangentially directed at 180 degrees from each other installed through the burner body in the vicinity of the pilots. The spin energy would be controlled by, more or less, jet air. This could be accomplished by adding ATP technology to regenerative burners.

(*b*) The walls and floors should have piers to allow hot gas to flow behind and under the load pieces. (See fig. 6.21.) The top-to-bottom temperature differential could be reduced by applying very small high velocity burners between the bottom piers which support the slabs. These burners would provide a small amount of heat to the pit bottom and would increase the combustion gas flow down the pit walls even to a point of recirculating pit gases. With these additional gases, plus burner heat, the temperature difference top to bottom should be less than 40°F (22°C).

(*c*) To increase the mass of gas in the pits at or near soak conditions, it is recommended that the regenerative burners be fired direct (cold air firing) to avoid the need to increase excess air to keep the slabs uniform in temperature. With cold air firing, we believe scale volume will not increase as it would with excess air.

### 6.9.2. Problems with One-Way, Top-Fired Soak Pits

In the late 1930s, the steel industry began a trend toward one-way, top-fired soak pits to get more space under the cranes. They were a great improvement over regenerative pits. The very expensive scrapping of a burned ingot was practically eliminated, and ingots had much more uniform temperatures. Prior to that time, heaters fired a pit until they could not see the ingots through a peep sight, because their color (temperature) and that of the background were so close to identical.

The problems of the one-way, top-fired pits were not recognized until new mills had only this type of pit to supply them with heated steel. The overall problem was the $U$-shaped combustion gas flow pattern, which created large temperature differences between the top and bottom and far wall to near wall at both the bottom and top of the ingots. The actual temperature differences lengthwise along the top of a pit varied from 140°F (78°C) with a hot charge to 300°F (167°C) with a cold charge. With these very large temperature differences, the time at maximum firing rate was very short—for example, heating hot heats $\frac{3}{4}$ hr $\pm$ $\frac{1}{4}$ hr. The time from arrival at the temperature setpoint to fuel input arrival at minimum input was 7 hr, $\pm 1$ hr. Therefore, the cycle time for a hot heat, with 2-hr out time, was just less than 8-hr—instead of the nominal 3 to 4 hours (a longstanding rule of thumb of the industry).

By the 1950s, the problem was widely known. Dr. Schack, a renowned authority from Germany, set up a test to study the problem and suggested a possible solution using water model studies. His solution was to increase the forward energy of the burner to increase recirculation, bottom to top, at the burner wall. The idea was excellent, but because of the dissimilarity of water and gas densities, the problem became worse when applied. The poc "$U$-flow" pattern had to be changed by varying the spin of the combustion gases. A fixed spin burner was developed, but the spin was either too little or too much in nearly all cases.

Then, burner manufacturer North American Mfg. Company of Ohio produced a burner that controlled the temperature to $\pm 10$°F (5.6°C) by a lot of spin or no spin (on/off control). The result was that the high-fire period was lengthened and the cutback period was reduced. A hot heat was ready in about 5 hr instead of 8 hr. Temperature measurements were taken with five thermocouples along the length of the pit bottom. When the pit temperature was thought to be uniform and the ingots ready to be rolled, the front-to-back temperature difference was 175°F (97°C).

To correct this temperature differential, a proportionally controlled spin of the poc was needed to automatically control temperature in the sidewalls, front, and back of the pit.

Such a proportionally controlled spin burner and control system were developed in the early 1980s and installed on six pits in Dunkirk, France, with excellent results. The top-to-bottom differential was only 40°F (22°C). The high-fire period was very long, and the cutback period was 40 min, with a cycle time of about 3 hr. Instead of the combustion chamber being uniform from front to back of the pit, the burner wall was now 80°F (44°C) hotter than the opposite wall. As the pit temperature reached setpoint, the differential at the ingot tops began to disappear. With the cutback to minimum fuel input, the combustion chamber temperature differential was near zero, but the front wall temperature began to drop, requiring the use of a forward gas jet (supplied within the burner) to move the peak heat flux closer to the front wall, giving even ingot temperatures.

At minimum fuel and air input, the ingot top-to-bottom temperature differential was again about 40°F (22°C). This difference was caused by the heat losses of the pit bottom. The basic reasoning for this is that with a smaller mass of gas flowing, the temperature drop of the gas must be greater to supply the bottom heat loss. Example 6.6 below illustrates this.

**Example 6.6:** A pit furnace is being fired with natural gas and 10% excess air, and has a 2400 F (1589 C) flue gas exit temperature. The wall, hearth, and roof losses are calculated to be 1.55 kk Btu/hr. With cold air firing, there is a 40°F (22°C) temperature difference from top to bottom of the ingots. Predict the corresponding temperature difference when using 1300 F (704 C) combustion air, and when using oxy-fuel firing.

From Figure 3.15 of *reference 51,* the available heat will be 36% with cold (60 F, 16 C) combustion air, or 56% with 1300 F (704 C) preheated combustion air. Thus, with pit losses of 1.55 kk Btu/hr, the gross input rate would be

$$1.5/0.36 = 4.2 \text{ kk Btu/hr when using cold combustion air,}$$

$$\text{or } 1.5/0.56 = 2.7 \text{ kk Btu/hr when using 1300 F combustion air.}$$

If cold air firing has a 40°F (22°C) temperature drop from top to bottom of the pit, the same pit with 1300 F combustion air would have a temperature drop of

$$40°\text{F (4.2 kk Btu/hr/2.7 kk Btu/hr)} = 74°\text{F}$$

to balance the heat loss of the pit bottom.

With the use of oxygen for combustion instead of air, the thermal drop would be perhaps three times the 40 F due to the much smaller quantities of flue gas (theoretically one-third of ambient air firing) to carry energy to the pit bottom. In fact, one-way, top-fired soaking pits are a very poor application for oxygen firing due to the small volumes of poc gases available to carry heat to the ingot bottoms. Other temperature differences in the pit might be as much as three times as great if air were replaced with oxygen.

Some engineers attempt to counter this problem with increased recirculation. They could spin the combustion products to reduce temperature differentials along the length of the pit, but the top-to-bottom temperature differentials would remain approximately three times as great as those with ambient air firing (120°F or 67°C). Even this possibility is unlikely because the volume of poc is so small and because convection heat transfer is proportional to velocity to the 0.7 power. The result is that oxygen combustion in soaking pits is not a wise choice when the quality of rolled material is temperature-uniformity-sensitive.

Almost any effort to reduce fuel cost will result in less air flow and correspondingly less poc circulation, so temperature differentials increase. When these differential increases result in either product rejects or excess slag formation, any fuel saving is far outweighed by the cost of metal loss.

**_6.9.2.1. Atmosphere in Soaking Pits and its Effects._** Tests of scale formation with different oxygen levels indicate that the curve looks like an "*S*" where the rate of scale formation rises about five times from slightly reducing to slightly oxidizing. However, these curves are often generated at temperatures below any scale melting or softening, which may change the results. For example, when heating silicon steel for direct rolling to strip, reducing the oxygen in the atmosphere from 3.0 to 0.5% improved the yield from 55 to 69%. At temperatures above the scale melting points, the liquid state immediately flows to the pit bottom, offering no further protection from oxidation of the newly exposed iron.

If there were no free oxygen, and only $CO_2$ and $H_2O$ available for oxidization, the rate of scale formation would be significantly less, improving yield.

The use of a reducing atmosphere (with some combustibles) is not without difficulty. Scale formed with a slightly reducing atmosphere sticks to the ingot surfaces and may be rolled in, creating pits. To remove the scale, the soaking pit atmosphere has been returned to 3% $O_2$ for a short period to remove the sticky scale by melting. In a way, this scenario gives some proof to the hypothesis that the melting of the scale changed the rate of scale formation because of the oxidizing furnace atmosphere.

### 6.9.3. Heating-Soaking Slabs

To heat slabs uniformly with regenerative burners, the following steps are necessary and should not be compromised:

1. Add ATP technology to the regenerative burners.
2. Add bottom and sidewall piers with small tempest burners through the long walls to fire under the bottom piers to pump the combustion gases down the long walls.
3. Below some firing rate, for example, 10 kk Btu/hr, the burners should fire direct to increase mass flow to improve temperature uniformity, by firing direct, bypassing the regenerative beds. (The poc of these burners should exit through flue openings below the burners.)

**Example 6.7:** Compare fuel requirements for a slab-soaking furnace fired with regenerative burners, and with and without added burners for 'pumping' (stirring, circulation). (See fig. 6.21.)

*Given:* Heat 60 tons per 5-hr cycle of steel slabs 7' × 7' × 10" (2.13 m × 2.13 m × 0.178 m) to 2100 F (1150 C); furnace size = 25' × 10' × 12' high (7.62 m × 3.05 m × 3.66 m high); two main regenerative burners firing at a total of 20.6 kk Btu/hr (21.6 GJ/h); 16 'stirring' burners firing a total of 1.6 kk Btu/hr (1.69 GJ/h). Each main burner has two tangential air lances for spin control, feeding 5 to 10% of the total air. Figure 6.21 is an endwise cross-sectional view of the furnace, showing the piers, circulation patterns, burner, and T-sensor locations.

*Operating information:* 2.9 hr at high fire; 0.3 hr cutback, 0.8 hr delay, 1 hr charge and draw—losing 0.02 kk Btu/ft$^2$hr (0.227 GJ/m$^2$h), Total cycle = 2.9 + 0.3 + 0.8 + 1 = 5.0 hr.

*Calculations:*

| | |
|---|---|
| High-fire fuel input, main burners = 2.9 hr × 20.6 kk Btu/hr | = 59.7 kk Btu. |
| High-fire fuel input, stirring burners = 2.9 hr × 1.6 kk Btu/hr | = 4.6 kk Btu. |
| Cutback fuel input, main burners = 0.3 hr × 20.6 kk Btu/hr | = 6.2 kk Btu. |
| Cutback fuel input, stirring burners = 0.3 hr × 1.6 kk Btu/hr | = 0.5 kk Btu. |
| Charge/draw input, cover open 1 hr with estimated gross loss | = 7.7 kk Btu. |
| TOTAL INPUT w/REGENERATIVE & STIRRING BURNERS | = 78.7 kk Btu/cycle. |

Fuel consumed = 78.7 kk Btu/cycle/(60 tpc) = 1.3 kk Btu/ton

= 78.7 kk Btu/cycle/(60)(2000) lb/cycle = 656 Btu/lb.

From figure A.7 in *Reference 51* or figure A.14 in *Reference 52*, read 370 Btu/lb as the heat content of steel heated to 2400 F (1316 C); therefore, the heat to the loads is:

12 tons/hr × 2,000 lb/ton × 370 Btu/lb = 8.88 kk Btu/hr

or 88.8 kk Btu/hr × 5 hr/cycle = 44.4 kk Btu/cycle.

Thus, the overall efficiency of the 5-hr cycle is (44.4/78.7) × 100% = 56%.

or (370/656) × 100% = 56%.

An alternative to the bottom-stirring-burner arrangement of example 6.7 would be going back to bottom-firing main burners (as with the Amsler-Morton pits of years ago), which achieved good bottom circulation without the added capital and operating costs of the extra little stirring burners. Piers would be required on the hearth and sidewalls to allow hot poc gases to circulate horizontally beneath and up behind the slabs. In that case, the calculations corresponding to example 6.7 might be:

**Alternative Example 6.7:** Bottom-fired main burners only.

| | |
|---|---|
| High-fire fuel input, main burners = 2.9 hr × 20.6 kk Btu/hr | = 59.7 kk Btu. |
| Cutback fuel input, main burners = 0.3 hr × 20.6 kk Btu/hr | = 6.2 kk Btu. |
| Charge/draw input, cover open 1 hr with estimated gross loss of | 7.7 kk Btu. |
| TOTAL INPUT w/REGENERATIVE & STIRRING BURNERS | = 73.6 kk Btu/cycle. |

Fuel consumed would be 73.6 kk Btu/cycle/(60tpc) = 1.23 kk Btu/ton

= 73.6 kk Btu/cycle/(60) (2,000) lb/cycle = 613Btu/lb.

Overall efficiency of a 5-hr cycle would be (44.4/73.6) × 100% = 60%.

or (370/613) × 100% = 60%.

The operating cost would be less as shown in the alternative example, and the first cost might be less because of no stirring burners. Some managers may wish to try for the traditional horizontally fired, top-fired burners without the stirring burners, but experience has shown that will be unable to accomplish even heating without prolonged soak times, which cost higher fuel bills and lower productivity. Accepting the poor temperature uniformity means accepting poorer product quality, which costs loss of customers or paying the fuel bill twice to do the job over correctly.

## 6.10. UNIFORMITY CONTROL IN FORGE FURNACES (for forging small steel pieces, see sec. 3.8.7)

The forging industry's customers demand increasingly tight temperature standards that require close temperature control throughout each forged piece. Often, the furnace must be certified, using a grid of test T-sensors in an empty furnace. Such certification without load(s) in the furnace may have been an improvement over no testing, but the addition of loads changes firing rates, gas movement, and heat transfer at nearly all locations in the furnace. If uniform product temperature is required, better means must be developed for internal furnace temperature control while heating products. Essentially, the problem is twofold: control of the temperature *above* the load(s) and control of the temperature *below* the load(s).

Loads should not be placed directly on a hearth or leaned against the furnace sidewalls because both surfaces have heat losses, which will be supplied by the loads and, in the process, also chill them.

### 6.10.1. Temperature Control Above the Load(s)

With the advent of fuel-directed, ATP burners, two temperature locations can be held at the same temperature or a constant difference in temperature, a nearly flat temperature profile regardless of the load size or location.

In addition to the two-point temperature control, other temperature measurements and control loops can be added in each zone to act as control monitors. When used with low-select devices on their output signals, these monitors can automatically take control of energy input to prevent overtemperature in the sensor locale. With sufficient monitors, overtemperatures at all potential hot spots of the load can be eliminated.

With the previous type control system and burners, the temperature control above the loads can be excellent, provided sufficient zones are installed. For batch furnaces, the minimum number of zones should be three—one for each end wall and one for the main body of the furnace. If there are two side-by-side doors, five zones are desirable—one for each side wall, two for furnace body, and one behind the center doorjambs.

### 6.10.2. Temperature Control Below the Load(s)

Temperature control below the load(s) depends on load piece location. If a product is placed on the hearth, the top-to-bottom temperature difference will never be uniform, and the magnitude of the top-to-bottom $\Delta T$ will depend on the following variables:

(a) load thickness—greater thickness yields greater $\Delta T$,
(b) load shape—rectangular pieces are a greater problem than round
(c) hearth heat loss—more heat loss causes more $\Delta T$ in the load pieces
(d) scale thickness on hot faces of load pieces
(e) exposed heat transfer area—a greater number of equivalent sides exposed will mean smaller temperature differentials
(f) thickness of scale on the hot face(s) of the product

Every effort should be made to position loads on piers or stools (preferably of low mass construction), especially for load pieces more than 4 in. (0.10 m) thick. Material more than 6 in. (0.15 m) thick should never be placed on the hearth unless the distance between centerlines of the pieces is at least twice the product thickness. Under no circumstance should pieces be piled on top of one another.

For truly uniform temperature across the bottom of the product, essentially equal clearances under and above the product must be provided, along with equal firing treatment. Because equal treatment, above and below, is often impractical at high temperatures, the clearance should be no less than necessary to accommodate the flames of a small, very high velocity burner without flame impingement. Those burners must be stable with at least 150% excess air (to reduce the concentration of triatomic gases that drives heat from the gas blanket into the loads). For example, if the burners are on 30-in. (0.76 m) centers, firing across an 8 ft (2.4 m) wide hearth, a 1 000 000 Btu/hr (1.055 GJ/h) burner with maximum velocity of combustion products leaving the burner tile of 200 mph (322 km/h), or a tile pressure of at least 4 in. wc (100 mm of water) generally will be satisfactory. Figure 6.22 depicts a suggested configuration of product relative to burners and T-sensors.

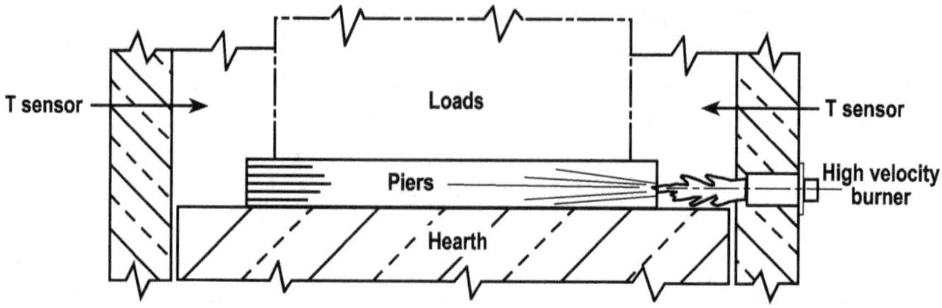


**Fig. 6.22.** Enhanced heating. Suggested arrangement with a row of high-velocity burners (type H, fig. 6.2) firing under the loads.

To assure a low temperature difference across the furnace width, T-sensors must be located on each side of the furnace. One sensor should be 1 to 3 in. (25 to 75 mm) above the pier in the wall opposite the burner(s) that controls the fuel input, with the combustion air flow held constant. When the furnace arrives at setpoint, the other sensor (in the burner wall at the same elevation) will be within ±6°F (3.3°C) of the opposite wall temperature. (See fig. 6.23, also refer to figs. 2.21 and 3.26.)

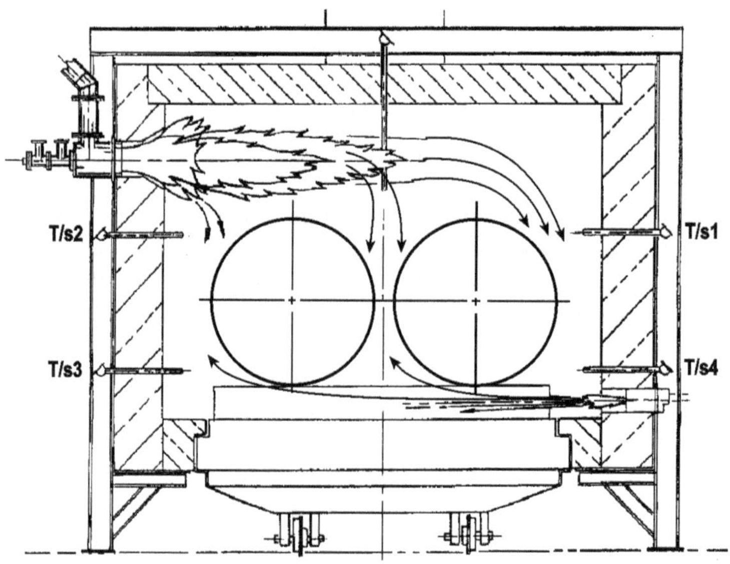


**Fig. 6.23.** Car-hearth forging furnace with enhanced heating, using overfiring ATP burners and underfiring high-velocity burners. T-sensor 1 adjusts the top burners' input and T-sensor 2 setpoint. The various gas flow paths from the upper burners are adjusted automatically, by T-sensor 2 controlling the degree of flame spin. T-sensor 3 controls input to the underfiring high-velocity burners by holding maximum air flow at all times and reducing fuel. The T-sensors should be replicated at each temperature control zone along the length of the load(s). The top center T-sensor is for high-limit shutdown. The roof flue has a cap damper for automatic furnace pressure control.

***An anomaly!*** To keep the temperature differences from one end to the other of the load(s) across the furnace width very small requires that gases flowing under the loads have nearly the same temperature from side to side of the furnace, which means that they *should not transfer much heat* to the load(s), hearth, or piers. That requires (1) high mass flow, (2) low concentration of triatomic gases (excess air, but no oxygen enrichment), and (3) minimum gas beam width (cloud thickness, pier height). This minimizing of the temperature drop of the gases flowing across the hearth means that the heat transfer from the gases between the piers, hearth, and loads must be kept small. The heat transferred must be supplied from a temperature drop in the gases moving under the load. To reduce that gas temperature drop and thereby maintain temperature uniformity, gas beam (thickness) must be kept small (8 to 12 in., 0.203 to 0.304 m), and the percentage of triatomic gases in the circulating gases must be kept low.

The mass of the piers should be kept small to minimize the heat absorbed by them because that heat would have to be supplied by the gases moving below the product, adding to the temperature loss of those gases. This scheme requires the location of flues to minimize interaction between zones. By following these practices, the across-the-furnace temperature profile above and below the loads will be very flat, providing very small temperature differences in the load(s) regardless of the loading pattern.

The previous control method will not provide uniform temperatures if the charge is improperly placed on the piers. Neither ingots nor small pieces should be piled on top of one another, which restricts heat transfer to one or more of the load pieces or surfaces. Carelessly placed load pieces will be heated very slowly because not all sides may be exposed to heat transfer so they will not pass quality control, and fuel will be wasted to heat them all over again. Another problem is having one or more loads too close to a sidewall where there is very little hot gas movement, leaving a very cold side for those pieces. The people charging furnaces must be made aware of the importance of their efforts in producing quality products via uniform heating.

If the management cannot be convinced to fire under the loads, a minimum of 4 in. (0.10 m) vertical clearance between the loads and the hearth will provide considerably better temperature uniformity and productivity. However, the clearance must be maintained open by frequent removal of accumulated scale.

## 6.11. CONTINUOUS REHEAT FURNACE CONTROL

### 6.11.1. Use More Zones, Shorter Zones

To improve reheat furnaces, many operators have invested in improved controls hoping to reducing fuel costs and improve product quality. Results have been disappointing because the heating zones were too long. For example, consider a top- and bottom-fired 100 ft (30.5 m) long furnace. When heating 8.5 to 10.0 in. (216 to 254 mm) thick load pieces, the top and bottom soak zones should be 25 to 30 ft (7.6 to 9.1 m) long, thus leaving 70 to 75 ft (21.3 to 22.9 m) for the top- and bottom-fired heating zones. With such an arrangement, the balance of the furnace normally would be divided into three top zones and three bottom zones—possibly 30 ft (9.1 m) top and bottom heat zones, 30 ft (9.1 m) top and bottom preheat zones, and 15 to 20 ft (4.6 to 6.1 m) top and bottom (unfired) charge zones. Except for the soaking zones, these zones are far too long to adequately control the furnace, especially after productivity adjustments. For example, after a delay, the newly charged pieces would have to move through the unfired zone and 50 to 60% of the preheat zone before the control temperature measurement would sense the newly charged, much colder material. This happens in both the top and bottom preheat zones and again in the heat zones, with the result that the new material is discharged too cold to roll.

The cause of the problem is much-extended heating time during the delay for all material in the furnace from charge door to soak zone. With this scenario, all material is much more uniformly heated, top to core and bottom to core, to *temperatures above design*. After the end of the delay, several pieces should be discharged to check the gauge. After the gauge is satisfactory, rolling can begin at about 80% of maximum rate. The product charged at the time of gauge checking may be rollable without difficulty. However, when the mill gets to 80% of full speed, loads entering the unfired top and bottom zones will be heated at very low rates, and the same will occur in the first 50 to 60% of the heat and preheat zones.

If the temperature measurements in the heat and preheat zones are sensitive, the firing rates of the heat and preheat zones, top and bottom, will reach 100% for the balance of the time that new material is in those zones. With these 100% instead of 80% firing rates, the load pieces then entering the furnace with firing rates at 100% will be heated above the uniform conditions desired. When this instability (too high firing followed by too low firing) begins, it is almost impossible to achieve uniform heating. This is the "domino" or "wave" effect mentioned relative to other furnaces throughout this book and in section 6.11.2.

If the heating zones from the charge door to the soak zone were much shorter and more numerous, for example, seven instead of three top zones, and seven instead of three bottom zones (including added firing in the normally unfired zone), the furnace program would enter the correct action as the second or third piece is extracted, and firing would be consistent with the actual mill supply of hot pieces from the furnace. The instability of the firing rates would be avoided, fuel rates reduced, and product quality improved.

With the authors' recommended six top heating zones and six bottom heating zones, the temperature measurement would control each small zone as the heating curve directs and would not get out of step as has been the case with very large zones. A furnace with the many zones recommended would probably be a roof-fired or side-fired furnace. Side firing would need ATP technology to control the loads' temperatures evenly from end to end across the furnace width.

Another reheat furnace problem that could be avoided by having more heating zones would be having charge zones hotter during low productivity than during high productivity. This occurs in many instances with large zones. For example, a program calls for the loads leaving the heat zone at 2200 F, but after a mill productivity upset (delay), the loads are leaving at only 2100 F. The control opens the input to 100%. As a result, the exit gas temperature leaving the heat zone will be very high, contributing to high fuel rates. If the furnace were configured with short zones, only the short zone needing a higher firing rate would fire harder; so the flue gas temperature would rise only slightly.

In the previous chapter, figure 5.10 illustrates a longitudinal reheat furnace with regenerative burners. *The following applies to each half of the furnace:* Two T-sensors through the roof of each of the two center soak zones to 2" (50 mm) above the thickest load and two T-sensors through each sidewall and 2 in. (50 mm) above the hearth control the three soak zones. Two sidewall T-sensors, 2 in. (50 mm) above the hearth control the top heat zone. Two T-sensors about 12 in. (0.3 m) below the skid rails control the bottom zone. Two T-sensors about 12 in. (0.3 m) below the top zone roof provide remote setpoints for the bottom zone's two controlling T-sensors. Sidewall T-sensors protruding into the zone are more responsive, but vulnerable, so flush installation in large recessed cups are often used.

The top preheat zone (fig. 5.10) has a high-limit *controlling* T-sensor near the hearth and near the loads' exit from the preheat zone, set to take over control of that zone if it senses more than 2200 F. At this location, the T-sensor indicates *load temperature* well (which is preferred over furnace temperature). The next zone (top heat zone) could be affecting the load temperature in the preheat zone, which would have a setpoint [T-sensor high, and 6 ft (1.82 m) from the load entrance] of 1600 F to 1800 F (870 C to 980 C). Load temperature entering any zone should be controlled to prevent it from rising above the setpoint of the next zone, which would waste fuel and prevent heat transfer in that next zone, which happens with light loading. Similarly, a zone's exit temperature may be too low with heavy loading.

### 6.11.2. Suggested Control Arrangements

Figures 6.24 and 6.25 show control arrangements found by coauthor Shannon to minimize the hunting 'domino effect' or 'accordion effect' mentioned in section 6.11.1, after a delay in a loaded multizone continuous furnace. Reviewing that effect, when a delay occurs, loads just 'sit' in each zone, soaking toward thermal equilibrium with that zone, with some heat radiating to or from adjacent zones. By the time the delay ends, the normal temperature gradient through the furnace length will be somewhat leveled, depending on the delay length. Load pieces near the discharge end of the furnace may be too cool, and those near the charge end, too hot.

After the delay, as the conveyor, pusher, or walker resumes operation, new cold pieces will be moved into the charge zone, causing the automatic temperature control to turn the burners there to high fire while most of the other zones will be idling because of pieces being overheated during the delay. Theoretically, automatic temperature controls should bring all the zones into proper temperature pattern. But the problem is that pieces with appreciable mass have center temperatures considerably different fromtheir surface temperatures. This creates an 'inertia' effect that we term a 'domino' or 'accordion'* wave action of the temperatures through the furnace length.

*Similar to the phenomenon that highway air patrol pilots observe after a driver slows suddenly, then speeds up. From the airplane, the spacing between cars looks like the side pleats of an accordion—gradually enlarging and contracting waves.

| SENSORS | CONTROL | MEASURE | LOCATED |
|---|---|---|---|
| T/s1 | Top heat zone | Early load & exit gas | Below roof to see load and sense gas temp |
| T/s2 | Top heat zone | High limit load temp | 2" above thickest load, if through roof; 2" above hearth if through side. |
| T/s3 | Provides set point for T/s4 | Top heat zone temp | 12" below roof |
| T/s4 | Bottom zone | | 12" below skid rails |
| T/s5 & T/s6 | Soak zone | Load temp | (Same as 1 and 2 above) |

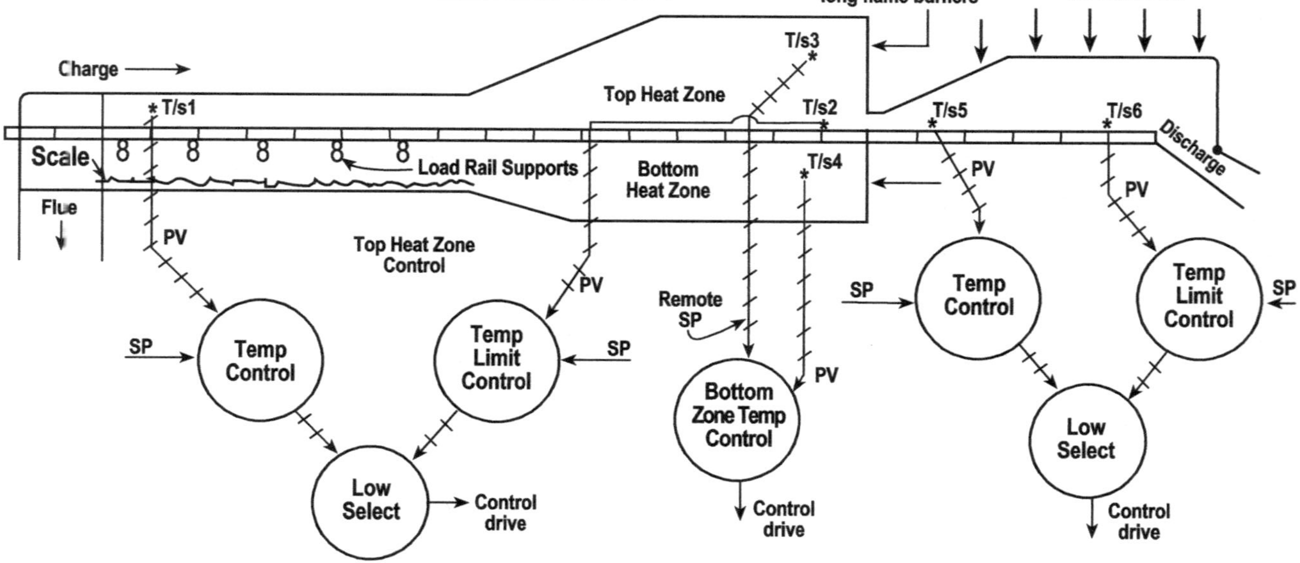


**Fig. 6.24.** Three-zone reheat furnace temperature control for best productivity, least fuel rate. This control system minimizes scale formation by preventing overheating. Scale accumulation forces bottom zone gases to top zone, reducing bottom side heating. PV = process variable; SP = setpoint; T/s = temperature sensor.

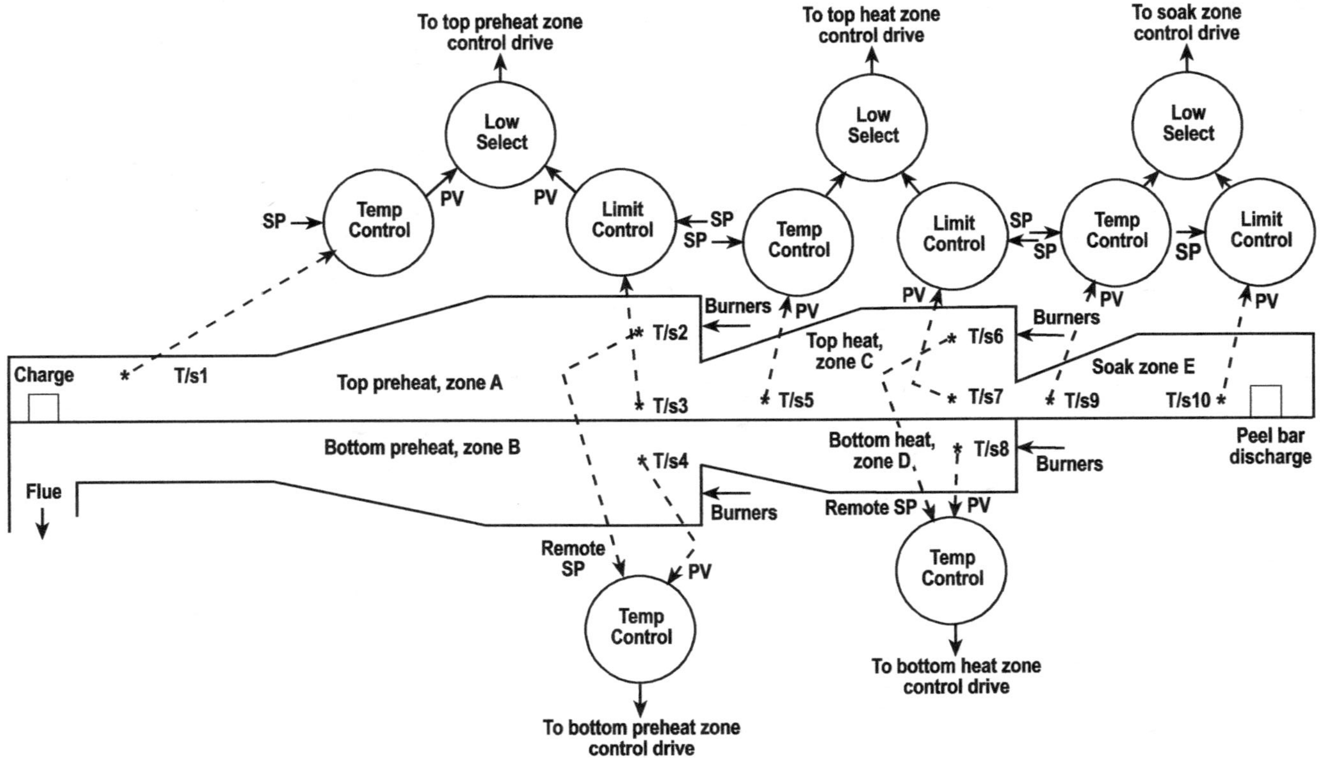


**Fig. 6.25.** Five-zone reheat furnace temperature control for best productivity, lowest fuel use. This control scheme allows quick recovery from production delays. PV = process variable; SP = setpoint; T/s = temperature sensor.

To prevent that problem, coauthor Shannon exhorts furnace owners to use more and shorter zones, and to locate control T-sensors low in the furnace sidewalls so that they can more promptly detect changes in *load temperature (not furnace temperature)*, and thereby react more promptly. T-sensors must be installed no higher above the furnace hearth than the thickness of the load pieces.

***6.11.2.1. Walking Hearth Furnace Control.*** The design of steel reheat furnaces has developed to such an extent that many early problems have been solved or at least remedied. However, the following are some difficulties that still cannot be estimated accurately enough to prevent concerns in final designs.

1. Slot losses in walking hearth and rotary furnaces due to infiltrated air and refractory condition
2. Actual excess air to be used in predicting %available heat
3. Actual reduction in heat transfer in bottom zones caused by skids
4. Accurate calculation of dropout losses
5. Determination of door losses due, largely, to infiltrated air

***6.11.2.2. Comparisons of Four Heating Modes.*** Heating capacities and fuel consumption rates were compared by developing heating curves[†] for 6" × 6" × 24' (0.152 m × 0.152 m × 7.32 m) steel blooms being heated to normal rolling temperatures in a walking hearth reheat furnace using air preheated by (*a*) regenerator, (*b*) a recuperator, (*c*) a regenerator with enhanced heating, and (*d*) a recuperator with enhanced heating. The same losses were used for all comparisons (see table 6.5 and figs. 6.26 to 6.29.).

To keep fuel consumption reasonable with recuperative air heating, it was necessary to keep the final poc exit temperature very low by keeping furnace capacity moderate. This is not necessary with regenerative air heating because the regenerative air heating beds lower the exit gas temperature, thus reducing fuel rates to a minimum. With recuperative air heating or with cold air, the furnace and the furnace gas exit temperature would have to have been 650 F (343 C) to compete with regenerative air heating's low fuel rates. Furnace heating capacity and fuel rate can vary if the charge zone temperature or load charging temperature varies.

A profound difference will occur in fuel rates when delays happen. With recuperation, the furnace exit gases may rise to 2000 F (1093 C) and more during the delay, then be diluted to 1500 F ± 250°F (816 C ± 139°C) by infiltrated air from many causes resulting in very low air preheat. Regenerative air heating depends only on the regenerative bed, and therefore, as the furnace gas temperature rises, the air preheat rises. The result is that the available heat of the combustion reaction falls during a delay with a recuperator, but may even rise during a delay with a regenerator. For these reasons, regenerative air heating and furnace capacity can be very high and still maintain low fuel rates while recuperative and cold air firing can have low fuel rates only with very low charge end furnace temperatures at all times, if coupled

[†]by the Shannon Method, explained in chap. 8.

**TABLE 6.5 Comparisons of heating curves for 6 in. (0.152 m) square steel blooms in a continuous reheat furnace, spacing = 1.6:1, with or without enhanced heating**

| Figure Description | tph | mtph | Time (min) | Fuel rate, (kk Btu/ton) | $\Delta T$ at end °F | $\Delta T$ at end °C | Maximum furnace temperature F | Maximum furnace temperature C |
|---|---|---|---|---|---|---|---|---|
| 6.26 regenerator | 115 | 104 | 81.6 | 1.07 | 40 | 22.2 | 2360 | 1293 |
| 6.27 recuperator | 100 | 91 | 105.6 | 1.32 | 50 | 27.8 | 2320 | 1271 |
| 6.28 regenerator w/enhanced heating | 136 | 123 | 69.5 | 1.13 | 20 | 11.1 | 2360 | 1293 |
| 6.29 recuperator w/enhanced heating | 119 | 108 | 88.8 | 1.32 | 30 | 16.7 | 2360 | 1293 |

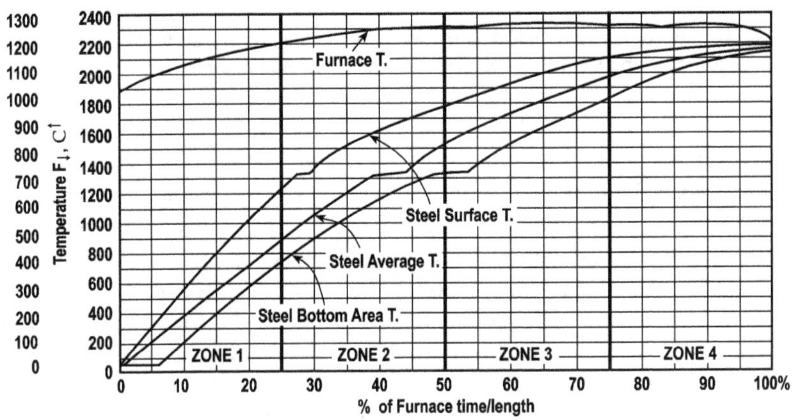


**Fig. 6.26.** Heating curves for 6 in. (0.152 m) square steel blooms in a 96 ft (29.3 m) long continuous reheat furnace, spaced 1.6:1, with air preheat by *regenerator*. 115 tph (104 mtph).

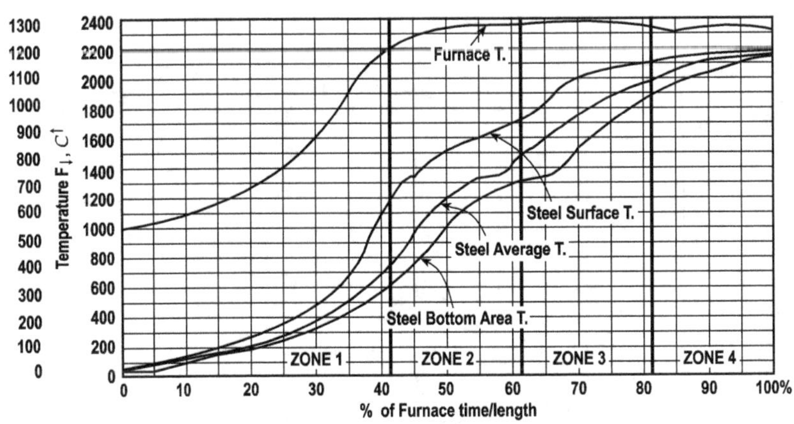


**Fig. 6.27.** Heating curves for 6 in. (0.152 m) square steel blooms in a 96 ft (29.3 m) long, continuous reheat furnace, spaced 1.6:1, with air preheat by *recuperator*. 100 tph (91 mtph).

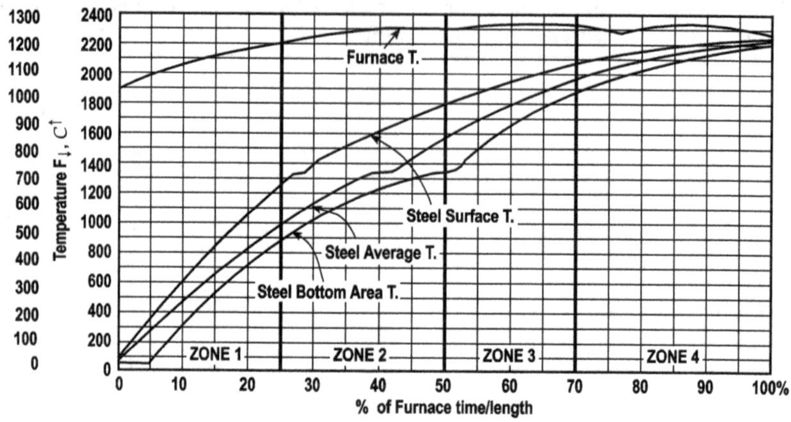


**Fig. 6.28.** Heating curves for 6 in. (0.152 m) square steel blooms in a 96 ft (29.3 m) long continuous reheat furnace, spaced 1.6:1, *with regenerator, enhanced heating.* 136 tph (122.9 mtph).

with very low air infiltration. From the temperature curves, one can conclude that for products spaced out on the hearth, and with enhanced heating, regeneration can raise productivity by 25% while raising fuel rates by only a small amount. Careful evaluation of flue gas exit temperature is critical when estimating fuel rates. (See sec. 2.4 and 5.1.) Some erroneously assume flue gas exit temperature is the same as furnace temperature. If the exit gas temperature had fallen that low, it could not deliver heat to the furnace! A $\Delta T$ is necessary to drive heat flow from the combustion gases to the furnace. Some specific cases are: about 1600 F (871 C) flue gas for a 1200 F (649 C) furnace, ~1900 F (1038 C) flue gas for a 1600 F (871 C) furnace, ~2200 F (1204 C) flue gas for 2000 F (1093 C) furnace, and ~2550 F (1400 C) flue gas for a 2400 F (1316 C) furnace.

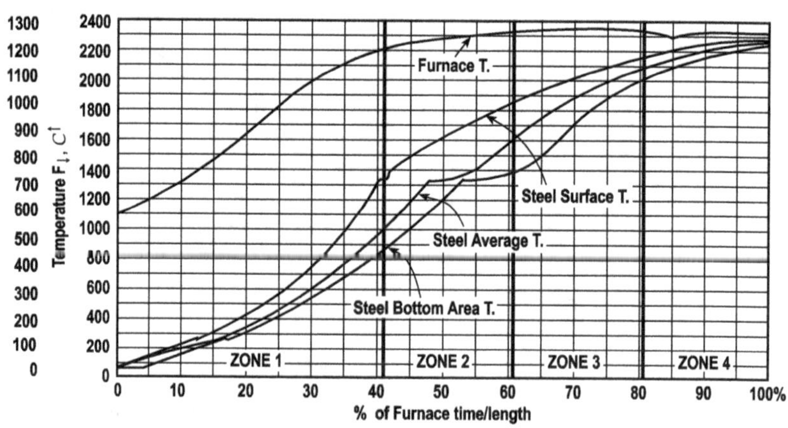


**Fig. 6.29.** Heating curves for 6 in. (0.152 m) square steel blooms in a 96 ft (29.3 m) long, continuous reheat furnace, spaced 1.6:1, *with recuperator, enhanced heating.* 119 tph (108 mtph).

> The industrial furnace field's real-life equivalent of Marmaduke Surfaceblow (world-famous serviceman and problem solver), Larry Hawersaat, Sr., used to say, "Cheap—cheap—cheap is for the birds!"

### 6.11.3. Effects of (and Strategies for Handling) Delays

***6.11.3.1. Effects of Delays.*** Sections 6.4 and 6.5.1 showed the effects of production delays on continuous steel reheat furnaces. As new cold loads are brought into the preheat zone after a delay, the heating and soak zones have yet to get the message that a massive cold load is about to enter their areas. That starts an overcorrection with sudden jumps to maximum input, followed by an oscillating accordianlike wave action going through several cycles of too-cold/too-hot/too-cold/too-hot output resulting in inability to roll quality product. This is brought on by inadequate ability of T-sensors to "feel" changing load temperatures promptly because of incorrect T-sensor locations, not enough short zones to avoid overcorrections, and not enough burner input near the charge end of the furnace to accommodate sudden changing needs after delays.

Suggested corrections include: (*a*) adding burners in top and bottom preheat zones, (*b*) shortening the top heating zone(s) or dividing them into more zones, (*c*) shortening the bottom heating zone(s) or dividing them into more zones, (*d*) relocating control sensors nearer the level of loads, and (*e*) programming control sensors to make top and bottom zones work as pairs.

All of the previous problems are aggravated by the "roller coaster"-like swings of the flue gas exit temperature changing a recuperator's output air preheat, and possibly damaging the recuperator, especially if lowest bidder favoritism has resulted in an induced draft fan of inadequate pressure and volume. The life of that fan also may be shortened.

**Warning:** Do not count on any continuous furnace always running at a continuous rate. Every furnace, oven, dryer, heater, boiler, and incinerator has to start up from cold or cool down from hot occasionally; therefore, designers and operators should build in flexibilities that will avoid damage to equipment and product during noncontinuous situations.

**Strategies for Handling Delays:**

A. If a 30-min delay is expected:
   1. Thirty min before, lower top and bottom heat zone setpoints to 2250 F (1204 C);
   2. Ten min before the delay, reset soak zone setpoints to 2250 F (1204 C);
   3. Ten min before the mill is to resume production, raise soak zone setpoints to normal;
   4. as soon as the delay ends and fresh material is charged, increase the firing rates of the two heat zones by increasing their setpoint to normal, taking care not to trip the furnace due to inadequate dilution air capacity and pressure.

B. If a 30-minute delay begins without prior knowledge:
   1. reduce soak zone setpoints to 2250 F (1204 C), *quickly*!
   2. lower heat zones setpoints to 2250 F (1204 C);
   3. ten min before the mill is to start, raise soak zone setpoints to normal;
   4. as fresh material enters the furnace, raise heat zone setpoints to normal, being careful not to trip the furnace due to inadequate dilution of air capacity and pressure.

C. If a delay of 2 hr is expected:
   1. Thirty min before the expected delay is to start, reduce the heat zones' setpoints to 2150 F (1177 C).
   2. Ten min before the delay is to start, reduce the setpoints of the soak zones to 2200 F. (1204 C);
   3. Forty-five min before the mill is to resume, raise the heat zones' setpoint temperatures to 2250 F (1232 C);
   4. Thirty min before the mill is to start, raise the soak zone's setpoints to 2250 F (1232 C);
   5. Ten min before the mill starts, raise soak zones to normal setpoints;
   6. as fresh material enters the furnace, raise the heat zone setpoints to normal again. Be aware of flue gas temperature levitation. Do not allow it to exceed the trip setting;
   7. it is highly recommend that the furnace trip temperature be reset to 1650 F ± 50°F (900 C ± 28°C) to assist the operator in proper operation of the furnace. Also recommended is early replacement of the dilution air fan or at least an increase in its output capacity and pressure all possible by a larger impeller and motor. Without these changes, the furnace will be difficult to operate correctly because the furnace priorities will be compromised by dilution air inadequacies.

D. Unexpected 5-hr delay:
   1. reduce soak zones' setpoints to 2200 F (1204 C) quickly as the delay begins;
   2. reduce heat zones' setpoints to 2150 F (1177 C) quickly as the delay begins;
   3. Forty-five min before the mill is to start, raise the heat zones' temperature setpoints to 2250 F (1232 C);
   4. Thirty min before the mill is to start, raise the soak zones' temperature setpoints to 2250 F (1232 C);
   5. Ten min before mill restart, raise the soak zones to their normal temperature setpoints;
   6. as fresh material begins to be charged, raise the heat zone setpoints to normal, being wary of a recuperator flue gas temperature furnace trip, by firing only enough fuel to hold the flue temperature below the trip setting. A better solution may be to manually control the fuel to the two heat zones so that the recuperator flue gas temperature does not trip off the furnace.

E. Also recommended:
   1. reset the furnace trip due to flue temperature between 1500 F to 1650 F ± 50°F (816 C to 890 C ± 28°C);
   2. redesign the dilution air system to increase the ambient air flow into the flue upstream of the recuperator entry to automatically prevent the temperature of the flue gas from tripping off the furnace;
   3. relocate the control T-sensors in the heat and soak zones as follows:
      *a*) top heat zone and control sensor should be between the first and second burners, 8" (0.2 m) above the hearth;
      *b*) add a second T-sensor, 3 to 4 ft (0.9 to 1.2 m) before the soak zone entry and 8" (0.2 m) above the pass line in the top heat zone to guide the operator as to the heating effect in the top heat zone;
      *c*) add a third temperature measurement in the top heat zone to act as a remote setpoint for the bottom zone. In fact, the present control temperature measurement in the top heat zone could be used for this purpose;
      *d*) the bottom control T-sensor should be located at about the same distance from the discharge of the bottom heat zone as the remote setpoint sensor is from the discharge of the top heat zone;
      *e*) change the location of the control T-sensors in the top soak zones to 3 ft (9 m) into the top soak zones 8" (0.2 m) above the pass line with an additional T-sensor 8" (0.2 m) above the pass lines 3 to 4 ft (0.9 to 1.2 m) from the zone discharge, for operator knowledge;
      *f*) use the present top soak zone measurements as remote setpoints for the two bottom soak zones.

By following the previous menu, delays can be managed smoothly, with the least possible trouble. The following were recommended for a new furnace that was inadequately designed for a new mill in 2001: (1) Redesign the dilution air system. (2) Replace the recuperator with one of much larger capacity and built for a higher inlet gas temperature. (3) Install a temperature control system operated from two heat zones and two top zone T-sensors. The top preheat zone control T-sensors should be placed in a sidewall 6 to 10 ft (1.8 to 3 m) from the charge door, limited by the T-sensor near the pass line before the soak zone. The bottom zones should receive this remote setpoint from the T-sensor high in the top zones and several feet from the soak- or heat-zone entry. With the new dilution air system, the control concept will require only soak-zone setpoint changes for delays.

***6.11.3.2. Heating Curves Showing Effects of Delays and Corrections.*** To understand the process of heating billets after a delay, see figure 6.30, which shows the normal furnace temperature profile (*top* curve) and the billet heating curve (*lower* curve) before a 30-min delay. Then, figure 6.31 shows the furnace temperature and the load heating curve for billets that stayed in the furnace during a 30-min delay. Figure 6.32 shows the inadequate heating of the second and third billets to enter the furnace after the delay if customary T-sensor locations are used.

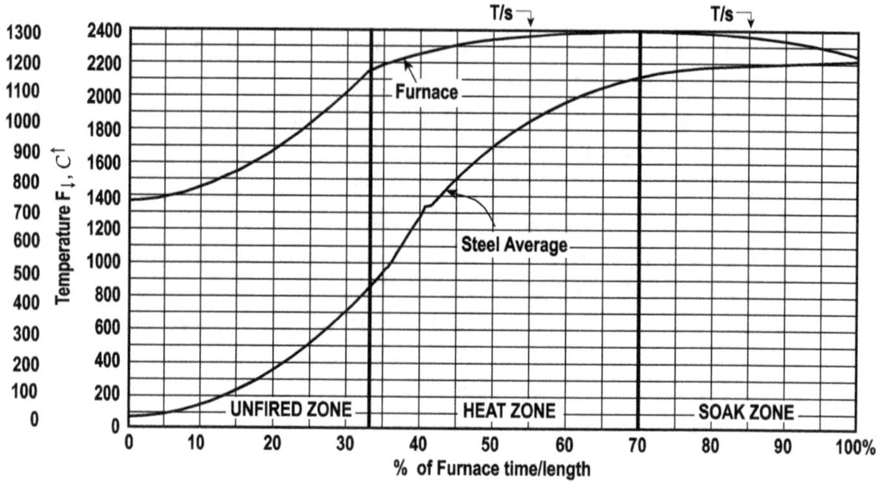


**Fig. 6.30.** Heating curve for a three-zone steel reheat furnace (*top* curve) and loads (*lower* curve) in normal operation **(without any delay).** The billet discharge temperature is 2220 F (1215 C).

In contrast, figure 6.33 shows the furnace temperature and the steel heating curves for the third billet charged after the end of the delay, when using coauthor Shannon's temperature control system for alleviating the problems of figure 6.32. This arrangement (shown across the top of fig. 6.33 and in figs. 6.24 and 6.25) has T-sensors located in a fast-moving furnace gas stream through the sidewall or roof where they

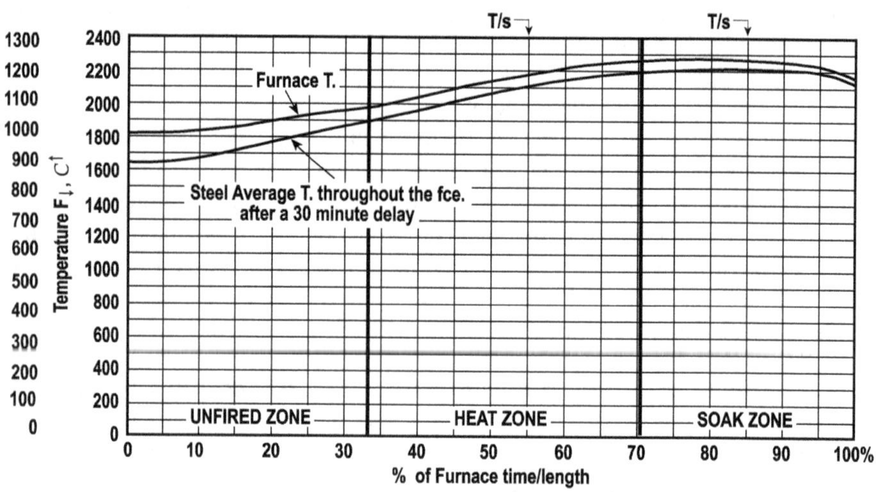


**Fig. 6.31.** Heating curve for a three-zone steel reheat furnace (*top* curve) and loads (*lower* curve) **after a 30-min delay.** Loads will be badly scaled from too early and too long exposure to high furnace temperature. (See example 8.3.1.)

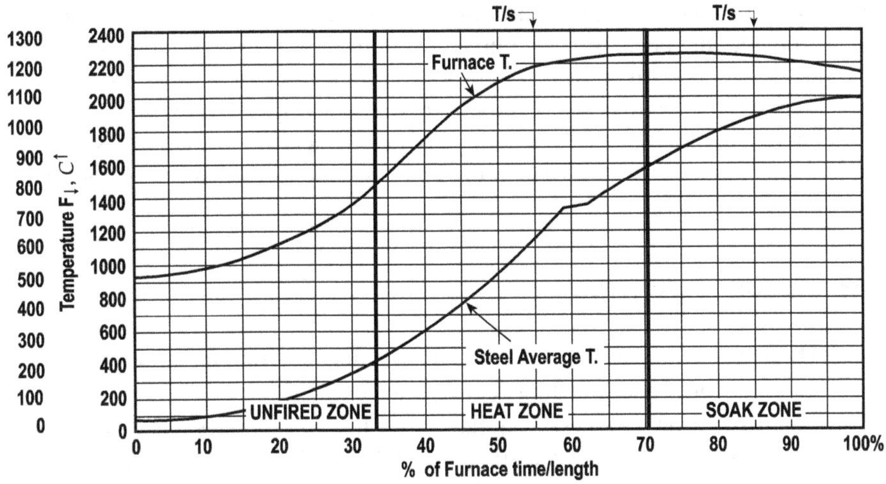


**Fig. 6.32.** Heating curve for a three-zone steel reheat furnace (*top* curve) and of the **third billet to enter the furnace at the end of a 30-min delay** (*lower* curve). Discharge temperature of this third load piece is only 2000 F (1093 C)—too cold to roll. Note that the furnace temperature at the charging entrance has cooled from 1360 F (738 C) in figure 6.30 to 920 F (493 C); and furnace temperature at the entrance to the heat zone has dropped from 2140 F (1171 C) in figure 6.30 to 1450 F (788 C) in this figure 6.32.

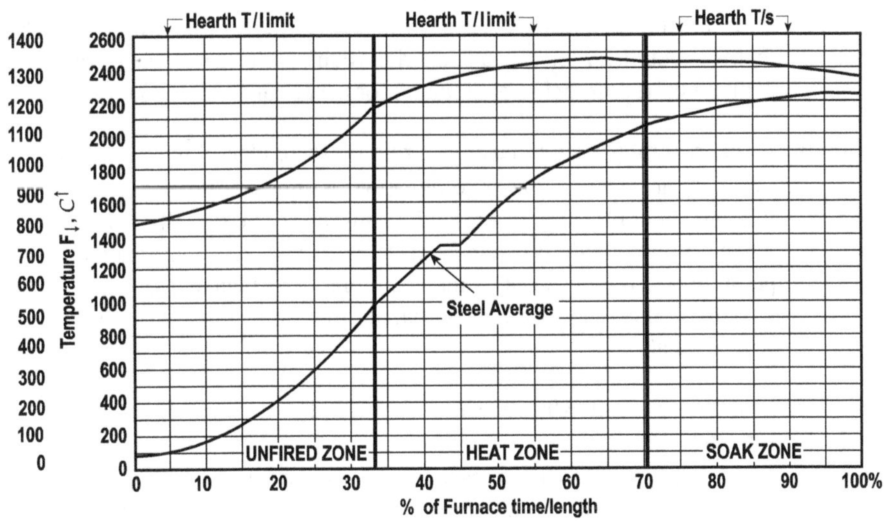


**Fig. 6.33.** Heating curve for a three-zone steel reheat furnace (*top* curve) and of third billet (*lower* curve) to enter the furnace after a 30-min delay and **with coauthor Shannon's system of T-sensor locations** (nearer hearth for *load* temperature sensing and control, instead of furnace or flame). Steel discharge temperature is 2240 F (1227 C)—good for rolling, *and* the furnace can resume its usual productivity more promptly after the delay.

can sense *load* temperature, but where the sensors cannot lose heat by radiation into the flue or charging opening, which may be (relatively) "cold holes." The sensor in the preheat zone is (limited by) the sensor near the hearth a few feet from the heat zone discharge. Those two sensors and controls have their signals pass through a "low select" device to prevent load overheating because the temperature control is located earlier in the billet's exposure history. Figure 6.33 shows such two-sensor control in the soak zone.

## 6.12. REVIEW QUESTIONS

**6.12Q1.** Is it better to have an air or fuel distribution manifold for a row of burners built curvy and streamlined or big and boxy?

**A1.** Big and boxy, unless you can afford time and money for a computer-designed and fabricated streamlined design that can assure uniform distribution to all burners at all firing rates. A big plenum box is ideal.

**6.12Q2.** Are the requirements for combustion the same as the requirements for an explosion?

**A2.** No, but almost. An explosion has all the requirements of combustion, except that it is not steady state, and instead requires accumulation of a combustible mixture of fuel and air, and sometimes confinement.

**6.12Q3.** How does air/fuel ratio affect product quality?

**A3.** Air/fuel ratio determines whether the atmosphere in a furnace is rich, lean, or neutral. Different load materials require different atmospheres (and sometimes at different temperatures) for best final product quality.

**6.12Q4.** Is the 'neutral pressure plane' (or 'zero pressure plane') really a plane?

**A4.** Probably not, because flows (circulation) within the furnace cannot exist without slight pressure differentials. Thus, the plane is really only a plane when all burners are off, flues and doors closed, and no horizontal temperature differentials exist. It may be more like a blanket that someone is shaking in the wind. But realize that all differentials within a large space will be small.

**6.12Q5.** Is there any reason why you should not specify a high turndown capability for a new furnace?

**A5.** Yes. Higher turndown requires higher blower pressure, which can increase the cost. You must find a compromise turndown ratio between cost and flexibility.

**6.12Q6.** If you cannot see the flow arrows from the designer's diagram when looking into a newly operating furnace, how can you know if the actual flow patterns are correct?

**A6.** Finished product quality is the test. You can infer some flow results by careful study of visible or measured temperature patterns. It is difficult to tell someone how to develop good heating judgment. You can help yourself develop good heating judgment by studying fluid dynamics and heat transfer, and by listening to experienced operators.

**6.12Q7.** How can the temperature difference from burner wall to opposite wall above the load(s) be held to a minimum?

**A7.** By controlling the spin of the combustion gases. A second method (not as good) is to alternate burners side to side, above the load, preferably with no greater than 2.5-ft center-to-center spacing.

**6.12Q8.** What should be the firing rate of a soaking pit that is to heat a 90-ton charge of 0.23% carbon steel ingots in a total of 9 hr? Assume a 25 ft long × 10 ft wide pit with heat losses of 1.5 kk Btu/hr. The average waste gas temperature over the 9 hr is estimated to be 2000 F. The ingot discharging temperature should average 2300 F.

**A8.** From figure 5.1 at average 2000 F flue gas with 10% excess air, read 40% available heat as an average over the 9-hr period. From figure 2.2, estimate the heat content of the steel at 2300 F as 364 Btu/pound. Therefore,

| | |
|---|---|
| heat to loads = (90 tons) (2,000 pounds/ton) (364 Btu/pound) | = 65.5 kk Btu in 9 hr. |
| heat losses = (9 hr) (1.5 kk Btu/hr) | = 13.5 kk Btu in 9 hr. |
| Total 'heat need' = required available heat = 65.5 + 13.5 | = 79 kk Btu in 9 hr. |
| Gross heat input required = 79/0.40 | = 198 kk Btu in 9 hr. |
| Firing rate required over 6 hr actual firing time = 198/6 | = 33 kk Btu/hr |

**6.12Q9.** Where should control T-sensors be located for shortest heat cycles with protection for the product in a continuous reheat furnace?

**A9.** In both sidewalls of the furnace at the height of the tops of the loads.

# 7

# GAS MOVEMENT IN INDUSTRIAL FURNACES

## 7.1. LAWS OF GAS MOVEMENT

Temperature uniformity involves *improvement by movement* of radiating triatomic gases as well as convection poc. (See also chap. 5 of *reference 51*.) Concepts of this chapter will be facilitated by the following review of the laws of gas movement concerning buoyancy, velocity head, fluid friction between gases and solids, and flow induction.

### 7.1.1. Buoyancy

A column of hot air (fig. 7.1) weighs less than an equally tall column of cold air, which is shown dotted to form a U-tube manometer. The dotted column corresponds to the atmosphere outside a stack or chimney. The difference in weights of the columns creates a pressure difference ($\Delta P$) known as "draft" (see glossary), expressed in inches or millimeters of water column on a manometer. The draft is proportional to the height of the gas column and to the difference in densities of the hot and cold gas columns. The densities of air and other gases depend on their pressures and temperatures, thus: **density, $\rho = p/RT$**, where density is pounds per cubic foot (US) or kg/m$^3$ (SI), $T$ is absolute temperature rankine (US) or kelvin (SI), and $R$ is a constant = 53.3 fp/pound mol °R for air (US), or 287 joules-kg-mol °K for air (SI). Densities are tabulated in *references 51* and *52*.

The theoretical draft (lift, suction) of a tall column of hot gas, as in a furnace, vertical duct, or stack is:

$$\Delta P_{"wc} = \frac{7.63 h_{ft}(P_{b,atm})}{(T_{aF}+460)}\left\{1 - G\left[(T_{aF}+460)/(T_{gF}+460)\right]\right\} \qquad (7.1)$$

where

$\Delta P_{"wc}$ = pressure difference "wc between a cold air and a hot gas column

$h_{ft}$ = height in feet of the hot gas column

$P_{b,atm}$ = barometric pressure, in atmospheres
$G$ = gas gravity = density$_g$/density$_a$
$T_{aF}$ & $T_{gF}$ = average temperatures of air & gas columns, respectively, Fahrenheit

$$\Delta P_{\text{mmH}_2\text{O}} = \frac{635.9 h_m (P_{b,atm})}{(1.8 T_{aC} + 492)} \{1 - G[(T_{aC} + 273)/(T_{gC} + 273)]\} \qquad (7.2)$$

where

$\Delta P_{\text{mmH}_2\text{O}}$ = pressure difference in mm of water, cold air to hot gas column
$h_m$ = height in meters of the hot gas column
$P_{b,atm}$ = barometric pressure, in atmospheres
$G$ = gas gravity = density$_g$/density$_a$
$T_{aC}$ & $T_{gC}$ = average temperatures of air & gas columns, respectively, Celsius

As you wade into the water at the beach to a point where the water is 1 m deep, consider a cubic meter of water, which has a density of 999 kg/m$^3$. The pressure on the square meter of beachbottom at your feet would be 999 kg/m$^2$. If you wade into the water at a beach where the water is 1 ft deep, think of a cubic foot of water, which has a density of 62.4 lb/ft$^3$. The pressure on a square foot of sand at your feet would be 62.4 lb/ft$^2$. That same pressure would be pressing down on the lower leg of a 1 foot high column of water in a U-tube manometer (see fig. 7.1).

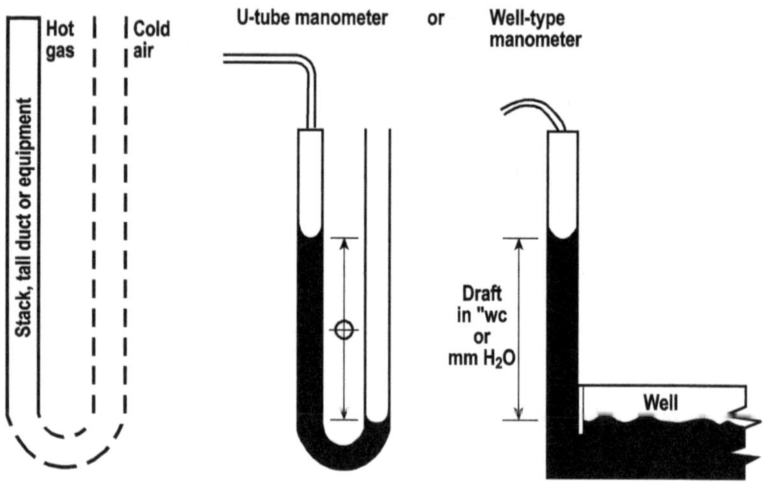


**Fig. 7.1.** Diagrams showing the cause of stack draft by analogy with a U-tube manometer. Solid lines represent a duct or stack of hot gas; dashed lines represent an adjacent column of cold air. The "well," or short, fat leg of the *far right* manometer, has a cross section so many times larger than the left leg that the change in elevation of the right leg can be ignored.

That same intensity of suction (vacuum, draft) could be pulling up on the top of the other leg of a U-tube manometer if connected to the bottom of a column of hot flue gas, and if the other leg of the manometer was open to atmospheric pressure. We measure draft (negative pressure) and other small pressures in units of "wc or $mmH_2O$. The aforementioned 12 in. wc = 62.4 lb/ft$^2$, or 62.4 lb/ft$^2$ × (1ft$^2$/144 in$^2$) = 0.433 psi (pounds per square inch), or 0.433 lb/in$^2$ × (× 16 oz/lb) = 7.17 osi (ounces per square inch).

The draft from equation (7.1) is plotted in figure 7.2 for a range of mean column air temperatures. For industrial heating fuels with high C/H ratio, the curve may be as much as 7% higher, but the usual excess air will bring the draft value back to very close to the plotted curve for hot air. The draft will be less during bad weather, and at high elevations when and where the barometer reading will be less than at sea level, in proportion to the ratio of actual barometric pressure to standard, both in the same units.

For tall columns of hot gas, the average temperature may be taken as the arithmetic mean between top and bottom. If the hot column is closed at the top and open at the bottom, the "draft" becomes an excess pressure in the hot column, that excess pressure being greatest at the top, with atmospheric pressure at the open bottom end. If a hot gas column is closed at the bottom and open at the top, atmospheric pressure will exist at the open top, with pressure less than atmospheric at the bottom of the column.

If the temperature of the hot column is constant and if the hot column is open at both ends, but contains a resistance to flow, then the draft will cause a flow through the column in such a manner that the draft will be balanced by the resistance to flow, which is the sum of all velocity heads plus friction heads.

### 7.1.2. Fluid Friction, Velocity Head, Flow Induction

Fluid friction is covered by information on pressure losses in pipes, ducts, orifices, valves, and fittings in pt 5 of *reference 51*.

As a current of air or jet of fluid (such as the poc from a burner) passes through a space (such as a furnace), it gathers unto itself molecules of the surrounding fluid, imparting velocity to them by viscous friction, or drag. The main stream slows down in such a manner that the total momentum of the two streams (Moving Mass × Velocity) is conserved. The total (included) angle of the cone that envelops the combined moving mass varies with the initial velocity and density of the jet. In cold air, it is about 16 degrees for slow jets traveling at 10 fps (3 mps), increasing gradually to about 25 degrees for jets at more than 1,000 fps (305 m/s).

When a jet of cold air induces hot air or combustion gases, the jet expands at greater angles than in cold air. The velocity at the edge of the jet is near zero, but the velocity at the center of the jet stream is approximately twice the average velocity. Care must be taken in applying these generalities to furnace jets, to use them only for currents in which combustion has been completed, (*a*) because changes of specific volume due to combustion affect the result considerably and (*b*) because the combustion process may be quenched by the induced cold air. Jet induction is discussed again in sec. 7.4.

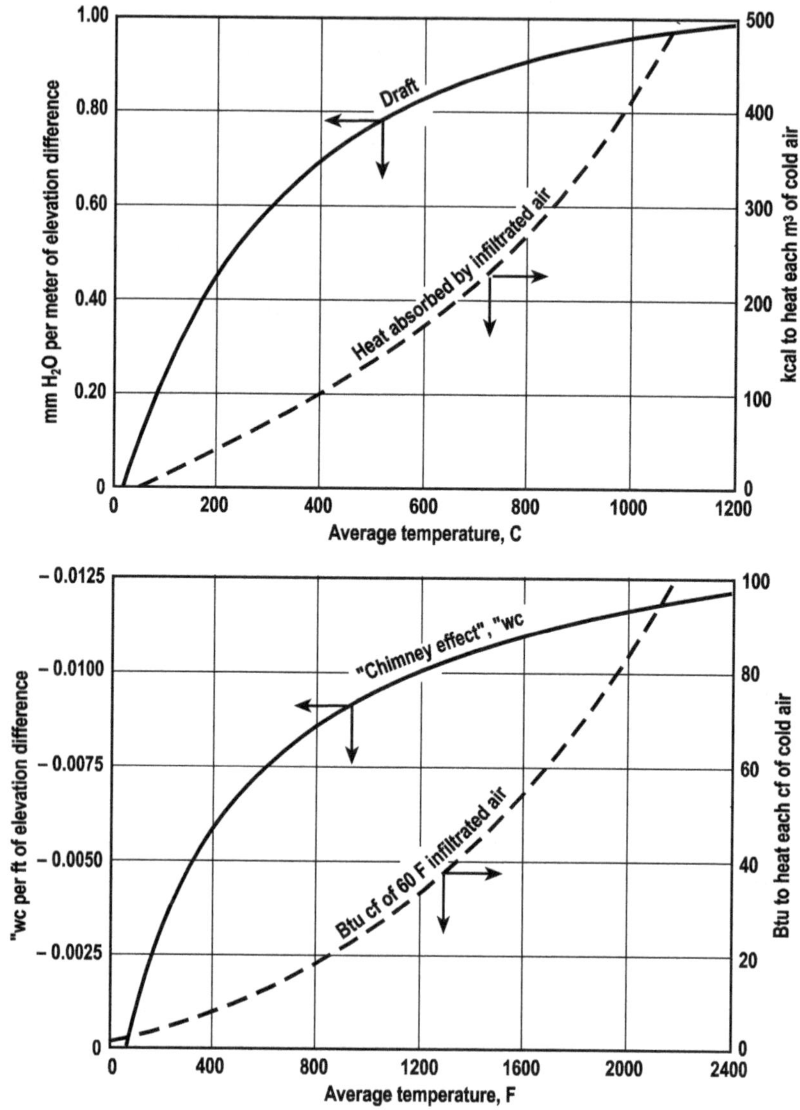


**Fig. 7.2.** Draft developed in a hot chamber, and fuel input required to heat infiltrated air. The vertical scale is the difference in height between a cold air inlet (crack, door opening) and a hot gas outlet at the top (flue, stack top, top of door opening). (Courtesy of *reference 52.*)

If a gas or air current passes along a furnace wall or load surfaces, it is retarded by both viscosity and turbulence. The retardation due to turbulence grows with the roughness of the surface of the wall. By the law of conservation of momentum, flow deceleration causes a rise in pressure.

In passing through tall ducts or tall apparatus, hot gases cool, contract in volume, and move more slowly. This is equivalent to a gradual enlargement of the stream cross section, as in a constant-temperature Venturi tube; thus, some of the kinetic energy is transformed into pressure energy. The maximum amount of pressure recoverable in a frictionless tube corresponds to the difference in the velocity heads of the initial and final velocities. But, the pressure recovery from this effect is so small compared to frictional pressure drop that it is negligible in most practical cases. Velocity heads (velocity pressures) are tabulated on p. 133 of *reference 51*.

## 7.2. FURNACE PRESSURE; FLUE PORT SIZE AND LOCATION (see also *references 51* and *59*)

Two good guidelines for pressure conditions in furnaces are:

1. In most industrial process heating, the pressure in the heating chamber should be atmospheric, or only very slightly positive, at all firing rates.
2. The lower the temperature to which the material is to be heated, the greater the necessity for thorough gas circulation in the heating chamber, especially if loads are placed compactly in the furnace or oven (e.g., piled or coiled material that is to be heated rapidly and uniformly. (See sec. 6.6.)

If furnace pressure is much greater than atmospheric pressure, flame or hot gases will leak out of all openings—wasting fuel, harming people and materiel near the leaks, and shortening the life of doors, doorframes, conveyors, seals, and refractories. If the furnace pressure is less than atmospheric pressure, cold air will be drawn in around doors, observation ports, conveyors, seals, and cracks—chilling parts of the load and wasting fuel.

In a tall furnace, it is impossible to have the same pressure at all levels because the furnace acts as a chimney, with its internal pressures increasing with elevation within the furnace. Depending on the magnitude of (*a*) pressure created by a forced draft fan or blower or (*b*) suction created by an induced draft fan, eductor, or natural chimney draft, the furnace may have any of the following situations:

| | | Situation | | | | |
|---|---|---|---|---|---|---|
| | | 1 | 2 | 3 | 4 | 5 |
| top pressure | = | +++ | ++ | + | 0 | - |
| center pressure | = | ++ | + | 0 | - | -- |
| bottom pressure | = | + | 0 | - | -- | --- |

How should an engineer select situation 1, 2, 3, 4, or 5 (i.e., automatic furnace pressure control setpoint) for the pressure sensor location? And is the pressure sensor located properly for the process?

Assume that the furnace has or will have cracks, and leaky seals around doors, peep sights, sensors, and car hearth or conveyor. Establish an ongoing inspection and repair program to minimize these possible sources of inleakage or outleakage.

Encourage operators to be proud of the prime condition of their furnaces. Keep an automatic furnace pressure controller, and its sensing lines, clean and in good operating condition.

**Objectives:**

1. Protect the loads from unwanted cooling by infiltrated air.
2. Keep out tramp air, which wastes fuel.

An engineer or operator can detect air inleakage by holding a smoldering wood splinter very close to the bottoms of doors and other suspected leak points, and observe the direction of smoke flow.

To prevent the entrance of tramp air, furnace pressure situation 1 or 2 is required. Some people suggest keeping the zero-pressure-plane below the lowest load, but it is safer to keep it below the lowest possible leak. Although it may be physically impossible to locate a sensor below the lowest possible leak, the furnace pressure sensor can be located higher if the setpoint pressure is purposely increased to bias it to control at a higher pressure level corrected for the sensor's higher elevation. (See table 7.1.)

Section 6.6.2 gives recommended details and locations of furnace pressure control sensors and their compensating (room pressure) taps.

Of the previously mentioned tabulated five situations, situation 1 is probably most desirable for industrial heat-processing furnaces. If the hearth is tight so that there can be no inleakage from below, the pressure at hearth level should be controlled at +0.02 in. wc (0.51 mm $H_2O$). For conveyor furnaces and car-hearth furnaces, there may be a chance of a leak below the hearth level (as at a water seal or sand seal), in which case the +0.02 in. wc (0.51 mm $H_2O$) pressure should be the setpoint for that lowest leak level. The control sensor should be just high enough above the hearth to avoid blockage by accumulated scale or refractory crumbs, and the control setpoint biased upward per table 7.1 for the difference in elevation between the sensor and the lowest leak. This will achieve the three objectives listed previously.

The desirable slightly positive pressure at hearth level is easily maintained if the poc exit via a hearth-level flue or under a door. This "downdrafting" arrangement has the advantage that relatively cool poc near the loads are swept out, and more of the hot gases contact the load(s) and the hearth, reducing temperature differentials.

When furnace gases are vented through the roof, they usually leave at a higher temperature; thus, the thermal efficiency will be reduced.

**TABLE 7.1 Elevation bias corrections for furnace pressure control setpoint when the furnace pressure sensor is above desired control level**

| | | | | |
|---|---|---|---|---|
| US units | | | | |
| Furnace temperature | 1200 F | 1600 F | 2000 F | 2400 F |
| Add "wc/foot of height | 0.0101 | 0.0110 | 0.0115 | 0.0120 |
| SI units | | | | |
| Furnace temperature | 700 C | 900 C | 1100 C | 1300 C |
| Add mm $H_2O$/m of height | 0.858 | 0.920 | 0.964 | 0.997 |

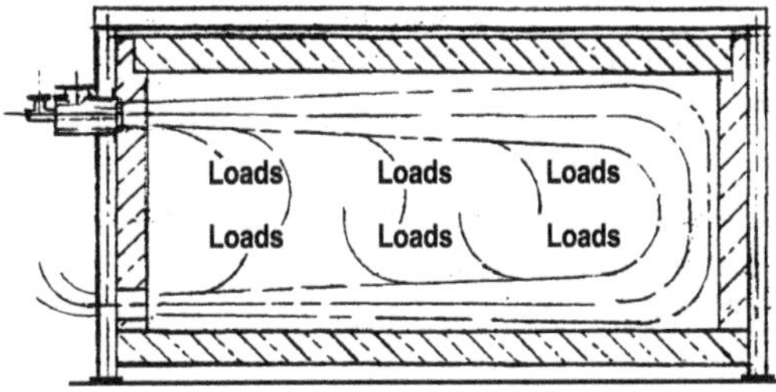


**Side view of a batch car-hearth furnace with top-firing, bottom-fluing.**

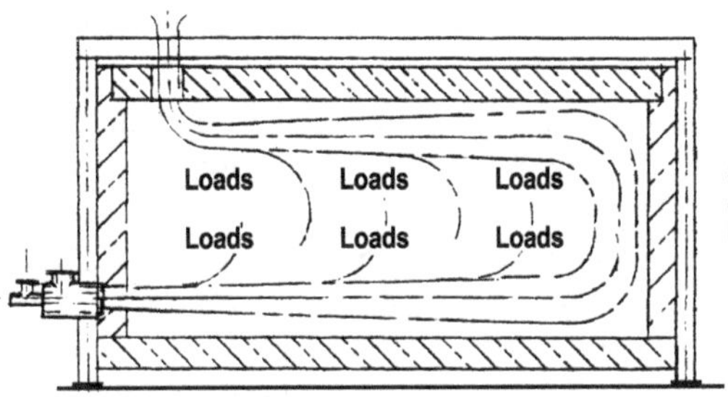


**Side view of a batch car-hearth furnace with bottom-firing, top-fluing.**

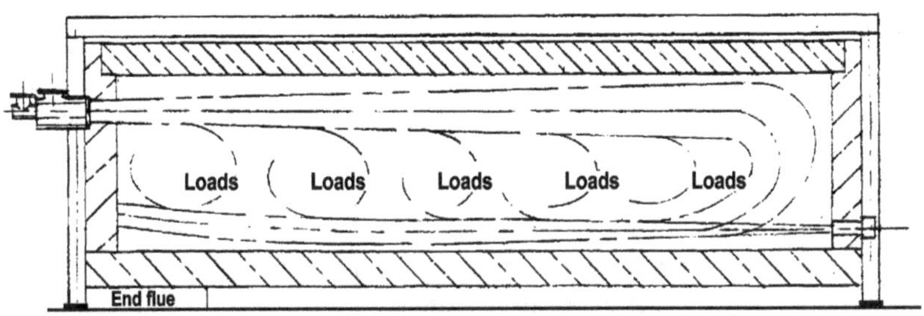


**End view, continuous reheat furnace with enchanced heating. A bottom flue is located at one end of the furnace.**

**Fig. 7.3.** Gas flow patterns must be carefully controlled in all types of furnaces to assure effective heat transfer, fuel efficiency, productivity, and product quality.

Bottom firing* (i.e., burners below the loads) delivers heat to the usually cooler hearth, making up for hearth losses that otherwise would be taken from the loads or from the gas blanket. (See fig. 7.3.) Bottom firing is sometimes used with roof vents, but roof flues can be undesirable because at low-firing rates, the gases may short-circuit direct to the roof flues (giving poor temperature uniformity and poor fuel economy). Roof vents also can cause negative or low furnace pressure; therefore, oversize vents should be avoided, and furnace pressure should be controlled with a stack closure. Tall furnaces are especially susceptible to this problem.

*Bottom firing and top fluing = updrafting; Top firing and bottom fluing = downdrafting. (Avoid using terms such as "overfiring" and 'overfired,' which mean overdone. Similarly, avoid the terms "underfiring" and 'underfired,' which also can mean insufficiently heated.)

> Dense loading requires prolonged heat cycles to achieve temperature uniformity throughout the pack. This was learned early in updrafted periodic brick kilns. Wherever there was a slightly wider vertical space between columns of bricks, the hot poc from the bottom-fired burners would follow the path of least resistance, and thus all the inner surfaces of that column would get hotter, creating more "chimney effect," which became a runaway effect. This produced overburned bricks around any hot columns, and underburned bricks everywhere else.

Roof flues can be used with top firing if the flames have sufficient momentum (even at low firing rates) that they will fly past the flues and not up the stack. (See fig. 7.12.)

In long *batch furnaces*, good temperature uniformity requires that each zone have at least one flue. Otherwise, changes in the firing rate in one zone can adversely affect other zones. It is possible to have one flue located between two adjacent zones.

Furnaces have been built with one flue in the end wall by the charge door (to supply a recuperator). The zone closest to the flue can operate over setpoint if the products to be heated are located near the discharge door. This is a very serious problem when ±25°F temperature variation is specified to be held at all times. Temperatures 100°F over setpoint have been witnessed.

To determine the *flue port size*, the firing rate should be calculated from a heating curve (chap. 8). However, the required firing rate can be calculated if the following information is known: (*a*) weight of loads to be heated per hour, (*b*) final load temperature required, (*c*) rate of temperature rise, (*d*) heat losses expected, (*e*) a conservative flue gas temperature expected, and (*f*) a conservative air/fuel ratio.

**Example 7.1:** *Given:* A car furnace (batch) 10' × 20' × 9' high inside is to heat 40 tons of steel loads from 60 F to 2250 F at a rate of 250°F per hour. Specific heat of steel, from p. 275 of *reference 52* is 0.165 Btu/lb°F. Average flue gas exit temperature will be 2200 F. The fuel will be natural gas with 10% excess air. Average losses, in Btu/ft$^2$hr are: roof 900, walls 500, door 1100, and car 600.

*Calculate:* (*a*) heat needs, (*b*) %available heat, (*c*) gross heat required, (*d*) design burner input, (*e*) flue gas volume at flue temperature, and (*e*) flue size.

*Solution:* (*a*) The average specific heat of steel, from table A.16$_{US}$ of *reference 52* is 0.165 Btu/lb°F. Heat to steel = wc ΔT = (40 ton) (2,000 lb/ton) (0.165 Btu/lb°F) (250°F/hr) = 3.3 kk Btu/hr

LOSSES:

roof = (20 × 10) (900) = 180 000

walls = (2) (20 × 9) (500) + (10 × 9) (500) = 225 000

door = (10 × 9) (1100) = 99 000

car = (20 × 10) (600) = 120 000

TOTAL = 624 000 Btu/hr = 0.624 kk Btu/hr.

(*b*) Heat needs = heat to steel + losses = 3.3 + 0.624 = 3.924 kk Btu/hr. %available heat at 2200 F at 10% excess air is 37% (from fig. 5.1).

(*c*) Gross heat required $= \dfrac{\text{(heat needs)}}{\text{(\%available heat/100\%)}} = \dfrac{3\ 924\ 000 \text{ Btu/hr}}{0.37} =$ 10.6 kk Btu/hr.

For abnormal conditions, a security factor of 1.2 is advised, or perhaps 1.4 for extra wall heat for a cold startup. 1.4 × 10.6 kk = 15 kk Btu/hr. ("Rules of thumb" may be very case specific or overly safe, but can be assuring "ballpark" guides; thus coauthor Reed prefers to call them "thumb guides." One such is 80 000 Btu/hr ft$^2$ of hearth for large high-temperature car furnaces, which gives 80 000 × 20 × 10 = 16 kk Btu/hr for the job in this example).

(*d*) A convenient thumb guide is the average of 11 natural gases on pp, 36 to 38 of *reference 51* is 11.4 scf of flue gas (with 10% excess air) per 1000 gross Btu. From that thumb guide, (15 000 000 Btu/hr) (11.4 cf fg/1000 Btu) (2200 + 460) / (60 + 460) = 875 000 acfh (actual ft$^3$/hr of 2200 F flue gas in this example.

(*e*) Assuming that the flue has a double ell refractory stub stack to protect personnel and to reduce radiation loss from the furnace, pp. 225 to 227 of *reference 51* imply that a flue velocity at temperature might be 20 fps. The flue opening in the roof should be

$$\frac{875\ 000 \text{ ft}^3/\text{hr}}{(20 \text{ ft/sec})\ (3600 \text{ sec/hr})} = 12.15 \text{ ft}^2$$

which would be a 3.5' ID square or a 3.93' ID round, flue opening. QED

It is possible to calculate the dimensions of ports and flues so that the resistance of ports and flues will be balanced by the draft (suction) plus furnace pressure. However, good practice in automatic furnace pressure control usually necessitates a stack damper that always takes a minimal pressure drop. Therefore, the real balance is: stack draft + furnace pressure = $\Delta P$ furnace exit orifice + $\Delta P$ stack skin friction + $\Delta P$ damper. Tables 7.2 and 7.3 from Prof. Trinks' fifth edition list information for a few specific cases that illustrate points mentioned earlier and equations 7.3, 7.4, and 7.5 below.

$$\text{Flue area} = \text{flue flow/flue velocity} \tag{7.3}$$

$$\text{Flue area, ft}^2/\text{ft}^2 \text{ of hearth} = \frac{\text{flow, ft}^3/\text{hr ft}^2 \text{of hearth}}{\text{(velocity, ft/sec)(3600 sec/hr)}} \tag{7.4}$$

$$\text{Flue area, in}^2/\text{ft}^2 \text{ of hearth} = \text{(eq.7.4)} \times (144 \text{ in.}^2/\text{ft}^2) \tag{7.5}$$

Table 7.2 shows that, for a very small furnaces (low flue, small cross section) and for low temperatures, the velocity through the flues and ports must be low (14 fps) if excessive furnace pressure is to be avoided. It also shows that in large furnaces with high temperature, velocities up to 40 fps may be practical. It appears impractical to formulate a simple rule for flue port size that is applicable to all furnaces. For quick estimates, however, it may be helpful to conclude from table 7.2 that velocities of 19, 23, and 27 fps are good averages for 1200 F, 1600 F, and 2200 F furnaces, respectively. On that basis, the figures of table 7.3 were derived using equations (7.3), (7.4), and

**TABLE 7.2 Velocities in flues and stacks**

| | Stack Temperature | | | | | | | | |
|---|---|---|---|---|---|---|---|---|---|
| | 1200 F | | | 1600 F | | | 2200 F | | |
| Stack Height→ | 3' | 8' | 20' | 3' | 8' | 20' | 3' | 8' | 20' |
| Flue size | Maximum velocity with reasonable furnace pressure (fps) | | | | | | | | |
| 4.5" × 4.5" | 13.8 | 18.0 | 22.8 | 15.8 | 20.7 | 26.4 | 18.3 | 24.1 | 30.7 |
| 4.5" × 9" | 14.1 | 18.6 | 24.5 | 16.1 | 21.5 | 28.3 | 18.7 | 25.2 | 33.0 |
| 9" × 9" | 14.4 | 19.4 | 26.4 | 16.4 | 22.4 | 30.5 | 19.1 | 26.2 | 35.8 |
| 18" × 18" | 14.6 | 20.2 | 28.6 | 16.7 | 23.3 | 33.0 | 19.4 | 27.3 | 38.9 |

*Note:* ' feet. " inches.

(7.5). These figures are necessarily approximate. Deviations have been found both up and down. The figures do not apply to continuous or recuperative or regenerative furnaces.

Table 7.2 is based on a heating rate of 100 lb of steel per hour for each square foot of hearth whereas 40 lb/ft$^2$hr is more reasonable for low-temperature furnaces. However, sometimes a furnace that was designed for low temperature is pushed into service at a higher temperature, in which case a damper or large piece of hard refractory can be used to partially block off an oversize flue. In smaller furnaces, the ratio of flue port area to hearth area must be larger.

If a flue poc carries heavy particulates and has ells (elbows) or horizontal sections where particles may be deposited, flues must be made even larger, and clean out doors must be provided, and used! For some forge furnaces and for bolt heading furnaces, all the poc are purposely forced out the slot through which the stock is charged.

For continuous furnaces, the previous suggestions for sizes of vents and flues are not applicable. The multiplicity of designs is so great that each type and rate of heating requires a separate calculation. The fuel consumption, rate of flow of poc, and temperature at which they leave the furnace are determined either by calculation or by comparison with existing, similar furnaces.

**Concluding reminders about furnace pressure:**

1. Negative furnace pressure increases fuel consumption. A recent complaint about a car furnace that could not reach capacity was found to be a problem with a $\frac{1}{2}$ in. gap all around the large car that admitted so much cold tramp air

**TABLE 7.3 Thumb guide generalizations relative to table 7.2. The first row is calculated as in example 7.2. The last row is via equations (7.3), (7.4), and (7.5) using approximate velocity figures from the center three rows of table 7.2.**

| | Stack temperature | | |
|---|---|---|---|
| | 1200 F | 1600 F | 2200 F |
| Flow, ft$^3$/hr ft$^2$ hearth | 900 | 2040 | 5000 |
| ft$^2$ of flue/ft$^2$ of hearth | 2.0 | 3.5 | 7.4 |

(excess air) that the %available heat had dropped far below its design level. Car seals have a good purpose, and need to be maintained!

2. Negative furnace pressure diminished product quality by admitting cold drafts that cause temperature nonuniformity, and may change the metallurgically required atmosphere in the furnace. Poor product quality raises fuel, labor, and material costs because the job has to be done all over again. It may cost loss of business and customers.
3. The only gain from a negative furnace pressure is lowered fan or blower costs (operating and capital).
4. Excessive positive pressure—more than about 0.02 in. wc (0.5 mm)—endangers people nearby, and shortens the life of furnace components.

## 7.3. FLUE AND STACK SIZING, LOCATION (see *references 51* and *59*)

### 7.3.1. The Long and the Short of Stacks

Most modern industrial heat-processing units are equipped with forced draft. Therefore, they do not need stacks for draft creation—only stub stacks to deliver hot gases away from where they might harm people, equipment, or the building that protects them from the weather. The poc can be discharged directly from the flues into the workspace, where a ceiling fan or a hood with a vent through the roof (monitor) delivers them to the atmosphere. Some large regenerative furnaces and steam power-generating boilers still depend on stacks for draft, but use of stacks is now mostly limited to need to deliver poc out of buildings or to high elevations for dispersal.

A slight positive pressure is usually desirable in the furnace, so the stub stack can be whatever height is needed to reach through the roof and sufficiently high above the surrounding buildings to prevent backdrafts or eddies from blowing down into it. The need to carry gases above surrounding buildings often makes them too high, therefore, a damper must be used to reduce excess draft.

Many furnace stacks are not only too tall but also too large. This may be because the steel shell of the stack often needs a protective refractory lining, which may be difficult to install in a small-diameter stack. Stack dimensions should be determined by calculation for each individual case.

A thumb guide for determining stack cross-sectional area (inside the lining) is to make it equal to about 60% of the sum of the areas of all exhaust ports or flues, provided that they were properly sized. This reduction to 60% is reasonable because the gases cool down on their way through the stack and because one large duct creates less frictional resistance than many small ducts of the same total cross-sectional area.

The method of calculation of stack size varies with local conditions, but one must first picture the pressure pattern through the combustion system and the furnace, as suggested in figure 7.4.

From figure 7.4 it is possible to write an equation of pressure balancc, similar to balancing one's checkbook or applying the law of conservation of energy (1st Law of Thermodynamics) in a heat balance.

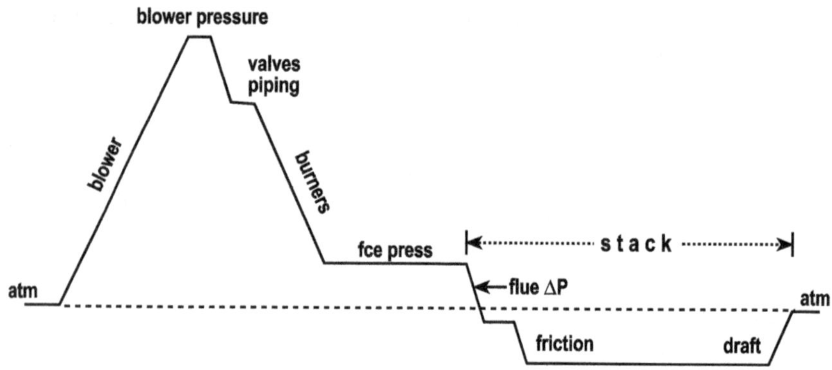


**Fig. 7.4.** Typical pressure-pattern picture for a combustion system and furnace. The vertical pressure drops are not to scale. The pressure drop across the burner's nozzles might be of the order of 20 to 25 in. wc whereas the furnace pressure should be about 0.02 in. wc.

$$(1 \text{ atm} = 0 \text{ gauge pressure}) \tag{7.6}$$

(*a*) 0 + blower pressure − valve & pipe drop = pressure to burners;
(*b*) burner pressure − burner $\Delta P$ = furnace pressure;
(*c*) furnace pressure − $\Delta P$ across flue = stack entrance pressure;
(*d*) stack entrance pressure − stack friction $\Delta P$ + stack draft = 0.

The following is a listing of where to find numbers to fill equation (7.6):

Blower pressure from the blower manufacturer's data.
Valve pressure drop from the valve manufacturer's data.
Pipe $\Delta P$ from tables or formulas in handbooks (e.g., *reference 51*).
Burner pressure drop from burner manufacturer's data.
Furnace pressure by the furnace engineer, cooperating with operators and managers responsible for quality, energy, and safety. (approximately +0.02 in. wc).
$\Delta P$ across flue as per Example 7.1.
Stack friction from pipe-friction formulas in *reference 51.*
Stack draft from pp. 221 to 225 of *reference 51*, or suction of an ID fan.

### 7.3.2. Multiple Flues

Multiple flues are difficult to balance, whether individual dampers are used for every flue or a single damper is positioned beyond where they merge into a single stack. The idea of downdrafting (flues at furnace bottom) is good for furnace circulation and efficient use of fuel. It has sometimes been done with a row of flues at hearth level. However, designers have often connected bottom flues to refractory stacks within thick furnace walls to protect persons around the furnace from burns by hearth-level openings. This defeats the purpose of downdrafting because each of the tall in-the-wall stacks creates a different suction effect.

Using a long shaft to operate many dampers in parallel at the tops of in-wall stacks presents a balancing-problem nightmare. Air dampers (sec. 6.6.3) also may be difficult to balance with multiple flues. A better way to protect personnel is to simply erect open-bottomed stacks as barometric dampers at each flue, positioned to shield anyone from the hot flues.

With multiple flues, if anything (scale, refractory crumbs, misplaced loads) partially blocks one or more of the hearth-level flues, that flue's low flow will cause it to cool and other hotter flues will carry more flue gas load, causing them to get hotter. This results in irregular heating of the loads in the furnace, and may eventually cause runaway overheating of the hotter flues. This same sort of unbalance of flue loads can be caused by different firing rates in adjacent zones or by burner locations that create localized positive or negative pressure on one flue entrance more than on another.

To avoid the aforementioned upsets of the furnace designer's intended furnace circulation pattern, simple *air dampers* are advised at the base of each in-the-wall stack. These can be simple holes, almost the size of the vertical stack cross section, in the bottom of each in-the-wall stack. On furnaces without in-the-wall stacks, personnel can be protected from low-level flues by mounting round vertical sheet

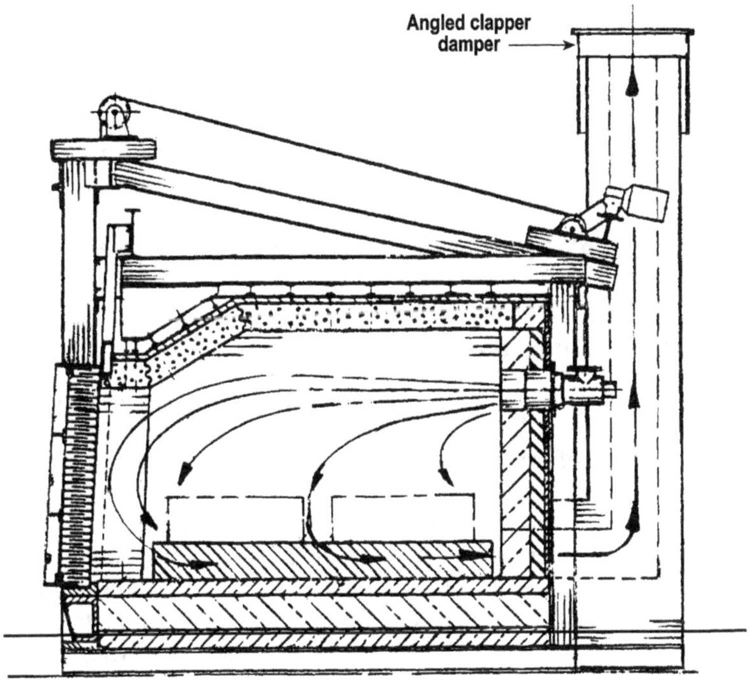


**Fig. 7.5.** Back-wall-fired in-and-out furnace. Stacks without bottom openings (without barometric dampers) must have automatic furnace pressure control.

> When asked where to locate the burners on a furnace, revered furnace man, Lefty Lloyd, replied: "Put the burners where you want. . . . Just let me decide the locations of the flues and the loads." Modern furnace builders would probably prefer to decide locations of all three—burners, flues, and loads, but the ideal decision would be for builders and operators to discuss and cooperate on all such matters.

metal ducts lined with ceramic fiber close to the outside of the furnace. These ducts should be wide open at top and bottom. Each should have a flue entry cut in its sidewall facing the horizontal low-level flue through the furnace sidewall.

For the reasons cited earlier, and to save construction costs, modern practice leans toward one or a few flues. This, however, complicates the problem of achieving uniform heat transfer to all loads, and emphasizes the need for thorough study of circulation for each furnace. (See fig. 7.5.)

With modern adjustable flame burners and with high-momentum burners, there is no such thing as a "neutral pressure *plane*." It is more like a wrinkled, billowing sheet. This effect also is exaggerated by the desire to counteract the "shadow problem" of straight-line radiation heating by using enhancing convection and radiating hot gases. The latter cool quickly, and therefore must be replaced constantly, causing ripples in the neutral pressure "plane."

Design, control, and operating engineers must think through furnace circulation patterns when locating pressure and T-sensors (*a*) where they will read representative answers and (*b*) where they can effectively measure changes (signals) that need to be detected for effective pressure or temperature control. (See sec. 6.6.2.)

## 7.4. GAS CIRCULATION IN FURNACES (more improvement by movement)

### 7.4.1. Mechanical Circulation

Mechanical circulation can be accomplished internally by *plug fans* (usually in the roof) with the driving motor outside the furnace and a drive shaft extending through the roof to an axial set of blades within the furnace. Materials limitations restrict this method to rather low temperature furnaces.

External means of mechanical circulation are *induced draft fans* and *forced draft fans*. Neither can do as thorough a job of in-furnace circulation as well-planned and strategically placed burner jets, but these draft fans or blowers do assist in overall transport or movement of gases out of and into a furnace. Induced draft fans have their inlet connected to the furnace, and therefore create a suction or negative pressure; forced draft fans and blowers have their outlet connected to the furnace, and therefore create a positive pressure. Large power boilers often have both induced and forced

> *Draft* can refer to a chilling breeze on grandma, a pulling force as with a team of draft horses, the depth of a ship, a weather pattern involved with local atmospheric pressure, and (in this book) the difference in pressure that moves air and poc through a furnace. These all seem slightly related . . . except, draft beer?

draft fans, thereby creating a *push-pull* system with *balanced pressure* somewhere in the boiler furnace between them. For further details on fans and blowers, consult *references 29* and *51.*

### 7.4.2. Controlled Burner Jet Direction, Timing, and Reach

Oxygen firing lowers the volume for circulation and raises the gas temperature, both of which may exacerbate nonuniformity. Excess air improves the circulation volume with lower gas temperature.

Pulse-controlled- and stepped firing has attracted many adherents. Burners are cycled on and off systematically in all portions of the furnace. *Pulse firing* uses less fuel than excess air firing. By operating the burners only at full high-fire or off, a maximum gas blanket temperature and maximum velocity for high convection heat transfer are attained whenever the burners are firing. Related to this is maximum mass flow, yielding minimum temperature drop along the gas path, providing maximum temperature uniformity for the loads along the paths of the jet gases.

*Stepped firing* alternates the positions of the burners that are on and those that are off in a programmed timing pattern to further even out temperatures, positionwise and timewise. This is the best method currently available for *small* burners for obtaining both excellent temperature uniformity and low fuel cost.

Most conventional burners have different temperature profile shapes and lengths at high fire rate than with low fire rate. These variations cause load temperature variations with respect to position in the furnace and with respect to time. Furnace engineers must try to locate burners and operate them to average out these temperature discrepancies. One solution is to use a combination of alternated small and large burners along the side of a continuous furnace. A better solution is burners with changeable temperature profile. In car-hearth furnaces, another means for providing side-to-side temperature uniformity is by firing from alternate sides.

*ATP burners* can control their thermal profile by by varying their spin to change the directions and lengths (reach) of their jets while maintaining near-stoichiometric air/fuel ratio. They are the best method currently available with *large* burners for obtaining both low fuel cost and excellent temperature uniformity because two T-sensor locations can be controlled by one burner (discussed in several places within this book). Regenerative burners with flame profile control will be the answer for excellent uniformity *and* fuel economy.

### 7.4.3. Baffles and Bridgewalls

Baffles and bridgewalls can sometimes be used to deflect hot furnace gas streams for the betterment of circulation, thereby improving load temperature uniformity and efficiency. However, they may be awkward and reduce the furnace versatility for a variety of load sizes and shapes.

### 7.4.4. Impingement Heating

Impingement heating, or direct flame-contact heating, has been used for some metal heat-treating operations involving long runs of identical load pieces because they achieve fast throughput rates for small pieces and take less floor space, but they have not achieved good fuel efficiency. They require close/consistent timing, position, and temperature control. *Skelp heating* for welding tube uses impingement heating. (See sec. 4.5.)

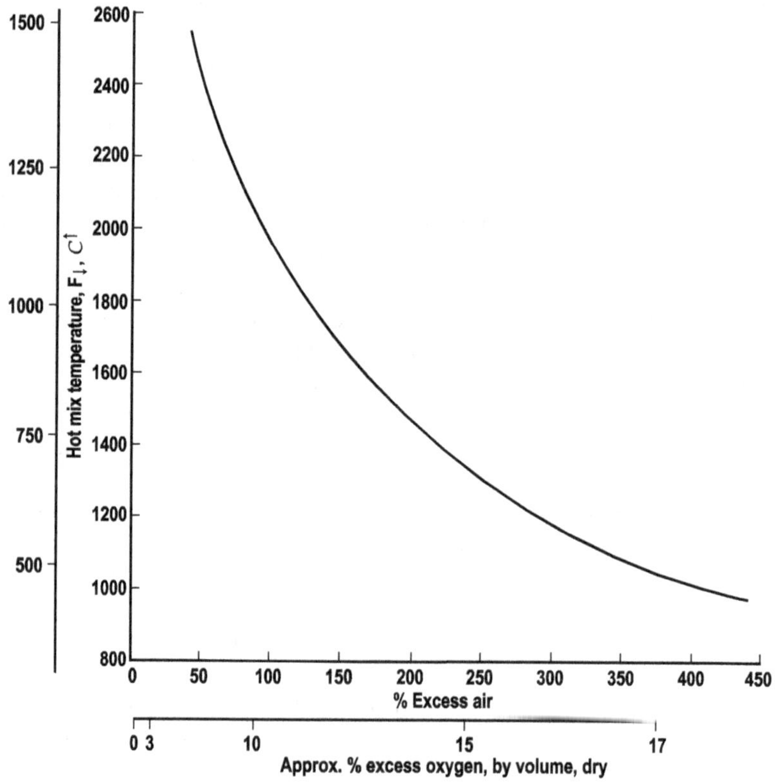


**Fig. 7.6.** Percent excess air necessary to maintain a required hot mix temperature when burning natural gas or distillate oil with cold air. (See also figure 3.18.) Example: To find the amount of excess air necessary to keep the hot mix below 2400 F, enter the vertical scale at 2400 F. Then move right to the curve, then down. Read 75% excess air.

Skelp heated in this way begins a very rapid scaling when scale softening temperature (2320 F or 1271 C) is exceeded. The heat liberation (2850 Btu/pound, 1583 kcal/kg) sustains the same burning reaction as with a burning torch. (See sec. 4.5.) Visible flame may contain some pic, so if it contacts some load materials it could react with the load surface, thus affecting quality, by forming a very tight scale, particularly if there is even a slight quantity of nickel in the steel.

Directing flames into or between load pieces (as in some enhanced heating situations (see sec. 7.5.1) can result in overheating and scaling of their surfaces. When such *nearly contacting flames* raise a steel surface above 2320 F (1271 C), the scale turns shiny, reducing the load's ability to absorb heat—a condition that must be avoided. This can be prevented by using enough excess air to keep the hot-mix temperature (adiabatic flame temperature) below the 2400 F (1315 C) level. Figure 7.6 is useful in planning this operating capability. Figure 7.7 is helpful in using an oxygen analyzer to monitor the actual operation.

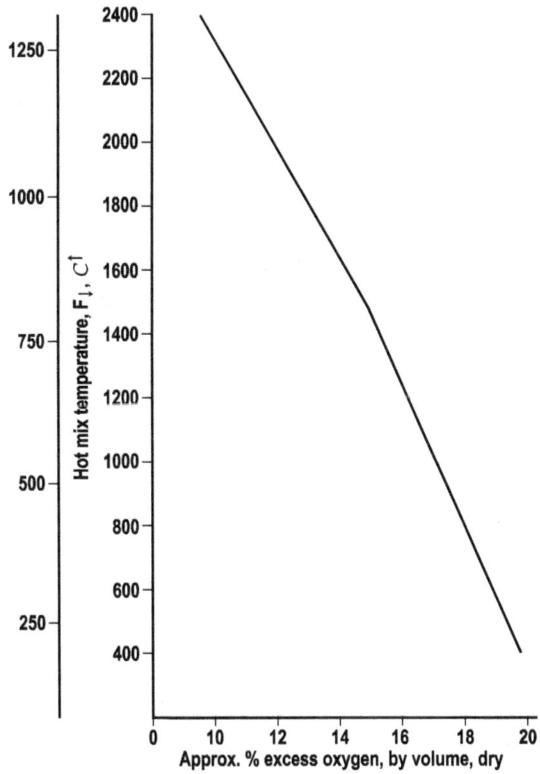


**Fig. 7.7.** Percent excess oxygen needed to maintain a required hot mix temperature when burning natural gas or distillate fuel oil using nonpreheated air.

### 7.4.5. Load Positioning Relative to Burners, Walls, Hearth, Roofs, and Flues (sec. 7.3 discusses flue location)

Operators and managers must understand the following general principles:

*Principle 1.* Loads should be placed on piers or stools.

*Principle 2.* Loads must not be positioned so that they obstruct outlet ports (flues) in the hearth or sidewalls. Likewise, loads must not obstruct inlet ports (burners) or their flames, or their intended paths supplying circulation to the loads in distant parts of the furnace.

*Principle 3.* Loads must not be placed so close together that gases cannot easily pass between them.

*Principle 4.* Loads should be placed where they can be "seen" (radiated to) by furnace walls, hearth, ceiling, flames, and hot gases, and so that all load pieces receive nearly equal exposure. Each load piece should be positioned so that as many sides as possible are exposed to radiation and convection.

The following discussions of specific furnace situations, some derived from actual case histories, illustrate the fact that after an engineer becomes familiar with (1) burners and their possible gas flow patterns in furnaces, (2) furnace equipment and load-handling equipment, and (3) the specific load characteristics and heating process, he or she can apply common sense to modernizing the industrial heating process for gains in productivity, quality, and economy. (Other goals that a furnace engineer must always keep in mind are safety and pollution control.)

***7.4.5.1. Heat Treatment of Railway Wheels.*** This treatment requires a toughness that combines a very long wheel life with a tire that must be much harder than the rest of the wheel. This requires that the tire be quenched and then tempered to prevent brittleness and to have the proper hardness.

*Hardening Heat Treatment.* To harden a 0.50% to 0.70% carbon tire, the wheel first must be heated to 1550 F ± 50°F to assure that the crystals of iron are austenitic when quenched. A manipulator is used to place the wheels two-high onto a special pier device in a rotary hearth *hardening* furnace. Three-high stacking is not recommended because thermal interaction with the top and bottom wheel may give the center wheel a heating curve very different from the other two. The interaction between the wheels may even impair the heating cycle of the top and bottom wheels.

Railroad wheel plants have separate hardening and tempering furnaces to provide better quality wheels than would be possible with dual-purpose furnaces. Enhanced heating should be able to help them increase throughput of wheels as much as 30%. In a hardening furnace, if the wheels are stacked two-high and separated from each other by 8 to 12 in., the heating process can be enhanced by installing small high-velocity burners in the wall at the centerline of the space between the wheels to drive hot poc and pull hot furnace gases between the wheels, thereby increasing heat transfer to both wheels and improving the temperature uniformity of both wheels. If the bottom wheel rests on its pier without burners directing gas under it, small high-velocity burners should be installed below the bottom wheel to improve heat transfer and furnace capacity.

If building a new furnace, a long continuous furnace is suggested, with a jig holding the wheels in a vertical position crosswise to the furnace centerline. As the wheels are moved through the furnace, small crosswise, high-velocity burners could provide hot gas movement between the wheels to increase their heat exposure and thereby the capacity of the furnace, or reducing the furnace size and capital cost.

*Quench and Temper Heat Treatment.* After the wheels are heated above the A-3 line on an iron–carbon/cementite phase diagram, quick *quenching* in a facility reduces tire temperature below 200 F to transform austenite to martensite, which is very hard and brittle. To toughen the martensite so that it can resist wear and accept shock, it is necessary to *temper* the load by raising its temperature to somewhere in the range of 1000 F to 1290 F (538 C to 699 C) depending on the final product use.

As the wheel exits the quench chamber, its average temperature can be 150 F ± 50°F, and it is then placed in a *tempering* furnace. In the tempering furnace, the wheel is brought to the desired temperature as quickly as possible. There are many types of tempering furnaces. These furnaces should be able to heat the whole wheel to a very uniform temperature to provide wheels that wear well without failing. Using enhanced heating in the tempering furnaces can significantly increase the production rate and the uniformity of the wheels being treated because it can double the heat transfer rate. In temper furnaces, it is necessary to look at the position of the wheels for opportunities to apply high-velocity burners to increase capacity and improve temperature uniformity.

Enhanced heating is accomplished using small high-velocity burners set far back from the wheels to pull large volumes of dilute hot furnace gases between the wheels. This technology can help many heat-treating operations. Increasing the heat transfer by enhanced heating can save the price of another furnace or allow a production increase in the range of 30 to 100%, depending on how the burners are applied and the effect on the exposure factor of the wheels. Figure 7.8 suggests how high-velocity burners might be applied for enhanced heating in both the hardening and tempering furnaces.

***7.4.5.2. Soaking Pits.*** (*See also* sec. 6.9.1; see example 3.3 in sec. 3.6.) The importance of circulation in gaining uniform heating is discussed in sections 6.9 and 8.3.1. A difficulty with soaking pits is the accumulated scale on the hearth, which impedes circulation around the bottoms of the ingots or slabs. Even without scale accumulation, the lower parts of the loads are difficult to heat as quickly as the rest of each tall standing load. Raising the loads on piers is difficult because of the loads' tremendous weight. Firing tunnels between piers might be easily plugged with accumulating scale. Of course, one of the objectives of more uniform heating is to minimize scale formation, thus, maybe a combination of better firing practice and better housekeeping would help one another. These also would help minimize metal loss and improve ingot/slab surface quality.

Figure 7.9 shows a desired circulation pattern with slabs stacked four-high. Leaning ingots against the sidewalls would hinder this flow pattern. Operators of all kinds of furnaces must remember that placing loads against any outside wall or hearth is bad

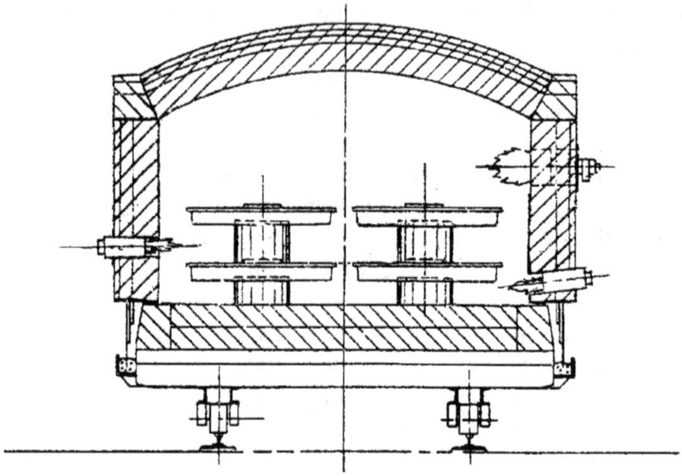

**Fig. 7.8.** Better heat treatment of railroad wheels with high-velocity burners and 50% higher exposure in a conventional furnace by stacking the wheels two-high on special piers. This sectional view could be of a rotary hearth or longitudinal continuous furnace, or a car-hearth furnace.

A 100% higher exposure factor may be possible in a suggested new continuous furnace with high-velocity burners with the wheels held vertically on jigs.

practice because those surfaces tend to be at lower temperatures, and they themselves need access to flowing gases to help them receive heat from circulating furnace gases and then to retransmit that heat to the loads. The closely piled slabs in figure 7.9 have less than 65% of their surface exposed to heat transfer. If the loads were free-standing spaced-out ingots, they would have close to 90% of their surface exposed to heat transfer. Obviously, either way, the top surfaces could be overexposed, perhaps 'washed' (see Glossary), and their bottom ends will be the first portion to become too cold to roll. (However, with a T-sensor below each ATP burner, and overtemperature control, washing can be completely avoided.

On a one-way, top-fired soaking pit, with conventional type 1, 6, or 7 forward or long flames (fig. 6.2), the hot poc gas path would pass over the tops of all the ingots, then flow down the end wall opposite the burner(s) and find its way across the hearth to a flue under a burner. At maximum firing rate, with 35% hearth coverage, the temperature difference between the ends of the pit might be 140°F to 300°F (78°C to 167°C). In these circumstances, the high-fire period will be very short, and the cutback time (between maximum and minimum firing rate) may be as long as 7 hr.

Some operators erroneously think that temperature equalization occurs because the flow path changes to a shorter U-shape (short-circuiting midway down the pit length from pit top to pit bottom), but they have cause and effect interchanged. The flow changes to the shorter path because the T-sensor at the far end gets so hot that it signals the burners to cut back to a lower input rate. Then, the gases have less

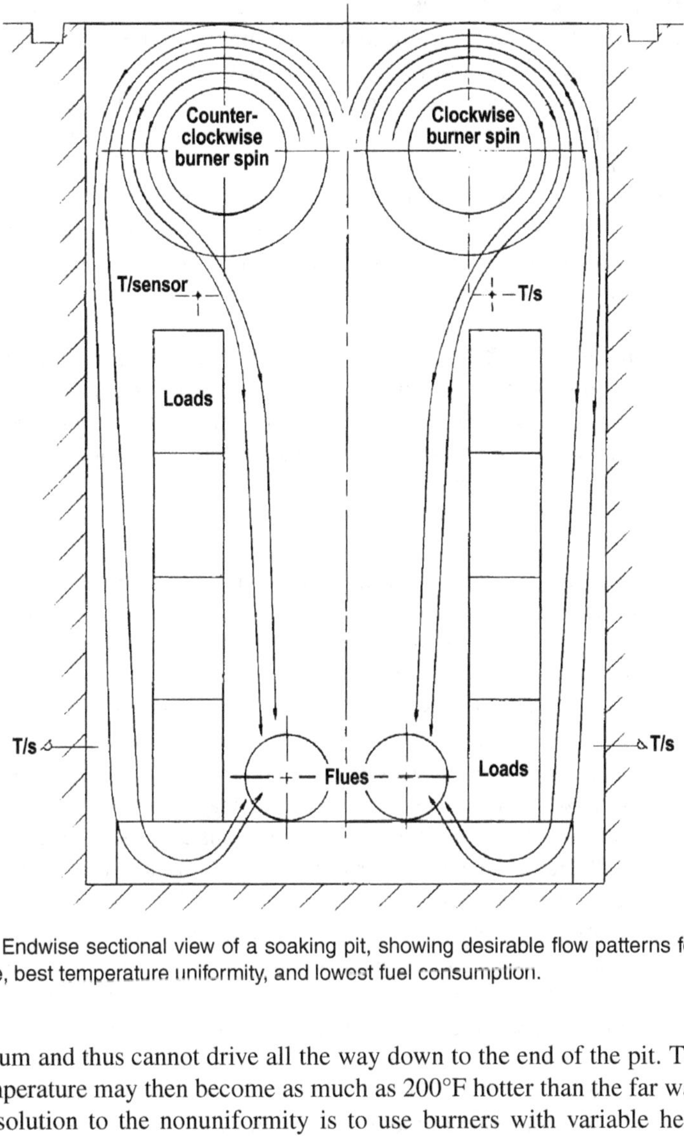


**Fig. 7.9.** Endwise sectional view of a soaking pit, showing desirable flow patterns for shortest firing time, best temperature uniformity, and lowest fuel consumption.

momentum and thus cannot drive all the way down to the end of the pit. The burner wall temperature may then become as much as 200°F hotter than the far wall.

The solution to the nonuniformity is to use burners with variable heat-pattern capability, which vary the spin by adjusting the ratio of tangential gas flow to axial gas flow. The spin is controlled with T-sensors at opposite ends of the pit approximately 3 ft (0.9 m) above the pit bottom, and is successful in keeping those two T-sensors within 10°F (2.8°C) of one another. A high-limit T-sensor in the burner end wall below the burner protects against "washing"* (melting slag) on the ingot tops. A soaking pit installation with this arrangement was heating 23.6 in. (0.6 m) square ingots with a cutback period of 40 min.

*"washing" = overheating, forming oxide (slag), and melting it. (See glossary.)

***7.4.5.3. Batch Forge Furnaces.*** (See example 3.1 in sec. 3.2, example 3.2 in sec. 3.4, plus sec. 3.4, 3.5, and 7.5.) Bottom-firing minimizes uneven heating of loads (*a*) by keeping the hearth hotter by balancing conduction losses through the hearth and (*b*) by enhancing circulation for convection and gas radiation close to the lower sides of the lowest load pieces. Use of piers or posts to elevate load pieces above the hearth is advised.

Slot forge furnaces are wasteful of fuel and prone to uneven heating because of the tremendous heat loss through the slot. They must have movable flaps for easy opening to add or withdraw pieces, *and operators must be convinced that they must close them promptly after every use.*

***7.4.5.4. Continuous Reheat Furnaces.*** Continuous reheat furnaces may be rotary or linear. Either can be side fired or top fired. Top firing may be done with conventional type A, F, or G forward thrust flames (fig 6.2) in a sawtooth roof or with type E flat flames in a flat roof. End firing alone can be used only in small linear reheat furnaces, but it is sometimes used in combination with roof- or side-firing in all sizes. (See also sec. 3.8.5.) For donut rotary hearth furnaces, much detail is discussed in section 6.4.1.

Gas flow in a round furnace is very different from flow in a rectangular furnace. With the flue located near the charge door, the gas flow in a rectangular furnace is from the discharge end of the furnace to the charge door. In a round furnace, the gases can move either of two ways. With this situation, there can be a large area somewhere in the furnace where there is no hot gas flow, and therefore little heat transfer. In addition, any gas that moved through the soak zone toward the flue will be very hot, increasing the combined flue gas temperature and thereby increasing fuel consumption. Another problem with gas flows in rotary furnaces is that the major portion of the gas travels near the inner wall, the shortest distance to the flue. This can result in the inner wall being 400°F (222°C) hotter than the outer wall, causing poor temperature uniformity and poor thermal efficiency.

More load pieces can be placed in a large rotary furnace, if they are placed near the outer wall to take advantage of the greater hearth area (preferably not closer than about 1 ft, 0.3 m). With side firing, the outer wall will have nearly twice as many burners as the inner wall because of the greater available space for locating them and because of the need for more energy input to heat more hearth and loads. With the temperature profiles of conventional burners at high fire favoring high heat release away from the burner wall, there should be more inner wall burners than outer wall burners to avoid a large temperature differential across the hearth (inner wall much hotter). Therefore, the outer wall burners should be a type that releases energy quickly whereas inner wall burners can be of conventional design.

Rotary furnaces are generally less efficient than rectangular furnaces, but they can better handle rounds and varying short lengths. In the United States, most continuous furnaces have been built for labor economy. If fuel economy is desired, it has to be attained by adding recuperation or regeneration.

A *recent installation of enhanced heating* in Ohio increased a furnace capacity from 30 tph to 40 tph. The primary physical process for increasing heat transfer with enhanced heating is the movement of the cold stagnant gases from between the furnace loads and replacing them with hotter furnace gases from above the product. This replacement gas movement was provided by small high-velocity burners that can move five to seven times their mass flow. With this still hotter replacement gas between the loads, an added heat source for the loads and hearth is available to provide more uniform heating, requiring less time waiting for uniform heating and thereby increasing furnace heating capacity. The heat transfer changes are the result of: (1) number of stirring burners, and (2) their firing rate; (3) gas velocity, (4) temperature, and (5) beam between loads; and (6) load, (7) hearth, and (8) roof temperatures being nearly the same. The increase in heating capacity depends on the gas blanket thickness, gas temperature, gas velocity, hearth temperature, and load temperature—all of which are increased by enhanced heating (adding stirring burners at or near the hearth level). The hearth between the load pieces runs hotter, providing additional heat transfer by radiation and conduction to the pieces resting on it. Another bonus from the enhanced heating burners is the heat remaining in their gases, which exit the "tunnel" between the load pieces and add temperature to the triatomic gases in the space above the loads, further increasing their heat transfer ability to the top areas of the loads. The next example attempts to evaluate the magnitude of the previously mentioned gains.

Circulation problems often occur in bottom zones of steel reheat furnaces with pusher and walking beam conveying systems. The problem is inadequate clearance for flow space beneath the loads. The many insulated structural crossover supports and water risers for the skid rails impede longitudinal poc flow under sides of the loads. Hot gases (that are supposed to transfer heat to the undersides of the loads) escape into the top zone, making that zone too hot and leaving the bottom side too cold. Suggestions are (*a*) keep bottom clear of scale pileup, (*b*) design the clearance (flow depth, $H_{bz}$ between bottoms of crossover beams and the top of scale on the hearth to be equal to the top zone clearance, $H_{tz}$, between the lower face of the roof refractory and top surface of the loads (fig. 7.10). Added advantages are (1) a thicker Triatomic gas cloud 'beam' for gas radiation to undersides of the loads, and (2) easier access for bottom zone cleanout, repairs, and replacements.

**Example 7.2:** Estimate the possible increase in furnace capacity by addition of gas radiation to refractory radiation. Consider a 2:1 space-to-thickness ratio for 8 in. rounds in a furnace with a 36 in. high space above the rounds filled with 2250 F gases (see fig. 7.11). Divide the periphery of each round into quadrants of 25% area each.

**Step 1.** Figure the radiation from hot refractory only. From figure 8.3, the normal exposure factor for rounds positioned with a spacing factor* of 2.0 is 48% of the total peripheral surface area. Each of the side quadrants receives half of the refractory radiation into the 8 in. hearth space between rounds, so the effective refractory radiation receiving area of each side quadrant is only 25% × 0.48/2 = 6%. The bottom quadrant has 0% effective area; thus, the total effective refractory radiation receiving area for the four quadrants is 25 + 6 + 6 + 0 = 37%.

*"Spacing factor" is the center-to-center 'pacing' divided by piece width.

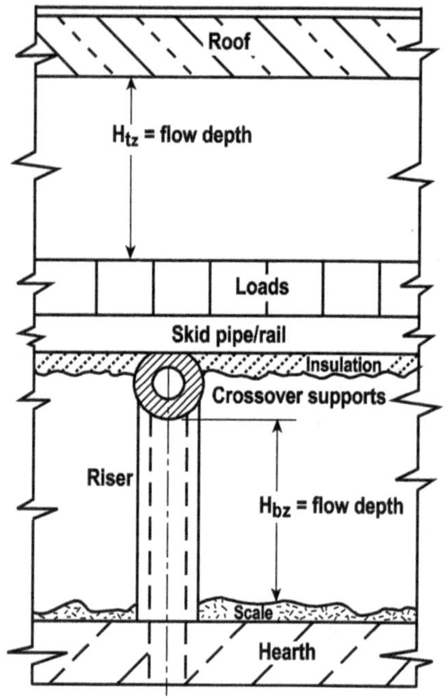


**Fig. 7.10.** Avoid scale accumulation. Redesign $H_{bz}$ equal to $H_{tz}$.

**Step 2.** Calculate the added hot gas radiation from the 36 in. thick gas blanket above the top quadrant and the 8 in. wide blanket between the loads. With enhanced heating, the blanket between the loads will be boosted back up to at least the 2250 F temperature assumed for the 36 in. blanket above the loads.
The coefficient of heat transfer from figure 2.13 drops from 22.5 (for the 36" beam above the top quadrant) to 8.1 (for an 8" beam at the side quadrant). The gas radiation between the rounds to each side of each round amounts to (8.1/22.5) $\times$ 25% = 9% effective area (compared to 25% for the top quadrant). The bottom quadrant has 0% effective area, thus, the effective *gas* radiation receiving area for the four quadrants is $25 + 9 + 9 + 0 = 43\%$.

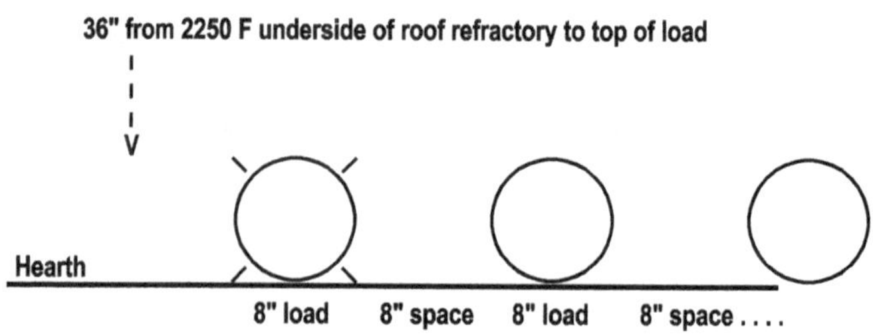


**Fig. 7.11.** Radiation geometry.

**Step 3.** Determine the total %gain from adding gas and refractory radiation = $\frac{25+6+6+9+9}{25+6+6}$ = 55/37 = 1.48 times as much heat actually transferred, compared to refractory alone.

If we add the 9% increase in heat transfer (to the hearth, between the rounds) and the increase in hearth temperature with enhanced heating, the gain would be (55 + 9)/37 = 1.72 times the original heat transfer. This does not include the smaller increase from convection heat transfer by the enhanced heating gases.

Obviously, some of these increases overlap; therefore, a conservative figure of only 25% increase has been used. A recent installation of enhanced heating to only 40% of a furnace resulted in an output 1.29 times the original. A bonus will be elimination of "barber poles" in seamless mill rounds leaving the first piercer by using enhanced heating in the last 15 min of their heating time in a rotary furnace.

A second bonus benefit, capacity-wise and quality-wise, from enhanced heating can occur for loads that are tight together, as in a pusher furnace. When such material is being heated, the temperature profile is uniform from the roof down to about 6" above the load. From there to the load piece, the temperature drops quickly to load temperature. With enhanced heating, the roof temperature would be maintained almost all the way to the load's surface, increasing heat transfer significantly. This also is true in bottom-fired zones, where the temperature is maintained almost constant from the furnace bottom to 6" below the lower surface of the load, where it drops quickly to the load surface temperature. In cases where it is possible to direct gases against this lower load surface, heat transfer will be increased significantly.

### 7.4.6. Oxy-Fuel Firing Reduces Circulation

Oxy-fuel firing reduces circulation because the poc do not contain all the nitrogen that came with air-fuel firing, thus, convection heat transfer is reduced. However, the concentration of triatomic molecules is greatly improved by the elimination of the inert nitrogen molecules, resulting in more than a 300% increase in gas radiation heat transfer. Although the new poc stream has a net improvement in its heat transfer capability, oxy-fuel firing may have a problem with nonuniform heating because the much-reduced gas stream volume may not provide the necessary circulation to deliver its heat to all surfaces of the loads—particularly the bottoms of ingots in soaking pits. A similar problem with *integral regenertor/burners* makes them impractical with soaking pits until small sizes and remote regenerator beds become available to locate the flues at hearth level.

Inadequately heated ingot bottoms in soaking pits may cause someone to increase input to the burners, overheating the ingots' tops, resulting in "washing" of the ingots. If without velocity effects, washing begins above 2490 F (1365 C). With high velocity, washing begins slightly above the softening of scale, about 2320 F (1271 C). For washing to occur, the gases flowing over the steel must contain 1 to 3% excess oxygen. At only 0.5% oxygen, the iron is competing with CO and $H_2$ for the remaining oxygen, and therefore, the oxidation rate of the iron is much slower. With more than 1%

oxygen, this competition does not exist, so the slag-making oxidation proceeds at a much higher rate. The heat release from oxidation of the iron further raises the temperature of the iron, sustaining the reaction. As the temperature falls, the slag-making reaction rate slows. It is believed that slag formation will cease at about 2250 F (1332 C).

As scale melts and runs off the steel surface, it exposes more virgin iron to the rapid oxidation (ablative melting). The exothermic heat release makes the reaction almost self-sustaining, similar to the reaction accomplished by a cutting torch.

With burners that do not direct the combustion gases at the steel surface, the oxidation of the iron takes place without the velocity stimulant, so less oxygen contacts the hot surface. Without as much oxygen available, the reaction slows, and the exothermic heat of the reaction is not available to sustain the washing—similar to the effect of shutting off the oxygen to a cutting torch.

## 7.5. CIRCULATION CAN CURE COLD BOTTOMS

The ideal way to achieve uniform heating would be to locate equally large burners below as above the load, but this creates design and material problems for supporting or suspending heavy loads. (Loads should not be placed directly on a hearth, which is inherently colder than the sidewalls or "ceiling" of a furnace.) To counter the nonuniformity problem, a row of small burners firing through "tunnels" (formed by piers or posts supporting the loads) was used on the bottom or hard-to-heat sides. If the furnace is wide (so that the tunnels are long), there can be a nonuniformity problem between the two ends of each tunnel. This does not affect product quality as seriously as the nonuniformity with the load on the hearth, or even on piers with no bottom-firing, but it is often not uniform enough for current high-quality standards.

A perfect heating situation would have each load piece completely surrounded (360 degrees in all planes) by equally high heat transfer rates to all its surfaces. That is often impossible or impractical because of (*a*) load shape and size, (*b*) handling and support problems, and (*c*) lack of appropriate piers, posts, or kiln furniture. The resultant uneven heating necessitates a long soak time to let the temperatures "even out" within the load, with possible increased fuel costs. Long soak times may cause excessive surface oxidation, and they surely cause lowered furnace productivity.

### 7.5.1. Enhanced Heating

Enhanced heating is a practical answer to the nonuniformity problem. It increases convection and radiant gas heat transfer by raising the temperature of the gases between load pieces by perhaps 500°F. Enhanced heating uses a row of small high-velocity burners, aimed under and between the load(s) through "tunnels" formed by piers or posts supporting the loads. This also counterbalances heat loss through the hearth. Correcting the cold hearth problem alone may increase productivity by 50%, with improved product temperature uniformity. Using enhanced heating in the last 15 min of heating rounds in a rotary hearth furnace will often raise the hearth temperature enough to eliminate the cold bottom quadrants on the rounds that cause "barber poling" in seamless mill rounds leaving the first piercer.

Enhanced heating burners are often fired with excess air (fuel-only control) to get higher mass flow. The high-velocity burners can "reach" farther across a wide furnace, and they are in the tunnels for a shorter time, giving them less time to cool, which creates a crosswise nonuniformity. But therein lies an anomaly. The crosswise cooling is still too much for good temperature uniformity in the load, so engineers purposely *lower the heat transfer rate* in the tunnel (by supplying high input mass flow through tunnels of small cross section), thereby reducing the cooling of the gases, maintaining a more "level" side-to-side temperature in the furnace. An added aid is alternately firing high-velocity burners from each end of every other tunnel, thereby allowing each left burner's temperature pattern (arcing down, like a trajectory) to be averaged out by downward temperature patterns from the right in adjacent tunnels.

Because of the perplexing anomaly that arises with enhanced heating, it is important to understand its principles and how it evolved. Convection and gas radiation heat transfer can both deliver heat at quite high flux rates, but both also result in fast cooling of the source itself—the poc gases. Luckily, the high-velocity burners induce (or pump) high mass flows of furnace gases through each tunnel; otherwise, steep temperature drop would occur along their gas paths. This is one of the basic reasons why furnace gas circulation is so important—and the reason why high-momentum (high-velocity) burners have been such a boon in industrial process heating.

Increasing the input through high-velocity burners can result in high flue gas exit temperatures with poor fuel efficiency. The best arrangement would involve: (1) burners firing first into a high-heat chamber and (2) gases passing into a load preheating chamber, where they would be allowed to slow, cool, and finally exit at a reasonably low temperature, resulting in an acceptable fuel efficiency. This implies a continuous furnace wherein the loads and furnace gases move counterflow (in opposite directions). However, the three-ingot batch forge furnace of figure 7.12 illustrates a case where the gases exiting from the ends of underload tunnels have time and distance in which to slow down, and give off more heat before finding their way out the flue. If they get caught up in the inspirating effect of the big main burner flames, they will "go around again," adding to the effectiveness of the main burners.

When a high-velocity jet leaves a burner nozzle, it inspirates inert poc from the surroundings. If the surrounding poc are 100° to 200° hotter than the walls, and if the jet gas is 800° hotter than the surrounding gas, the two streams would mix, and that mixture might be 300° hotter than the walls. With higher jet gas momentum (Velocity × Mass), the jet would inspirate more of the surrounding gas, mixing with it, resulting in less than 300° above the wall temperature (see fig. 7.13).

The fact that the jet gas has its temperature moderated by its inspiration of surrounding gases decreases its ability to transfer heat by gas radiation. This is a way that enhanced heating helps temperature uniformity. If the mixture of jet and entrained gas moving under the load cools only 15°, then the load will have only about a 15° side-to-side $\Delta T$.

If the jet gas passageway (tunnel) were reduced from a 2 ft (0.61 m) crosswise gas beam to half as wide, figure 2.13 shows that the ability of 2200 F (1204 C) gas

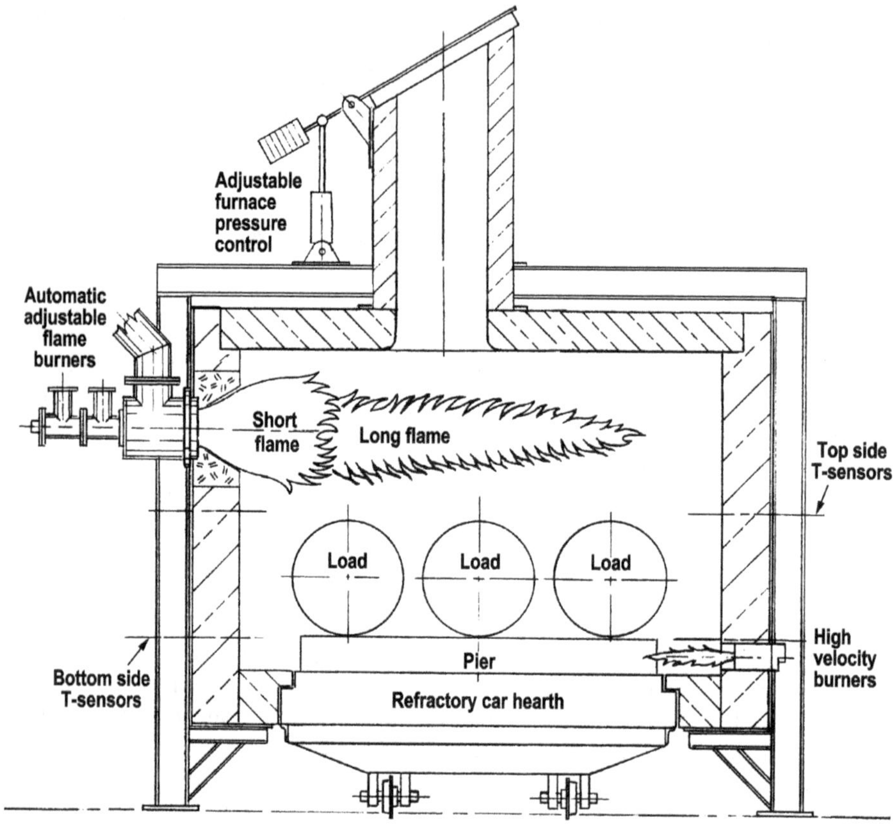


**Fig. 7.12.** Top-side temperature uniformity is assured with adjustable thermal profile burners and two-sensor control. High momentum, underload burners (enhanced heating) with two-sensor control improve bottom temperature uniformity.

to radiate to the loads would be reduced from 17 to 11, or a reduction of about 35%. This means that narrower tunnels under the load which force the poc through faster also cool less, improving crosswise temperature uniformity.

**Anomaly Summary:** *For good product temperature uniformity*, the underpassages on a batch furnace must have minimum temperature difference from end to end. The following suggestions relate to underfiring where gas underpassages are much smaller than those above the loads.

The heat transfer rate from the poc gases to the loads must be moderate because the load temperature will reflect the poc temperatures. Therefore,

A. The entry gas/flame temperatures should be moderated by dilution with excess air or recirculated furnace gases or both. This has a two effects:

1. With lower gas-to-product temperature differences, both radiation and convection heat transfer rates will be slower.

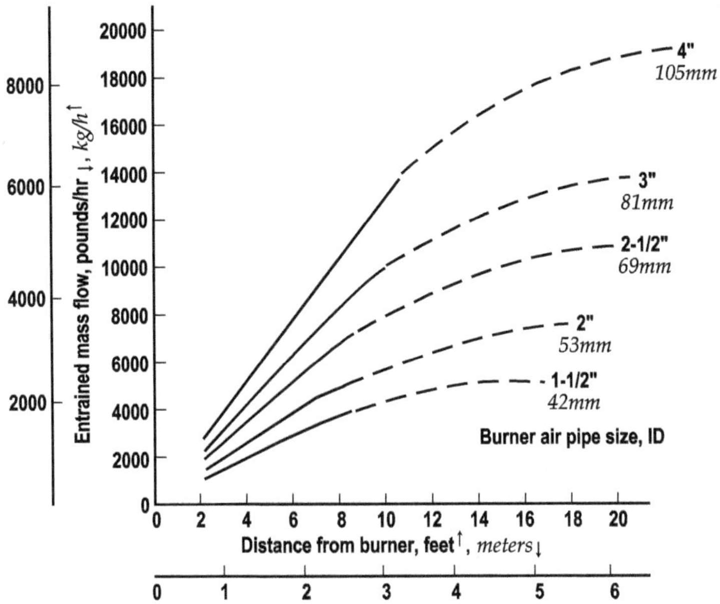


**Fig. 7.13.** Entrained furnace gas is estimated to have 500 fps (152 m/s) port velocity at 1700 F (927 C).

2. The increased mass flow of gas in the passages below the loads becomes a stabilizing factor in holding a near-constant temperature across the furnace load's width.

B. The gas passage cross section (for minimum temperature change, to limit heat transfer by gaseous radiation) should be less than 12 in. (<305 mm) high.

[In contrast, *for a high producton rate—just the opposite*—the underproduct passages should be at least 2 ft (0.61 m) high to nearly reflect the cross section above the loads, where control of the heat release pattern by the burner practically eliminates cross-furnace temperature differences in the product.]

## 7.6. REVIEW QUESTIONS

**7.6Q1.** How does recirculation improve temperature uniformity?

**A1.** Very high temperatures and very low temperatures are moderated (diluted) by the increased mass flow brought about by recirculation. In the heat transfer formulas, these effects are present in the mass flow velocity of the convection formula and in the volumes of triatomic molecules affecting radiation.

**7.6Q2.** Under what circumstances does one want to design for *less* heat transfer from the poc?

**A2.** When product temperature differentials are above those specified, especially the temperature drop of poc (and consequently of product) from one end to the other end of a between-the-piers firing tunnel, use furnace gas recirculation or excess air to level the end-to-end temperature drop. Greater mass flow at lower inlet temperature is needed to level out the temperature pattern from end to end. A consequence of this will be a higher exit poc temperature (lower fuel efficiency).

**7.6Q3.** Why aren't regenerative burners or oxy-fuel firing applicable to soaking pits?

**A3.** The poc gas mass flow is less with regenerative heating and much less with oxy-fuel firing because of much higher efficiencies. That means the poc gas stream cannot carry or deliver as much heat, so the temperature profile is much steeper, resulting in greater temperature differences. In the case of oxy-fuel firing, the higher percentage of triatomic molecules in the poc further increases heat transfer, resulting in even greater temperature differentials. These problems are worse after passing the cutback point in the firing sequence. With the ingot top-to-bottom temperature differentials possibly exceeding 200°F (111°C), the ingot bottom surface will crack as it is rolled.

**7.6Q4.** Where should temperature control sensors be located for uniform cross-furnace temperature control with enhanced heating?

**A4.** As close as possible to the loads so that they will be more sensitive to changes in load temperature than those of wall, crown, or hearth temperatures.

**7.6Q5.** How can you minimize the temperature drop from side to side under the load in a furnace?

**A5.** Limit the size of the piers to 8" to 12" high, use excess air, or use high-velocity burners with fuel turndown only, and use piers of minimum mass and with many openings. Heat requirements will be minimum, and heat transfer rates will be low (desirable) due to the minimum gas blanket thickness. Low heat transfer is desired to minimize poc cooling as the poc move across the furnace width.

**7.6Q6.** How is draft created in furnaces?

**A6.** (*a*) Natural draft (no mechanical energy) is created by a difference in furnace gas density and ambient gas density (outside the furnace).

(A thumb guide for furnaces at or above 2000 F (1093 C) is that each foot (0.3 m) of furnace height will cause about 0.01 in. wc (0.25 mm wc) less pressure inside the hot furnace than In the surrounding room.)

(*b*) Forced draft is generated by pressure or suction from fans, blowers, air jets, or gas jets.

**7.6Q7.** Why is oxygen firing fundamentally less uniform?

**A7.** There are two reasons for less uniformity:

(*a*) The volume of the poc with oxy-fuel firing is only 28% as much as with the same heat release with air-fuel firing, so the combustion reaction is at a much higher temperature with oxy-fuel, and the poc stream is therefore capable of tranferring heat more rapidly.

(*b*) Without the presence of nitrogen (from air), the poc sream is almost 100% triatomic molecules versus only about 26% with air-fuel firing. Therefore, the oxy-fuel flame is hotter, and the thermal profile of its poc stream is much steeper, making nonuniformity more probable.

**7.6Q8.** How can reasonable uniformity be achieved with top firing only in a batch furnace?

**A8.** Flues must be provided near the hearth in each zone because gas movement is necessary wherever loads are located. This is difficult without external energy directing the gases. A recommended solution is placing the loads on 8 in. to 12 in. high piers and applying enhanced heating with small high-velocity burners firing between the piers.

**7.6Q9.** Why is the cycle time shorter when firing batch furnaces with both top and bottom firing?

**A9.** Heat transfer area is nearly doubled with top and bottom firing, except for the "shaded" areas caused by piers or rails. If only one-side heating can be justified, choose bottom-side heating even though its exposed area will be less because its temperature uniformity will be better than it would be with top-side-only heating.

**7.6Q10.** How do enhanced heating burners increase the effective heat transfer area of the product when there is space between the product pieces?

**A10.** When the spaces between the load pieces are perpendicular to the furnace gas flow, the gases between the loads are practically stationary, so their temperature will stay very near that of the loads. With essentially no temperature difference between these gases and the loads, little if any heat transfer takes place. If energy can be supplied to the stagnant area between the loads by small high-velocity burners (enhanced heating), the effective heat transfer area between the loads *and the hearth* will increase by more than 25%.

**7.6Q11.** When heating a load such as a rolling mill roll, why is it desirable to have at least four zones of temperature control above and four zones below the load?

**A11.** The two end zones above and the two end zones below are required to control the temperatures at the furnace ends, where heat losses are greater so that the ends of the loads do not "see" cooler surfaces. The functions of the middle four zones (two above the load, and two below the load) are to provide temperature uniformity in the areas immediately around the loads. Even more zones could be effective in preventing the small bearing journal ends of the rolls from being over- or underheated because of the different mass of the main cylinder section of the roll. That might require five top and five bottom zones, but ten zones have been judged excessive when limiting the control temperature rise to 25°F to 35°F (13.9°C to 27.8°C). From this lengthy answer, one can see why a gas movement study is so important in a batch furnace in preventing out-of-specification temperatures in the product!

**7.6Q12.** Where should the flues be with top and bottom firing, and what is the best number of flues?

**A12.** With top and bottom firing, the flue exits are normally installed in the furnace roof. If more than one flue is to be used, they should be placed to avoid gases from one zone moving through another zone. With three top and three bottom zones, two flues are necessary—on centerlines between zones.

**7.6Q13.** When designing a flue system, what security factor should be used to make future productivity adjustments possible?

**A13.** A security factor of 1.3 is suggested, applied to the maximum burner firing rate and with flue gas exit temperatures 200°F (111°C) above the furnace running temperature at maximum rates. Some furnace designers may be irritated by these specifications, but they are needed to recover a furnace's normal temperature profile quickly. These specifications are more necessary for a mill with many delays to provide the versatility needed. It is important to be aware of different goals—furnace designers want to build an inexpensive furnace so that they can get the order, but operators want versatility to be able to heat and roll as many tons as possible.

# 8

# CALCULATIONS/ MAINTENANCE/QUALITY/ SPECIFYING A FURNACE

## 8.1. CALCULATING LOAD HEATING CURVES

The objective of this exercise is to develop a set of time/temperature curves such as shown in figs. 6.26 to 6.33 and figure 8.1. In this book, the authors frequently urge the readers to use this "Shannon Method" to develop such curves for their own specific loads, processes, and heating equipment so that they can better analyze their furnace capabilities and requirements, and make good engineering judgments relative to their control.

On figure 8.1, the 20 abscissa units = 100% of time or distance in the furnace. For sample problem 8.1.1, with 890 ft (24.4 m) inside furnace length, each division therefore represents 880/20 = 4 ft or 1.22 m. Other given data are 2068 #/pc; 0.668'/pc center to center; 200 000 #/hr.

The total time for each load piece in the furnace =

$$\frac{(80' \text{ fce length})\ (2068\ \#\ \text{wt each load piece})\ (60\ \text{min/hr})}{(0.668' \text{ ctr to ctr of load pieces})\ (200\,000\ \#/\text{hr to be heated})} = 74.3\ \text{min.}$$

Furnace heating curves are not just for furnace designers. Furnace users also need to be able to calculate heating curves to purchase a new furnace or improve an existing furnace to reduce concerns about receiving proper value. Plant engineering departments too often are interested in advice that reduces capital costs without regard for results. When operators cannot produce, engineering departments may have failed to examine the facts thoroughly to determine the root cause so that the operator is assisted or the supplier questioned to correct the deficiency. Heating curves help in making these and other decisions.

If engineering departments calculated heating curves specifically for their furnaces and loads, they would be able to determine correct specifications for the furnace to meet their specific needs. In addition, when required to reduce costs, they could be aware of the results and inform plant management of the limitations imposed on the

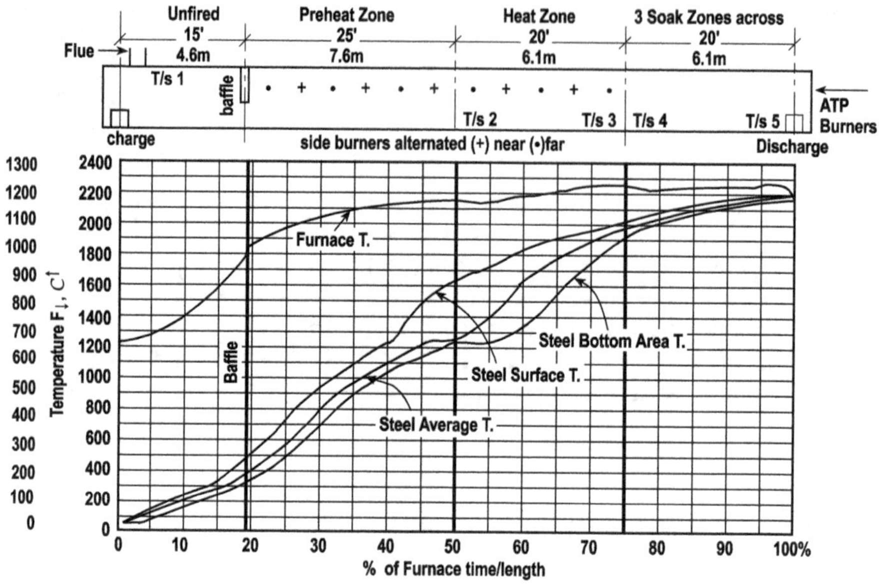


**Fig. 8.1.** Typical temperature-versus-time curves for a steel reheat furnace.

operators. The heating curve calculation may reveal other cost savings and procedures needed for long-term good results.

Areas where someone* might cut corners are:

1. Practically eliminating design security margins† on design firing-rate capabilities;
2. Underestimating the flue gas exit temperature, or measuring the flue gas temperature with a sensor that "sees" the cold tubes of the recuperator;
3. Lowering excess air too much;
4. Building or selecting a recuperator with less than the furnace firing capacity;
5. Reducing the flue system capacity below that of the total furnace firing rate;
6. Calculating the dilution air capacity to handle less than the total possible flue gas entering the recuperator;
7. Designing the system with insufficient fan energy for mixing the dilution air and flue gases.
8. Ignoring the need for design security factors to allow for abnormal situations such as additional air from infiltration.
9. Underestimating furnace heat losses, including increases with furnace age.

*Particularly someone trying to establish a low price for a proposed new unit.

†See the glossary, under safety factors, about security factors and margins.

Coauthor Shannon designed a system considering all the normal deficiencies and with a 20% security factor[†], but he found that the system was just large enough to control the flue gas temperature entering the recuperator. This emphasizes the need to play it safe with expensive long-term equipment design and selection.

In another situation involving a recently built new furnace, Consultant Shannon found that after a delay, the operator had to further delay return of the furnace to operation because the flue gas temperature entering the recuperator was too high. To try to remedy the situation, the operator lowered the dilution air setpoint temperature from 1650 F (900 C) to 1300 F (704 C), which reduced the preheated air temperature during low firing rates by several hundred degrees F. This particular furnace was so under fired (with all zones at maximum firing rate) that it limited the maximum production rate for the mill. The furnace designer may not be the only cause of these problems. Other reasons are clients who (1) are not knowledgeable or (2) have no consultant to provide the knowledge, or (3) purchase from the lowest bidder, regardless of past results.

These problems are the primary reasons why the authors felt the need to produce a sixth edition of this book. It is hoped that clients, through their engineers and this book, will gain sufficient knowledge to write strict specifications and insist on adherence thereto. Then, the knowledgeable engineers can convince others not to cut corners, thus protecting their plant from undersized recuperators, fans, flue systems, and dilution air systems. Those who accept such "corner cuttings" will forever raise operating costs, but lower productivity and product quality. These problems harm not only the particular plant, but the whole industry, which is always seeking to lower costs, raise productivity, and improve quality for its customers.

### 8.1.1. Sample Problem: Shannon Method for Temperature-Versus-Time Curves

*Given:* 200 000 pounds/hour of 0.4% carbon steel to be heated to 2150 F ± 25°F for rolling. The 4.5" sq × 30 ft long billets are spaced 8" center-to-center, so the spacing-to-thickness ratio = 8"/4.5" = 1.78, on a walking hearth.

*Preliminary Decisions:* Walking hearth four-zone reheat furnace, with all zones longitudinally or side fired. Zone 1 (charge end) is to be unfired. Zones 2 and 3 are to be side fired, and zone 4 (soak) is to be fired longitudinally, using ambiet air in all burners. Fuel = natural gas. Hearth width should include 2 ft clearance on each end of 30 ft long billets = 34 ft.

*Find:* Hearth area and length—first try = 80 ft.

Plot: Temperature versus time curves.

Later: Determine input rates to all zones.

*Look-up data:* Load density = 489 $lb/ft^3$ (*reference 51*, table 4.4b). Load emissivity = 0.85 (from *reference 51*, table 4). Estimated possible hearth loading = 83.3 $lb/ft^2$, from figure 4.21, considering space-to-thickness ratio, number of zones, whether with bottom heating, and/or with enhanced heating.

[†]See the glossary, under safety factors, about security factors and margins.

*Procedure—Phase A—Datasheet.* Table 8.1 = blank datasheet. Table 8.2 = the datasheet filled in for this problem to find heat transfer factor, $H$. If $H$ is less than 0.4 with time-lag (line 11) greater than 6 min, increase the furnace length by 10% and perform a second iteration. Continue iterations with increased furnace lengths until $H$ is greater than 0.4. If $H$ is greater than 0.47 with time-lag (line 11) less than 6 minutes, try an iteration with 10% less furnace length.

#### *8.1.1.1. Exposure Factor as a Function of Space-to-Thickness Ratio*

Refer also to chapter 2. For two-side heating, Figures 3.7 and 8.2 show a maximum of 83%. It could only be 100%, if the side and end areas could receive radiation at the same rate as top and bottom (four-side heating).

**TABLE 8.1 Blank preliminary datasheet for steel temperature-versus-time curves**

| | | Iteration #___. | |
|---|---|---|---|
| #1) | *Furnace*: type_______ Number of zones = ___top, ___bottom | | |
| #2) | *Load*: Material__________ production rate = | ___lb/hr | ___kg/h |
| a) | Thickness | ___feet | ___meter |
| b) | Length | ___feet | ___meter |
| c) | Width | ___feet | ___meter |
| #3)d) | Weight, = (a) (b) (c) [ 489(lb) or 7834(kg)] = | ___pounds | ___kg |
| e) | Grade: carbon content, stainless, other | ___%C | ___%C |
| #4)f) | Discharge temperature | ___F | ___C |
| g) | Temperature variations allowed | ± ___°F | ± ___°C |
| #5)h) | Furnace inside heating width = load length b | ___feet | ___meter |
| i) | Estimated possible hearth loading, from figure 4.21 | ___$lb/ft^2hr$ | ___$kg/m^2h$ |
| j) | Furnace inside length* | ___feet | ___meter |
| k) | Effective hearth area = (b) (j) | ___$ft^2$ | ___$m^2$ |
| #6) | Production rate/unit hearth area = (#2) / k | ___$lb/ft^2$ | ___$kg/m^2$ |
| #7) | Load spacing, centerline to centerline, $\Delta\phi$ | ___feet | ___m |
| #8) | Spacing / thickness ratio = #7 / a | ___ft/ft | ___m/m |
| #9) | Load exposure – % of 4 sides, from fig. 8.3 | ___% | ___% |
| #10) | Effective weight / exposed area = d / (b) (2a + 2c) (#9/100%) = | ___$lb/ft^2$ | ___$kg/m^2$ |
| l) | Lag factor, $F_1$† for exposure, from fig. 8.3 = | ___ | ___ |
| #11) | Lag time = ($a^2$) ($F_1$) (144 / 10) = | ___minutes | ___min. |
| #12) | Total heating time = (J/#7) (60) (#3d) / #2 = | ___minutes | ___min. |
| #13) | Emissivity or absorptivity = | ___ | ___ |
| m) | Number of time increments on selected plotting paper | ___ | ___ |
| #14) | Time increment = #12 / m = | ___minutes | ___min. |
| #15) | Heat transfer factor, $H$ = ( #13 × #14 × 1000 ) / (#10 × 60) = | ___ | ___ |
| #16 | If #15 is not above 0.43, try a new iteration, with a new j = above j × (0.43/#15) | ___ | ___ |

Permission is granted owners of this book to copy this blank datasheet.

*Shorter length may save on capital investment, but will raise operating costs.

†With 1-side heating, $F_{1,\text{one side htg}} = 8$; $F_{1,\text{two side htg}} = 2$; $F_{1,\text{four side htg}} = 1$. To find $F_1$ between these values, first use fig. 3.6 or fig. 8.2 to find the % of full exposure ignoring end areas; then read $F_1$ from fig. 3.3 or 8.3.

**TABLE 8.2 Preliminary datasheet for steel temperature-versus-time curves showing numbers for sample problem 8.1.**

| | | | Iteration #1. |
|---|---|---|---|
| #1) | *Furnace (fce)*: type = walking hearth. Number of zones = 4 top, 0 bottom | | |
| #2) | *Load*: 0.4% C steel; production rate = | 200 000 lb/hr | 90 700 kg/h |
| a) | Thickness 4.5/12 = | 0.375 feet | 0.114 meter |
| b) | Length | 30 feet | 9.15 meter |
| c) | Width | 0.375 feet | 0.114 meter |
| #3)d) | Weight = (a) (b) (c) [489(lb) or 7834(kg)] = | 2063 pounds | 936 kg |
| e) | Grade: carbon content, stainless, other | 0.4 %C | 0.4 %C |
| #4)f) | Discharge temperature | 2150 F | 1177 C |
| g) | Temperature variations allowed | ±25° F | ±14° C |
| #5)h) | Furnace inside heating width = load length b | 30 feet | 9.1 meter |
| i) | Estimated possible hearth loading, from fig. 4.21 | 156 $lb/ft^2hr$ | 763 $kg/m^2h$ |
| j) | Fce inside length* = 1st iteration try | 80 feet | 24.4 m |
| k) | Effective hearth area = (b) (j) = (30) (80) = | 2400 $ft^2$ | 223 $m^2$ |
| #6) | Production rate/unit hearth area = (#2) / k | 83.3 $lb/ft^2$ | 124 $kg/m^2$ |
| #7) | Load spacing, centerline to centerline | 0.668 feet | 0.204 m |
| #8) | Spacing / thickness ratio = #7 / a | 1.78 ft/ft | 1.78 m/m |
| #9) | Load exposure – % of 4 sides, from fig. 8.2 | 41 % | 41 % |
| #10) | Effective weight /exposed area = d / (b)(2a + 2c) (#9/100%) = 2063/{(30) [4 (0.375)] [41/100%] = | 112 $lb/ft^2$ | $kg/m^2$ |
| l) | Time-lag factor, $F_1{}^{\dagger}{}_{\text{for exposure}}$, from fig. 8.3 = | 3.05 | 3.05 |
| #11) | Lag time = $(a^2)$ $(F_1)$ (144 / 10) = = $(375)^2$ (3.05) (144) /10 = | 6.18 minutes | 0.10 h |
| #12) | Total heating time = (j/#7) (60) (#3d) / #2 = = (80 / 0.668) (60) (2063) / 200 000 = | 74.3 minutes | 1.23 h |
| #13) | Emissivity or absorptivity = | 0.85 | 0.85 |
| #14) | Time increment, in minutes = #12 divided by number of time units on graph paper = 74.3 minutes/20 units = | 3.71 minutes | 0.062 h |
| #15) | Heat transfer factor, H = (#13) (#14 in hr) (1000) / #10 = (0.85) (3.71/60) (1000) / 112 = | 0.47 | 0.47 |
| #16) | If #15 is not above 0.43, try a new iteration, with a new j = (1st iteration j) + 10% = | —— | —— |

*Shorter length may save on capital investment, but will raise operating costs.

†$F_{1,\text{one side heating}} = 8$; $F_{1,\text{two side heating}} = 2$; $F_{1,\text{four side heating}} = 1$. To find $F_1$ between these values, first use figure 3.7 or figure 8.2 to find the % of full exposure ignoring end areas; then read $F_1$ from figure 3.8 or 8.3.

The curve of space-to-thickness ratio with two-side heating has been questioned by many for not rising above about 83% of the full surface area minus the end areas. To study this, compare two-side heating of a 6" billet with a 3:1 space-to-thickness ratio versus four-side heating with 2200 F gas cloud (blanket) thickness. Even at a space-to-thickness ratio of 3:1 with two-side heating, the sides receive heat approximately as in table 8.3, with space between the sides instead of a gas blanket above and below the load.

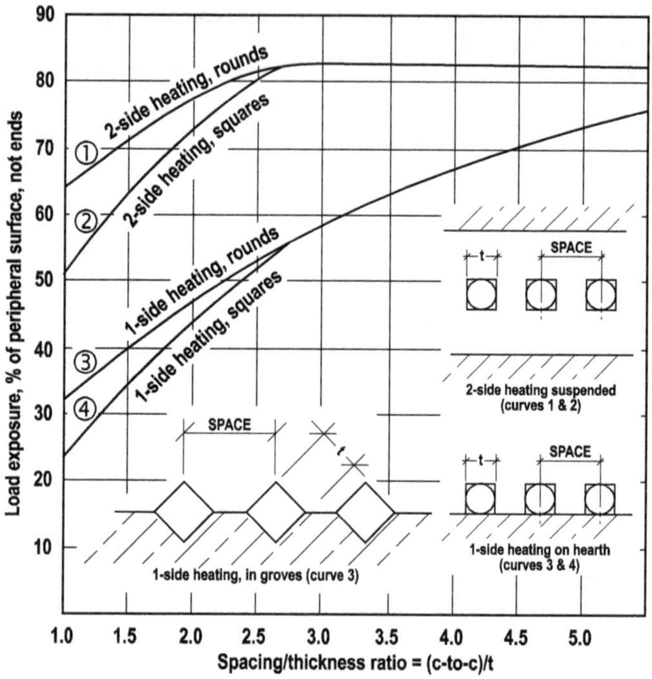


**Fig. 8.2.** %Exposure versus workpiece spacing ratio. (Same as figure 3.7.) Billet "spacing ratio" = centerline to centerline distance, C, divided by billet width or diameter, W. Using a centimeter scale facilitates interpolating. Use the answer from this graph as the input to the abscissa of fig. 8.3.

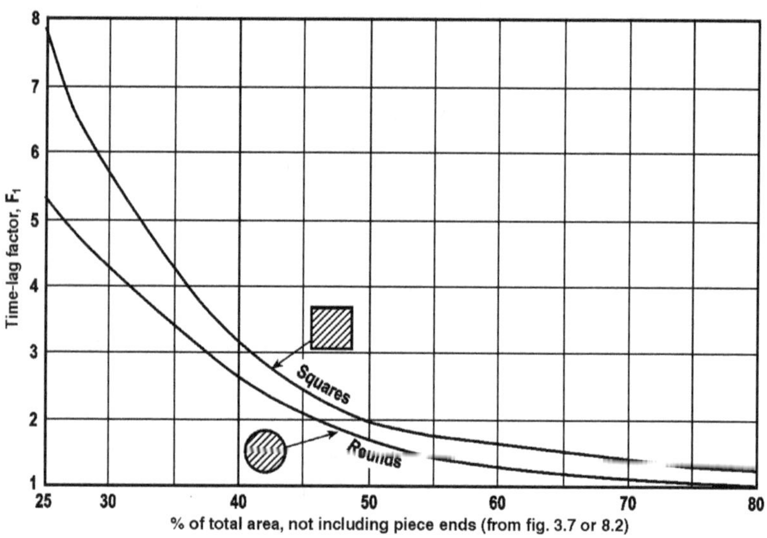


**Fig. 8.3.** Exposure factors, for squares and rounds with various sides exposed, or various percentages of total area exposed. For square sections with all four sides exposed, F1 = 1.0. (See eq. 3.1 and 3.2.) Use a centimeter scale to interpolate. (See example 3.2 and table 8.2.)

**TABLE 8.3 Comparison of two-side heating and four-side heating**

| | Two-Side | Four-Side |
|---|---|---|
| Radiation coefficient from gas blanket between pieces vs. gas blanket above pieces | 14 | 17.3 |
| Convection coefficient between pieces with 1 fps vs. 15 fps above = $(1/15)^{0.78} = 1/8.28$ | 1 | 8.28 |
| Estimated gas-to-load temperature difference | ~100°F | ~1000°F |
| Angularity of exposure – weighted 30° vs. 90° | 1 | 3 |

It is difficult to weigh the relative importance of each of the above four influencing factors, but the tabulated comparisons would indicate that the 83% figure should be conservatively acceptable.

Table 8.4 compares heat transfer rates for 6" (152 mm) square billets in a Curve 2 versus Curve 4 situation, both with spacing ratios of 3:1 and 2000 F (1090 C) furnace gas. Gains from wider spacing have diminishing returns (especially for four-side heating). All curves droop at low spacing-to-thickness ratios because all radiation is less with narrower spacing.

Round loads have smaller lag-time exposure factors than rectangular loads because radiation at a low angle, as from a nearby sidewall, has a better chance of reaching an adjacent load piece because round pieces make less shadow on an adjacent piece— if both have the same spacing ratio. The percent of full peripheral exposure also influences the lag-time, which becomes more important with thicker loads. Figure 8.3 gives the lag-time exposure factor F1 versus percent of full peripheral exposure.

***Procedure—Phase B—Draw a longitudinal cross section*** of the furnace interior, showing zone boundaries; burners, flue, and baffle locations; sensor locations; charge, discharge, and hearth. This side-sectional furnace drawing will be referred to as

**TABLE 8.4 Comparison of heating rates from curves of figure 8.2**

| | Curve 2 | Curve 4 | Curve 2/Curve 4 |
|---|---|---|---|
| Gas beam, B, in ft; in m | B = 4T<br>2 f; 0.6 m | B = 3T<br>1.5 f; 0.5 m | |
| Gas radiation flux, from fig. 13.13 of *ref. 52,* Btu/hr ft$^2$; kW/m$^2$ | 7700; 24 | 6200; 20 | 7700/6200 = 1.24 |
| Estimated furnace gas velocity over load piece surfaces, f/s; m/s | 15 | 3 | |
| Convection heat transfer coefficient hc from p. 91, *Reference 51* | $(15)^{0.75}$ | $(3)^{0.75}$ | 7.62/2.28 = 3.34 |
| Convection heat transfer = hc$\Delta T$<br>(Estimated effective $\Delta T$ for convection = 200°F) | 1524 | 456 | |
| Combined effect of gas radiation and convection | 7700 +<br>1524 | 6200 +<br>456 | 9224/6656 = 1.38 |

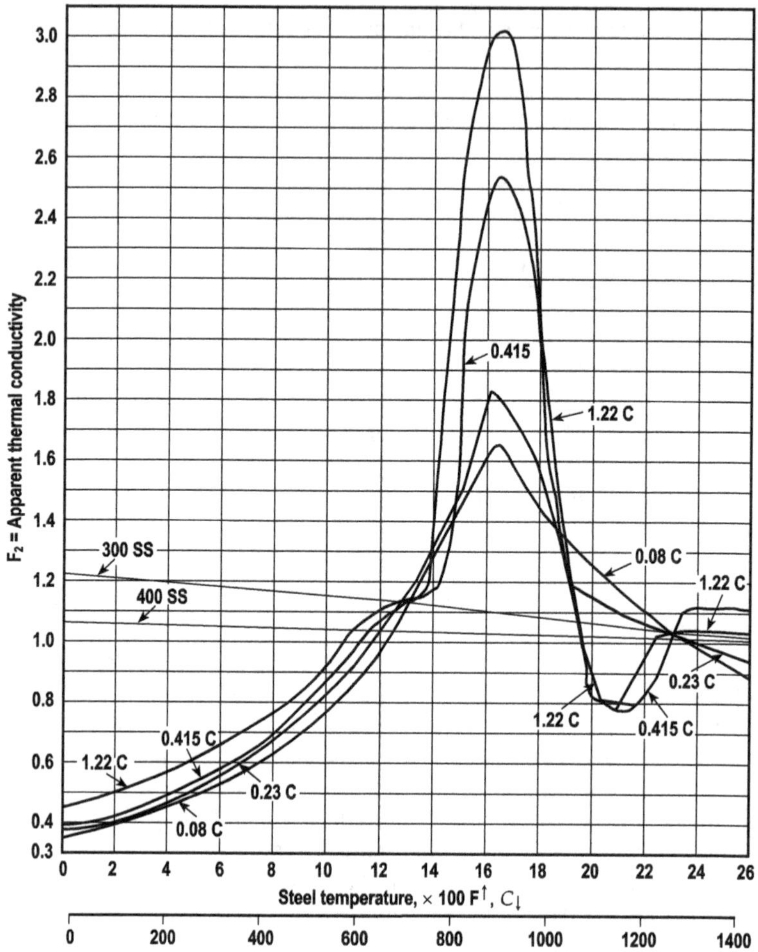


**Fig. 8.4.** Effect of carbon content in various steel grades on heat absorption is shown by these "grade factors" used in the last steps of table 8.3 (worksheet) for the Shannon Method for plotting steel heating curves. The peaks in this graph show the effect of the dramatic increase in heat absorption for steels containing various percentages of carbon, C, during the crystalline phase changes between 1200 F and 1900 F (650 C and 1038 C). SS = stainless steel.

figure 8.5, but it appears as the top of figure 8.1 and figure 8.5. T-sensor 1, in the first (unfired) zone, controls the input to the second (preheat) zone.

### 8.1.2. Plotting the *Furnace* Temperature Profile, Zone by Zone on Figs. 8.6, 8.7, and 8.8

***Procedure—Phase C—Preparing to plot a furnace temperature curve.*** (Plotting *load* temperature curves will follow in sec. 8.1.3.). Using 11 in. × 17 in. (0.28 m × 0.43 m) graph paper, lay out a vertical temperature scale and a horizontal

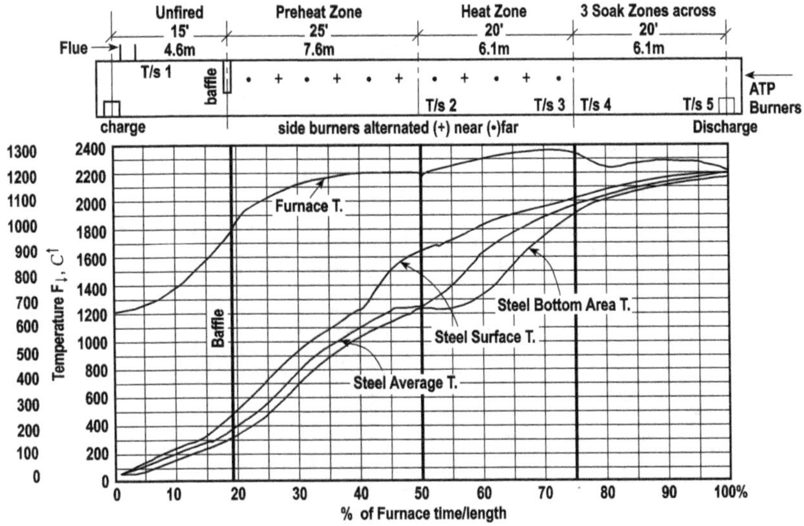


**Fig. 8.5.** Typical time-versus-temperature curves for a steel reheat furnace, with a side-sectional drawing aligned above the curves.

scale for time (or distance through the furnace). Enlarge or reduce the drawing of Phase B to align with the 3-piece graph, hereafter referred to as Figures 8.6, 8.7, and 8.8.

Three load temperature curves, for the load surface, load average, and load core (or load bottom in the case of one-side heating) will be assembled in section 8.1.3. Identify the job with a title box containing information such as owner, furnace identity, load description, design production rate, graph number, furnace type, process, load spacing, expected fuel rate, emissivity, person making the calculation, and date.

Divide the temperature profile sheet [11 in. × 17 in. (0.28 m × 0.43 m) graph paper = figures 8.6, 8.7, and 8.8] into 20 units and number them. At the right end (furnace discharge) of the bottom scale of the graph, mark (*a*) 100%, (*b*) total time the load will be in the furnace, and (*c*) total effective furnace length. Divide each of these scales into appropriate units (%, ft or m, hr and min). Draw vertical lines to show zone interfaces—aligned with the sketch (from Phase B), now at the top of this graph, Figure 8.6.

***Procedure—Phase D1—Soak Zone.*** Begin drawing the expected temperature profile of the furnace walls and roof (top curve on Fig. 8.11), starting with the discharge (right) end of the soak zone. Deciding zone temperatures is difficult—not an exact science. Some engineers are tempted to assume flat zone temperature profiles, but that cannot be because the furnace interacts with the flame temperature profile, charging rate, and heat transfer to the load. The furnace temperature drops slowly from the discharge to the beginning of the soak zone, to the point where the higher heat zone temperature raises the inlet soak zone temperature from 2230 F (1220 C) to 2340 F (1280 C). The authors suggest some guidelines in the following paragraphs.

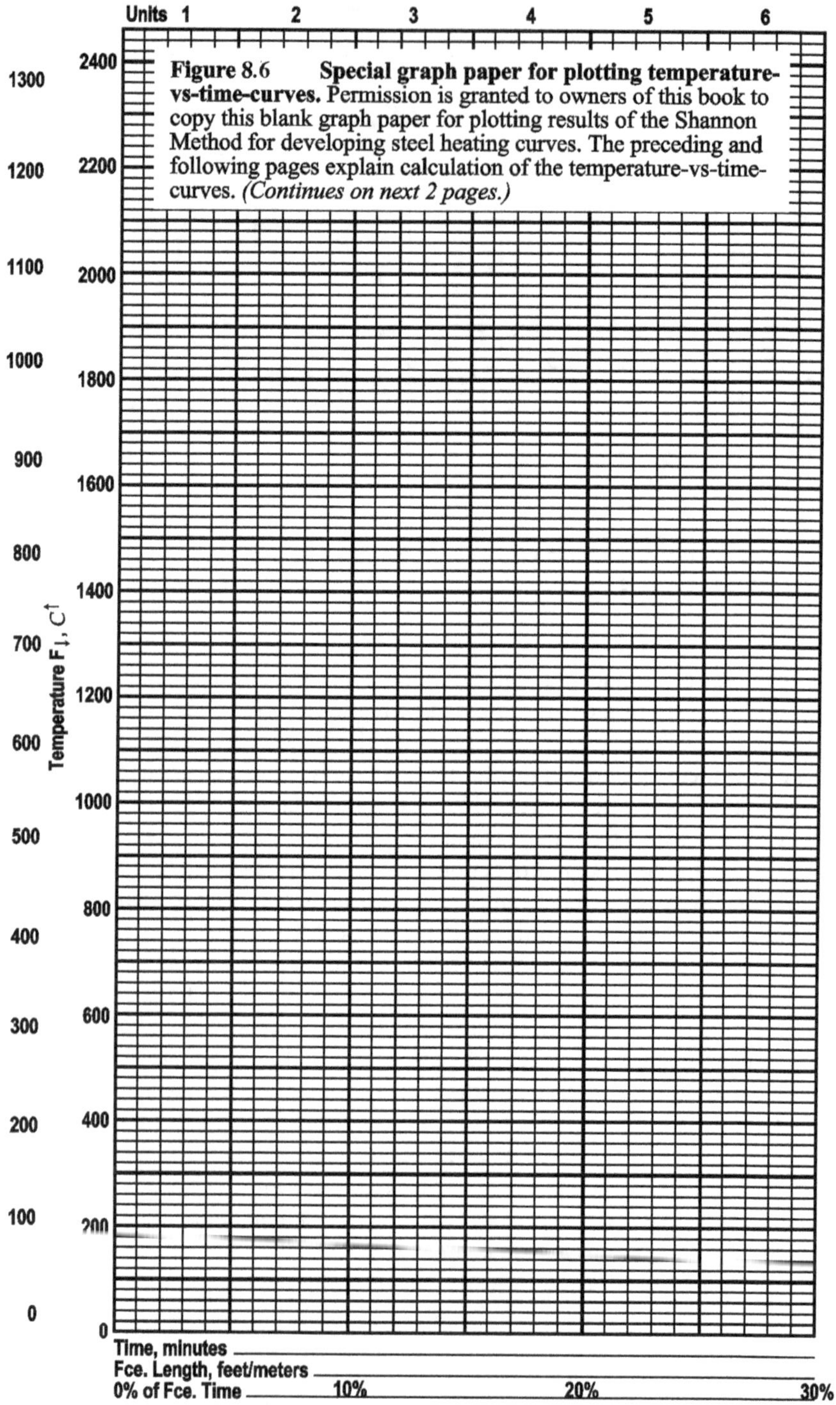


**Figure 8.6 Special graph paper for plotting temperature-vs-time-curves.** Permission is granted to owners of this book to copy this blank graph paper for plotting results of the Shannon Method for developing steel heating curves. The preceding and following pages explain calculation of the temperature-vs-time-curves. *(Continues on next 2 pages.)*

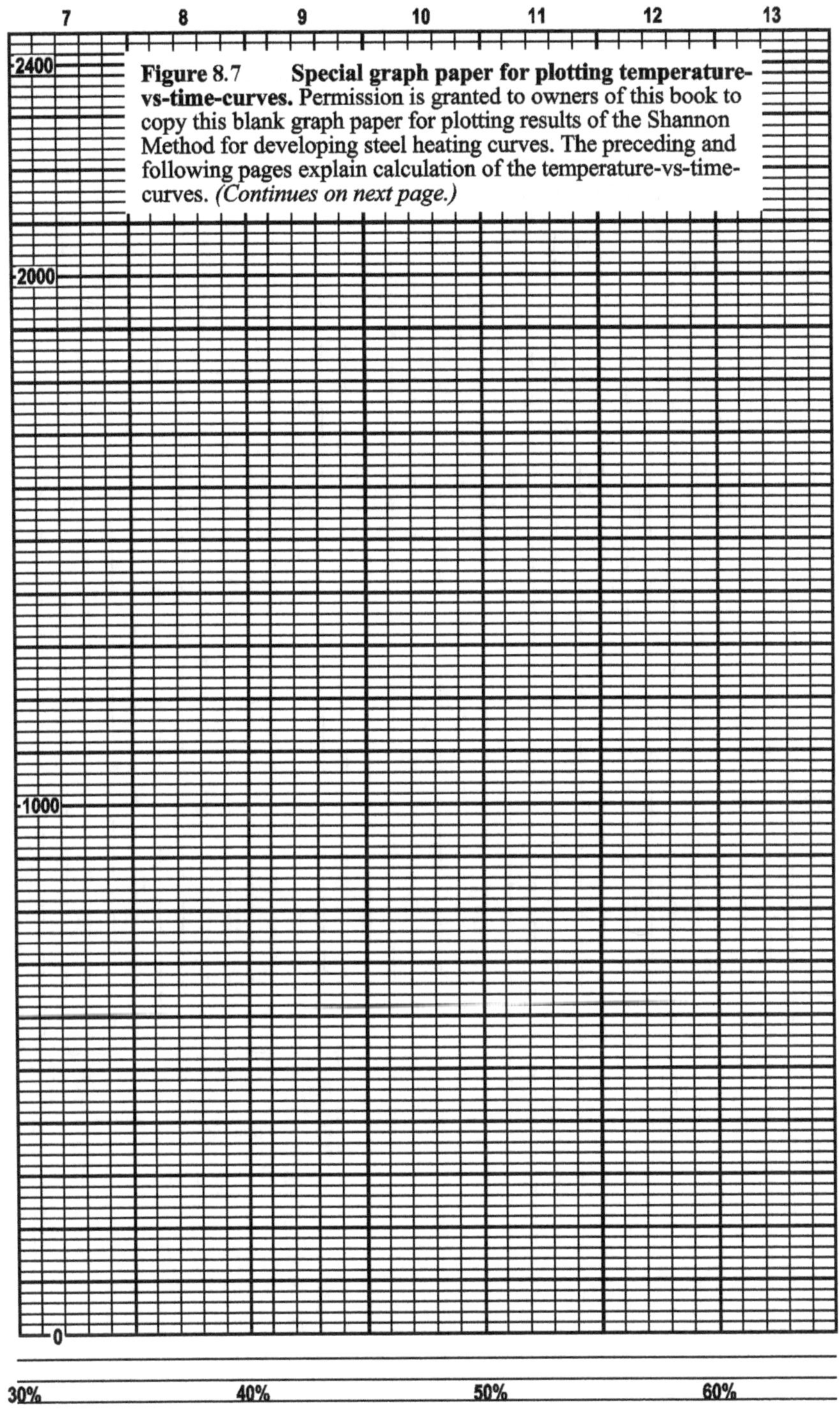


**Figure 8.7 Special graph paper for plotting temperature-vs-time-curves.** Permission is granted to owners of this book to copy this blank graph paper for plotting results of the Shannon Method for developing steel heating curves. The preceding and following pages explain calculation of the temperature-vs-time-curves. *(Continues on next page.)*

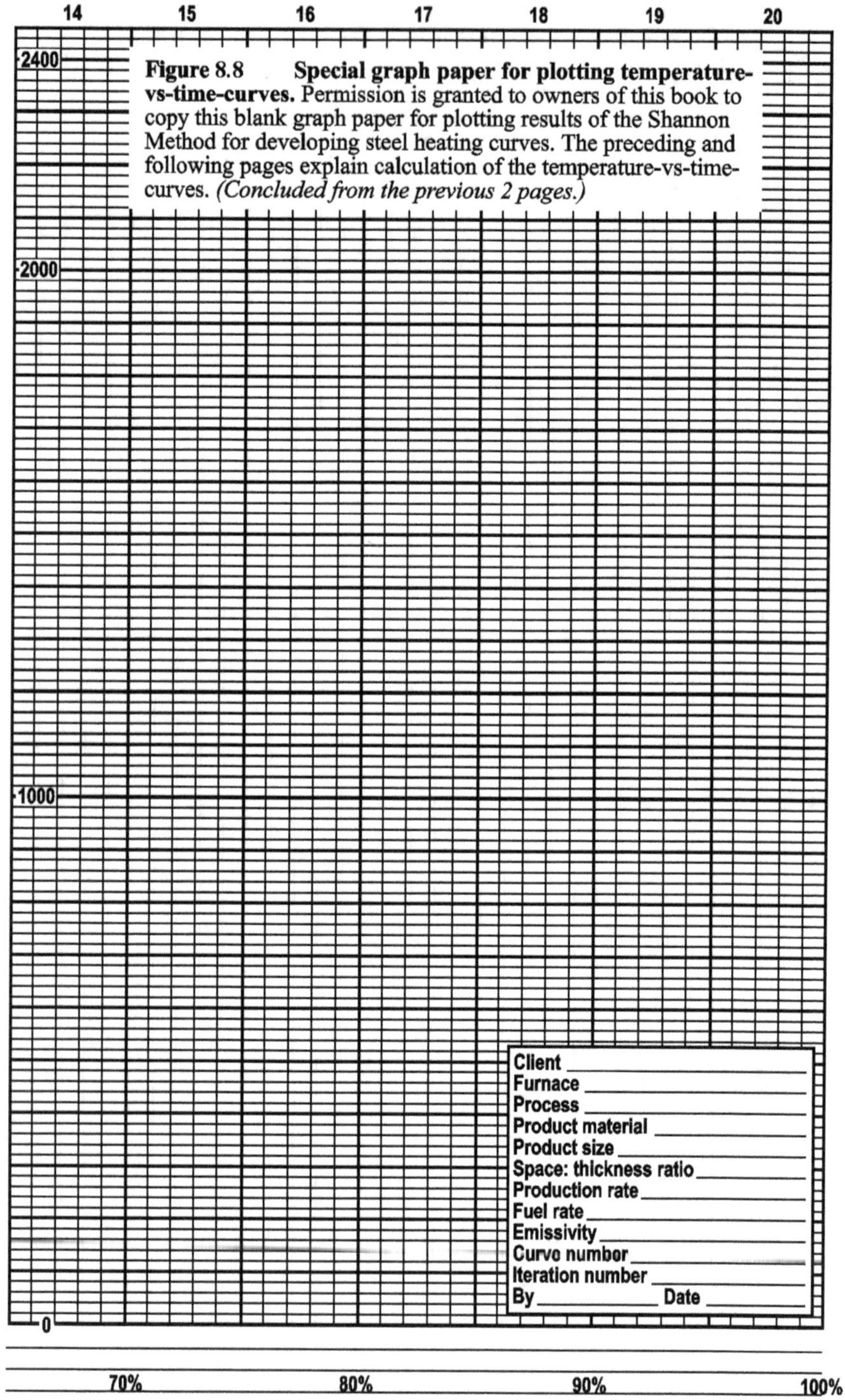


**Figure 8.8 Special graph paper for plotting temperature-vs-time-curves.** Permission is granted to owners of this book to copy this blank graph paper for plotting results of the Shannon Method for developing steel heating curves. The preceding and following pages explain calculation of the temperature-vs-time-curves. *(Concluded from the previous 2 pages.)*

*Soak Zone Guidelines—if the soak zone is end fired with conventional burners,* the discharge end-wall temperature will be about the expected rolling temperature. At 4 ft (1.2 m) from the discharge wall hot face, furnace temperature will be about 70°F (39°C) above rolling temperature. At soak zone entry, the product will be close to rolling temperature. The average zone wall temperature should be 50° to 70°F (28° to 39°C) above the normal metal rolling temperature.

*Soak Zone Guidelines—Whether the soak zone is side-fired, roof-fired, or longitudinally fired,* the discharge end wall temperature may be 20°F (11°C) below the maximum soak temperature. If the soak zone is fired with side burners, roof burners, or longitudinal ATP burners, the discharge end wall temperature may be 20°F (11°C) above the maximum soak temperature. If the load pieces are discharged through end wall openings with large heat losses, the whole range of soak zone temperatures should be plotted as 25 to 50°F (14 to 28°C) below the just-mentioned pattern, allowing for large heat losses of the door and extractor or dropout.

***Procedure—phase D2—Heat Zone.*** For this example, assume a radiation shield curtain wall between the soak and heat zones. The design steel rolling temperature is 2150 F (1177 C), so it is reasonable to plan for a heat zone temperature of 2350 F; certainly no higher than 2400 F.

With a heat zone longitudinally top fired, the burner wall temperature would be 100°F (56°C) above the product discharge temperature and 100°F below the peak temperature of the zone at high fire. With side firing, the heat zone curve raises the zone entering temperature quickly to a peak of 2340 F (1280 C). The heat zone temperature then falls with greater slope than the soak zone to 2180 F (1193 C) just before the preheat zone starts to rise to a maximum of 2180 F (1193 C).

*Heat Zone Guidelines.* Typically, furnace roof/side temperatures peak about 15 ft (4.6 m) from the burner wall, then slowly fall to 2100 F to 2300 F (1149 to 1266 C) depending on zone length, firing rate, flame length, and the value of the heat transfer factor, H (A high H value will increase the slope of the zone temperature). The temperature at the charge end of a 20 ft (6.1 m) long heat zone will probably be 150°F below the peak zone temperature.

*Heat Zone Fired From the Sides or Roof.* The discharge wall would be at peak temperature, and its temperature would begin to fall about 10 ft from the zone discharge. The downhill slope would be shallow near the discharge, but steeper near the charge end of the zone because of changing heat flux and product temperature. Continuing energy input to the charge end of the zone, and lower heat flux from the flame profile, will cause the zone temperature change differential (peak to charge end) to be 100°F to 150°F (56°C to 83°C), depending on the H value.

***Procedure—phase D3—Preheat Zone.*** If longitudinally fired, this zone would have a peak temperature of about 2250 F (1252 C) at a point 5 to 10 ft (1.5 m to 3.0 m) from the burner wall. The burner wall temperature would probably not be more than 2200 F (1204 C). The entry end of this zone is cooler because the product at the entry is generally at ambient temperature; therefore, the temperature difference is greatest at that instant. The load temperature then rises rapidly because of the 4th power effect of radiant heat transfer. If roof fired or side fired, the slope of the temperature curve will be more moderate because in that case combustion takes place to the entry of the fired zone.

The preheat zone thermal profile slopes to a minimum of 1800 F (982 C) at the entry baffle (between unfired and preheat zones) where fuel stops burning. Note the more rapid drop in temperature after (*left of*) the baffle.

If a furnace has one or more bottom zones, use the same thinking to do the temperature profiles. Before the next phase, check to see if any of the following need reconsideration: (1) Is the exposure factor still applicable? (2) Is the time-lag correct? (3) Is the time in the furnace still correct?

***Procedure—phase D4—Unfired Charge Zone.*** To minimize the flue gas exit temperature from the furnace, use of an unfired zone is generally wise (unless using regenerative burners, which create a low exit gas temperature leaving the beds). An unfired zone of 15 to 25% of the furnace length would start at the charge door, allowing the furnace gases to supply all the heat in that zone. To make that zone most effective, a radiation heat shield (baffle) should be placed between the discharge end of the unfired zone and the beginning of the preheat zone. There will be no heat input in this zone other than the sensible heat from the poc of other zones, therefore, the zone temperature drops 300°F to 450°F (166°C to 250°C). That lowers the exit gas temperature, raising the fuel efficiency. The unfired zone temperature profile has a steeper slope than the preheat zone, but not as steep as with regenerative burners positioned almost to the charge door.

*Charge Zone Guideline: Check the furnace curve slope.* When doing a heat balance of an unfired preheat zone, it is possible to check on the slopes of the temperature curves of preheat and unfired zones. If the slopes are too steep, excess energy will be available, and furnace temperature will be higher than estimated. If insufficient energy was available at the beginning of the unfired zone, the slope was not steep enough.

Drawing a furnace temperature profile is not easy. With practice, engineers can use common sense and this method to make a reasonably correct estimate of the furnace temperature curve that will serve them well. As with any calculation, engineers should note factors influencing the outcome or that may affect the next step in the iteration—and modify their design accordingly. For example, they should now check to see if the charge zone rise in furnace temperature and load temperature are actually possible from the falling furnace gas temperature and resultant change in available heat. **Warning:** In a furnace temperature profile, the temperature in the first 30% of the furnace length should not exceed 2300 F, where scale begins to soften. Softened scale has a very smooth, reflective surface that will not absorb heat, resulting in lower load temperature at the discharge.

Many who calculate heating curves draw straight lines for the zone temperature. With *longitudinally fired furnaces*, others attempt to estimate an ascending, then flat, and finally a declining temperature profile. With several longitudinally fired zones (sawtooth roof), the ascending-flat-declining pattern may repeat in each zone. The combustion reaction begins in the burner tile (quarl) of a conventional longitudinally fired burner. As the air and fuel emerge from the tile at the burner wall, the reaction is just starting, and therefore the energy released and the temperatures are low.

As the gases move away from the burner wall, their reaction accelerates, providing more and more energy for transfer to walls, roof, and load. As the temperature rises, more and more heat is transmitted to the product directly, and indirectly by way of the refractory. The temperature profile begins at the burner wall 100°F to 150°F (56°C to 83°C) below the zone temperature as typically measured from the roof 15 ft (4.6 m) beyond the burner wall. Depending on the type of burner, the rate of temperature rise to the location of the control T-sensor may or may not be rapid. If the burner has a lot of combustion spin, the temperature will rise rapidly, beginning at the burner wall.

Generally, the rate of heat transfer is low near the burner wall because the temperature differences are very small. (Load movement is counterflow to flame movement; thus, the flame reactants are coolest as they leave any one zone whereas the load pieces are hottest as they leave any one zone.) As the distance from the burner wall increases, the load surface is colder and the flame temperature is hotter because the combustion reaction rate accelerates. However, a control T-sensor 15 ft (4.6 m) from the burner wall limits the furnace temperature at that point. (This temperature is held to a setpoint determined by the operator or by a model.) With *high-spin burners*, as one follows the temperature profile away from its maximum and in the direction of flame reactant flow, the furnace temperature declines quickly to the setpoint, and thereafter drops rapidly to the exit.

With *nonspin burners*, the furnace temperature at the control sensor will probably be the highest in the zone. Nonspin burners may have a location in the heating zone where the combustion reaction is increasing at a rate almost the same as the rate of increase in energy requirement of the product. In this case, the zone temperature profile would be flat. However, beyond the completion of the combustion reaction (a variable distance, depending on the firing rate), the flame temperature profile declines because the heat source has ended, and cold loads continue to enter the zone, absorbing more energy. Because the location of the end of the combustion reaction is unknown, accurate calculation of the slope of the temperature decline curve is very difficult.

In a longitudinally fired zone with all multiple burners firing at 20 kk Btu/hr (586 MW) maximum in the nonspin mode, the temperature profile may begin to decline 25 to 30 ft (7.6 to 9.1 m) from the burner wall because of completion of the combustion reaction and of the cooling effect of cold, heavy loads entering the zone. With spin-type burners, the temperature profile decline would begin much earlier, perhaps 10 ft (3 m) from the burner wall. Because the furnace temperature near the burner wall would have been hotter than the zone setpoint at 10 to 15 ft (3 to 4.6 m), production output of that zone would have been greater because more heat would have been transferred. In addition, the available heat will be higher because the temperature of the gases leaving the zone will be lower.

A two-sensor zone control, with sensors at the elevation of the top of the product, is recommended. A spin burner will give the best production rate and best (minimum) fuel consumption. To take maximum advantage of this, more and shorter zones should be used.

**Warning**: *Beware of a hot charge (entrance) in the charge zone.* There are cases where the actual temperature at the charge end of a zone appears to be very hot, and yet the furnace productivity is low and the product too cold for good rolling quality.

The cause may be (1) operating the furnace in batch mode or (2) reflective scale on the surface of the load, interfering with heat transfer. The second case begins with a zone setpoint of 2300 F (1260 C) or above in the charge area, which causes rapid scale formation that insulates the product so that the scale surface itself reaches its softening temperature (about 2320 F or 1270 C), creating a reflective surface that lowers heat transfer. Radiation cannot heat a mirror, so the zone temperature becomes very hot but cannot transfer energy to the load. Although logic might indicate the need for higher temperature to increase product temperature, the zone temperature must be reduced to prevent reflective scale formation. This should be done by lowering the setpoint to 2200 F to 2250 F (1200 C to 1230 C), thus preventing the scale from reaching 2320 F (1270 C). After the load is again absorbing heat without the reflective scale, the operator may slowly raise the zone temperature toward 2300 F. If the charge end of the zone again becomes very hot, the setpoint was raised too high.

These cases show that calculating an accurate zone temperature profile is difficult. A flat temperature profile for part of a zone may be correct, but with most zones and firing rates, the temperature profile must ascend or decline to reflect the dynamic heat exchange rates in furnace zones.

Both *side firing* and *roof firing* add energy along the furnace length. If the burners are duplicates in every way, the temperature will rise from the charge end and peak at the discharge end of the each zone. For maximum productivity, the zone charge end burners should be larger, as directed by heating curves, if productivity is of more concern than fuel efficiency.

*Regenerative burner firing* is much like other side-fired furnaces (except oxy-fuel firing) in that maximum production necessitates installing burners as close to the flues as possible to hold the furnace temperature up almost to the charge door. The reason is that with regenerative burners, the mass of gas moving to the flues is very small because 80 to 100% of the flue gases are used to preheat air in each burner's heat exchange bed to provide very low fuel rates.

To use *oxy-fuel firing* (near-pure oxygen instead of air) in industrial furnaces to improve productivity, furnace designers must be aware of the major changes this can cause in the furnace temperature profile, and (*a*) the mass of the combustion gases is reduced by about 67%, (*b*) the percentage of triatomic gases in the poc increases from 26 to near 100%, and (*c*) the best possible efficiency goes from 35 to 70% available heat in many heat zones.

The furnace thermal profile starting at the burner wall (longitudinally fired) increases much more rapidly with oxy-fuel firing than with air-fuel firing because there is only one-third the mass of gas to absorb the same heat release from the same chemical reaction. Additionally, the temperature decline is even more rapid than with ATP burners because of higher heat transfer from the small mass of gas containing 100% triatomic gases versus 26%. Earlier higher available heat release changes the profile. Because of these changes, oxy-fuel's thermal profile is much more sensitive. The burner design may modify some of these differences.

To maximize productivity, more regenerative burners (and sometimes side-fired burners) should be installed as near as practical to the flues; otherwise the unfired area will be at lower temperatures, and thus will not be able to transfer much heat. With a large, unfired preheat zone, delay difficulties are magnified because the zone T-sensor allows the newly charged material to move a long way before it affects the firing rates.

If low fuel use is more important than productivity, correct engineering would be to have a long, unfired section to remove maximum possible energy from the flue gas. The location of the first fired-zone T-sensor should be near the flue. However, if saving productivity is more important than saving fuel, an unfired zone should be fired.

### 8.1.3. Plotting the *Load* Temperature Profile

Plotting the load temperature profile on a graph requires the use of a worksheet, tables 8.5 and 8.6. Now you will begin to work back and forth between the graph and worksheet. Whereas section 8.1.2 worked from right to left (decreasing temperatures) when plotting *furnace zone* temperature curves, this section 8.13 will now work from left to right (increasing temperatures) in plotting the *load* temperature curves. To begin the process of drawing the load temperature rise curve, estimate an average load surface temperature in the first group of three time units and record it on line [d] of your worksheet, table 8.7.

***Overview of the method:*** (Letters correspond to worksheet lines, tables 8.5 to 8.9.)

[b] From the estimated furnace temperature curve (fig. 8.11), read the average temperature of the first group of three increments. [d] Estimate an expected product surface temperature.

[e] From table 8.9, at temperature [b], find the black body radiation heat flux, Btu/ft$^2$ hr, from the *furnace* in the first group of 3 increments (first 15% of total inside furnace length or time in the furnace).

[f, g] Not applicable unless both top and bottom firing, or very thick load.

[h] From table 8.9, at temperature [d], find the black body radiation heat flux, Btu/ft$^2$ hr, from the *load* in the first group of 3 units.

[i] Subtract [h] from [e] for net radiation heat flux rate, furnace to load.

[j] Multiply the net radiation [i] by 3H (for a group of 3) to get the Btu/lb heat content rise in the group of 3 units, or 2H for a 2-unit group.

[l] Use Table 8.9 again, but this time to look up the new average load temperature corresponding to the new heat content. This is the average load piece temperature for the first group of 3. On figure 8.10 and 8.11, plot this temperature at the right end of the 3rd unit in the 1st group of 3.

[m] Look up the grade factor, $F_2$, from figure 8.4, at the new average temperature at the discharge end of the section. This is for use in calculating time-lags [n] and [o], which are functions of the thermal conductivity of the load material, and the Btu/pound change for each new average group-of-3 temperature. These time-lags determine when bottom and top temperatures of the load piece arrive at the

calculated average load temperature of step [1], when the top surface reaches the [l] temperature, and where (on figs. 8.10 and 8.11) to plot for the bottom* surface temperature arriving at [l].

[n] Use formula here for minutes for heat to diffuse from top to average.

[o] Experiments have shown that the time-lag for heat to diffuse from average to bottom is about 0.62 of [n, the time-lag from top to average].

As the final steps for the first group of 3 units, on figure 8.5 or 8.11, *plot furnace temperatures* [b, c] if not already done in phases C and D. Then, *plot load average temperature* [l] at the end of the 3rd increment as the first point on the *average temperature curve*. Next, *plot load bottom temperature* [l again] at [o] minutes to the *right* of $p_1$ as the first point on the *bottom temperature curve.* Finally, *plot load top temperature* [l again] at [n] minutes to the *left* of $p_1$ as the first point on the *top temperature curve.*

NEXT ITERATION: Visually extrapolate the average temperature curve to estimate a new [d] in the next group of 3 units on table 8.5.

***Procedure—phase E.*** (This is a repetition of the 'overview' above, but with more detailed explanations.) On copies of the blank worksheets from tables 8.5 and 8.6, at line [b] enter the average furnace temperature for each of the 7 groups of 3 increments that you plotted on your graph, figure 8.5, as a result of procedures C and D. Because our example is for one-side heating, skip lines [c] and [f]. Estimate the average load surface temperature for the first group of 3 increments, and enter it on line [d].

In table 8.7, enter the difference between the black body radiation rate for furnace temperature [b] and load temperature [d], on line [i]. Multiply [i] by 3H, for the 3 unit group, and enter the resulting Btu/pound heat content rise of the load on line [j]. The Shannon method's H factor reduces black body radiation by the effect of emissivity (absorptivity), In succeeding columns, use line [k] to totalize the cumulative Btu/pound. In figure 8.9, convert the new Btu/pound heat content to a new average temperature throughout the load (270 F for the first three time units), and record it on line [l].

**Example:** A 100 F piece of oxidized steel (emissivity = 0.79) has a flat surface parallel to a nearby 1600 F kaolin insulting refractory (emissivity = 0.49). From table 2.3, $F_a = 1$ and $F_e = 1/[(1/0.79) + (1/0.49) - 1] = 0.433$. From table 8.9 above, the net $q_{bb}$ = (30 960 Btu/hr ft$^2$ for the refractory) − 168 Btu/hr ft$^2$ for the steel) = 30 790 Btu/hr ft$^2$. Therefore, net radiation heat flux between the two surfaces (by equation 2.6) = $q_{bb}\ F_eF_a$ = 30 790 (1) (0.433) = 13 300 Btu /hr ft$^2$. *(Continued detailed explanation of the Shannon method from before Table 8.5.)*

On your own copy of figure 8.6, plot the average load temperature for the first group of 3 units, from line [l] of table 8.7, by marking a point at 270 F at the

*Bottom temperature for top-only heating, but center temperature if using top and bottom heating. *(This detailed explanation of the Shannon Method for plotting steel heating curves continues several pages later, after the worksheets and table 8.9.)*

right end of the 3rd unit. (See enlargement in figure 8.10.) Assume starting load temperatures (top, average, and bottom) to be 60 F (16 C) for all three curves that you will eventually draw. Do not connect the dots until you have at least 3 points along each of the 3 curves.

To plot the load top surface temperature, it is necessary to determine a time-lag between when the top surface arrives at 270 F and when the average (core) load temperature arrive at 270 F. (We have already plotted the 270 F core temperature at

**TABLE 8.5 Blank heat transfer calculation worksheet**

Client__________ Furnace size & type:__________ Curve #______
tph =______ Load dimensions & grade__________ Date______
H = ______ 3 H = ______

| | | | | | | | | | | |
|---|---|---|---|---|---|---|---|---|---|---|
| [a] Units | 1 | 2 | 3 | 4 | 5 | 6 | 7 | 8 | 9 | 10 |
| [b] Furnace temperature, top average for group of 3 units | | | | | | | | | | |
| [c] Furnace temperature, bottom avg[a] | | | | | | | | | | |
| [d] Load surface temperature | | | | | | | | | | |
| [e] Furnace black body radiation, from table 8.9, at top temp[a] [b] | | | | | | | | | | |
| [f] Furnace black body radiation, from table 8.9, at bottom temp [c] | | | | | | | | | | |
| [g] Avg[a] fce[a] top & bot[a] radn[a], [e+f]/2. If more zones, add $g_2$, $g_3$, etc.[a] | | | | | | | | | | |
| [h] Load black body radiation, from table 8.9, at temperature [d] | | | | | | | | | | |
| [i] Net radiation between fce at b temp and load at d temp = [g] − [h] | | | | | | | | | | |
| [j] Btu/# rise = [i] (3H), or [i] (2H) for last group, of 2 | | | | | | | | | | |
| [k] Cumulative Btu/#. $k_1 = 0 + j_1$; $k_2 = k_1 + j_2$; $k_3 = k_2 + j_3$; etc | | | | | | | | | | |
| [l] Average load temperature, from figure 8.9 | | | | | | | | | | |
| [m] Lag factor $F_2$, from figure 8.4 at temperature [l] | | | | | | | | | | |
| [n] Time lag, in % of total fce time, from average to top = #11[d] (0.6[e]) [m] / (#12[d]/100 spaces) = 5 [m] | | | | | | | | | | |
| [o] Time-lag, %, from average to bottom[c] = 0.62[e] [n] | | | | | | | | | | |

Permission is granted to owners of this book to make copies of blank worksheets, tables 8.5 and 8.6

[a] See glossary for abbreviations. [c]to bottom if 1-side heating; to center if 2-side heating. [d]table 8.2. [e]From experimental evidence

avg = average. betw = between. bot = bottom. col = column.

etc = et cetera = and so forth. fce = furnace. radn = radiation. temp = temperature.

**TABLE 8.6 Blank heat transfer calculation worksheet**

Client__________ Furnace size & type:__________ Curve #______
tph =______ Load dimensions & grade__________ Date______
H = ______ 3 H = ______

| | 11* | 12* | 13 | 14 | 15 | 16 | 17 | 18 | 19 | 20 |
|---|---|---|---|---|---|---|---|---|---|---|
| [a] Units | | | | | | | | | | |
| [b] Furnace temperature, top average for group of 3 units | | | | | | | | | | |
| [c] Furnace temperature, bottom avg[a] | | | | | | | | | | |
| [d] Load surface temperature | | | | | | | | | | |
| [e] Furnace black body radiation, from table 8.9, at top temp[a] [b] | | | | | | | | | | |
| [f] Furnace black body radiation, from table 8.9, at bottom temp [c] | | | | | | | | | | |
| [g] Avg[a] fce[a] top & bot[a] radn[a], [e+f] / 2. If more zones, add $g_2$, $g_3$, etc.[a] | | | | | | | | | | |
| [h] Load black body radiation, from table 8.9, at temperature [d] | | | | | | | | | | |
| [i] Net radn between fce at [b] temp and load at [d] temp = [g] − [h] | | | | | | | | | | |
| [j] Btu/# rise = [i] (3H), or [i] (2H) for last group, of 2 | | | | | | | | | | |
| [k] Cumulative Btu/#. $k_1 = 0 + j_1$; $k_2 = k_1 + j_2$; $k_3 = k_2 + j_3$; etc. | | | | | | | | | | |
| [l] Average load temperature, from figure | | | | | | | | | | |
| [m] Lag factor $F_2$, from figure 8.9 at temperature [l] | | | | | | | | | | |
| [n] Time-lag, in % of total fce time, from average to top = #11[d] (0.6[e]) [m] / (#12[d]/100 spaces) = 5[m] | | | | | | | | | | |
| [o] Time-lag, %, from average to bottom[c] = 0.62[e] [n] | | | | | | | | | | |

Permission is granted to owners of this book to make copies of blank worksheets, tables 8.5 and 8.6
[a]See glossary for abbreviations. [c]to bottom if 1-side heating; to center if 2-side heating. [d]table 8.2. [e]From experimental evidence.
**Note:* Units 11 and 12 on this page are part of the same group of 3 as is Unit 10 (last column on the previous page); so the first column of calculated figures to be inserted on this page should be the same as those of the last column of table.
avg = average. betw = between. bot = bottom. col = column.
etc = et cetera = and so forth. fce = furnace. radn = radiation. temp = temperature.

the end of the 3rd unit.) Use figure 8.4 to read $F_2$, the time-lag factor for the grade of steel. In this case, for 0.04% carbon at 60 F, interpolate $F_2 = 0.44$, so record this on line [m].

Calculate the lag time, in percentage of total time in the furnace, for the same 270 F to diffuse from top surface to core of a load piece = 6.18 (0.6) [m]/(1% of 743

**TABLE 8.7 Heat transfer calculation worksheet (continues on table 8.8)**

Client (sample). Furnace size & type: 80' × 34's walking hearth. Curve # 2.
tph = 100. Load dimensions & grade 4.5" × 4.5" × 30' 0.4% C steel. Date 70202.
H = 0.47 3 H = 1.41

| | 1 2 3 | 4 5 6 | 7 8 9 | 10 |
|---|---|---|---|---|
| [a] Units | 1 2 3 | 4 5 6 | 7 8 9 | 10 |
| [b] Furnace temperature, top average for group of 3 units | 1350 F | 1840 | 1910 | 2150 |
| [c] Furnace temperature, bottom avg[a] | b | b | b | b |
| [d] Load surface temperature | 200 F | 590 | 1100 | 1540 |
| [e] Furnace black body radiation, from table 8.9, at top temp[a] [b] | 18.4 | 48.1 | 54.2 | 79.8 |
| [f] Furnace black body radiation, from table 8.9, at bottom temp [c] | b | b | b | b |
| [g] Avg[a] fce[a] top & bot[a] radn[a], [e+f] / 2. If more zones, add $g_2$, $g_3$, etc.[a] | 18.4 | 48.1 | 54.2 | 79.8 |
| [h] Load black body radiation, from table 8.9, at temperature [d] | 0.325 | 2.09 | 10.2 | 25.7 |
| [i] Net radn between fce at [b] temp and load at [d] temp = [g] − [h] | 18.1 | 46.0 | 44.0 | 54.1 |
| [j] Btu/# rise = [i] (3H, or 2H for last group of 2) = (18.1) (3) (0.47) = | 25.5 | 64.9 | 62.0 | 76.3 |
| [k] Cumulative Btu/# = 0 + $j_1$ = 0 + 25.5 = | 25.5 | 90.4 | 154 | 229 |
| [l] Average load temperature, from figure 8.9 | 270 F | 770 | 1160 | 1430 |
| [m] Lag factor $F_2$, from figure 8.4 at temperature [l] | 0.44 | 0.72 | 1.13 | 2.19 |
| [n] Time-lag, %, average to top = (6.18[d])(0.6[e])[m]/(0.743[d]) = 5 [0.44] = | 2.2 | 3.6 | 5.7 | 11 |
| [o] Time-lag, %, from average to bottom[c] = 0.62[e][n] = 0.62 [2.2] = | 1.4 | 2.2 | 3.5 | 6.8 |

[a]See glossary for abbreviations.
[b]Not applicable.
[c]to bottom if 1-side heating; to center if 2-side heating.
[d]Table 8.2.
[e]From experimental evidence.
avg = average. betw = between. bot = bottom. col = column.
etc = et cetera = and so forth. fce = furnace. radn = radiation. temp = temperature.

minutes) = 5 [m] = (5) (0.44) = 2.2%. Record this as [n], and plot your first point on the top surface temperature curve at 270 F and 2.2% to the left of the average temperature point. Then calculate the lag time, in percentage of total time in the furnace, for the same 270 F to diffuse from core to bottom, which is 62% of [n] = 0.62 (2.2) = 1.4%. Record this as [o], and plot the first point on the bottom surface

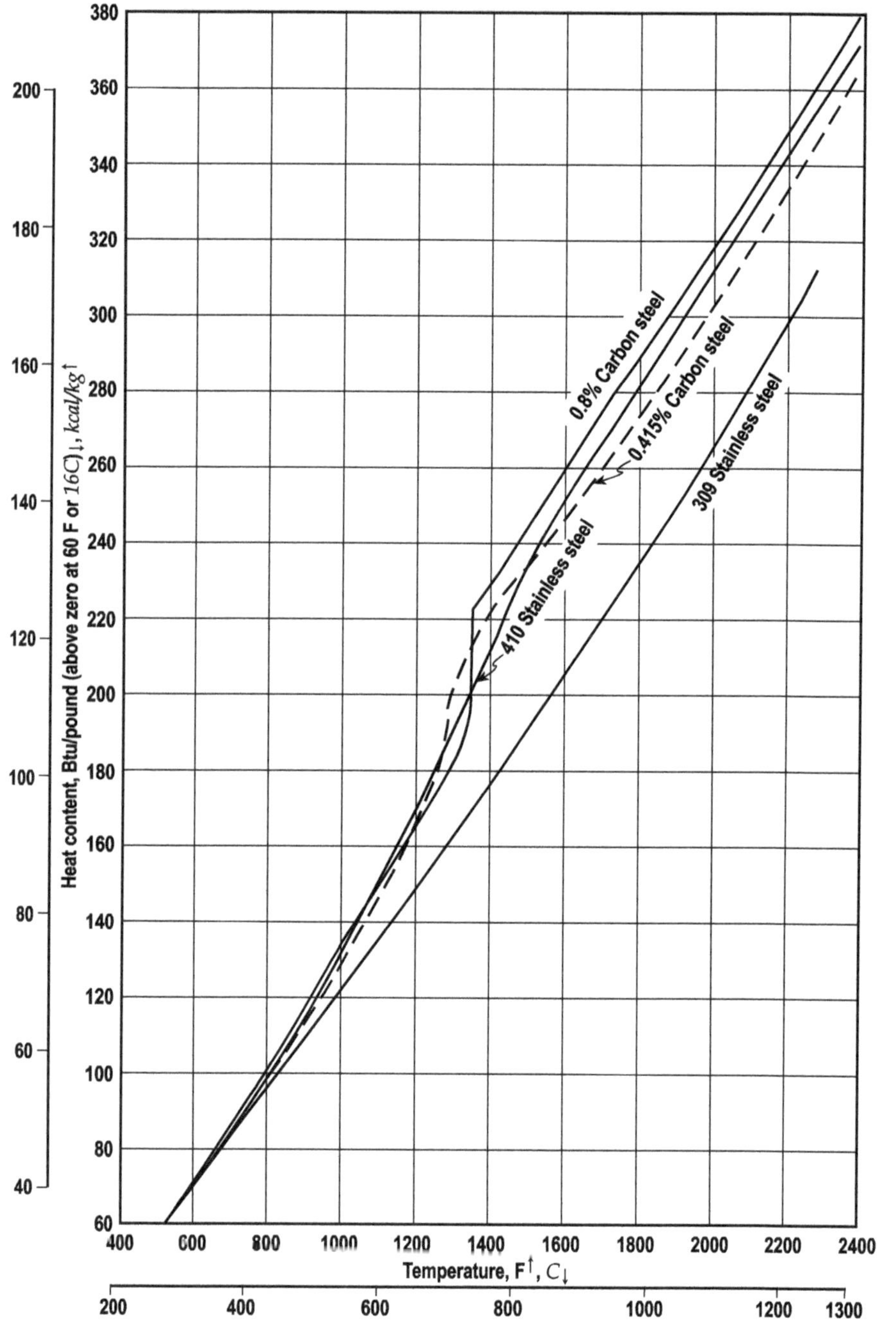


**Fig. 8.9.** Heat contents of four steels in normal working temperature ranges. For heat contents of other metals, consult pp. 260–263 of *reference 52.*

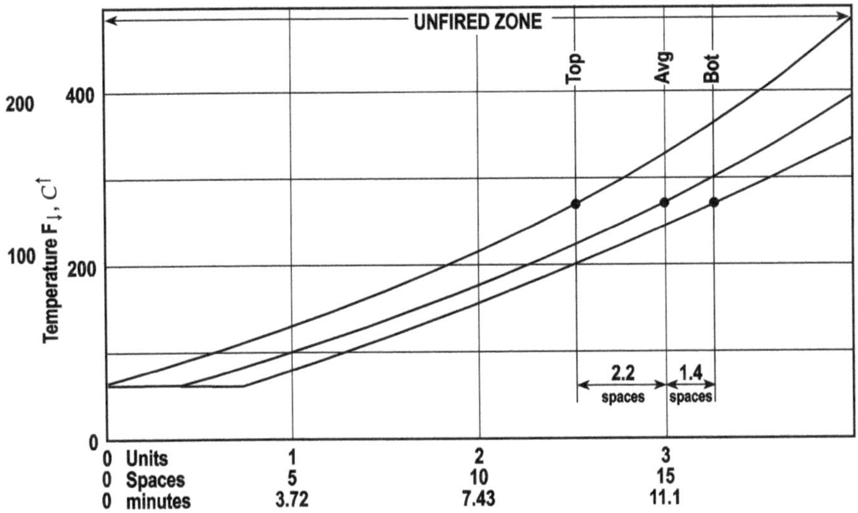


Fig. 8.10. Enlargement of plotting top, average (avg), and bottom (bot) temperatures at 270 F, from [I].

temperature curve at 270 F and 1.4% to the right of the average temperature point. (See fig. 8.10—enlargement of plotting for the first points of the 3 curves.)

Return to step [d], and use the two points that you now have on the top surface temperature curve (at 0 and 3 units) to estimate the average load surface temperature for the next group of three units. Proceed down the second column of numbers on table 8.7. The only bumps or humps in the curves should be at the 1300 F to 1400 F

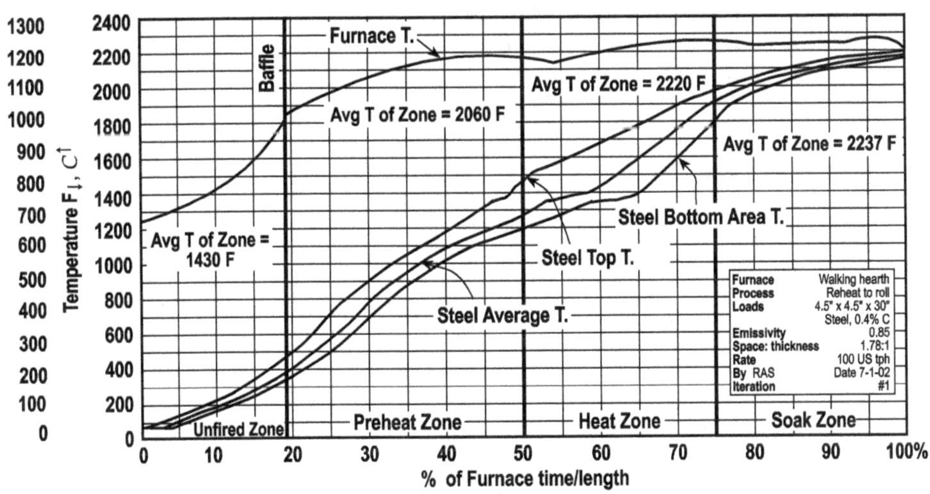


Fig. 8.11. Temperatures-versus-time graph: Results of sample problem 8.1. Preceding text explains the calculation of these curves.

**TABLE 8.8 Heat transfer calculation worksheet (continued from table 8.7)**

Client (sample). Furnace size & type: 80' × 34's walking hearth. Curve # 2.
tph = 100. Load dimensions & grade 4.5" × 4.5" × 30' 0.4% C steel. Date 70202.
H = 0.47 3 H = 1.41

| | | 11* 12* | 13 14 15 | 16 17 18 | 19 20 |
|---|---|---|---|---|---|
| [a] | Units | | | | |
| [b] | Furnace temperature, top (average for group of 3 units) | 2150 F | 2240 | 2230 | 2240 |
| [c] | Furnace temperature, bottom avg[a] | b | b | b | b |
| [d] | Load surface temperature | 1540 F | 1820 | 2060 | 2160 |
| [e] | Furnace black body radiation, from table 8.9, at top temp[a] [b] | 52.3 | 44.9 | 20.7 | 10.4 |
| [f] | Furnace black body radiation, from table 8.9, at bottom temp [c] | b | b | b | b |
| [g] | Avg[a] fce[a] top & bot[a] radn[a], [e+f] / 2 If more zones, add $g_2$, $g_3$, etc.[a] | | | | |
| [h] | Load black body radiation, from table 8.9, at temperature [d] | | | | |
| [i] | Net radn between fce at [b] temp and load at [d] temp = [g] − [h] | | | | |
| [j] | Btu/# rise = [i] (3H, or 2H for last group, of 2) = 52.3 (3) (0.47) = | 73.7 | 63.3 | 29.2 | 9.8 |
| [k] | Cumulative Btu/# = [previous k] + [new j] = | 226 | 290 | 319 | 329 |
| [l] | Average load temperature, from figure 8.9 | 1430 F | 1900 | 2100 | 2160 |
| [m] | Lag factor $F_2$, from figure 8.4 at temperature [l] | 2.19 | 0.84 | 0.88 | 1.07 |
| [n] | Time-lag, minutes avg to top = $(6.18^d)(0.6^e)/(0.743^d)$[m] = 5 [2.19] = | 11 | 4.2 | 4.4 | 5.4 |
| [o] | Time-lag, minutes, avg to bottom[c] = = $0.62^e$ [j] = 0.62 [11] = | 6.8 | 2.6 | 2.7 | 3.3 |

*Note: Units 11 and 12 on this page are part of the same group of 3 as unit 10, the last column of table 8.7; so the first column of calculated figures on this page duplicates those in the last column of table 8.7.
[a]See glossary for abbreviations.
[b]Not applicable.
[c]to bottom if 1-side heating; to center if 2-side heating.
[d]Table 8.2.
[e]From experimental evidence.
avg = average. betw = between. bot = bottom. col = column.
etc = et cetera = and so forth. fce = furnace. radn = radiation. temp = temperature.

(700 C to 760 C) crystalline change for carbon steels (fig. 8.9). If curves are not about as smooth as those of figs. 8.1 and 8.11, try a new iteration, with different estimates for furnace and load surface temperatures.

You are on your way. It is a long job, but rewarding. You will not only get answers to many questions but information needed to conduct a realistic heat balance AND a

**TABLE 8.9 Black body radiation heat flux rates, in thousands of Btu/hr ft² from equation 2.6. Example: For 150 F, read 0.253 = 253 Btu/hr ft².**

| Temperature, F | 0 | 10 | 20 | 30 | 40 | 50 | 60 | 70 | 80 | 90 |
|---|---|---|---|---|---|---|---|---|---|---|
| 0 | 0.076 | 0.083 | 0.091 | 0.098 | 0.107 | 0.116 | 0.125 | 0.135 | 0.145 | 0.157 |
| 100 | 0.168 | 0.181 | 0.194 | 0.207 | 0.222 | 0.237 | 0.253 | 0.270 | 0.287 | 0.306 |
| 200 | 0.325 | 0.345 | 0.367 | 0.389 | 0.412 | 0.436 | 0.461 | 0.487 | 0.514 | 0.543 |
| 300 | 0.572 | 0.603 | 0.635 | 0.668 | 0.703 | 0.739 | 0.776 | 0.815 | 0.855 | 0.896 |
| 400 | 0.939 | 0.983 | 1.030 | 1.078 | 1.127 | 1.178 | 1.231 | 1.285 | 1.341 | 1.395 |
| 500 | 1.458 | 1.520 | 1.585 | 1.651 | 1.718 | 1.787 | 1.859 | 1.933 | 2.009 | 2.088 |
| 600 | 2.169 | 2.525 | 2.338 | 2.425 | 2.515 | 2.608 | 2.703 | 2.801 | 2.902 | 3.005 |
| 700 | 3.111 | 3.220 | 3.332 | 3.446 | 3.563 | 3.684 | 3.807 | 3.933 | 4.062 | 4.194 |
| 800 | 4.330 | 4.469 | 4.612 | 4.758 | 4.908 | 5.061 | 5.217 | 5.377 | 5.540 | 5.707 |
| 900 | 5.878 | 6.053 | 6.232 | 6.415 | 6.602 | 6.792 | 6.987 | 7.186 | 7.390 | 7.597 |
| 1000 | 7.808 | 8.024 | 8.245 | 8.470 | 8.700 | 8.934 | 9.173 | 9.417 | 9.665 | 9.919 |
| 1100 | 10.18 | 10.44 | 10.71 | 10.99 | 11.27 | 11.55 | 11.84 | 12.13 | 12.43 | 12.74 |
| 1200 | 13.05 | 13.37 | 13.69 | 14.02 | 14.35 | 14.69 | 15.04 | 15.40 | 15.76 | 16.12 |
| 1300 | 16.49 | 16.87 | 17.25 | 17.64 | 18.04 | 18.45 | 18.86 | 19.28 | 19.70 | 20.13 |
| 1400 | 20.57 | 21.02 | 21.47 | 21.93 | 22.40 | 22.87 | 23.35 | 23.84 | 24.34 | 24.85 |
| 1500 | 25.37 | 25.89 | 26.42 | 26.95 | 27.50 | 28.06 | 28.62 | 29.19 | 29.77 | 30.35 |
| 1600 | 30.96 | 31.56 | 32.18 | 32.80 | 33.43 | 34.07 | 34.72 | 35.38 | 36.05 | 36.73 |
| 1700 | 37.42 | 38.11 | 38.82 | 39.54 | 40.26 | 41.00 | 41.75 | 42.51 | 43.28 | 44.06 |
| 1800 | 44.85 | 45.65 | 46.46 | 47.28 | 48.11 | 48.95 | 49.80 | 50.67 | 51.55 | 52.43 |
| 1900 | 53.33 | 54.23 | 55.16 | 56.09 | 57.04 | 57.99 | 58.96 | 59.95 | 60.94 | 61.94 |
| 2000 | 62.96 | 63.99 | 65.03 | 66.08 | 67.14 | 68.23 | 69.33 | 70.44 | 71 56 | 72.69 |
| 2100 | 73.84 | 75.00 | 76.17 | 77.36 | 78.56 | 79.78 | 81.01 | 82.25 | 83.50 | 84.78 |
| 2200 | 86.07 | 87.37 | 88.69 | 90.02 | 91.37 | 92.73 | 94.10 | 95.49 | 96.90 | 98.32 |
| 2300 | 99.8 | 101.3 | 102.7 | 104.1 | 105.6 | 107.2 | 108.7 | 110.3 | 111.8 | 113.4 |
| 2400 | 115.1 | 116.7 | 118.3 | 120.0 | 121.6 | 123.3 | 125.0 | 126.7 | 128.5 | 130.2 |
| 2500 | 132.0 | 133.7 | 135.5 | 137.4 | 139.2 | 141.1 | 143.0 | 144.9 | 146.9 | 148.8 |
| 2600 | 150.8 | 152.7 | 154.7 | 156.7 | 158.8 | 160 8 | 162.9 | 165.0 | 167.1 | 169.2 |
| 2700 | 171.4 | 173.6 | 175.8 | 178.0 | 180.3 | 182.6 | 184.9 | 187.2 | 189.5 | 191.8 |

better "feel" for what your furnace can and cannot do. Do not just think about the end results, but as you calculate your way through your furnace, think about what factors make the curves rise more or less rapidly, and what you could do (operation-wise, design-wise) to make your process more productive, quality effective, and efficient.

**Batch furnace heating curves** can be calculated in a manner very similar to that for continuous furnaces. Note that the horizontal scale or abscissa is labeled distance *or time*. The resulting curves may show some differences. For example, the length of the 'cutback time,' which depends on (*a*) the length of the gas flow path from when it first begins to give up its heat until it exits via the flue and (*b*) the lag time of the products being heated (see the definition of 'cutback period' in the glossary).

**Example:** A 25' long × 10' wide soaking pit heating 36" × 36" × 90" high ingots (33 000 pounds each) can be heated from cold to ready to roll in 10 hr, with a cutback time of 2.2 hr with burners and controls for spin control. Without spin-control burners and with only one control T-sensor, the job took 12 hr and had a cutback time of 4 to 7 hr—the main reason for the long cutback time of 4.3 hr before versus 2.1 hr after the modernization. Furthermore, at the beginning of the cutback time, the prior case had a bottom temperature difference from the wall opposite the burner to the burner wall of more than 180°F, versus near zero with modern spin control. The previous way still had this differential when the ingots were drawn. If using cold air, the top-to-bottom difference on ingots was 20°F (11°C) with no spin control, but 40°F (22vv°C) with spin control. If oxy-fuel firing were used, this bottom temperature difference from end to end would be as great as 180°F to 400°F (100°C to 222°C), even with a long cutback.

With the usual U-shaped gas flow pattern, the cutback period can be shortened by high/low or on/off firing. To illustrate this, assume high and low firing rates of 20 kk Btu/hr and 6 kk Btu/hr, respectively, a turndown ratio of 3.33:1. Therefore the ratio of sensible heat flow rates to the furnace gas is 3.33 to 1. This means that the gas temperature passing the last ingot will be much hotter than when at low fire. This last ingot before the flue is the most difficult to bring to rolling temperature, and it determines the pit'sproductivity and total fuel use.

### 8.1.4. Heat Balance—To Find Needed Fuel Inputs

Whether you are designing a new furnace or evaluating an existing furnace, after completing the Shannon Method for calculating steel temperature-versus-time curves (sec. 8.1), the next logical step is determining fuel inputs required for each of the furnace zones. The gross heat input required is given by equation 2.1, repeated here as equation 8.1:

$$\text{Energy input} = \frac{\text{`heat needs' for load and furnace}}{\text{available heat, as a decimal)}} \qquad (8.1, 2.1)$$

The 'heat needs' for a *continuous furnace* ***after heat-up*** are: heat to the loads; heat losses to the walls, hearth, and roof; and heat losses to cooling water and openings. (See all in a Sankey diagram, sec. 5.11.)

Ways to minimize losses are discussed in chapter 5. The following text and worksheet (table 8.1) explain the methods for evaluating heat to the load and heat losses for the furnace of sample problem 8.1. Furnace dimensions and other furnace data are not presented at the beginning of this sample problem 8.1, but rather looked up or presented at the point of need during the progress of the following solution.

***8.1.4.1. Refractory Heat Loss*** *Sample Problem 8.1—Required Fuel Inputs.* An added aspect of sample problem 8.1 (the same continuous walking beam steel reheat furnace): calculate the required gross heat input to each zone. (See worksheet tables 8.14 to 8.17.).

*Heat balance worksheet guide.* {Numbers in this type parentheses refer to line numbers of tables 8.14 to 8.17}.

{1} Relates to batch furnaces; leave blank for this continuous furnace.

{3 through 7} Determine heat absorbed by the load in each zone.

{8 and 12} Determine wall, roof, hearth, and closed-door heat losses by using equivalent inches of firebrick thickness—in tables 8.14 and 8.15. For an existing furnace, check the furnace drawings and specifications for refractories used and their thicknesses. In general practice, equivalent firebrick thicknesses are: 40 in. fb to 50 in. fb for roofs, 65"fb for sidewalls, and 40"fb for hearths. (See fig. 8.12 and pp. 100–114 of *reference 51.*)

{10, 11} Apply only to batch furnaces. (Consult pp. 103–106 of *reference 51,* or the refractory manufacturer for heat storage data.)

{12, 13} For losses through slots and open doors, see figs. 5.7 and 5.8, and *reference 51,* pp. 114–117, Vol. I, *Combustion Handbook,* (*reference 51*).

{14} There is little heat loss from rolls conveyors that stay within the hot furnace chamber all of the time. For conveyors that move in and out of the furnace, calculate loss/hr = (weight/hr) (specific heat) ($T_{max.} - T_{min.}$).

{15} Summation of {8} through {14}.

{16} See fig. 5.3 or eq. 5.1, and adjacent discussion.

{17} Actual measured combustion air temperature entering the burner.

{18} From fig. 5.1 or 5.2 for natural gas, or ask North American Mfg. Co. or fuel supplier for a "Stoic" printout on your specific fuel with hot air.

{19} Work from right to left, starting with the available heat of the last column (unfired zone): $Co_u = Ah_u - Ah_{p,} Co_p = Ah_p - Ah_{h,} Co_h = Ah_h - Ah_{s,} Co_s =$ zero, where $Co$ = carryover, $Ah$ = available heat from {18}, subscript$_u$ = unfired zone, $_p$ = preheat zone, $_h$= heat zone, $_s$ = soak zone.

{20} Sum of heat to loads and losses = {7} + {15}

{21} Heat from one zone flowing to and being absorbed by the loads in the next zone.

{22} Sum of loads and losses, minus carryover = {20} – {21}.

{23} Gross heat input required = (heat needed)/(% available heat/100) = {22}/{18} = fuel rate in each zone.

{24} Summation of gross inputs for soak, heat, and preheat zones.

{25} Safety factor—See glossary and the discussion at the end of this chapter.

{26} Zone design gross input = {23}{25} = amount of burner input capacity to be supplied to each zone.

**8.1.4.2. Heat Losses to Cooling Water** For water-cooled doors and doorframes, include those losses with the heat balance tabulation for door looses. The engineer doing a heat balance must take responsibility for double-checking that no heat losses have been overlooked. Water-cooled surfaces absorb furnace heat at such an intense rate that they cannot be overlooked.

Cooling-water heat losses must be tallied, especially from bottom-fired zones, that is: (*a*) skidrails & pipes—insulated + uninsulated, (*b*) crossovers & pipes—insulated + uninsulated, (*c*) riser pipes—insulated + uninsulated.

See figure 8.13 for cooling-water heat losses for the previous components of a typical skid pipe system—all in Btu/sq ft of *bare* pipe surface, even for cases where the bare pipe is covered with insulation.

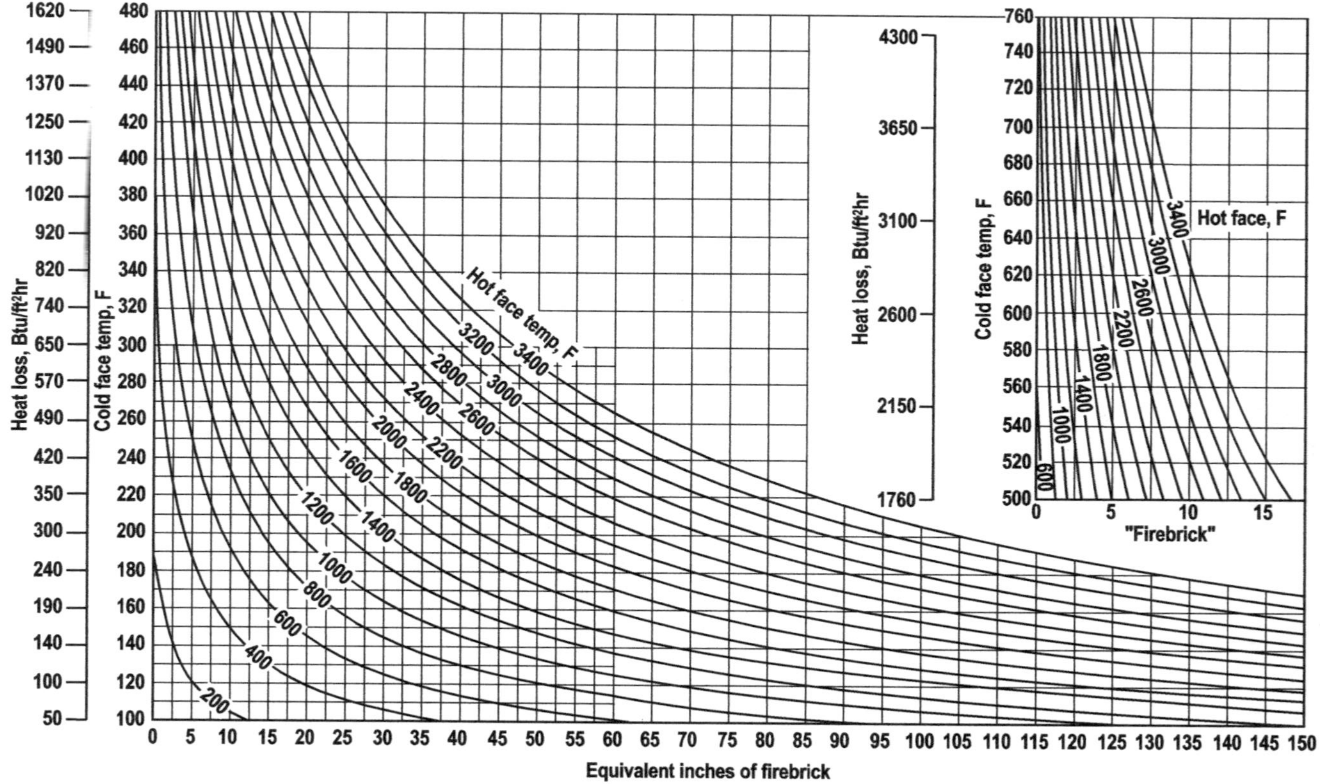


**Fig. 8.12.** Heat losses for various equivalent firebrick thicknesses of vertical walls, with no wind and 70 F ambient air. Losses will be slightly higher from roofs; slightly lower for hearths and bottoms. To interpolate, use an engineer's scale at 20 graduations per inch on the vertical scales; at 50 graduations per inch on horizontal scales. (Courtesy of *reference 51*.)

**TABLE 8.10 Heat balance. Main worksheet**

Client __________. Date ________. Iteration ____. By __________
Furnace: __________ Zones = _ top, _ bottom. Piece weight ________
Load size: ______________ Load material: ______________
*Fce IDs*: ____ w × ____ l × ____ h. *Rate*: ____lb/hr, ____tph, ____kg/h

| ZONE → | Soak | Heat | Preheat | Unfired |
|---|---|---|---|---|
| {1} Time interval, units on fig. 8.11 | | | | |
| {2} Avg zone temp, from fig. 8.11 | | | | |
| {3} Load temp, $T_{out}/T_{in}$ | | | | |
| {4} Btu/lb, $\{h_o\}$, $\{h_i\}$, from fig. 8.9 | | | | |
| {5} Btu gain/pound = $\{h_o\} - [h_i\}$ | | | | |
| {6} Pounds heated per hour | | | | |
| {7} **Heat to loads**, kk Btu/hr | | | | |
| {8} Wall + roof + bottom refr heat loss, kk Btu/hr | | | | |
| {9} Water heat loss, kk Btu/hr | | | | |
| {10} Wall + roof + bot heat storage[s], kk Btu/hr | | | | |
| {11} Pier, car, kiln furniture storage[s], kk Btu/hr | | | | |
| {12} Door loss, kk Btu/hr | | | | |
| {13} Slot loss, kk Btu/hr | | | | |
| {14} Roll or conveyor loss, Btu/hr | | | | |
| {15} **Total losses and storages**[s] = Σ{7 through 14} | | | | |
| {16} Zone exit gas temp, F | | | | |
| {17} Air preheated to, F | | | | |
| {18} %available heat/100 (figs. 5.1 or 5.2) | | | | |
| {19} AvHt carryover, from previous zone | | | | |
| {20} **Total loads, losses, storages**[s] = {7 + 15} | | | | |
| {21} Carryover from adjacent zone = {19}{24} | | | | |
| {22} Heat needed = {20} − {21} | | | | |
| {23} Gross heat input required = {22}/{18} | | | | |
| {24} Cumulative of {23} | | | | |
| {25} Safety factor (see last page this chapter) | | | | |
| {26} Zone design gross input = {23}{25} | | | | |

**Total input for all zones** = $\{23_{soak} + 23_{heat} + 23_{preheat}\}$ =
= ________ Btu/hr or /tph = ________ kk Btu/ton

Permission is granted to owners of this book to make copies of this blank worksheet, table 8.10. (See also table 8.14 and 8.15.)
[s]storage or tare applies only to batch (non-continuous) furnaces.
avg = average. betw = between. bot = bottom. col = column. eqn = equation, formula. etc = (et cetera), and so forth. fce = furnace. kk =millions. radn = radiation. refr = refractory. temp = temperature. Σ = total sum. (See glossary.)

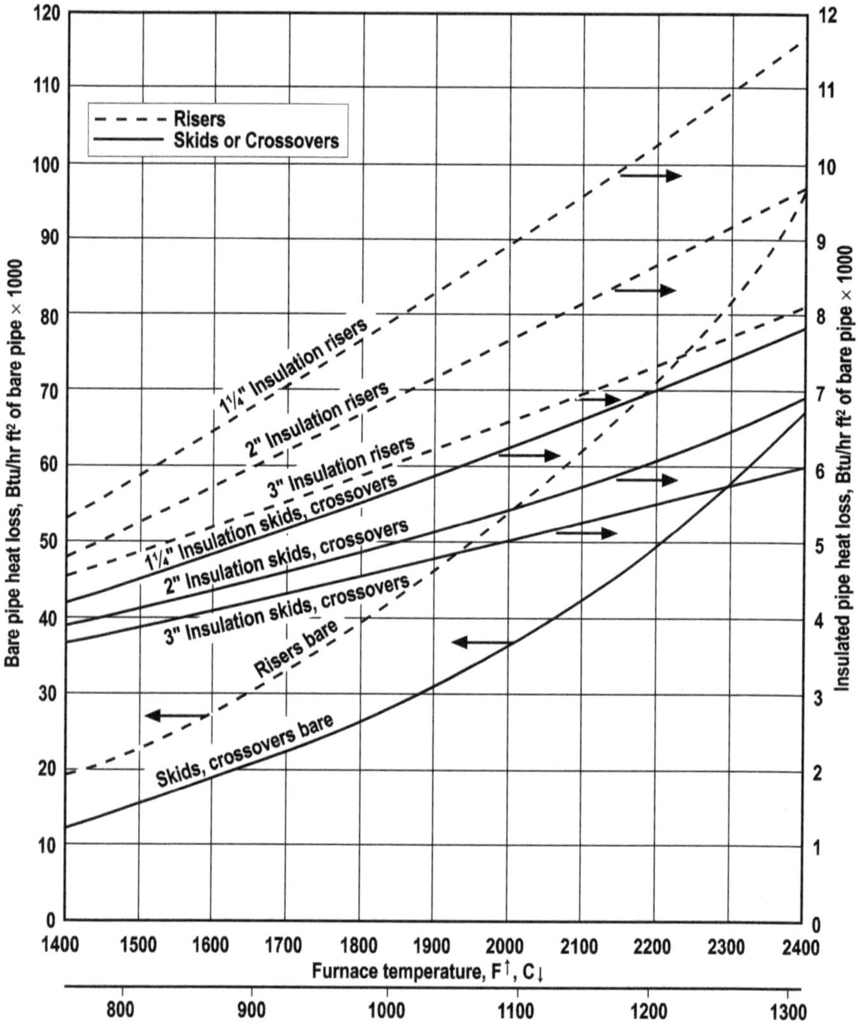


**Fig. 8.13.** Cooling-water heat losses to skid pipe systems. All but the 3" insulation curves are courtesy of Bloom Engineering Co., Inc.

Water heat loss per zone = (8.2)
(total bare pipe surface ft$^2$/zone) (loss, Btu/ft$^2$ of bare pipe, by fig. 8.13)
where (total bare pipe surface ft$^2$/zone)
= 3,142(bare pipe length, ft) (bare pipe OD"/12).

Heat losses to water for water-cooled doors and doorframes should be included with the tabulation for door losses. The engineer doing a heat balance must take responsibility for double-checking that no heat losses have been overlooked. Water-cooled surfaces absorb furnace heat at such an intense rate that they cannot be overlooked.

**TABLE 8.11 Heat balance. Refractory loss worksheet**

Client ________. Date ________. Iteration ______. By ________
Furnace: ________ Zones = _ top, _ bottom. Piece weight ________
Load size: ________ Load material: ________
*Fce IDs*: ______ w × ______ l × ______ h. *Rate*: ______lb/hr. ______tph. ______kg/h

Equivalent "firebrick ("fb) is from table 4.18b and 4.18c or fig. 4.15d of *Reference T48.*
Total "fb is the sum of "fb for all the layers in a wall, roof, or hearth.
Heat loss, Btu/ft$^2$hr, is from fig. 8.12 for the total "fb at zone hotface temp.

TOP SOAK ZONE. Hot face temp = ______F. ID length____ft, width____ft, height____ft

**Roof** layer 1. Thickness = ______in. Refractory = ________in. fb = ______.
" layer 2. Thickness = ______in. Refractory = ________in. fb = ______.
" layer 3. Thickness = ______in. Refractory = ________in. fb = ______.
Total 3 layers "fb = ____. Roof heat loss thru 3 layers, from fig 8.12 = ______Btu/ft$^2$hr.
Top soak zone roof loss = (roof loss) (area, ft$^2$) = (Btu/ft$^2$hr) (w) (l) = ______Btu/hr.

**Wall** layer 1. Thickness = ______in. Refractory = ________in. fb = ______.
" layer 2. Thickness = ______in. Refractory = ________in. fb = ______.
" layer 3. Thickness = ______in. Refractory = ________in. fb = ______.
Total 3 layers in. fb = ____. Wall heat loss thru 3 layers, from fig 8.12 = ______Btu/ft$^2$hr.
Top soak zone wall loss = (wall loss) (wall ft$^2$) = (Btu/ft$^2$hr) (2w + 2l) (h) = ______Btu/hr.

**Bottom** layer 1. Thickness = ______in. Refractory = ________in. fb = ______.
" layer 2. Thickness = ______in. Refractory = ________in. fb = ______.
" layer 3. Thickness = ______in. Refractory = ________in. fb = ______.
Total 3 layers in. fb = ____. Bottom heat loss thru 3 layers, from fig 8.12 = ______Btu/ft$^2$hr.
Top soak zone bot loss = (bot loss) (bot ft$^2$) = (Btu/ft$^2$hr) (w) (l) = ______Btu/hr.
TOTAL TOP SOAK ZONE LOSS = roof + walls + bot = + + = ________.

TOP HEAT ZONE: Hot face temp = ______F. IDs: l = ____ft, w = ____ft, h = ____ft

**Roof** layer 1. Thickness = ______in. Refractory = ________in. fb = ______.
" layer 2. Thickness = ______in. Refractory = ________in. fb = ______.
" layer 3. Thickness = ______in. Refractory = ________in. fb = ______.
Total 3 layers in. fb = ____. Roof heat loss thru 3 layers, from fig 8.12 = ______Btu/ft$^2$hr.
Top heat zone roof loss = (roof loss Btu/ft$^2$hr) (roof ft$^2$) = (____) (____) = ______Btu/hr.

**Wall** layer 1. Thickness = ______in. Refractory = ________in. fb = ______.
" layer 2. Thickness = ______in. Refractory = ________in. fb = ______.
" layer 3. Thickness = ______in. Refractory = ________in. fb = ______.
Total 3 layers in. fb = ____. Wall heat loss thru 3 layers, from fig 8.12 = ______Btu/ft$^2$hr.
Top heat zone wall loss = (wall loss Btu/ft$^2$hr) (wall ft$^2$) = (____) (____) = ______Btu/hr.

**Bottom** layer 1. Thickness = ______in. Refractory = ________in. fb = ______.
" layer 2. Thickness = ______in. Refractory = ________in. fb = ______.
" layer 3. Thickness = ______in. Refractory = ________in. fb = ______.
Total 3 layers in. fb = ____. Bottom heat loss thru 3 layers, from fig 8.12 = ______Btu/ft$^2$hr.
Top soak zone bot loss = (bot loss Btu/ft$^2$hr) (bot ft$^2$) = (____) (____) = ______Btu/hr.
TOTAL top heat zone loss = roof + walls + bot = + + = ________Btu/hr.

Permission is granted to owners of this book to make copies of this blank worksheet (see also tables 8.14 and 8.15).

**TABLE 8.12 Heat balance. Refractory loss worksheet**

Client ____________ Date __________. Iteration _____. By ____________
Furnace: ____________ Zones = _ top, _ bottom. Piece weight ____________
Load size: ____________ Load material: ____________
*Fce IDs*: _____ w × _____ l × _____ h. *Rate*: ______lb/hr. ______tph. ______kg/h

TOP PREHEAT ZONE: Hot face temp = ____F. ID length ___ft, width ___ft, height ___ft

**Roof** layer 1. Thickness = _____in. Refractory = ____________in. fb = _____.
" layer 2. Thickness = _____in. Refractory = ____________in. fb = _____.
" layer 3. Thickness = _____in. Refractory = ____________in. fb = _____.
Total 3 layers in. fb = _____. Roof heat loss thru 3 layers, from fig 8.12 = _____$Btu/ft^2hr$.
Top preheat zone roof loss = ($Btu/ft^2hr$) (roof $ft^2$) = (_____) (_____×_____) = _____Btu/hr.

**Wall** layer 1. Thickness = _____in. Refractory = ____________in. fb = _____.
" layer 2. Thickness = _____in. Refractory = ____________in. fb = _____.
" layer 3. Thickness = _____in. Refractory = ____________in. fb = _____.
Total 3 layers in. fb = _____. Wall heat loss thru 3 layers, from fig 8.12 = _____$Btu/ft^2hr$.
Top preheat zone wall loss = ($Btu/ft^2hr$) (roof $ft^2$) = (_____) (_____×_____) = _____Btu/hr.

**Bottom** layer 1. Thickness = _____in. Refractory = ____________in. fb = _____.
" layer 2. Thickness = _____in. Refractory = ____________in. fb = _____.
" layer 3. Thickness = _____in. Refractory = ____________in. fb = _____.
Total 3 layers in. fb = _____. Bottom heat loss thru 3 layers, from fig 8.12 = _____$Btu/ft^2hr$.
Top preheat zone bot loss = ($Btu/ft^2hr$) (roof $ft^2$) = (_____) (_____×_____) = _____Btu/hr.
TOTAL preheat zone loss = roof + walls + bot = + + = _____Btu/hr.

TOP UNFIRED ZONE: Hot face temp = ____F. IDs: l = ____ft, w = ____ft, h = ____ft

**Roof** layer 1. Thickness = _____in. Refractory = ____________in. fb = _____.
" layer 2. Thickness = _____in. Refractory = ____________in. fb = _____.
" layer 3. Thickness = _____in. Refractory = ____________in. fb = _____.
Total 3 layers in. fb = _____. Roof heat loss thru 3 layers, from fig 8.12 = _____$Btu/ft^2hr$.
Top unfired zone roof loss = ($Btu/ft^2hr$) (roof $ft^2$) = (_____) (_____×_____) = _____Btu/hr.

**Wall** layer 1. Thickness = _____in. Refractory = ____________in. fb = _____.
" layer 2. Thickness = _____in. Refractory = ____________in. fb = _____.
" layer 3. Thickness = _____in. Refractory = ____________in. fb = _____.
Total 3 layers in. fb = _____. Wall heat loss thru 3 layers, from fig 8.12 = _____$Btu/ft^2hr$.
Top unfired zone wall loss = ($Btu/ft^2hr$) (roof $ft^2$) = (_____) (_____×_____) = _____Btu/hr.

**Bottom** layer 1. Thickness = _____in. Refractory = ____________in. fb = _____.
" layer 2. Thickness = _____in. Refractory = ____________in. fb = _____.
" layer 3. Thickness = _____in. Refractory = ____________in. fb = _____.
Total 3 layers in. fb = _____. Bottom heat loss thru 3 layers, from fig 8.12 = _____$Btu/ft^2hr$.
Top unfired zone bot loss = ($Btu/ft^2hr$) (roof $ft^2$) = (_____) (_____×_____) = _____Btu/hr.
TOTAL unfired zone loss = roof + walls + bot = + + = _____Btu/hr.

Permission is granted to owners of this book to make copies of this blank worksheet (see also tables 8.14 and 8.15).
Repeat preceding segments, relabeled for other zones.
TOTAL REFRACTORY LOSSES = Summation of all above zone heat losses = + + + + + + + = _____Btu/hr.

**TABLE 8.13 Heat balance. Water loss worksheet**

Client ______________. Date __________. Calculation ______. By ______________
Furnace: ________________ Zones = _ top, _ bottom. Piece weight ____________
Load size: ___________________________ Load material: ________________________
*Fce IDs*: _____ w × _____ l × _____ h. *Rate*: ________ lb/hr. ________ tph. ________ kg/h

**Bottom __________ zone**[w]. Average temperature ________ F.

*Skids:*
1. Length bare ______. 2. Bare OD ______. 3. loss, Btu/hr (by fig. 8.13) __________.
4. Length insulated? ______. 5. loss, Btu/hr ("  " 8.13) __________.

*Crossovers:*
1. Length bare ______. 2. Bare OD ______. 3. loss, Btu/hr (by fig. 8.13) __________.
4. Length insulated? ______. 5. loss, Btu/hr ("  " 8.13) __________.

*Risers:*
1. Length bare ______. 2. Bare OD ______. 3. loss, Btu/hr (by fig. 8.13) __________.
4. Length insulated? ______. 5. loss, Btu/hr ("  " 8.13) __________.

[w]Repeat preceding segments, relabeled for other zones; then add together all cooling water losses and enter the sum in line {9} of table 8.10. Permission is granted to owners of this book to copy this blank worksheet, table 8.12.

See figure 8.13 for cooling-water heat losses for the previous components of a typical skid pipe system—all in Btu/square foot of *bare* pipe surface, even for cases where the bare pipe is covered with insulation.

It is necessary to perform this cooling-water heat-loss procedure for as many times as it takes to cover all water-cooled surfaces within the furnace.

***8.1.4.3. Heat Losses Through Open Doors, Slots, Other Openings.*** Figures 5.7 and 5.8 plus pages 114 to 117 of Volume I of the Combustion Handbook (*Reference 51*) provide good methods for evaluating these losses. In addition to the radiation heat loss out through slots, designers and maintenance personnel have another reason for keeping the slots small: *tramp air inleakage,* which must be considered in deciding how much excess air to use when entering the available heat chart for line {18} in tables 8.10 and 8.14. The following calculation applies a simplified method for evaluating slot radiation losses—applied to the slots between hearths of the walking-hearth furnace of sample problem 8.1. The slot lengths are the zone lengths, plus or minus a few feet at charge and discharge.

*Simplified slot radiation loss calculation.* A zone's total slot heat loss = (total slot area) multiplied by (black body radiation from the zone's refractory temperature inside to ambient temperature outside).

The *total refractory losses* = the sum of all preceding zone heat losses = 0.739 soak + 0.589 heat + 0.657 preheat + 0.328 unfired = 2.313 kk Btu/hr. Enter above zone totals in respective columns on line{8} of table 8.16.

| Zone | Temp | Radiation[r] | × | # of slots | × | Width | × | Length | = | Loss in kk Btu/hr[t] |
|---|---|---|---|---|---|---|---|---|---|---|
| Soak | 2240 F | 0.09137 | × | 6 | × | 1"/12 | × | 21' | = | 0.960 |
| Heat | 2200 F | 0.08607 | × | 6 | × | 1"/12 | × | 20' | = | 0.861 |
| Preheat | 2060 F | 0.06933 | × | 6 | × | 1"/12 | × | 25' | = | 0.867 |
| Unfired | 1430 F | 0.02193 | × | 6 | × | 1"/12 | × | 15' | = | 0.164 |

[r]Black-body radiation, in kk Btu/ft$^2$hr, from table 8.9. It is rationalized that no emissivity, no absorptivity, or any shape factor need be used here because narrow slots have immense radiating source and receiving areas relative to their slot area (like a pinhole camera).
[t]Record figures from this column on line {13} of table 8.14.

*Conclusions.* Lines {23} and {24} of table 8.16 are the sought-after end results of all the preceding heat balance work. These figures can be used to check whether an existing furnace has enough input to serve the jobs it is now expected to do. Alternatively, this information can be used to select gross Btu/hr burner inputs to each zone of a new furnace, or for modernization of an existing furnace.

The reader will discover many differing opinions on the size safety factors to use between the previous conclusions and the actual burner inputs to be applied to a furnace. The authors of this book feel that most current designers should use larger safety factors for the following reasons:

1. Too many engineers use furnace temperature as flue gas exit temperature when looking up %available heat. (See fig. 5.3.)
2. Too many furnace designers figure on only 5% excess air (1% excess oxygen), but most furnace zones end up operating with 15% to 20% excess air, which limits their capacity. The reason for this discrepancy is unknown, but it is necessary to face reality.
3. Too many companies use a safety factor of 1.15 or less. Coauthor Shannon uses 1.2, or preferably 1.4, mainly to hasten recovery after mill delays when newly charged cold loads need more than design input.
4. Furnace buyers may not be familiar with furnace technology, and they may be obligated to buy the least-expensive bid. For example, the energy need following a delay is much higher than this equilibrium design.
5. Specifications do not stipulate all parameters that should be followed.
6. Failure to allow for future business growth and changing product specifications.

An underfueled furnace is the most costly furnace in the long run; An under-air-capacitied combustion system, a close second. All the aforementioned problems and many sad cases of furnace inadequacy can be avoided by furnace users having a better understanding of their own needs. To make a product at the lowest possible cost, you need a thorough understanding of the relationships between fuel economy, product quality, and productivity.

**TABLE 8.14 Heat balance. Refractory loss worksheet[1] for sample problem 8.1**

Client (sample). Date 07 03 02. Iteration 2*. By RAS

Furnace: walking hearth Zones = 4 top, 0 bottom. Piece weight 2,068 pounds

Load size: 4.5" × 4.5" × 30 ft Load material: 0.4%C steel.

*Fce IDs*: 34' w × 80' l × 6 ft h. *Rate*: 200 000 lb/hr. 100 tph. 90 700 kg/h

Equivalent firebrick, "fb, is from table 4.18b and 4.18c or fig. 4.15d of *reference 51.*

Total "fb is the sum of "fb for all the layers in a wall, roof, or hearth.

Heat loss, Btu/$ft^2$hr, is from fig. 8.12 for the Total "fb at zone hotface temperature.

TOP SOAK ZONE. Hot face temp = 2240 F. ID length 22 ft, width 34 ft, height 6 ft.
**Roof** layer 1. Thickness = 5 in. Refractory = Kast-Set@2.4/" in. fb = 12.
" layer 2. Thickness = 4 in. Refractory = APG 22LGR@1.95 in. fb = 8.
" layer 3. Thickness = 2 in. Refractory = B-W 1900 block in. fb = 25.
Total 3 layers in. fb = 45". Roof heat loss thru 3 layers, from fig 8.12 = 400 Btu/$ft^2$hr.
Top soak zone roof loss = (roof loss) ( $ft^2$) = (400) (34) (22) = 0.299 kk Btu/hr.

**Wall** layer 1. Thickness = * in. Refractory = * in. fb = *.
" layer 2. Thickness = ____ in. Refractory = ____ in. fb = ____.
" layer 3. Thickness = ____ in. Refractory = ____ in. fb = ____.
Total 3 layers in. fb = 65*. Wall heat loss thru 3 layers, from fig 8.12 = 270 Btu/$ft^2$hr.
Top soak zone wall loss = (wall loss) ($ft^2$) = (270) (468[d]) = 0.126 kk Btu/hr.

**Bottom** layer 1. Thickness = * in. Refractory = * in. fb = *.
" layer 2. Thickness = ____ in. Refractory = ____ in. fb = ____.
" layer 3. Thickness = ____ in. Refractory = ____ in. fb = ____.
Total 3 layers in. fb = 40*. Bottom heat loss thru 3 layers, from fig 8.12 = 420 Btu/$ft^2$hr.
Top soak zone bot loss = (bot loss) ($ft^2$) = (420) (34) (22) = 0.314 kk Btu/hr.
TOTAL top soak ZONE LOSS = roof + walls + bot = 0.299 + 0.126 + 0.314 = 0.739 kk Btu/hr.

TOP HEAT ZONE: Hot face temp = 2240 F. IDs: l = 20 ft, w = 34 ft, h = 6 ft
**Roof** layer 1. Thickness = ____ in. Refractory = ____ in. fb = ____.
" layer 2. Thickness = ____ in. Refractory = ____ in. fb = ____.
" layer 3. Thickness = ____ in. Refractory = ____ in. fb = ____.
Total 3 layers in. fb = 45. Roof heat loss thru 3 layers, from fig 8.12 = 360 Btu/$ft^2$hr.
Top heat zone roof loss = (roof loss Btu/$ft^2$hr) (roof $ft^2$) = (360) (680) = 0.245 kk Btu/hr.

**Wall** layer 1. Thickness = ____ in. Refractory = ____ in. fb = ____.
" layer 2. Thickness = ____ in. Refractory = ____ in. fb = ____.
" layer 3. Thickness = ____ in. Refractory = ____ in. fb = ____.
Total 3 layers in. fb = 65. Wall heat loss thru 3 layers, from fig 8.12 = 270 Btu/$ft^2$hr.
Top heat zone wall loss = (wall loss Btu/$ft^2$hr) (wall $ft^2$) = (270) (240[+]) = 0.065 kk Btu/hr.

**Bottom** layer 1. Thickness = ____ in. Refractory = ____ in. fb = ____.
" layer 2. Thickness = ____ in. Refractory = ____ in. fb = ____.
" layer 3. Thickness = ____ in. Refractory = ____ in. fb = ____.
Total 3 layers in. fb = 40. Bottom heat loss thru 3 layers, from fig 8.12 = 410 Btu/$ft^2$hr.
Top heat zone bot loss = (bot loss Btu/$ft^2$hr) (bot $ft^2$) = (410) (680) = 0.279 kk Btu/hr.
TOTAL top heat zone loss = roof + walls + bot = 0.245 + 0.065 + 0.279 = 0.589 kk Btu/hr.

*For easier overview, authors skipped repetition of details in this solution, using current practice cited for lines {8–9} of the heat balance worksheet guide, namely 40–50 in. fb for roofs, 65 in. fb for sidewalls, and 40 in. fb for hearths.

[d]Area corrected for discharge wall.

**TABLE 8.15 Heat balance. Refractory loss worksheet$_2$ for sample problem 8.1**

Client (sample). Date 07 03 02. Iteration 2*. By RAS
Furnace: walking hearth Zones = 4 top, 0 bottom. Piece weight 2068 pounds
Load size: 4.5 in. × 4.5 in. × 30 ft Load material: 0.4%C steel.
*Fce IDs*: 34 ft w × 80 ft l × 6 ft h. *Rate*: 200 000 lb/hr. 100 tph. 90 700 kg/h

Equivalent firebrick, "fb, is from table 4.18b and 4.18c or fig. 4.15d of *reference 51.*
Total in. fb is the sum of in. fb for all the layers in a wall, roof, or hearth.
Heat loss, Btu/ft$^2$hr, is from fig. 8.12 for the Total in. fb at zone hotface temp.

TOP PREHEAT ZONE. Hot face temp = 2060 F. ID length 25 ft, width 34 ft, height 6 ft.
**Roof** layer 1. Thickness = ____ in. Refractory = ____________ in. fb = ____.
" layer 2. Thickness = ____ in. Refractory = ____________ in. fb = ____.
" layer 3. Thickness = ____ in. Refractory = ____________ in. fb = ____.
Total 3 layers in. fb = 50*. Roof heat loss thru 3 layers, from fig 8.12 = 315 Btu/ft$^2$hr.
Top preheat zone roof loss = (Btu/ft$^2$hr) (roof ft$^2$) = (315) (25 × 34) = 0.268 kk Btu/hr.

**Wall** layer 1. Thickness = ____ in. Refractory = ____________ in. fb = ____.
" layer 2. Thickness = ____ in. Refractory = ____________ in. fb = ____.
" layer 3. Thickness = ____ in. Refractory = ____________ in. fb = ____.
Total 3 layers in. fb = 65*. Wall heat loss thru 3 layers, from fig 8.12 = 275 Btu/ft$^2$hr.
Top preheat zone wall loss = (Btu/ft$^2$hr) (roof ft$^2$) = (275) (300) = 0.083 kk Btu/hr.

**Bottom** layer 1. Thickness = ____ in. Refractory = ____________ in. fb = ____.
" layer 2. Thickness = ____ in. Refractory = ____________ in. fb = ____.
" layer 3. Thickness = ____ in. Refractory = ____________ in. fb = ____.
Total 3 layers in. fb = 40*. Bottom heat loss thru 3 layers, from fig 8.12 = 360 Btu/ft$^2$hr.
Top preheat zone bot loss = (Btu/ft$^2$hr) (roof ft$^2$) = (360) (25 × 34) = 0.306 kk Btu/hr.
TOTAL preheat zone loss = roof + walls + bot = 0.268 + 0.083 + 0.306 = 0.657 kk Btu/hr.

TOP UNFIRED ZONE: Hot face temp = 1430 F. IDs: l = 17 ft, w = 34 ft, h = 6 ft
**Roof** layer 1. Thickness = ____ in. Refractory = ____________ in. fb = ____.
" layer 2. Thickness = ____ in. Refractory = ____________ in. fb = ____.
" layer 3. Thickness = ____ in. Refractory = ____________ in. fb = ____.
Total 3 layers in. fb = 45*. Roof heat loss thru 3 layers, from fig 8.12 = 215 Btu/ft$^2$hr.
Top unfired zone roof loss = (Btu/ft$^2$hr) (roof ft$^2$) = (215) (17 × 34) = 0.124 kk Btu/hr.

**Wall** layer 1. Thickness = ____ in. Refractory = ____________ in. fb = ____.
" layer 2. Thickness = ____ in. Refractory = ____________ in. fb = ____.
" layer 3. Thickness = ____ in. Refractory = ____________ in. fb = ____.
Total 3 layers in. fb = 65*. Wall heat loss thru 3 layers, from fig 8.12 = 160 Btu/ft$^2$hr.
Top unfired zone wall loss = (Btu/ft$^2$hr) (roof ft$^2$) = (160) (408) = 0.065 kk Btu/hr.

**Bottom** layer 1. Thickness = ____ in. Refractory = ____________ in. fb = ____.
" layer 2. Thickness = ____ in. Refractory = ____________ in. fb = ____.
" layer 3. Thickness = ____ in. Refractory = ____________ in. fb = ____.
Total 3 layers in. fb = 40*. Bottom heat loss thru 3 layers, from fig 8.12 = 240 Btu/ft$^2$hr.
Top unfired zone bot loss = (Btu/ft$^2$hr) (roof ft$^2$) = (240) (17 × 34) = 0.139 kk Btu/ft$^2$hr.
TOTAL unfired zone loss = roof + walls + bot = 0.124+0.065+0.139 = 0.328 kk Btu/hr.

**TABLE 8.16 Heat balance. Main worksheet for sample problem 8.1**

Client (sample). Date 07 03 02. Iteration $2^{p}$. By RAS
Furnace: walking hearth Zones = 4 top, 0 bottom. Piece weight 2068 pounds
Load size: 4.5 in. × 4.5 in. × 30 ft Load material: 0.4%C steel.
*Fce IDs*: 34 ft w × 80 ft l × 6 ft h. *Rate*: 200 000 lb/hr. 100 tph. 90 700 kg/h

| | ZONE → | Soak | Heat | Preheat | Unfired |
|---|---|---|---|---|---|
| {1} | Time interval, units on fig. 8.11 | 1–4.8 | 4.8–10 | 11–15 | 16–20 |
| {2} | Avg zone temp, from fig. 8.11 | 2240 F | 2200 F | 2060 F | 1430 F |
| {3} | Load temp, $T_{out}/T_{in}$ | 2200/2060 | 2060/1350 | 1350/490 | 490/60 |
| {4} | Btu/lb, $\{h_o\}$, $\{h_i\}$; via fig. 8.9 | 335, 314 | 314, 209 | 209, 53 | 53, 0 |
| {5} | Btu gain/pound = $\{h_o\} - [h_i\}$ | 21 | 105 | 156 | 53 |
| {6} | Pounds heated per hour | 200 000 | 200 000 | 200 000 | 200 000 |
| {7} | **Heat to loads**[bg] | 4.2 | 21 | 31.2 | 10.6 |
| {8} | Refractory (wall, roof, bottom) heat loss[b,g] | 0.74 | 0.59 | 0.66 | 0.33 |
| {9} | Water loss[b] | 0 | 0 | 0 | 0 |
| {10} | Storage[b] | 0 | 0 | 0 | 0 |
| {11} | Heat to piers[b] | 0 | 0 | 0 | 0 |
| {12} | Door loss[b] | 0.21 | 0 | 0 | 0.03 |
| {13} | Slot loss[b] | 0.96 | 0.86 | 0.87 | 0.16 |
| {14} | Roll or conveyor loss[b] | 0 | 0 | 0 | 0 |
| {15} | **Total losses and tare**[b] = Σ{8–14} | 1.91 | 1.45 | 1.53 | 0.52 |
| {16} | Zone exit gas temp, F, by eq. 5.1 | 2450 | 2350 | 2100 | 1830 |
| {17} | Air preheated to, F | 60 | 60 | 60 | 60 |
| {18} | %available heat/100 (= ah) | 0.28 | 0.31 | 0.38 | 0.45 |
| {19} | AvHt carry over, from next zone[c] | – | 0.03[c] | 0.07[c] | 0.07[cp] |
| {20} | **kk Btu/hr: loads, losses, tare** = {7 + 15} | 6.11 | 22.45 | 32.73 | 11.12[p] |
| {21} | Carryover from next = [19] [24] | – | 0.65 | 6.45 | 11.29[p] |
| {22} | Heat needed = {20} – {21} | 6.11 | 21.80 | 26.28 | [p] |
| {23} | Gross heat input required = {22}/{18} | 21.82 | 70.32 | 69.16 | – |
| {24} | Fuel rate, Cumulative of {23} | 21.82 | 92.14 | 161.30 | – |
| {25} | Safety factor | [z] | 1.4 | 1.3 | – |
| {26} | Zone design gross input = {23}{25} | 44.9[z] | 98.4 | 89.9 | – |

**Total input** for all zones = $\{26_{soak} + 26_{heat} + 26_{preheat}\}$ = 44.9 + 98.4 + 89.9 kk Btu/hr

**Max fce firing rate = 233.2 kk Btu/hr** (or 233.2/100 tph = 2.332 kk Btu/ton)
Furnace fuel rate{24}/100 tph, in kk Btu/ton = 1.613.

[b]Unless otherwise specified, heat units are in kk Btu/hr = millions of Btu/hr.
[c]Carryover %available heat ($cah_{unfired}$) = $ah_{unfired}$-$ah_{preheat}$; $cah_{preheat}$= $ah_{preheat}$-$ah_{heat}$; etc.
[g]From fig. 8.12.
[p]A previous iteration, not shown, found that {19 unfired} was 0.13, which resulted in a carryover to the unfired zone of 22.17 kk Bu/hr, which was much higher than needed. It was concluded that the temperature slope in the preheat zone was insufficient and the slope in the unfired was too steep; thus, the second iteration (above) was performed with steeper preheat zone slope and less steep unfired zone slope, which gave the reasonable {21} = 11.29 above. If {22} is less than 1 kk Btu/hr, that is close enough.
[z]For a soak zone, fall back on a rule of thumb of 60 000 Btu/ft$^2$ because a soak zone will need extra input to start up when filled with cold loads; therefore, (22') (34') (0.06) = 44.9 kk Btu/hr.

## 8.2. MAINTENANCE

Maintenance includes cleaning, lubricating, adjusting, inspecting, repairing, upgrading, *and safety*. Maintenance requires ongoing vigilance, just like safety, product quality, productivity, pollution control, economy—and ultimately, personnel relations, customer relations, community relations.

### 8.2.1. Furnace Maintenance

***8.2.1.1. Skid Systems.*** Inspect skid systems frequrently and make prompt corrections because they can be very vulnerable. The furnace should be taken off line four times per year to bring the skid insulation back to original condition. The water-cooling system for the skids should be flushed out and scale deposits removed by acid cleaning. If scale is found, improvements in the water recirculation and treatment systems should be installed or corrected. If pitting occurs, use more water treatment chemicals to lower the water's oxygen level.

***8.2.1.2. Burners.*** If at all possible, burners should have individual air/fuel ratio controls, with *air primary*, that is, air adjusted by heat demand (temperature), and fuel adjusted to follow air flow changes. If the air/fuel ratio control is *fuel primary*, the furnace might be accidentally filled with a rich mixture—a condition that is difficult to correct without crossing the explosive limit of the fuel. There should be a quick-shutoff fuel valve (reachable without a ladder) at the nearest exit.

*Burner tiles* must be inspected frequently, and replaced as soon as possible if damaged. Generally, cracks are not a major problem, but if pieces of tile are missing, replacement should have a high priority to avoid damage to the furnace and its loads. If burner block failure happens repeatedly, consult the burner manufacturer about another method of installation. Hot spots in a furnace shell around a burner may indicate that hot gas is leaking through a cracked tile or burner block, which will cause the shell to buckle outward, breaking the tile in tension. **Remember:** almost all refractories are strong in compression, but weaker in tension and shear.

*Burner Fuel Supply System.* Fuel line pressure regulators must have a manual shutoff valve on their upstream side. The gaseous fuel supply line to each furnace should have a *drip leg*, and perhaps filters or strainers. A drip leg is a vertical down-flowing gas supply pipe with a manual shutoff valve and then a side outlet tee to the burners. The continuing straight-down outlet of the tee should have a straight section about 1.5' (0.46 m) long, with a cap at the dead end to form a catch basin for liquids and solid particles. Allow space below the cap to permit its removal after placing a bucket below to catch accumulated liquid and dirt.

*Filters and Strainers.* The side offtake from the vertical fuel supply downcomer should have either two filters in parallel for dirty gaseous fuels or two strainers in parallel for liquid fuels. All strainers and filters must have shutoff valves both upstream and downstream, and these should be used to clean the filters and strainers

> Unplugging clogged fuel lines has led to fires, and even explosions. Use two filters (strainers) in parallel, with shutoff valves upstream and downstream of each, and clean them often with a nonflammable fluid. Remember, obey what your mother (and John Wesley, c. 1740) told you: "Cleanliness is next to Godliness." Otherwise you may end up next to devilish flames—sooner than you had planned!

frequently. Do not clean filters or strainers with any flammable fluid, and allow them to air-dry before replacement.

*Burner pilots* have much smaller passageways than main burners, thus they are subject to plugging. Clean them regularly, especially the tiny passageways in the pilot mixer. Care must be taken when cleaning pilots so that the 'cleanings' do not fall back into the cleaned parts or short out the pilot's spark gap. Reinstall the pilot assembly so that the pilot tip (nozzle) is only hand tight in the burner mounting plate—or you will never again get it out.

***8.2.1.3. Controls.*** Before a furnace is removed from operation, all three forms of its control—temperature (input), air/fuel ratio, and furnace pressure—should be checked for proper operation. Then, when the furnace is down, these controls should be calibrated and cleaned, especially the fluid flow measuring components. Actuators need cleaning and lubrication. Lost motion in the control valve linkage should be corrected.

***8.2.1.4. Seals, Doors, Hearth, Roof, and Walls.*** These all should be checked, cleaned, and repaired as part of regular preventive maintenance.

*Water seals* should get care similar to water-cooled skid systems (discussed earlier). The same applies for water-cooled doors and doorframes.

*Sand seals* need frequent filling and checks for trough damage. Ceramic fiber (fire-hose-like) seals for door bottoms need watching for tears.

*Doors* should be checked often and repaired promptly because hot gas leaks can lead to runaway ruin quickly. Seals around doors and car hearths need frequent repair or replacement. Doors should be checked for warpage and loss of refractory. Doors that are not used should be bricked up, but with addition of an observation port (with closure on a chain) and closure for monitoring furnace conditions during firing. If there are any gaps between doors and stationary furnace elements exceeding $\frac{1}{8}$ in. (3 mm), they should be adjusted for less leak.

*Hearth, roof, and walls* should be watched for buckling, hot spots, cold spots, and damaged or leaking refractories. In addition, look for signs of outleakage (hot spots, buckling) through the metal skin of the furnace, and especially around burners, doors, and peepsights. Rammed or blown patches should be installed and carefully dried/cured. Refractory hangers should be cleared of deep dust coverage, which can insulate them, causing their temperature to rise, reducing their strength. (Dust is a very good insulator, because it contains many tiny air spaces.)

### 8.2.2. Air Supply Equipment Maintenance

Air fans, blowers, compressors, and eductors must be monitored for vibration, change in sound, hot spots, lack of lubrication, and wear. Insist that inlet screens, filters, and silencers be kept in proper position, tightly, and that they be cleaned regularly with a nonflammable cleaning compound.

*Inlet vane controls* should be inspected for linkage or bearing looseness, and adjusted or replaced before they cause more trouble. The minimum air flow should be reset to 10% of maximum to protect recuperators and any other air-cooled devices.

*Fans, impellers, and motors* need clearances checked regularly, and reset if greater than specified by the supplier. Re-balance fan and motor assemblies regularly as preventive maintenance. Make sure that the fan is not in surge when balancing it. Clearances between stationary and moving parts should be checked regarding the supplier's recommendations—generally not larger than $\frac{1}{8}$ in. (3 mm), except for very large units. If the cost of a furnace going cold and ruining a load of products is greater than the cost of backup impellers and motors, buy both backups. Carefully label them accordingly and make sure that both maintenance and operating people know that standby replacements are on site, and where.

Inlet vane controls on blowers and fans should be checked for looseness of linkages and bearings, and corrected or replaced before they cause more trouble. Make sure that inlet screens, filters, and silencers are in place, tight, and cleaned. Do not use any flammable cleaning compound.

*Flexible connectors* need constant observation to check for separations. They are designed to prevent transmission of vibration, but they themselves are not immune to vibration problems. Watch for tears, wear spots, and separations. Replace with less severe bends, or reposition equipment to minimize misalignment. Make sure that all pipe fitters and installers know that flexible connectors are not to be used instead of pipe fittings. Their only purposes are to absorb vibrations and to correct for *minor* misalignment.

*Vibration isolators* may need checking occasionally.

### 8.2.3. Recuperators and Dilution Air Supply Maintenance

Recuperator heat exchangers need regular inspection for cracked, torn, or broken tubes or tube sheets.

*Flue gas temperature measurement* needs scheduled inspection to be sure the T-sensor does not "see*" the cold tubes, which will 'fool' the overtemperature control into letting flue gases get too hot.

*Minimum air flow* should be at least 10% of maximum air flow, and this must be maintained 100% of the time—not 98% of the time. The maximum flow should

*That is, the sensor must not be in a position where it could emit straightline radiation to surfaces that are purposely cool. The dilution air temperature control sensor must not 'see' cold recuperator tubes because that would allow the flue gas temperature to be 100°F to 250°F (55°C to 139°C) above design, reducing recuperator life. Too many recuperators have been burned out on their first day of use. Engineers and operators (who have safely passed the first-day test) should redouble their vigilance from there on.

include the maximum possible firing rate of all burners. The velocity of this air stream will provide sufficient energy to assure mixing of the dilution air with the flue gas to keep the recuperatorcomponents at a sufficiently low level to prevent damage.

Prevent *combustibles burning in the recuperator*—a damaging situation. For long recuperator life, limit the flue gas temperature to 1600 F (871 C), and check the actual reading with a high-velocity thermocouple. Frequent preventive maintenance must include using a high-velocity thermocouple to check the automatic over-temperature sensor.

Inspect the *dilution air* system to be sure that it has adequate capacity to cool the flue gas for protection of recuperators and other equipment. Perform this check regularly and especially after delays, when all zones will be at maximum input, with the loads hot all the way back to the charge door, thereby raising flue gas exit temperatures considerably above normal.

In many cases, the dilution air fan and system are not adequate in either volume or pressure to cool the flue gas below the maximum allowable temperature. Therefore, the authors recommend that the system be redesigned by a consultant who has experience with such systems. As a general rule, the air velocity at maximum dilution air flow should be at least 160 fps (49 m/s), which requires a pressure of 10"wc (0.25 m $H_2O$ gauge). This flow should be designed for maximum firing rate of all burners with flue gas temperature at least 2000 F (1093 C). This velocity will provide sufficient energy to mix the dilution air with the flue gas, even at low-dilution air requirements. An air flow capacity safety factor of 1.2 should be used when dilution air systems are designed—with adequately increased dilution air fan discharge pressure to deliver *and to mix*.

### 8.2.4. Exhortations

All furnace and machinery operators should have a check list of items to check every time they come on duty. All operating personnel should be encouraged to be on constant lookout for wear and tear and things going wrong, and to report them promptly to the maintenance department. AND, to keep their confidence, the maintenance department must take prompt action, never ridiculing their concerns. Nothing runs down a plant worse than loss of employees' pride!

Maintenance requires ongoing vigilance, just like safety. If these two aspects of plant operation are not conscientiously practiced, they may affect profits and personnel, customer, and community relations.

## 8.3. PRODUCT QUALITY PROBLEMS

### 8.3.1. Oxidation, Scale, Slag, Dross

Oxidation of any product—steel, aluminum, copper, brass, or bronze—can be minimized by close control of air/fuel ratio to a minimum of about 5% excess air. Less than that may result in presence of pic, which can cause hydrogen absorption and other defects, pollution, and hazards.

Under average conditions, the weight of *scale on steel* surfaces can be expressed by the following empirical equation generated by original author W. Trinks, based on observation. It has no known theoretical foundation. Its accuracy is about ±25%. Composition of the steel and of the furnace gases, and method of circulation of gases have great effect on scale formation.

$$\text{Pounds of scale/ft}^2 = 0.4 \times (T/2200)^5 \times t^{0.5} \tag{8.3}$$

where $ft^2$ is exposed area of steel; $T$ is steel surface temperature, $F$, not absolute, and $t$ is hours of exposure time.

Steel scale begins to soften at 2320 F ±50°F (1271 C ± 28°C), depending on its composition. It melts near 2500 F (1371 C), but that also depends on its composition.

If thick steel (which stays in the furnace for a long time) is heated in a hot furnace, the scale becomes mushy, if not liquid. Semimolten scale has caused many erroneous temperature measurements in steel heating furnaces. Scale is an insulator. Its conductance is lower in its solid form, but the high reflectivity of the molten form causes it to act as an insulator. If the scale is not shiny or glossy, optical pyrometers and radiation pyrometers measure scale temperature, but not steel temperature; pyrometers indicate a temperature somewhere between furnace ceiling temperature and scale temperature, but not steel temperature. Shiny scale (semimolten) reflects radiation; nearly eliminating heat transfer to the load.

Scale on steel is many different oxides of iron combined with sulfur, silicon, and other alloys in the steel. The melting point of this mixture varies from 1650 F to 2500 F, with a normal softening temperature of 2300 F. With large quantities of sulfur in the steel or in the furnace atmosphere, the softening temperature can be as low as 1600 F, and scale formation may be twice normal. With large quantities of silicon in the steel, the softening temperature can be as low as 2150 F, and scale formation 30% higher than normal. If neither sulfur nor silicon is above normal, the melting temperature of the scale is 2500 F. If that temperature is reached on the steel surface, scale will run off the steel piece like water and give evidence of washing. If the melted scale is permitted to drop into a bottom zone, the scale will gradually fill that space, requiring jackhammers for removal.

If scale softening occurs, there will be a highly reflective surface on the hot face of the scale, backed by a very porous (poor conducting), dull material. If a reflective scale condition is generated in the charge area of a reheat furnace, heat transfer to the steel in the remainder of the furnace will be significantly reduced because one cannot heat by radiation mirrors! A reflective scale condition can be generated by holding a charge zone above 2300 F; therefore, charge zones should be limited to 2300 F maximum.

***8.3.1.1. Effect of Temperature, Time, Atmosphere, and Velocity.*** The variables that affect scale formation are: (1) temperature, (2) time, (3) atmosphere, and (4) gas velocity—discussed in order of importance next.

*Temperature of the Steel Surface.* From 1900 F to 2000 F (1038 C to 1093 C), the rate of scale formed increases by 30%. At 2500 F (1371 C), scale runs off the

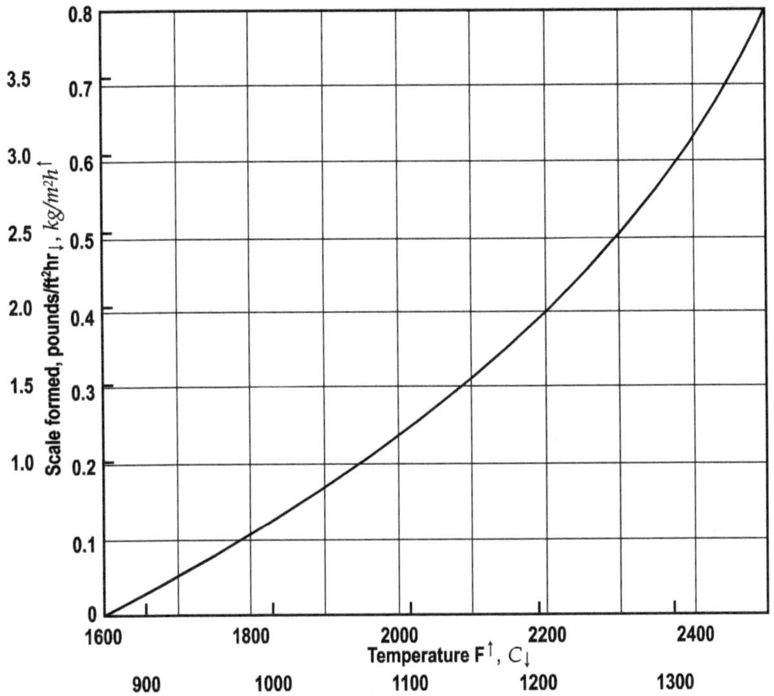


**Fig. 8.14.** Temperature effect on scale formation on steel.

load like water, again exposing the steel to furnace gases. Scale formation thereafter is largely controlled by the availability of oxygen in the furnace gases. (See fig. 8.14.)

*Time at Temperature.* If the time is doubled, the scale formed may increase by 40%. (See fig. 8.15.)

*Furnace Atmosphere.* If there is a reducing condition (a shortage of air for fuel combustion), the quantity of scale formed will be only about 20% as much as with a slight excess of air (oxidizing atmosphere). With only 50% of the air necessary to burn the fuel, almost no scale will be formed. If the combustion air were increased to just a little above the minimum to burn all the fuel, the scale formed per hour would increase about five times. If the combustion air were further increased, very little additional scale would be formed. (See fig. 8.16.)

*Silicon steels* may have to be heated to 2600 F (1370 C) to attain the desired characteristics and to control precipitation of grain boundary inhibitors. To limit costly scale loss at these high temperatures, holding the excess oxygen to 0.5% or less is very effective.

Heating under a reducing atmosphere forms scale that is almost impossible to remove, resulting in rolled-in scale in the finished product. Because rolled-in scale is intolerable, the last stage of the steel heating process is to hold the product at high

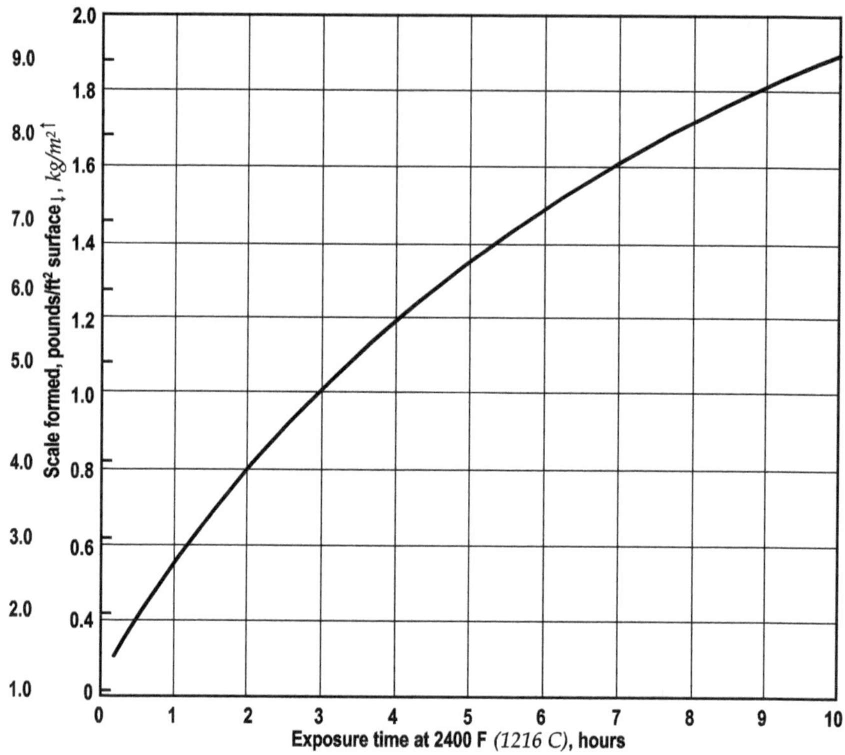


**Fig. 8.15.** Time effect on scale formation on steel.

temperature with at least 2% excess oxygen, or just enough oxygen to remove the tight scale in liquid form.

A furnace with no bottom soak zone can only correct the tight scale problem on the top side. This should cause management to provide a bottom soak zone, which also will improve productivity.

*Velocity of Furnace Gases Passing over the Steel Surface.* If the furnace gas velocity contacting the steel were increased, the inert gas at the surface of the steel would be stirred and enriched with more $O_2$, $CO_2$, and $H_2O$ (oxidizing agents), increasing scale formation. If the scale formed at 40 ft/second was 5 lb/hr, the scale formed at 80 ft/second would be about 60% greater or 8.12 lb/hr. The following are two examples of gas velocity increasing scale. (See fig. 8.15 and 8.17)

**Example 8.1:** A continuous weld pipe mill operated two turns a day, from 0800 to 2400 hour. At 2345 hour, the mill shut down, and the skelp was removed from the hot zone of the furnace. The water-cooled supports in the furnace also were removed. At 0800 hour the following morning, the skelp was replaced into the furnace on the furnace floor. Each bung top opening was uncovered and "L"-shaped hooks were inserted through the bung opening to lift the skelp off the floor.

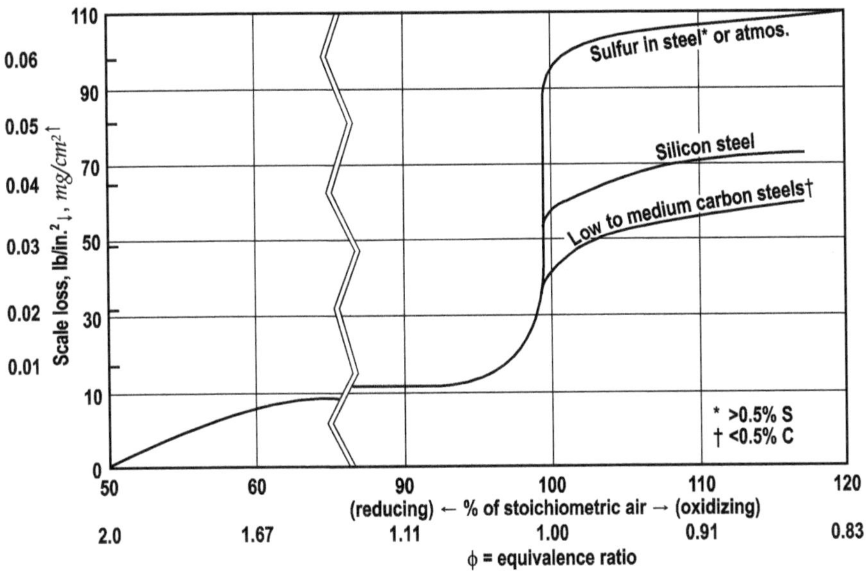


**Fig. 8.16.** Atmosphere effect on scale formation on steel. *The top curve is for steel containing more than 0.5% sulfur or for an atmosphere containing sulfur compounds. †The bottom curve is for steels having less than 0.5% carbon.

*Conclusion of example 8.1.* Another person installed a water-cooled support through a side opening under the skelp. The first man then removed the hook, and they repeated the procedure at the next bung. To rethread the furnace took a minimum of 30 min daily.

It was decided to try to keep the skelp in the furnace overnight at 1550 F (843 C) to save the rethreading time. At 2345 hr, the fuel was shut off, but the air for combustion was increased to maximum flow to increase the cooling speed of the skelp and furnace. With the very high velocity air flowing over the skelp, it scaled so rapidly that it disappeared within a minute—oxidized by the high velocity air.

At 2345 hr the following evening, both the fuel and air were shut off, the damper was opened fully, the bung hole closure tiles were removed, and the cinder drain openings were removed. Within 20 min the furnace temperature was 1550 F (843 C), and that temperature was held until 0800 hour the next morning, when the furnace was started up without rethreading. In this second case, the cooling air velocity was much lower; therefore oxidation was much slower.

**Example 8.2:** A blooming mill was to reroll 13 in. × 17 in. (0.28 × 0.43 m) blooms for a very critical application, so the soaking pits were to heat the blooms as uniformly as possible. Many pit loads were involved. Two pits were set up to fire with constant maximum air capacity to achieve best uniformity. (The other pits were fired with only 10% excess air.) The blooms in the pit using maximum air had more mass flow and therefore should have been more uniform and hotter, but they were uniformly colder! The blooms rested on the pit bottom, which lost heat through its hearth. The

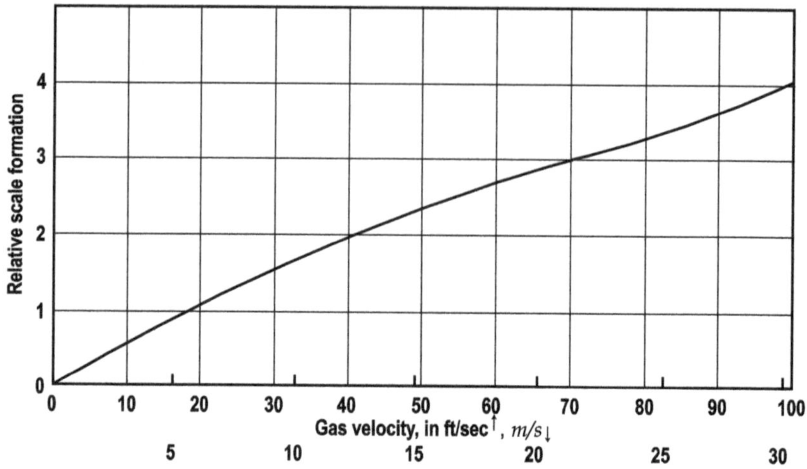


**Fig. 8.17.** Furnace gas velocity effect on scale formation on steel.

heat loss through the bloom bottoms had to be supplied by the heat transferred into the sides of the blooms. *BUT,* scale on the sides of the blooms was restricting the heat transfer, requiring a greater temperature differential to replace the loss. The thicker scale caused by the high-velocity gas flowing over the blooms reduced the bloom temperatures even though the flue gas temperatures indicated the whole pit was at a higher average temperature.

*Further Explanation of Scale Formation.* *Without high-velocity* gases flowing over a steel surface, scale melting begins at temperatures above 2249 F (1365 C). *With high-velocity* gases flowing over the steel surface, scale melting begins near the scale softening temperature, 2320 F (1270 C). Scale melting can proceed only if the high-velocity gases contain at least 1% more oxygen than needed for stoichiometric combustion. If the oxygen excess is less than 0.5%, carbon monoxide (CO) and hydrogen (H) will compete with the iron atoms for oxygen, lowering the scale formation rate to 20% of the rate with 1% excess oxygen.

At temperatures *below* about 2250 F (1232 C), iron diffusion is much slower than oxygen availability. Scale formation is controlled by the temperature and the rate of diffusion of iron atoms toward the scale surface and oxygen moving toward the load surface. At temperatures *above* 2250 F (1232 C), the iron diffusion rate is high enough that availability of oxygen controls the reaction rate.

With the combination of (1) higher temperature, (2) oxygen availability being the controlling factor, and (3) high velocity of furnace gases, spent gases are swept away, providing more oxygen to oxidize the iron atoms. If the velocity effect is great enough, the heat release from oxidation of the iron will raise the scale temperature to its melting point. The molten scale will flow off the steel surface, providing an unlimited source of iron atoms. Then, the burning iron provides heat to sustain the reaction, provided that heat conduction away from the steel load piece does not cool it enough to slow or stop the reaction (provided that the oxygen level of the flowing gases remains above 1% level, and the temperature level remains above 2250 F).

The previous interaction can provide a better understanding of differences in controlling the two different types of skelp-heating furnaces. In one type, the poc are directed at the edge of hotter-than-350 F skelp, initiating rapid iron burning if above 1% oxygen. Reaction heat melts the scale, and it falls off, exposing surface and giving the appearance of "washed steel." Width of the washed effect is controlled by the skelp body temperature which, in turn, is controlled by skelp line speed or furnace temperature.

Another type of skelp-heating furnace is fired to heat the furnace, avoiding direct impingement on the edges of the skelp. Iron burning does not begin until the skelp emerges from the furnace, where jets of oxygen or air (or both, one after the other) provide the oxygen for reaction. Again, the width of the washed area is controlled by skelp line speed or furnace temperature. Line speed control seems to be better because it is quicker to react to the changing furnace temperature. Both methods are satisfactory.

Coauthor Shannon believes that we have advanced in our understanding of scale formation in steel reheat furnaces, except for one problem—when a steel surface receives too much radiation too soon in a furnace.

**Example 8.3:** If steel at 1400 F is pushed into an area where the furnace temperature is 2250 F, from table 8.9 with $F_e F_a = 0.85$, the net radiation to the steel is (107 200 − 20 570) (0.85) = 73 600 Btu/ft$^2$hr, which is more than double the intensity in any other area of the furnace. Back when the steel was at 900 F, FeO scale began to form and accelerated at a rate about proportional to the 5th power of the steel surface temperature (in F). The temperature of the scale was accelerated at an even faster rate because its very porous (poor conducting) nature minimized heat transfer to the steel, trapping heat within the scale itself. With a compounding combination of high-temperature and high-velocity furnace gases flowing across the scale, excess oxygen in the furnace gases further oxidizes the FeO to $Fe_2O_3$. All these reactions release more heat, raising the scale temperature rapidly.

When the scale surface reaches 2320 F, the scale softens, forming a smooth surface that acts as a mirror. That reflective surface reduces heat flow into the steel; thus, the steel piece arrives at the furnace discharge at too low a temperature for proper rolling. One might wonder why the mirrorlike surface does not cause a problem in the hotter parts of the furnace. It does, but because the steel is hotter the temperature difference is less, giving less intense radiation. If the radiation heat flux were 45 000 Btu/ft$^2$hr instead of 73 610 Btu/ft$^2$hr, the heating time, and therefore the charge zone length, would be about 73 600/45 000 = 1.64 times longer. The shining scale has such high reflectivity that it has the effect of reducing the absorptivity (or emissivity), thus shortening the effective length of the furnace.

*Rolled-In Scale.* If steel alloyed with just a trace of nickel* is heated above 1500 F (816 C) with reducing conditions†, the scale will stick to the steel. The bond between

*Steel made from scrap will have at least traces of nickel because scrap invariably contains a small quantity of austenitic stainless steel, which contains nickel. Removal of nickel from steel is very difficult, so it is left in the steel.

†The reducing atmosphere that causes sticky scale is just barely reducing. In an experiment, 0.2% combustible formed a scale that was extremely thin but impossible to remove, even with a hammer.

the scale and the steel cannot be broken by descaling with water or even with a hammer, so when the steel passes through the rolls, the scale is rolled into the steel surface forming pits. Those pits must be ground out or cut out, or the steel will be scrapped.

*Trouble-shooting tips* for minimizing a harmful reducing atmosphere that can cause rolled-in scale:

1. An air/fuel ratio control system with fuel primary (fuel flow leading air flow) if (*a*) air supply system's design is inadequate, (*b*) maximum fuel flow limit is set too high, (*c*) designers assumed air flow resistances and fuel flow resistances in banks of burners in parallel are precisely equal, which they never are, and (*d*) operator adjusts fuel or air flow to a burner in a bank of burners controlled by a single air/fuel ratio control, thus causing some burners in the bank to go reducing.
2. Flame wherein coexisting reducing and oxidizing gases are delayed in mixing and burning until after they contact the surface of the steel.
3. Air/fuel ratio control errors due to flow or $O_2$ measurement problems.
4. Fuel with varying calorific value or density.

Solutions to some of these problems may require measurement of individual air and gas streams to individual burners and/or change of burner type to avoid slow mixing or flame impingement on the product.

### 8.3.2. Decarburization

The chemical removal of some of the carbon from the surface of steel is termed *decarburization*. The steels aversely affected by decarburization are generally those with 50 or more pints of carbon. The carbon is generally in a chemical combination with iron as $Fe_3C$, but it may be combined with other metals such as chromium. The combined carbon is easily oxidized by $CO_2$ and $O_2$ in the furnace gases, as is the iron in forming scale. However, unlike iron, the carbon under reducing conditions can react with hydrogen to form methane gas. Thus, holding a slightly reducing atmosphere in the furnace above 1500 F (816 C) does not lower the loss of carbon in the steel surface.

As steel temperature approaches 1500 F (816 C), the atoms and molecules of both solid and gas move faster, so the gas molecules are able to penetrate the solid more easily, resulting in significant chemical reactions. The surface carbon is oxidized or hydrogenated. As the steel temperature rises, the rate of decarburization increases at an accelerating rate to greater depths.

The only means for minimizing decarburization is by heating the steel to as low as possible a rolling temperature and holding the steel at high temperature for as short a time as possible. To salvage steel when much of the carbon has been removed from its surface is very costly and usually impractical. To meet a difficult decarburization depth specification, the following changes can help.

Change 1. To meet a difficult decarburization depth specification, roll to a finish size from the largest bloom possible. This spreads the decarburization the most, reducing its depth.

Change 2. Add tungsten or chromium to as high a percentage as the range allows for the specified grade because these elements form a tighter barrier to gas penetration of the steel than do other alloying elements.

Change 3. Fire with fuels having as little hydrogen as possible to minimize decarburization (but rarely is a fuel change an option).

Change 4. Increasing the heat transfer area of the steel to reduce heating time will reduce decarburization. A full walking beam furnace where the piece spacing can be increased to 2:1 or 2.5:1 provides for maximum heat transfer area on the billets; therefore, the resulting minimized heating time can result in minimized decarburization.

Change 5. Heat to as low a temperature as possible, and minimize heating time above 1500 F.

Change 6. Avoid delays. Remove loads from the furnace during delays.

Change 7. Add enhanced heating, combined with maximum space-to-thickness ratio, thus shortening heating time.

### 8.3.3. Burned Steel

Surface cracks in steel are the result of many problems that leave the steel surface looking broken up. Management nearly always calls this "burnt steel"—even when the furnace never was above 2450 F (1343 C). With true burnt steel, the crystal boundaries at the surface have been oxidized, which reduces the strength of the material and lowers its ability to be rolled. The material called "burnt" has often been rolled on a modern powerful mill when it was too cold to allow the elongation that the mill opening required.

When steel is really burnt, it has been heated to at least 2500 F (1371 C). In his long steel mill experience, coauthor Shannon has witnessed only one true case of burnt steel, and that was found to have experienced a pyrometer reading of 2600 F (1427 C). He has seen localized (spot) overheating (burnt steel) caused by flame impingement. If the steel has been "washed" with the very hottest gases, it may be burnt. Engineer Shannon also has witnessed cases where steel was scrapped as "burnt" because the surface had pits caused by rolled-in scale. As explained earlier, this sticky scale develops with steels containing a trace quantity of nickel when exposed to reducing atmospheres above 1500 F (816 C). Such scale is generally thin, but attached very, very tightly to the steel surface.

Higher carbon content in steel causes burnt steel at lower tempertures. Laboratory work has shown that steel with a carbon content of 0.2% can withstand 2650 F (1455 C) without burning, but 1.0% carbon steel will burn at slightly above 2450 F (1343 C).

### 8.3.4. Melting Metals

The major problems when melting aluminum (and some other low-temperature melting metals and their alloys) are usually oxide formation and hydrogen absorption. Both can seriously affect casting quality by causing oxide inclusions or porosity.

Electric heating has an advantage over fuel firing in that it avoids the hydrogen (from fuels).

## 8.4. SPECIFYING A FURNACE

### 8.4.1. Furnace Fuel Requirement

The fuel requirement of a furnace is the sum of all the heat uses and losses divided by the %available heat, expressed as a decimal. This calculation is made for each furnace zone. For batch heating from cold, it is necessary to add the heat restored in the furnace walls, hearth, and roof refractories with each furnace cycle. Storage heat can be quite a large sum if hard refractories are used. If lightweight or fiber-lining materials are used, the loss to heat storage will be less. Shuttle car configurations (sec. 4.3 and 8.11) reduce the heat lost from storage by shortening the time that the furnace door(s) are open.

The aforementioned summed heat requirements and losses of a furnace are called the "required available heat". The conversion to gross heat required or fuel required necessitates dividing by the decimal percent available heat for the flue gas exit temperature. Determining that flue gas exit temperature is a major problem. Most persons think all that is needed is to assume that a measured temperature at the flue connection is the flue gas exit temperature. This neglects the fact that the gas from which all the heat is supplied to the furnace is transferring heat to the product directly to the refractories and then to the product. For this heat transfer to take place, the poc must be hotter than whatever they are heating, and higher rates of heat transfer require higher source temperatures because heat always flows "downhill" from a high source temperature to a lower receiver temperature.

The Stefan-Boltzmann equations (2.6, 2.7, 2.8, and 2.9) show that heat transfer rate to most black or gray bodies varies as the difference in the 4th power of their absolute temperatures, which accentuates the difference between "furnace temperature" or "furnace wall temperature" and "poc gas temperature." *Case A:* In Figure 6.3, at 1000 F "furnace temperature" and 20 fps gas velocity, the temperature of the exiting poc gas is on the order of 1800 F. With a combustion air temperature of 600 F, if someone erroneously took the %available heat (from fig. 5.1) at 1000 F he would read 78%. He should have taken the %available heat at 1800 F, where he would read 57%. Therefore, if the required available heat were 100 kk Btu/hr (105.5 kJ/h), the gross heat required will be 100/0.57 = 175 kk Btu/hr (185 kJ/h), *NOT* 100/0.78 = 128 kk Btu/hr (135 kJ/h) as with the erroneous method. *Case B:* At 2500 F furnace temperature, with the same 20 fps, the poc gas temperature would be 2560 F. Corresponding figures are in table 8.16.

When specifying a new furnace, input calculations should be based on the true flue gas exit temperature—NOT ON FURNACE TEMPERATURE! Coauthor Shannon recommends adding a safety factor of 30% in general, but 40% in the charge zone to accommodate productivity expansion of the mill—the latter because inadequate charge-zone capacity can cause swings in input needs after delays. His experience

**TABLE 8.16 Comparisons of correct and erroneous ways of figuring furnace fuel requirement in example cases A and B, both at 20 fps velocity**

| | Furnace (fce) T | Flue Gas Exit (fge) T | %Available Heat | | Required Gross Input | |
|---|---|---|---|---|---|---|
| | | | w/fce T* | w/fge T | w/fce T* | w/fge T |
| A | 1000 F | 1800 F | 78%* | 57% | 128 kk Btu/hr* | 175 kkBtu/hr |
| | 538 C | 982 C | 78%* | 57% | 135 kJ/h* | 185 kJ/h |
| B | 2500 F | 2560 F | 30%* | 28% | 333 kkBtu/hr* | 357 kBtu/hr |
| | 1371 C | 1427 C | 30%* | 28% | 352 kJ/h* | 377 kJ/h |

*Erroneous—shown only for comparison.

has shown these extra fuel rates have paid huge benefits over the years for small first cost! Combustion airflow designs and ductwork should match these higher rates.

If the furnace is to use a recuperator, make sure the design uses the total maximum airflow for all zones to avoid running out of high-temperature air supply when it is most needed.

Beware of buying a furnace computer control whose designers lack an understanding of complex interactions of a furnace-and-mill system when delays occur. Operators must be able to understand a computer control model or they will become dependent on the computer supplier for help with every little glitch.

A two-sensor control, each with controller and with a low select device in each zone (except the entry zone) will be more effective and serviceable by mill operators. The entry zone will have one T-sensor located near the charge area in the flue gas flow. Its purpose is to follow productivity of the zone, especially after a delay. With this system, the additional zone T-sensors will keep the product heating on track without overheating. For best results, the sensors should be within a few inches of the load.

## 8.4.2. Applying Burners

Many engineers have applied new burners and found that they did not produce the desired effect or correct the problem for which it was purchased, or caused another problem. For example, the bottom heat zone (20 ft = 6.1 m) long) of a *steel reheating furnace* is fired longitudinally with several 10 kk Btu/hr burners. The temperature control sensor in the sidewall, 11 ft from the burner wall, provides reasonable heating as long as the mill is rolling steadily and the burners are operating at or near maximum firing rates. At the burner wall, the temperature profile is below setpoint temperature, but it rises to 20 F above setpoint at 13 ft from the burner wall. (See fig. 6.3.) If the furnace temperature had been higher in the first 6 ft from the burner wall, it would have transferred more heat, increasing productivity and lowering the flue gas exit temperature.

In addition to the lowering of heating capacity, another problem occurs when the mill stops and the firing rate is reduced—as shown by the 30 and 50% curves of figure 6.3. At 50% and smaller firing rates, the burner thermal profile changes, increasing the burner wall temperature and reducing the temperatures beyond about 3 ft (0.9 m) from the burner wall. In fact, the wall temperature could be over 75°F higher than the setpoint temperature. If the setpoint temperature was 2450 F when the mill stopped and the firing rate was reduced to 30% or less, the temperature control might hold the burner wall temperature at $2450 + 75 = 2525$ F. At this temperature and with movement of the load stopped, the surface of the load would soon be above 2490 F—the temperature at which the scale melts. The melted scale will drop into the heat zone bottom. After a fairly long time, the zone bottom will fill with solidified scale that will deflect the flame, interfering with heat transfer and gas flow patterns in addition to lowering yield.

The aforementioned problems occur because of the dynamics of combustion. As the firing rate is increased from minimum, the air $\Delta P$ needed to push the added air through the burner and tile must increase by the square of the pressure (because we are accelerating the air flow). The air in most burners provides the bulk of the energy for combustion gases, so as the firing rate increases, the air velocity increases, pushing the actual combustion and heat release zones farther and farther from the burner. Because of this dynamic, the flame's temperature profile (a measure of heat flux) changes longitudinally with firing rate, as shown in fig. 6.3.

To moderate the previous problem, a longitudinally fired zone in a reheat furnace can be fired with a combination of small and large burners designed to permit paralleling them. The small burners will have their peak heat release closer to the burner wall whereas the large burners will have a peak heat release farther from the burner wall. With such a combination, the zone temperature profile will be much flatter, regardless of the firing rate.

Another way to correct the "hunting" problem after a mill stoppage is to use burners with a controlled adjustable spin of the combustion products to keep two T-sensors, one close to the burner wall and one perhaps 10 to 15 ft (3 to 5 m) away, at the same temperature. At low-firing rates, this system may require a forward-firing gas lance to extend the heat flux to hold up the far thermocouple temperature. This lance can be turned on when the firing rate drops below a predetermined rate. The lance should be designed to pass 5 to 10% of the total fuel. Such a burner for controlling heat flux profile is now available. The same type of burner, with near and far T-sensors for control, is used to solve a crosswise temperature profile problem in cross-fired zones. (See sec. 3.8.5.)

### 8.4.3. Furnace Specification Procedures

When specifying a furnace for a new or existing facility with or without a consultant's input, the production rate for each product must be studied first. For example, on a mill that averages 60 tph, but with some production rates as high as 120 tph, a businessman would be inclined to buy a furnace for perhaps 80 tph. This example actually happened when a designer, realizing the businessman's folly, actually planned the furnace for 110 tph. After the furnace became operative, the mill averaged 100 tph, still with peaks of 120 tph. Furnaces that limit productivity are difficult to correct without a major expenditure and cause owners to avoid improving mill performance while the furnace is holding everything back. What appears to be saving money by building the furnace to meet less than current maximums can be a costly event of major proportions.

The furnace should be designed to at least the maximum rate that the mill ever produced. Designing for 20% above the peak is planning ahead to prevent future problems. It can give operators room to improve mill performance.

After a furnace design capacity is agreed upon, *product quality* must be addressed. The following quality problems must be considered:

1. Surface conditions: (*a*) unequal product dimensions due to poor temperature uniformity, (*b*) pits formed by rolled-in scale, (*c*) surface marks caused perhaps by the movement through the furnace, (*d*) loss of carbon in the product surface, and (*e*) cracks in the surface.
2. Hydrogen absorption
3. Scale loss reducing yield
4. Effect of furnace atmosphere
5. Mill cobbles

*Furnace fuel rate* must be addressed. The ideal furnace combustion system (to attain maximum efficiency and minimum fuel rates) is by preheating combustion air with a *regenerative burner system*, which requires more daily attention than does a recuperative system. With daily attention, a regenerative system's overall cost over a 5-year period will be less. The benefit occurs because the fuel rate depends on the heat exchange beds, not on furnace operating techniques. Fuel waste during delays is minimal with regeneration because the available heat is maintained at 70% + versus a drop of as much as 50% in available heat during delays with recuperation.

Some want to reduce costs of regeneration by using parallel burners in air and exhaust gas modes. Because of nonuniform packing of heat exchange materials, however, airflows and exhaust gas flows of regenerative burners are not identical, so each burner must have its own air/fuel ratio control and its own exhaust gas control system to provide near-maximum combustion efficiency.

Specifications should insist that the maximum-allowed firing rate of a burner should be limited to 6 in. (151 mm) of water-column pressure drop across the bed when the excess air is above 15% as measured by flow devices on the air and fuel streams. The reason for 15% rather than 10% excess air is because of air and exhaust valve leakage. This leakage of combustion air cannot be used to burn fuel, but as long as the air loss is not greater than 10%, all the fuel can be burned in the furnace.

If the capital cost of regeneration exceeds the available funds, *recuperative air preheating* should be used, but its payback is not so great because of its lower efficiency. With recuperation, the furnace should be sized to reduce the flue gas temperature to no more than 1700 F.* This usually means that the furnace temperature at the flue should not exceed 1300 F. Many will take exception to the 400°F between these two temperatures. They assume that the flue gas temperature is the same as the furnace wall temperature at the flue. The furnace gas temperature must be higher than the wall and load temperatures, or no heat can flow from heat source (furnace gas) to the wall and load.

To protect a recuperator from overheating and burnout, a *dilution air system* must be capable of reducing the flue gas temperature to 1500 F (816 C), and the design air velocity (for mixing the dilution air with the flue gas) must be at least 160 fps (49 m/s) at maximum furnace firing rate. This high velocity at maximum provides flow and enough air energy to mix with the flue gases at 10% rate. In addition, the maximum designed flow volume should be at least 25% greater than the calculated need. The reason for the additional dilution air is that the gas temperature may be higher than estimated. Many near-new recuperators and dilution air systems have had to be replaced because of premature burnout. Most of these occurred because the air flow was too low and the mixing energy too low, as a result of fan pressures less than 7 in. of water column (178 mm of water column) or maximum airflow velocity of only about 105 ft/s (32 m/s). The dilution air system (ducting) also must be considered so that the aforementioned required velocity and pressure can be delivered at the point of mixing just before the recuperator.

*Maintain a minimum airflow* of 10% of maximum recuperation design through the recuperator during all operating conditions to assure some coolant flow through all tubes to prevent them from being heated to flue gas temperature.

*Prevent unburned gases* from entering the recuperator.

*Flue gas temperature measurement* errors can cause difficulties in heat recovery systems. If a thermocouple can "see" cold recuperator tubes (i.e., if the T-sensor can radiate heat to cold recuperator tubes), it may read 100°F to 250°F (55°C to 139°C) lower than it actually is, so it will not be able to protect the recuperator tubes. The corrosion reaction rate of steel doubles with every 16°F to 18°F of temperature rise, so an error of 100°F in the flue gas temperature can reduce tube life to about one-third of its intended life.

*Furnace location* is important: There should be reasonable clearance around the furnace for future adjustments and modifications. A 20 ft (6 m) clearance on all six furnace sides is advised. Generous access space below and around the bottom zone is necessary, along with means for lowering and raising equipment to all parts of the furnace.

*Ambient conditions around a furnace* must be reasonable to allow quick repairs. Air movement from both inside and outside the building should be mandatory during construction, operation, and repairs.

*Guarantees* of fuel rate per ton of product, production rate, and minimum NOx emission rate should be included in the bids. If some reasonable way is available to

*This high-temperature limit has been rising over the years as better materials are employed and their cost can be justified. However, the advent of packaged regenerator-burners, which are more efficient and not dependent on high-temperature-conductive materials, has decreased interest in high-temperature recuperators.

specify a minimum scale formation and a minimum temperature variation within any one piece, those specs also would be desirable. On large furnaces, predicted thermal profiles for a variety of throughput rates should be expected.

If there is to be a *skid support system*, the heat transfer in the bottom zone must have a high priority, or skid marks can become a large problem. Anchoring of the skids must have attention to avoid difficulties.

*Cooling-water flow control*, along with a back-up system, are often necessary to protect sensitive parts from overheating.

If side firing is to be used downstream of longitudinal firing, *baffles* or other means must be used to prevent the longitudinal streams from deflecting the side-fired streams before they reach the furnace center. Otherwise, the product uniformity will suffer, and efficiency will be lower.

*Furnace control* should not be by a complicated modeling system that your operators cannot easily manage, or they will become a dependent on the installer much of the time. A simple system that can be understood by all concerned, including the management, will be the best.

The system installation engineer should explain how the control will react to controlling the product temperature of those pieces that were in the furnace during *delays* and those that were charged immediately after the delay.

Roof *heat losses* should be expected to be below 600 Btu/ft$^2$hr, sidewalls below 325 Btu/ft$^2$hr, and furnace hearth or bottom below 450 Btu/ft$^2$hr.

*Furnace pressure* should be controlled at a slightly positive level at the lowest leak elevation, preferably by a stack cap damper so that it can be seen when the system is in difficulty. When the plant manager can see that the damper is in trouble, correcting it becomes a priority. Where the damper is in the flue and unseen, repairs may never be performed.

*Air/fuel ratio control* should be by a very simple and reliable system, preferably one control per burner with fuel following air (air primary) so that lack of air reduces fuel. Zone temperature measurement should be by sensors near the product so that the product is the most important variable—not the furnace zone temperature, except in the entry zone where the sensor should "see" the product and "feel" the heating gases. Indexing of the load pieces helps to get the T-sensor to get a measurement as near to the product temperature as possible.

In summary, the authors wish to quote some wise points from Mr. Ralph Ruark's article in the July 2001 *Ceramic Industry* (pp. 27–30) on "What to Avoid when Buying a Kiln" (*reference 76*), much of which also applies to buying a furnace, oven, dryer, melter, incinerator, boiler, heater.

> "A kiln purchase should be achieved through a team effort. The team should include a kiln specialist, a ceramic engineer, a mechanical/electrical specialist, a quality assurance specialist, and someone intimately familiar with the production floor operation and product flow. One person simply cannot have the range and depth of knowledge to make sure that the perfect solution is achieved."
>
> "Innovative companies usually produce great results; those less innovative often survive by selling low cost products."

"There are components common to all kilns. Specifying certain materials and hardware by brand could minimize the spare parts necessary."

There are many versions of the following old saying: The Delight of Low Cost Will Soon Be Forgotten, But the Sadness of Poor Quality Can Embitter You (and Your Management) The Rest of Your Days!

## 8.5. REVIEW QUESTIONS AND PROJECT

**8.5Q1.** Regarding product quality, where is the one place in an oven or furnace that you do not want radiation?

**A1.** To or from T-sensor elements. If they emit radiation to any cooler surfaces, they will give an erroneously low reading. If they receive radiation from any hotter surfaces, they will give erroneously high readings. A theorist might argue that you want them to be sensitive to whatever might be received or emitted by the loads, but sensor elements have very small mass compared to loads; therefore, their temperature will rise or drop faster than that of the loads. The theorist's ideal location for a T-sensor would be embedded in the center of the hardest-to-heat part of a load.

**8.5Q2.** Regarding product quality concerns for industrial process heating operations, what is usually the most important process variable?

**A2.** Temperature uniformity, or more generally, temperature control.

**8.5Q3.** Arrange the following concerns in order of importance—in your opinion—for your furnaces:
Furnace productivity ___ Personnel productivity___ Product quality___ Fire prevention___ Fuel cost/Energy conservation___ Pollution minimization___ Safety___ Cleanliness___ Preventive maintenance___ Public relations___ Customer relations___ Training___ Employee relations___ Other___

## 8.5 PROJECT

Discuss the order of the previous concerns (8.5Q3) with associates, supervisors, and management. Then, agree on a consensus for your organization, put it in writing, and put it into practice!

# 9

# MATERIALS IN INDUSTRIAL FURNACE CONSTRUCTION

(See also the following on refractories:

Conductivity: *reference 51,* pt 4, pp. 81, 86–87.
Wall losses: *reference 51,* pt 4, pp. 100–115.
Burner tiles: *reference 52,* pt 6, pp. 10, 83–86.)

## 9.1. BASIC ELEMENTS OF A FURNACE

The basic elements of a furnace are (*a*) the heat-resistant lining with insulation; (*b*) the steel-supporting structure and casing; (*c*) heat-releasing, distributing, and control equipment, via fuel combustion or conversion of electric energy to heat, and including circulation of hot gases and provision for waste gas discharge; and (*d*) load-holding and load-handling equipment, including piers, skids, kiln furniture, hearth plates, walking beam structures, and roller and other conveyors.

Industrial heat-processing furnaces are insulated enclosures designed to deliver heat to loads for many forms of heat processing. The load or charge in a furnace or heating chamber is surrounded by sidewalls, hearth, and roof consisting of a heat-resisting refractory lining, insulation, and a gas-tight steel casing, all supported by a steel structure.

### 9.1.1. Information a Furnace Designer Needs to Know

In selecting materials for a furnace—new, rebuild, or maintenance—a furnace designer needs to know:

1. Temperature range required in production, including significant fluctuations and their intervals
2. Operating schedule—continuous or intermittent. Scheduled downtimes for maintenance, vacations, other

Firebrick was the dominant furnace material from about 5000 BC to the 1950s. Many years ago, man discovered that tufa (calcareous sinter, or solidified bubbled lava) is an excellent insulating material for high-temperature furnaces (maybe as in this book's frontispiece). Modern insulating firebrick is a man-made equivalent of tufa.

Firebrick originally provided load bearing walls, heat resistance, and containment. As steel framing and casing became more common, and as monolithic refractories were improved, furnaces were built with externally suspended roof and walls.

3. Material composition of loads to be processed, and effects of chemicals released on the furnace refractories, and metal structure
4. Fuel to be used, and their effects on the furnace refractories/structure
5. Probability of furnace damage by the loads as they are placed on the hearth, or as they move through the furnace
6. Advantages from using cooling water in the rails, lintels, other areas
7. External forces applied to the structure, for example, thrust exerted on the hearth and skids by a pusher
8. Nearby machinery that may transmit shock or vibrations to the furnace
9. Static and dynamic load on the foundation; nature of subsoil, drainage

## 9.2. REFRACTORY COMPONENTS FOR WALLS, ROOF, HEARTH

(See also further discussion of hearths in sec. 9.7.1.)

The linings of industrial furnaces require stable materials that retain their strength at high temperatures, have resistance to abrasion and to furnace gases, and have poor thermal conductivity (good heat-insulating capability).

Modern firebrick (from fireclay, kaolin) and silica brick are available in many compositions and many, many shapes for a wide range of applications and to meet varying temperature and usage requirements. High-density, double-burned, and super-duty (low-silica) firebrick have high-temperature heat resistance, but relatively high heat loss; thus, they are usually backed by a lower density insulating brick.

Insulating firebrick (kaolin) with many very small air pockets is a modern replacement for tufa.

### 9.2.1. Thermal and Physical Properties

The basic components of most refractories are oxides of various origins. Tables 9.1 and 9.2 list properties of some monolithic refractory materials.

**TABLE 9.1. Properties and analyses of five CASTABLE refractories (see also pp. 397–405 of *reference 26*, and pp. 95, 102–117 of *reference 51*.)**

| Characteristics (all hydraulic bond): | | | | | Special |
|---|---|---|---|---|---|
| Service range, 200 F to | 2600 F | 2800 F | 3000 F | 3100 F | 3100 F |
| Density, $lb/ft^3$ | 138 | 138 | 145 | 165 | 165 |
| Gallons water/100 lb dry | 1 | 0.94 | 0.88 | 0.88 | 0.81 |
| Cure time, hr | 24 | 24 | 24 | 24 | 24 |
| Abrasion loss[a] after 1500 F | 15 cc | 15 cc | 10 cc | 10 cc | 10 cc |
| Cold Modulus of Rupture[b], psi | | | | | |
| At 230 F | 1230 | 980 | 900 | 890 | 1000 |
| At 1500 F | 1155 | 1135 | 990 | 1025 | 1400 |
| At 2000 F | 1400 | 1210 | 1160 | 1090 | 1650 |
| At 2500 F | 1800 | 1450 | 1790 | 1375 | 2050 |
| At 3000 F | – | – | 3090 | 1500 | 2925 |
| Hot Modulus of Rupture[c], psi | | | | | |
| At 1500 F | 1250 | 1100 | 950 | 950 | 1350 |
| At 2000 F | 1660 | 1400 | 1670 | 1530 | 2400 |
| At 2500 F | 125 | 300 | 350 | 650 | 950 |
| Chemical analysis, % | | | | | |
| $Al_2O_3$ | 44.5 | 46.8 | 48.8 | 78.9 | 68.6 |
| $SiO_2$ | 47.2 | 46.1 | 46.4 | 16.3 | 26.9 |
| $Fe_2O_3$ | 1.1 | 1.3 | 0.8 | 0.9 | 0.8 |
| $TiO_2$ | 1.5 | 1.5 | 1.4 | 1.6 | 1.2 |
| CaO | 5.1 | 3.6 | 2.0 | 1.8 | 1.8 |
| MgO | 0.1 | 0.2 | 0.1 | 0.1 | 0.1 |
| Alk | 0.4 | 0.3 | 0.3 | 0.2 | 0.2 |

[a]ASTM C-704.
[b]ASTM C-133.
[c]ASTM C-583.

***9.2.1.1. Refractory Sizes and Shapes.*** Various refractory materials have been formed into numerous sizes and shapes, collectively termed "firebrick," evolving into standard sizes and shapes such as straight, small, split, soap, wedge, end skew, side skew, edge skew, neck, key, arch, featheredge, jamb, bung, circle, and block. Figure 9.1 shows a few of the many shapes available.

Furnace linings may be single or multilayer in form. Single layers usually suffice for furnaces operating at temperatures below 1400 F (760 C). Linings for modern high-temperature furnaces are almost always multilayer. The high-temperature layer, which forms the interior surface of the refractory, referred to as 'hotface,' is backed by one or more layers of less heavy-duty refractory and/or insulating materials, and then finally the outer metal shell or 'skin' (coldface). Furnace designers must make sure that the temperature at the interfaces between the various refractory and insulation linings does not exceed the safe temperature rating of the next layer. Most refractory suppliers have computer programs to check this for customers.

**TABLE 9.2. Properties and analyses of seven PLASTIC refractories (see also pp. 397–405 of *reference 26*, and pp. 95, 102–117 of *reference 51*.)**

| Characteristics (zero cure time) (A + H = air + heat bond) (Chem = chem bond) | | | | | | | |
|---|---|---|---|---|---|---|---|
| Service max., F | 2900 | 3000 | 3200 | 3100 | 3300 | 3200 | 3400 |
| Weight to place, lb/ft$^3$ | 153 | 159 | 157 | 160 | 174 | 180 | 188 |
| Weight use, lb/ft3 | 142 | 148 | 148 | 152 | 165 | 170 | 176 |
| Abrasion loss[a] after 1500 F | | 13 cc | – | 12 cc | – | 12 cc | 10 cc |
| Cold Modulus of Rupture[b], psi | | | | | | | |
| At 230 F | 510 | 1045 | 500 | 1140 | 510 | 1260 | 1495 |
| At 1500 F | 405 | 975 | 415 | 1190 | 400 | 1260 | 1755 |
| At 2000 F | 660 | 1230 | 585 | 1210 | 600 | 1780 | 2090 |
| At 2500 F | 700 | 1820 | 710 | 1990 | 490 | 1725 | 1870 |
| At 3000 F | – | 1830 | 1020 | 2310 | 870 | 1580 | 1400 |
| Hot Modulus of Rupture[c], psi | | | | | | | |
| At 1500 F | 575 | 1430 | 540 | 1770 | 620 | 1890 | 1960 |
| At 2000 F | 875 | 835 | 920 | 1260 | 915 | 740 | 1155 |
| At 2500 F | 175 | 540 | 145 | 750 | 390 | 750 | 1250 |
| Chemical analysis, % | | | | | | | |
| $Al_2O_3$ | 43.9 | 54.0 | 58.6 | 69.3 | 77.9 | 84.5 | 88.9 |
| $SiO_2$ | 48.6 | 36.9 | 34.5 | 22.1 | 17.0 | 7.9 | 4.8 |
| $Fe_2O_3$ | 2.1 | 1.1 | 1.4 | 1.0 | 1.5 | 1.2 | 0.5 |
| $TiO_2$ | 1.4 | 1.3 | 0.5 | 1.6 | 2.2 | 1.8 | 0.6 |
| CaO | 0.5 | 0.4 | 0.1 | 0.1 | 0.2 | 0.1 | 0.1 |
| MgO | 0.1 | 0.2 | 0.1 | 0.2 | 0.1 | 0.1 | 0.1 |
| Alk | 0.4 | 0.3 | 0.3 | 0.2 | 0.2 | 0.2 | 0.2 |

[a]ASTM C-704.
[b]ASTM C-133.
[c]ASTM C-583.

### 9.2.2. Monolithic Refractories

Monolithic refractories are classified by physical properties, consistencies, and grain sizing (e.g., powder, paste, clay). Construction methods have been developed to suit various installation procedures such as pouring, troweling, gunning, ramming, patching, gunning, blowing, slinging, vibrating, spraying, foaming, or injecting. The castable (poured), plastic (rammed), or blown (sprayed, foamed) forms of refractory materials are generally superior to layed-up, dipped refractory brick construction because they are less prone to leak, and they provide extended furnace life.

Monolithic material can be transferred by pumps over long distances and in large quantities for pouring in position. Much labor can be saved by selecting the right method for transferring and applying monolithic materials. Because the weight of monolithic refractory in a furnace is held by a large number of supports, small or large areas can be repaired or replaced wherever necessary without affecting the surrounding area. Monolithic refractory materials adhere well to surrounding materials.

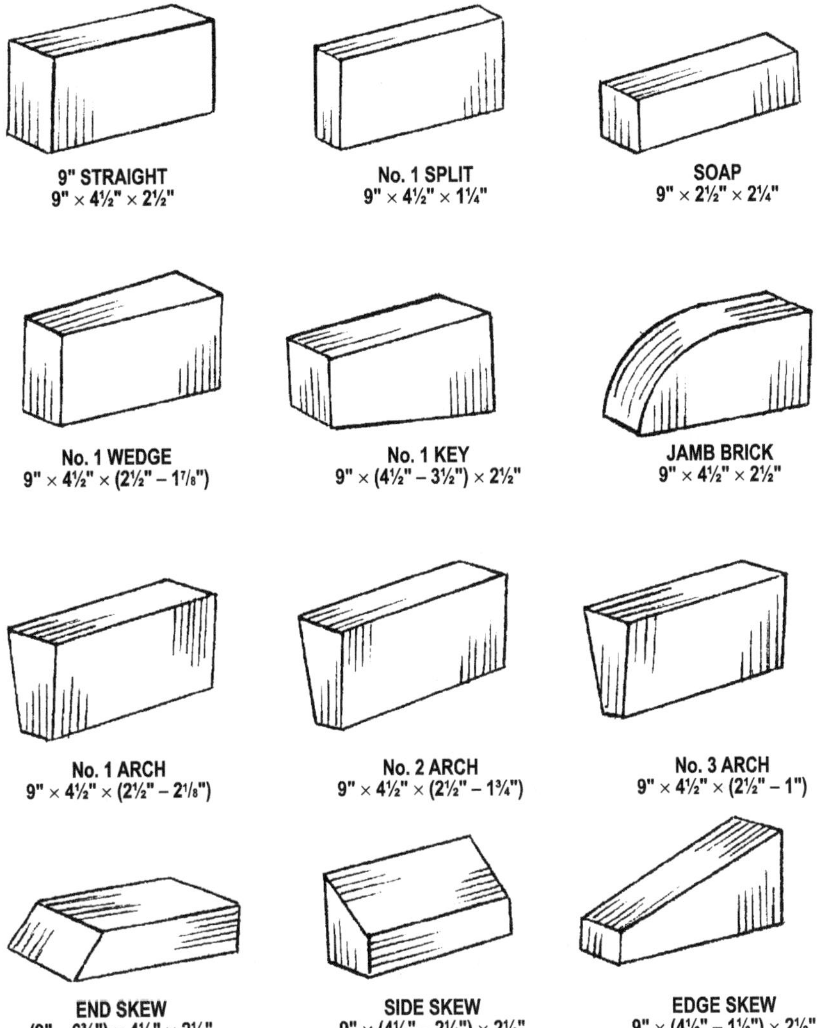


**Fig. 9.1.** Some typical refractory shapes.

Monolithic refractories are suitable for walls that must be gas tight. The weight of the furnace itself is sustained by supports that help the monolithic material adhere to the shell and prevent gas leakage.

Monolithic refractories have lower thermal expansion than most refractory bricks. Whatever small expansion does occur can usually be absorbed by the supports. Therefore, unlike refractory bricks, monolithic refractory walls do not require clearances for thermal expansion. Clearances required for brick construction may allow passage for furnace gas leaks out or air into a furnace. The superior sealing capability and reduced expansion of monolithic refractories make them suitable for higher furnace pressures and temperatures. Among the reasons for the growing use of monolithic refractories are versatility of the material and the flexibility of the self-supporting anchor system.

Some of the many variations of monolithic refractories are:

*Castable* refractories consist of course and fine grains with suitable bonding cement. After mixing with water, these are poured in place using molds or pouring forms.

*Trowelable* refractories are a kind of castable refractory mortar with a consistency that makes it easy to trowel into place—very useful for patching and for shaping complex surfaces.

*Plastic* refractories contain a binder material, and are tempered with water so that they have suitable plasticity for pounding or ramming into place.

*Ramming* refractories are similar to plastic, but somewhat more stiff.

*Patching* refractories, tempered with water and/or with a binder added for softer plasticity, which permits patching in place.

*Gunning* refractories have course and fine refractory grains and bonding agents, suitable for installation with a gunning machine.

*Injection* refractories can be injected in a slurry state into small places such as gaps and wide cracks, and for filling molds with narrow passageways.

*Vibratable* refractories are castable refractory materials that should be vibrated to fill all the voids in a mold.

*Slinging* refractories are for installation with a slinging machine.

*Coating* refractories are in the form of a thin slurry that can be brushed onto or otherwise coated on the working surface of other refractories.

*Refractory mortars* are finely ground refractory materials that, when tempered with water, become trowelable for bonding layed-up refractory shapes.

***Castable refractories*** are made in many compositions for specific uses, including insulating castables. Castables are generally formulations of heat-resisting aggregates and alumina cement that can be poured into forms. They also may be formulated for gunning or troweling. Castables are hydraulic or chemical setting. The degree of chemical setting varies considerable. Setting characteristics, including the ultimate strength of the refractory, vary with the bonding material.

With any material used in high-temperature applications, the effect of linear *thermal expansion*, and especially the permanent linear change, must be considered. Shrinkage of castables is less than that of plastic refractories; therefore, permanent linear change is less. Castable refractories are significantly superior to firebrick in permeability resistance and spalling resistance. Plastic refractories have better spalling resistance than either firebrick or castables.

*Thermal conductivity* of castable refractories is as much as 35% less than that of firebrick, that is, castables are better insulators. High alumina castables have high abrasion resistance, and are more durable at high temperatures.

**TABLE 9.3. Recommended minimum monolithic refractory thicknesses**

| Method | Vertical (sidewalls) | Horizontal (roofs, hearths) |
|---|---|---|
| Welling | 3 in./77 mm | 2.5 in./64 mm |
| Pouring | 4 in./102 mm | 6 in./154 mm |
| Gunning | 6 in./154 mm | 6 in./154 mm per layer |
| Ramming | 7.5 in./192 mm | 8 in./203 mm |

## 9.2.3. Furnace Construction with Monolithic Refractories

Furnace construction with monolithic refractories is determined by the method(s) to be used in installing the furnace lining, which may be dictated by furnace configuration, time limitations, or other local site conditions. The furnace designer must determine the minimum refractory thickness required. (See table 9.3.) Thicker-than-minimum linings are usually mandated by fundamental economic considerations such as fuel conservation (less heat loss), extended lining life, and reduced maintenance. Additional lining thickness also may be required because of workplace environmental considerations (e.g., external shell temperature or interal atmosphere).

Thermal expansion of monolithic refractories is less than that of refractory brick, but it must be considered. Monolithic refractories do shrink when cooled after heating. The following is a satisfactory method for determining the need and size of expansion joints. Determine the average temperature between the hotface and the junction with the next layer of lining. Multiply that average by the coefficient of expansion of the refractory, and by the longest dimension of the section to be installed. Deduct the shrinkage figured from the %permanent linear change, furnished by the supplier. If the result is positive, that number indicates the size of the expansion joint that must be supplied. Offset expansion joints are preferred. (See fig. 9.7.)

## 9.2.4. Fiber Refractories

Refractory materials can be melted, spun, and blown into fiber strands similar to "wool" or "blanket" insulations. They are used in many medium- and low-temperature furnaces and ovens furnaces, and for outer layers in multilayered refractory walls. Because of all their small air spaces, they are much better insulators than solid refractories, but they are more fragile, less durable, and more difficult to install so that they do not settle, shrink, or otherwise lose their good insulating property. Many of the suggestions in a later section on insulation installation can apply to fiber refractory installation.

A technique for use of fiber refractories in higher temperature furnaces is to fold and compress them in many horizontal layers, stacked one above the other, to form thick insulating walls. See the door and walls in Figure 3.5. Patented holders keep them in place and compacted. Abrasion, shrinkage, and porosity can be problems, but careful installation and use has proven them successful in specific applications. Installation can be faster and less expensive than monolithic and other rigid wall construction methods.

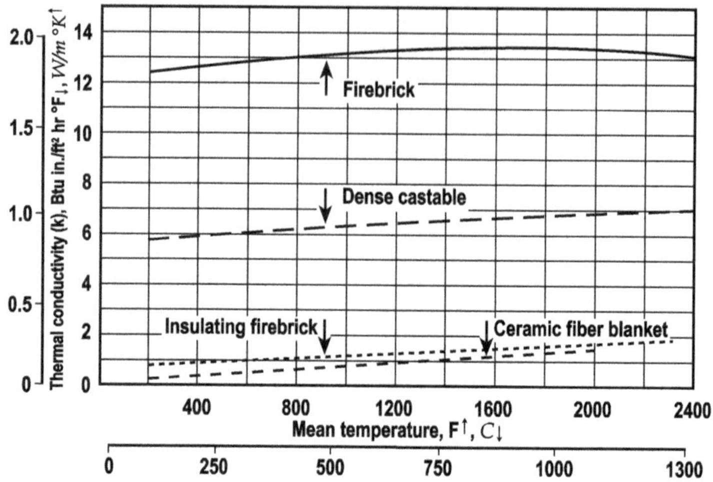


Fig. 9.2. Overview of fuel-saving characteristics of four classes of refractory linings. Lowest conductivity saves most fuel. Other considerations are weight, life, and ease of installation. Fuel savings with an added veneer of insulating refractory are usually greater if the furnace operation is cyclic than if continuous (Courtesy of *reference 13*).

## 9.3. WAYS IN WHICH REFRACTORIES FAIL

At temperatures above 2000 F (1367 C), refractories become more and more porous, allowing the hot furnace gases (poc, which may be several hundred degrees F above the zone temperature) to attack the chemistry of the refractory. In time, this attack reduces the surface strength of the refractories and causes their melting temperatures to be lowered. Examples follow. (See fig. 9.2.)

*Case 1: Hearths* In rotary-hearth steel-reheat furnaces, where load pieces are positioned directly on the hearth, the weight of the loads will cause depressions in the hearth after perhaps 6 mo. of operation. The cure for this problem is to build the hearth with stainless-steel rails built into the refractory hearth so that the "ball" of each rail protrudes above the top of the refractory surface 2 to 3 in. (5.8 to 7.6 cm). With this arrangement, loads are supported from deep in the hearth refractories where materials are cooler, and therefore stronger and not attacked by the furnace gases.

To also gain a heat transfer benefit from the rails mentioned previously, it is suggested that they be installed at an angle to the direction of load movement. Then, they can act as little piers between which hot poc gases from enhanced heating burners can travel to add to the effective heat transfer area on the bottom sides of the loads. That bottom area might have formerly had zero heat-transfer effectiveness. Even without enhanced heating, there will be some gain because the pieces will not be sitting directly on a relatively cold hearth.

The stainless-steel rails should have at least 22% chromium and 25% nickel. The ideal would be 28% chromium and 35% nickel, but the added cost may not be justifiable.

*Case 2: Roofs, Walls, Burner Tiles* If combustion gases are directed toward or across these surfaces, they become more porous, lose strength, slump, and even melt. Very dense refractories should be used at least near the surfaces exposed to gases hotter than zone temperature. Generally, a higher percentage of alumina makes a refractory more dense, and therefore less subject to the above problems. Strict attention to refractory installation instructions will minimize these problems. For burner tile installation, refer to the Appendix.

*Case 3: Thermal Stress, Vibratory Stress* Typical examples are burner quarls or tiles (which also are subject to Case 2 problems), but expand more on their inside (hottest) surfaces. Round ID and OD tiles have a slight advantage in this regard. Surrounding them with a "collar" of high-strength refractory is a sort of "brute force solution." Whatever surrounds them must be installed with a 360-degree-tight contact to prohibit leakage around the tile, which could overheat the furnace casing. Burner tiles in tall multilayered walls are subject to large cumulative expansion differences from floor to burner elevation.

*Case 4: Physical Wear, Some Atmospheres, Liquid Slag or Scale, Leaking Cooling Water* These also can be bad for refractories. After installation of castable, rammed, and gunned refractories, a long, slow dryout period is necessary to prevent spalling or explosions from steam formation within the refractories.

## 9.4. INSULATIONS

Most insulating materials achieve their low thermal conductance by virtue of the many small air spaces built into their structure. Nitrogen or other inert low-conductivity gases also can be used, but the cost of sealing in such alternate gases is usually prohibitive. The air spaces do not need to be small, but they must be narrow enough to prevent internal convection that would diminish their insulating effectiveness. Furnace refractory walls would have very dense material at the hotface (inside surface), followed by a layer of less dense refractory, then followed by a very porous or insulating material—for a "firebrick equivalent" of 55 in.

Soft, flexible "blanket" insulations are often the outer layer of a furnace or oven. To diminish outer surface heat loss, follow these admonitions:

1. Maintain a reflective or light-colored outer surface. Aluminum paint or foil is excellent on the outside metal "skin" if free of dust and oxide.
2. Keep insulating surfaces away from fans, drafts, winds, rain, and dirt.
3. Avoid dust-laden or fungal atmospheres.
4. Clean regularly by gentle blowing or brushing that will not change the surface reflectivity.

5. Insulations usually work better if not painted—unless already oxidized, in which case it is probably better to replace them frequently.
6. Prevent vibration which enhances heat loss and shortens insulating life.
7. Avoid puncturing, compressing, or touching. Do not walk on.
8. Perhaps add a protective sheet-metal skin, but with provision for easy opening for inspection. When the last layer of a composite wall is a fiber insulation, make certain that it is backed by a near-gastight "skin." Otherwise, the fiber will be of no value because hot gas will move through the fiber.
9. Keep all persons in the vicinity aware of these requirements.
10. Beware of health hazards for installers. They should wear breathing masks and eye and ear protection.

Rigid foamlike insulations are more durable, but still subject to crushing and to surface changes. Insulations made by spinning, weaving, knitting, braiding, blowing, or foaming refractory materials are generally preferred over animal, paper, plastic, metal, or glass fibers. All must be fireproof for industrial heating applications. New insulations must be tested carefully—not on a production line.

## 9.5. INSTALLATION, DRYING, WARM-UP, REPAIRS

Great care is necessary when installing refractory and insulating materials to assure a leak-proof enclosure. Outleaking hot gases can lead to runaway damage. Inleaking cold gases can be detrimental to product quality and raise fuel bills.

Do not compress insulating materials because their small air spaces provide greater insulating capability. Hooks, hangers, or shelves may help keep insulting materials from self-compressing with age and vibration. Both flexible blanket insulations and rigid insulating material such as solidified foams (refractory or organic) need to be carefully installed with no appreciable gaps between pieces, or between them and harder refractory or the metal "skin" of the furnace.

Installers must follow the supplier's instructions *very* carefully regarding mixing proportions, dryout time, and warm-up procedures. Failure to mix water with the as-received powder or granules exactly as specified in the supplier's instructions can lead to poor bonding or difficulty in applying the mixture. With some materials, a too-rapid drying or warm-up can result in a tight surface (a 'skin') that acts as a sealer to hold in remaining liquid. Continued heating will cause the trapped water within the undried mix to "flash" to steam, increasing its volume 1,600 times and resulting in many little explosions that rupture the "skin," often causing the product to be unacceptable, or subject to spalling.

An essential part of any drying operation is providing ample flow of the drying medium (usually warm air, not hot, air) to accomplish mass transfer, that is, to carry away the air that becomes saturated with moisture. This is a phenomenon similar to convection—very velocity dependent. Therefore, thoughtful positioning of circulating fans or high-velocity excess air burners during dryout is essential.

> **Warning:** When using burners for dryout or warm-up operations, do not skimp on adequate flame safety and programming just because you think this is a temporary operation. Too many new furnaces have never produced a nickel because of start-up explosions. The most dangerous times for furnaces are, as with airplanes, (pardon the expression) during takeoff and landing.

*Dryout times* generally recommended for large areas and thicknesses are 55 to 60 hr. Proportionally less time is reasonable for smaller areas and thicknesses, including patchings. However, if steam is noticed coming from the refractory, the drying process should be slowed by delaying any further temperature rise until steaming stops. Then, resume the temperature rise rate, but do not try to catch up to the original temperature profile. Allow the stopped period to extend the dryout time.

*Warm-up times* can be considerably less than dryout times, if no moisture needs to be driven off. Some warm-up time is important even for previously dried-out furnaces to minimize refractory spalling because of too-rapid or uneven thermal expansion of the dry, solid refractory.

## 9.6. COATINGS, MORTARS, CEMENTS

Patented coatings with high emissivity and absorptivity have been used successfully, but warrant careful investigation to be sure that the emissivity of the proposed new surface is sufficiently higher than the existing surface to warrant the investment. Will the better emissivity be permanent? Could it be subject to spalling, damage, or degradation because of furnace atmosphere?

Mortars and cements should be compatible with the chosen brick material. It is important to remember that simply dipping each brick in "slip" (very runny, thinned, less viscous mortar) may not provide sufficient bonding. A likely problem is judging that there has been sufficient curing or dryout time because the slip on the exposed surfaces of the bricks is dry, but not thinking about the much, much longer curing time required for the slip *between* bricks. Even a very experienced bricklayer for architectural brick may have inadequate judgment (feel) for when the mortar is not right for good furnace refractory work. Hurrying a furnace mason may be penny-wise and pound-foolish.

## 9.7. HEARTHS, SKID PIPES, HANGERS, ANCHORS

In continuous furnaces, cast or wrought heat-resisting alloys are used for skids, hearth plates, walking beam structures, roller, and chain conveyors. In most furnaces, the loads to be heated rest on the hearth, on piers to space them above the hearth, or on skids or a conveyor to enable movement through the furnace. The furnace interior can be observed through airtight peepholes or closeable sightports. (See sec. 9.8 for details on materials.)

### 9.7.1. Hearths

To protect the foundation and to prevent softening of the hearth, open spaces are frequently provided under the hearth for air circulation—a 'ventilated hearth.' Natural convection cooling of these spaces under a furnace is really not very effective—unless some forced flow cooling air is provided. Actually, a solid contact between furnace bottom and the earth may be better than still air cooling. If, however, the hearth is so hot that conducted heat might damage the furnace foundation, *forced underside ventilation* is necessary.

Because of possible abrasive damage during loading and unloading, hearths are often built up with extra layers of very dense refractories. Hard-fired brick shapes may be preferred over cast or rammed refractories. However, if the refractory shapes happen to buckle upward, loading of new pieces may catch on them and cause major damage. No matter now the hearth is constructed, operating personnel must be continually advised that: *Correct loading procedure on any type hearth is: (1) to let the load pieces down very gently in their final hearth location, (2) never lower a piece so that one corner or side touches the hearth surface before the entire bottom face contacts the hearth, and (3) never attempt to slide, push, or nudge pieces after they are in contact with the hearth surface. In other words, always save time and hearth by carefully doing it right the first time.*

In modern practice, hearth life is often extended by burying stainless-steel rails up to the ball of the rail to support the loads. The rail transmits the weight of the load 3 to 5 in. (0.07 to 0.13 m) into the hearth refractories. At that depth, the refractories are not subjected to the hot furnace gases that, over time, soften the hearth surface refractories. The grades of stainless rail used for this service usually contain 22 to 24% chromium and 20% nickel for near-maximum strength and low corrosion rates at hearth temperatures. With stainless-steel rails imbedded in a hearth, the hearth life can be extended by a factor of 1.5 to 3 times. Attempts to use other imbedding material have not been successful.

Hearths in high temperature furnaces, particularly in rotary hearth steel reheat furnaces, may suddenly fail with the steel load pieces sinking into the weakened refractory. This is caused by the long-term penetration of hot furnace gases into the refractory hearth material, changing its chemistry to lower its melting point. The aforementioned use of stainless-steel rails embedded in the hearth refractory extends the useful hearth life by supporting the furnace loads. The stainless rails extend the load deep into the refractory to a level where the softening point is still very high, so no deformation of the hearth occurs. Obviously, taller stainless rails will stretch the time to the next hearth rebuild.

### 9.7.2. Skid Pipe Protection

Modern full insulation reduces heat loss from pipes by more than 85%. The volume of cooling water required is less. Figure 9.3 shows a typical arrangement of skid pipes and supports for a pusher reheat furnace.

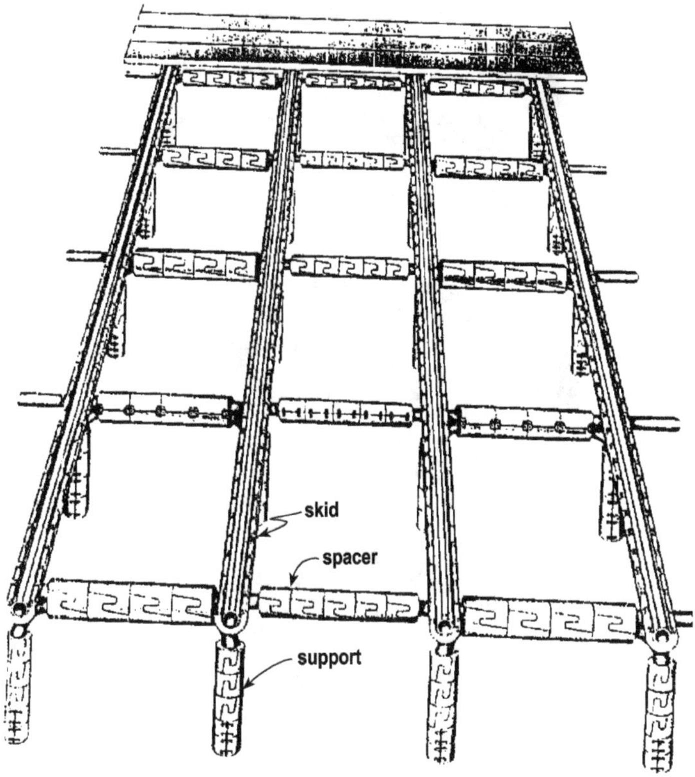


**Fig. 9.3.** Insulated water-cooled skid pipe and support arrangement for a pusher type furnace.

Figure 9.4 shows three types of support and skid pipe insulating covers. Type A is designed for risers, jacks, or crossovers. This has a lightweight insulating cover of interlocking segments having a flexible ceramic inner layer bonded to a rigid outer layer of formed ceramic fibers. Type B is similar to type A, but is designed for use in severe duty zones on risers, cross pipes, and jacks. It is welded to the pipe and finished in the same way as type C. Type C is used on the skid pipe, a severe service area. It is made from a 3000 F severe duty castable refractory and reinforced with stainless-steel fibers. The cover is welded to the skid rail though the openings as shown. The openings and all other voids are closed with a troweled castable refractory after the welding.

Figure 9.5 shows typical bake-out schedules for refractory construction, including skid and support refractory. The supplier's specific schedule must be used because there are so many different brands with varying ingredients and formulations. A 24-hr curing time should precede these. Line A is for new or major replacement refractory construction. Line B is for returning a furnace to operating temperature after it has cooled to the cure temperature. It is advisable to keep a furnace warm at curing temperature during vacations and other downtimes to avoid potentially damaging moisture accumulation in or on the refractories.

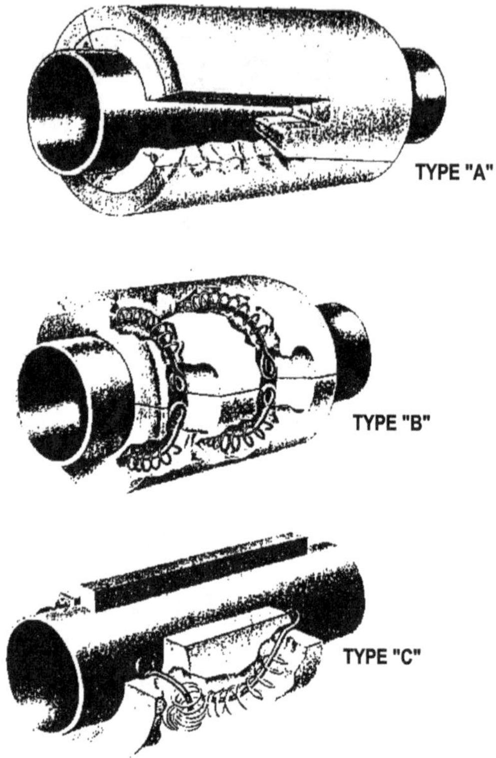


**Fig. 9.4.** Some types of skid pipe and support pipe insulators. Courtesy of Plibrico Company.

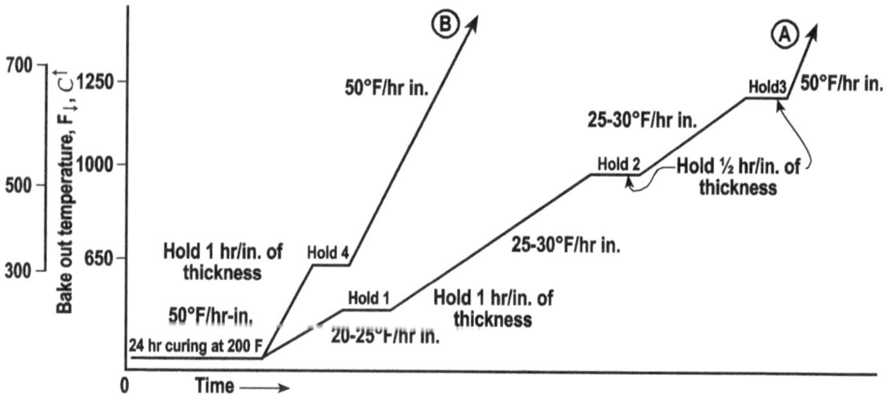


**Fig. 9.5.** Typical refractory bake-out schedules. *The specific schedule by the supplier must be used* because different designs use distinctively formulated materials. For multilayered linings, the hotface lining dictates the schedule.

Line A of the graph shows a 24-hr cure not to exceed 200 F (93 C) *after a new or major replacement refractory construction.*

1. From Cure to Hold 1, raise temperature 20°F to 25°F (11°C to 13°C) per hour for each inch of refractory thickness.
2. At Hold 1, maintain temperature at 350 F (177 C) for 1 hr for each inch of thickness. This critical period should be monitored closely.
3. From Hold 1 to Hold 2, increase temperature 25°F to 30°F (13°C to 16°C) per hour for each inch of thickness.
4. At Hold 2, maintain 1000 F (538 C) for $\frac{1}{2}$ hour for each inch of thickness.
5. From Hold 2 to Hold 3, again increase temperature 25°F to 30°F (13°C to 16°C) per hour for each inch of thickness.
6. At Hold 3, hold 1250 F (677 C) for $\frac{1}{2}$ hour for each inch of thickness.
7. From Hold 3 to operating temperature, increase temperature 50°F per hour for each inch of thickness.

Line B of the graph shows a 24-hr cure not to exceed 200 F (93 C) *after returning a furnace to operating temperature after it has cooled to cure temperature.*

1. From Cure to Hold 4, raise temperature 50°F (27°C) per hour for each inch of refractory thickness.
2. At Hold 4, maintain 350 F (177 C) for 1 hr for each inch of refractory thickness.

### 9.7.3. Hangers and Anchors

Although these two terms are sometimes used interchangeably, anchors are ceramic or high-temperature metal alloy shapes embedded in a monolithic refractory whereas hangers are usually the metal holders for the anchors. The hangers and anchors not only support the refractory wall or roof but do so while allowing slight expansion and contraction movements. (See fig. 9.6.)

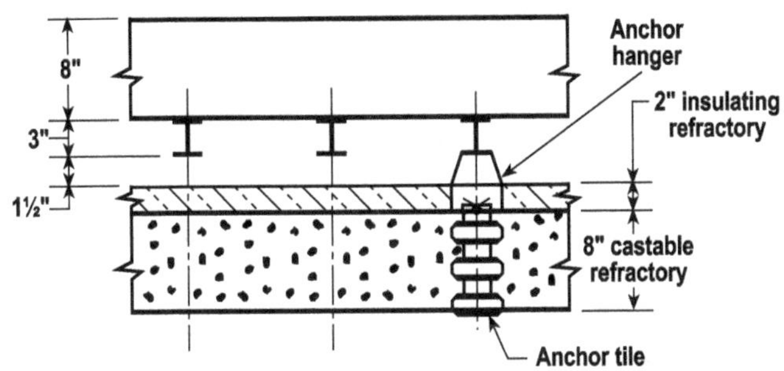


**Fig. 9.6.** Typical monolithic roof construction. Higher temperature operations may require thicker refractory, insulation, and cooling space.

Anchors and hangers must maintain their mechanical strength at the temperatures encountered. Data such as that developed in the example at the end of this chapter can provide the basis for determining the lining temperature gradient as well as guidance in selecting the type of support to be used. Alloy metal anchors that are to be imbedded in a monolithic refractory should have a flexible coating to allow for differences in the thermal expansions of the refractory and the metal.

After the type of support has been determined, spacing becomes a significant factor. There are two different ways to calculate the spacing, but they are contradictory in some respects. Method 1 is based on the premise that a thicker lining has more weight to support, so the supports should be closer together. Method 2 surmises that a thicker lining is stronger, so the supports can be farther apart. The conservative approach is to figure it both ways and select the way that results in the supports closer together.

Equation 9.1 assumes equal support spacings in both directions.

$$\text{Pounds load on one support} = (\text{Spacing, in.})^2\ (\text{lining thickness, in.})\ (\text{lining density, pounds/in.}^3) \tag{9.1}$$

Figure 9.7 illustrates an offset expansion joint in a monolithic wall.

Figure 9.8 shows some more typical monolithic refractory supports.

Another excellent application for anchors and hangers is in on-site rammed or cast refractory burner tiles for cases where the burner manufacturer does not provide a kiln-fired burner tile. These are usually for large burners. Figure 9.9 is a typical drawing provided by a burner manufacturer, with detailed dimensions and angles that

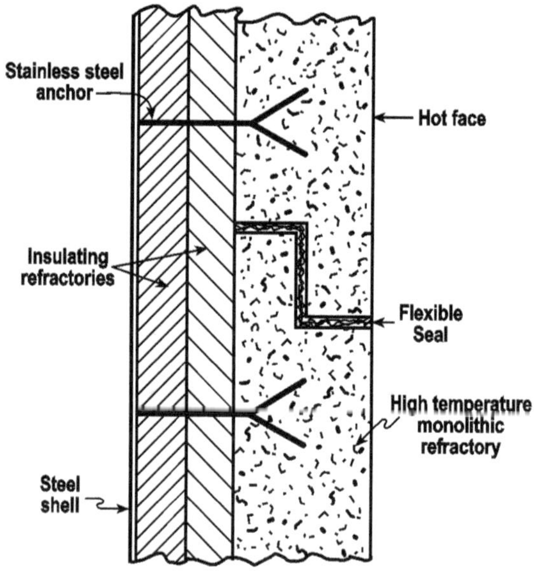


**Fig. 9.7.** An offset expansion joint in a monolithic wall with stainless-steel Y-anchors.

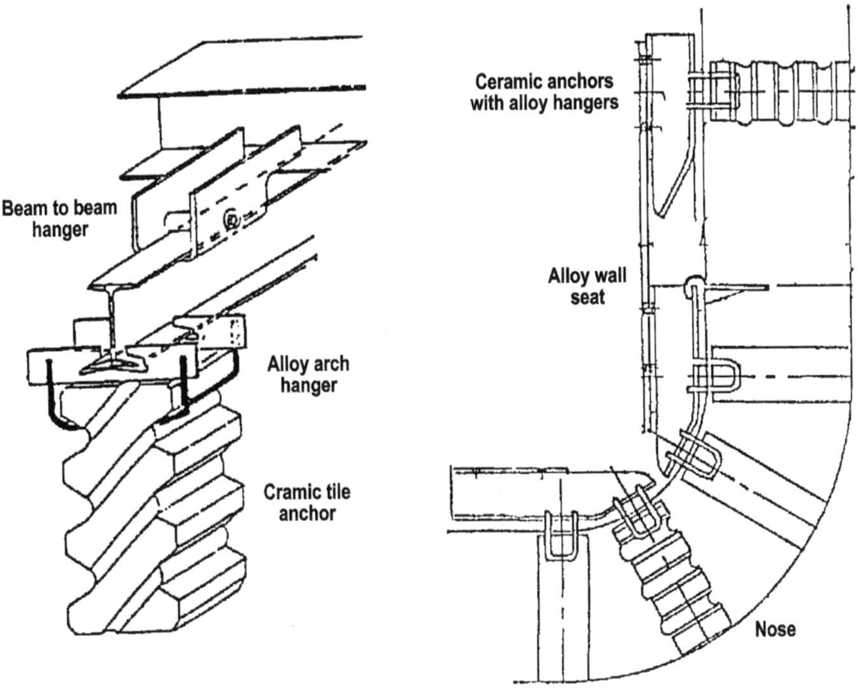


**Fig. 9.8.** Monolithic refractories in roof (arch) construction and in nose construction, using supports consisting of ceramic anchors held by alloy hangers.

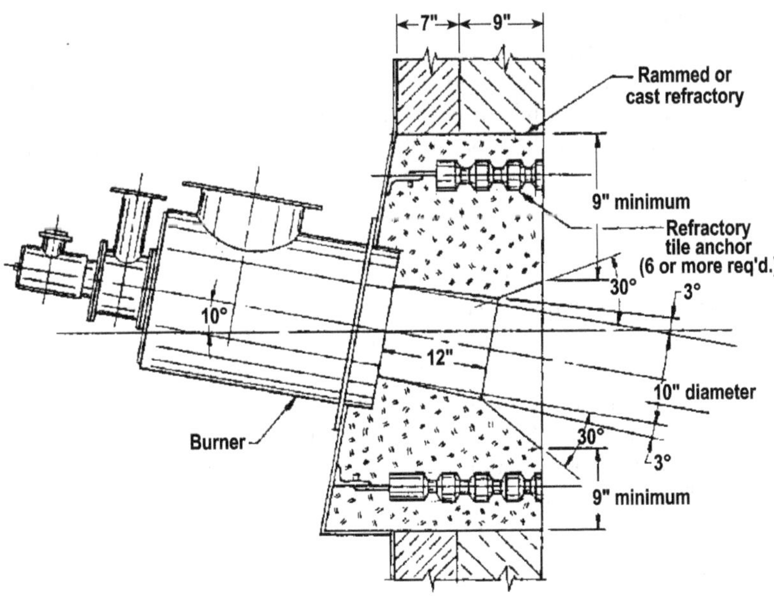


**Fig. 9.9.** Burner manufacturer's drawing with precise instructions for installation with rammed or cast monolithic refractory using ceramic anchors and alloy hangers.

must be observed precisely to assure that the burner flame will perform as promised. For these large installations, a rework results in a high cost in production time and labor; therefore, doing it right the first time must have a very high priority.

## 9.8. WATER-COOLED SUPPORT SYSTEMS

This section emphasizes water-cooled supports for skid rails and other conveying systems, but much of the information herein can be adapted to water-cooled doorframes and other equipment that needs cooling.

In furnaces with bottom zones, such as pusher or walking beam steel reheat furnaces, each *skid rail*, on which the loads rest or slide, consists of a schedule 160 pipe, 6.625" (0.1683 m) OD with 0.718" (18.24 mm) wall thickness, through which cooling water is circulated. A solid *skid wear bar* is securely welded onto the top surface of the pipe. The skid wear bars are often small diameter bars of heat-resisting, wear-resisting material. Their small diameter allows less contact area with the load pieces, thereby minimizing heat loss from the loads.

The water-cooled *skid rail pipe* supporting the skid wear bar is insulated with one or two different insulating materials to reduce heat gain (as these are subject to the same hot furnace gas heat transfer as are the loads). A group of crosswise water-cooled support pipes (*crossovers*) support the skid rail pipes from below and are attached to the furnace sidewalls. Vertical pipes (*risers*) support the crossover pipes. The outer surfaces of all the skid and supporting pipe structure must be capable of withstanding physical and thermal shock as well as chemical attack from the bottom-zone furnace gases.

The skid rail support system "shadows" some of the bottom-side heat transfer surface area of the loads (*a*) by its projected area and (*b*) by its gridwork of thick-walled "slots" that significantly reduce the radiation from bottom-zone refractories and gases. The degree of heat transfer reduction depends on the ratio of the skid spacing, $D$, to slot depth, $X$*. For an $X/D$ ratio of 4.5:1, figure 5.7 shows that with a rectangular opening having $W{:}D = 2{:}1$, the heat transfer to the undersides of the loads would be about 88% of what it would be if the slot thickness $X$ were zero.

Figure 9.10 shows a way to get more rigidity and strength in the skid pipe arrangement by stacking them two-high. This allows more horizontal space ($D$ dimension in fig. 5.7) between skid pipes, but adds to the depth ($X$ dimension)*. Equal spacing of all skid pipes having a large $D/X$ in figure 5.7 yields high radiation reception on the loads' bottom sides through the vertical slots. But in figure 5.7, the radiation rates drop off radically on the steep left part of the curves. Comparing the equal spacing with unevenly spaced skids (*bottom half* of fig. 9.10), the average of the high radiation of a wide $D$ and the low radiation from a narrow $D$ will be appreciably lower than the average from two slots of equal $D$. Equal spacing also will give better structural

*Figure 5.7 shows a horizontal slot as in the sidewall of a furnace, but for this case, with radiation shining up through one of the grids of slots formed by the skid rails and their crossover pipes, $X$ is vertical and $D$ is horizontal.

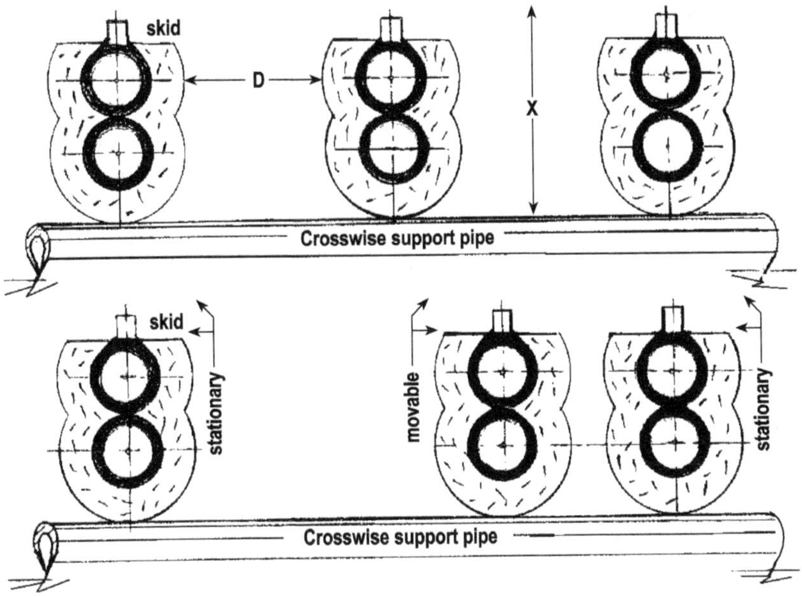


**Fig. 9.10.** Double-high skid pipes admit more radiation from the bottom zone to the loads than do single-high skid pipes, which must be wider for the same load bearing capacity. Equally spaced skid rails (top view) average more heat transfer to the load's undersides than do unevenly spaced rails (*lower view*)—by a ratio of 0.61:0.57 for one specific set of dimensions.

support. The number of skid rail pipes spaced across the furnace is determined by the load weight and a normal overhang of loads near the furnace walls, which should not exceed 18" (0.46 m). To find the optimum design requires careful evaluation of strength versus heat transfer and of capital costs versus operating costs.

In a walking beam furnace, the number of walking skids is one less than the number of stationary skids. They should be spaced out from one another as much as the load piece strength will allow because, as shown in the discussion earlier, bottom-zone heat transfer to the undersides of the loads suffers from narrow spacings (small $D$) and tall (high $X$) slots in the supporting gridwork. Evaluation of this effect should be recalculated for every combination of dimensions using figure 5.7.

When designing a skid system, the number of skid pipes and the number of crossovers should be kept to a minimum, the slot depth kept as small as possible, and insulation thickness as thin as reasonable with good strength. Generally, crossovers are limited to where there are riser supports. Wear bar thickness and height are compromises between minimizing cold streaking on the load bottoms because of too much heat loss to skid cooling water, and a reasonable wear time between wear bar replacements.

Recirculating water-cooling systems should have water treatment to control hardness to near zero and to prevent oxygen corrosion. If there is a steam boiler nearby, a common water treatment may be possible, but this should be explored with care. The cooling-water temperature *rise* should not exceed 20°F (11°C), and steaming should be avoided by keeping the maximum water temperature below 130 F (54 C). In determining the quantity of water to be circulated, it is important to realize that insulation may deteriorate, in which case the heat-carry-out capacity of the cooling-water system may have to increase tenfold.

Skid rail insulation warrants constant monitoring! An emergency second source of water is essential because the loss of cooling water can be very costly. Water should be *de-aerated* and leaks corrected promptly. If any air were to get into the cooling water, it would be swept along inside the top inner surface of a cooling water pipe. Air has lower thermal conductivity and heat capacity than water; thus, it will not pick up heat as water does; that is, air is a poorer coolant than water. The pipe will get very hot wherever there is air. Any overheated area on the pipe will therefore lose its strength, causing a support system failure that can be catastrophic. To prevent this, air must be bled out of the water from the top of the skid pipe and sloped continuously with no high spots all the way to the "bosh," a water-collecting container where air can be separated.

Scale formation in water-cooling systems weakens the pipes and reduces their heat-absorbing capability (like inside insulation), causing the outside surfaces to become very hot, reducing their strength and allowing them to bend, break, or burst. Oxygen corrosion from inadequately treated water will cause pits, which will become leaks into the furnace, requiring added fuel because of water's high latent heat of vaporization. Refractories will be harmed and short-lived if leaking water strikes them. If water leaks strike the furnace loads, the resultant temperature differentials may interfere with processing or cause rejection by quality control (or worse, the customer).

Load support system designers must realize that skids will never form an absolutely level pass line, nor will the loads be perfectly straight; therefore, the entire weight of any load piece may be on just two skids, the entire load weight of which might be on only two crossovers, the entire load weight of which may be on only two risers.

Top-quality welding is crucial for all water-cooling-system parts. A weld without full penetration is a crack, a failure. All welds must be sound tested. The welding of skids is critical and should have full penetration welds to succeed. A very successful way to reduce expansion problems is to have the skids be short bar pieces with bevels on the ends and about $\frac{1}{8}$ in. (3.2 mm) spaces endwise between them. To reduce heat transfer to the skids, it is advisable to use a high-temperature, low-conductivity (such as cobalt) wear bar on the skids in walking beam structures.

## 9.9. METALS FOR FURNACE COMPONENTS

Heat processing industries depend on materials that have strength at high temperatures.

Irons and steels have been the workhorses for holding industrial furnace refractory structures together. Metals that are to have extended life in furnaces with temperatures in excess of 1400 F (760 C) must meet the following requirements:

1. Not subject to rapid oxidation (scaling, slagging). (See table 9.4.)
2. Resistant to attack by mildly sulfurous atmospheres
3. Creep strength must be such that deformation will take place over an economically viable period of time when it can be repaired or replaced
4. Irreversible growth (by thermal expansion, grain change, oxidation) must not exceed the tolerance of the application

### 9.9.1. Cast Irons

Gray cast iron gives good service up to 1300 F (704 C). It has low tensile strength (fig. 9.11), so it should only be used in compression. It gives good service up to 1300 F (704 C).

Nodular cast iron has higher tensile strength than gray iron and will give good service up to 1600 F (871 C). It can be used in tension. Cast irons oxidize quite rapidly at high temperatures, although they are not as susceptible to oxidation as is steel.

**TABLE 9.4. Scaling temperatures of typical steel alloys**

| | Chromium | Nickel | Scaling Temperatures | | | |
|---|---|---|---|---|---|---|
| | | | Long-Term | | Intermediate | |
| Type | %Cr | %Ni | | | | |
| 301 | 17 | 7 | 1700 F | 927 C | 1600 F | 871 C |
| 302 | 18 | 8 | 1700 | 927 | 1600 | 871 |
| 302-B | 18 | 8 | 1800 | 982 | 1650 | 899 |
| 303 | 18 | 8 | 1700 | 927 | 1400 | 760 |
| 304-S | 18 | 8 | 1700 | 937 | 1600 | 871 |
| 305 | 18 | 8 | 1700 | 937 | 1600 | 871 |
| 308 | 20 | 10 | 1700 | 937 | 1600 | 871 |
| 309 | 25 | 12 | 2000 | 1093 | 1800 | 982 |
| 310 | 25 | 20 | 2100 | 1149 | 1900 | 1038 |
| 314 | 25 | 20 | 2100 | 1149 | 1900 | 1038 |
| 316 | 18 | 8 | 1700 | 937 | 1600 | 871 |
| 317 | 18 | 8 | 1700 | 937 | 1600 | 871 |
| 321 | 18 | 8 | 1700 | 937 | 1600 | 871 |
| 347 | 18 | 8 | 1700 | 937 | 1600 | 871 |
| 403 | 12 | – | 1300 | 704 | 1500 | 816 |
| 405 | 12 | – | 1300 | 704 | 1500 | 816 |
| 410 | 12 | – | 1300 | 704 | 1500 | 816 |
| 414 | 12 | 2 | 1250 | 677 | 1400 | 760 |
| 416 | 12 | – | 1250 | 677 | 1400 | 760 |
| 418 | 12 | – | 1300 | 704 | 1500 | 816 |
| 420 | 12 | – | 1200 | 649 | 1400 | 760 |
| 440 | 17 | – | 1400 | 760 | 1500 | 816 |
| 442 | 21 | – | 1800 | 982 | 1900 | 1038 |
| 446 | 28 | – | 2000 | 1093 | 2150 | 1177 |

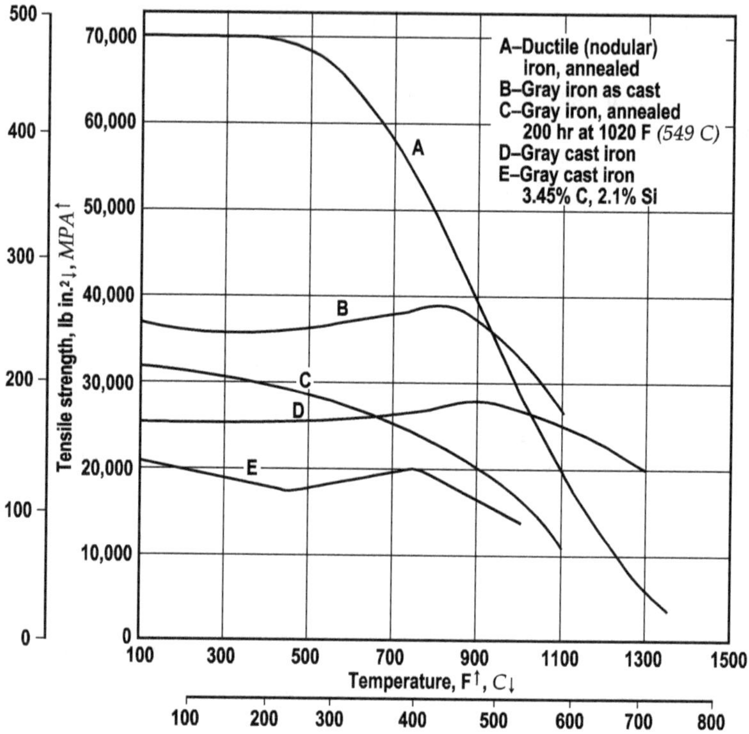


Fig. 9.11. Tensile strengths of cast irons at elevated temperatures.

***9.9.1.1. Growth Problems.*** Expansion of cast iron is not reversible and continues to grow at temperatures of 1000 F to 1500 F (538 C to 816 C). Additives, such as chromium and silicon, reduce growth somewhat. (See figs. 9.12a and b.) Tests have shown 3.5% growth for plain cast iron during 35 cycles totaling 320 hr at 1472 F (8900 C). It is evident that repetitive heating as well as temperature must be avoided to minimize growth, and that otherwise, ample space must be provided to accommodate this growth. Any cast iron can be used below 1300 F (704 C). Ductile (alloy) iron is serviceable up to 1600 F (871 C).

Steels also exhibit permanent growth after repeated heating to 1500 F (816 C) and hotter, but steel's growth is less than that of cast iron.

## 9.9.2. Carbon Steels

Structural quality shapes and plate (ASTM 36) usually provide satisfactory service for external furnace supports, shells, and external conveyor and walking beam components (see figure 9.13.)

Heavy wall water-cooled and insulated carbon steel pipe (ASTM 53) is used for rails, walking beams, and their supports. Effects of thermal expansion must be considered.

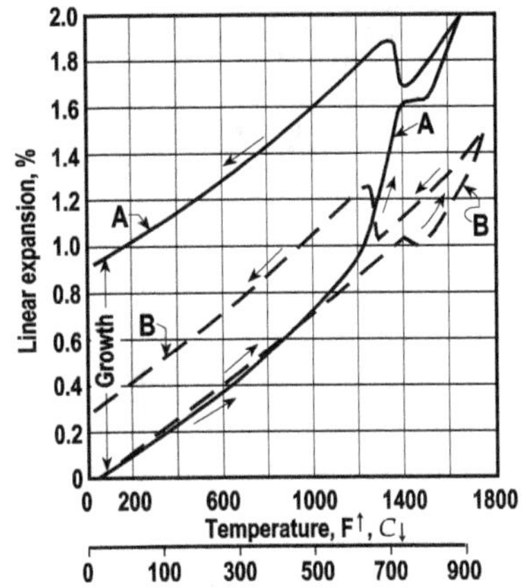


Fig. 9.12(a). Expansion and growth of cast iron, after a single heating. Curve A is for cast iron of 3.08% C, 1.68% Si. Curve B is for cast iron of 3.99% C, 1.60% Si, ss heated in 0.5 hr, then cooled in 2.5 hr.

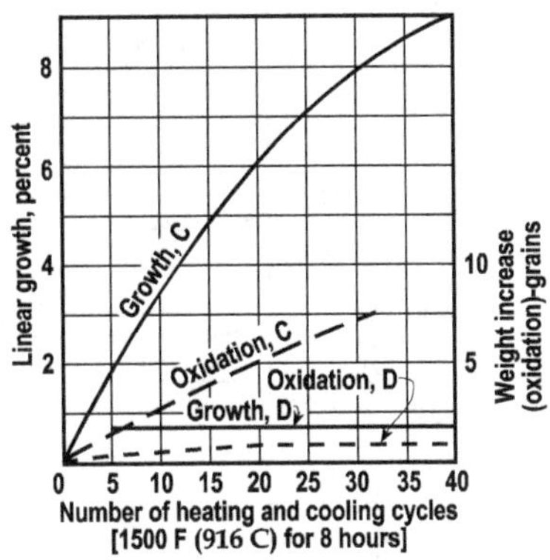


Fig. 9.12(b). Growth and oxidation of cast iron after repeated heating and cooling. Curves C are for plain cast iron, 3.26% C, 2.02% Si. Curves D are for cast iron containing 3.04% C, 1.62% Si, 14.31% Ni, 5.37% Cu, 3.26% Cr.

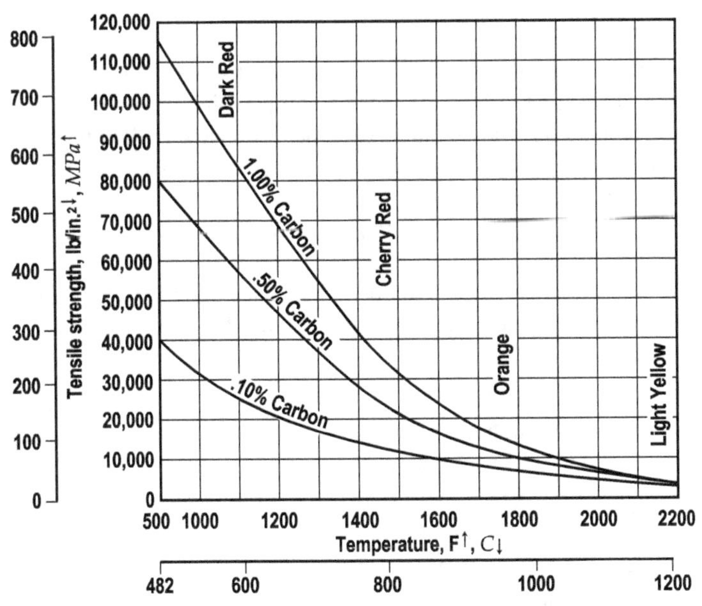


Fig. 9.13. Tensile strengths of carbon steels at various temperatures.

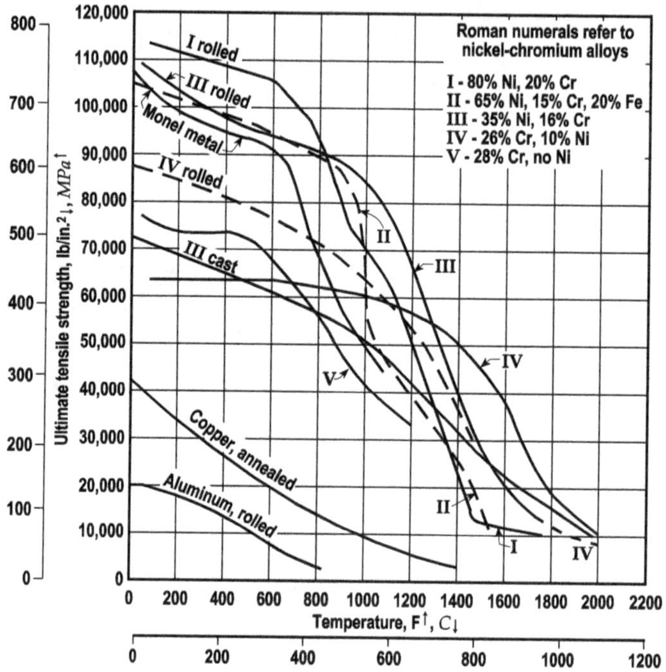


Fig. 9.14. Strength vs. temperature relations for various metals and alloys. Quick pull tests.

### 9.9.3. Alloy Steels

Iron–carbon–chromium–nickel alloy steels are used extensively in furnace applications such as heat treat containers, hearth components, drive chains, carburizing boxes, recuperators, regenerative burners, burner parts, and radiant tubes. The metal selection must consider the fact that the expansion rate of austenitic stainless steels is nearly twice that of ordinary steel. (See fig. 9.14.)

Below is a list of stainless steels used in process furnace design.

309 Austenitic stainless steel—excellent resistance to oxidation. High tensile and good creep strength at elevated temperature. Satisfactory for service in selected applications to 2000 F (1093 C).

310 Somewhat higher resistance to oxidation and higher creep strength.

316 Resistive to corrosion from most chemicals, particularly sulfuric acid. Superior tensile and creep strength at elevated temperatures.

442 A straight chromium ferritic steel. Corrosion resistant. Low propensity to scaling. Low tensile strength.

446 Heat resisting to 2150 F (1177 C). Resists oxidation better than 310, but has much less tensile and creep strength than 310 at high temperature. Sulfurous gases can be a problem. (See table. 9.4.)

**TABLE 9.5. Properties of steels for high-temperature uses (see also pp. 260–289 of *reference 52*)**

| Grade number | 304 | 309 | 310 | 316 | 410 | 430 |
|---|---|---|---|---|---|---|
| Heat resistance, $temp_{max}$, intermittent | 1600 F | 1800 | 1900 | 1600 | 1500 | 1600 |
| continuous | 1700 F | 2000 | 2100 | 1700 | 1300 | 1500 |
| Thermal conductivity, Btu ft/ft$^2$hr°F | | | | | | |
| at 212 F (100 C) | 9.4 | 9.0 | 8.0 | 9.4 | 14.4 | – |
| at 952 F (511 C) | 12.4 | 10.8 | 10.8 | 12.4 | 16.6 | – |
| Mean coefficient of thermal expansion | 11.0 | 10.9 | 10.9 | 10.7 | 6.4 | 6.6 |
| (in./°F) $(10)^{-4}$ at 68 F to | 1600 F | 2100 | 2100 | 1600 | 1300 | 1500 |
| Creep strength, lb/in.$^2$, at 1000 F (538 C) | | | | | | |
| 1% flow in 100 000 hours | 10 800 | 12 000 | 17 000 | 15 000 | 11 000 | 6500 |
| Yield strength[a], lb/in.$^2$ minimum | 30 000 | 30 000 | 30 000 | 30 000 | 32 000 | 35 000 |
| Ultimate strength[a], lb/in.$^2$ minimum | 80 000 | 75 000 | 75 000 | 75 000 | 60 000 | 60 000 |
| %Elongation[a] in 2 in. minimum | 50 | 40 | 40 | 40 | 20 | 20 |
| %Reduction in area[a] minimum | 60 | 50 | 50 | 50 | 50 | 40 |

[a]annealed.

## 9.10. REVIEW QUESTIONS, PROBLEM, PROJECT

**9.10Q1.** What refractory materials have been used to build furnaces for centuries?

**A1.** Fireclay (kaolin) brick and tufa (solidified bubbled volcanic lava).

**9.10Q2.** Why have water-cooled furnace doors, doorframes, and other parts fallen out of favor for industrial furnaces?

**A2.** Because they ultimately spring a leak, and the water causes costly damage to the furnace and its load, resulting in much downtime.

**9.10Q3.** What is the difference between dryout time for a newly installed refractory and warm-up time for a previously dried furnace?

**A3.** The difference is many more hours for dryout than for warm-up because dryout must slowly cause moisture to migrate to the surface and evaporate without sudden steam formation below the refractory surface, which could cause small explosions that can blow off the surface.

**9.10Q4.** At temperatures above 1200 F (650 C), why is it wise to use ceramic thermocouple wells in sidewalls instead of protruding alloy tubes?

**A4.** Because a metal protective tube will slowly yield to creep, bending downward against the wall, giving a poor reading. It will be very difficult to remove for replacement.

**9.10Q5.** When heating for dryout, what should be done when some areas begin to steam?

**A5.** Reduce the heat input to hold the temperature constant until steaming stops, then resume the temperature rise program. Do not try to catch up. The cycle will have to be lengthened by the amount of time that it was necessary to hold, to finish steaming.

**9.10Q6.** When the hearth of a rotary furnace begins to have grooves, what is the cause, and what can be done to increase the hearth life when replacing the hearth?

**A6.** The cause is hot furnace gas altering the refractory chemistry, lowering its softening temperature. When replacing the hearth, bury stainless-steel rails in the hearth so that they can support the load from deep in the refractory where it is unaffected by hot poc.

**9.10Q7.** What can cause roof support hangers to fail?

**A7.** When dust (from the flue or elsewhere) accumulates on the hangers, it will act as a layer of insulation, holding in heat conducted to them from the furnace. This will lower the hangers' strength; and can drop the roof.

**9.10Q8.** Recuperator tubes and tube sheets have failed, but their thickness has not been thinned. Why?

**A8.** Heating and cooling of the materials has work-hardened it, causing it to become brittle and fail.

**9.10Q9.** What can be done if you cannot find a T-sensor location for dilution air temperature control where it cannot radiate heat to the cold air tubes, and thereby give a false reading?

**A9.** Make a hemispheric depression in the refractory upstream of the recuperator and install the T-sensor recessed in that depression so that it cannot 'see' the cold tubes.

## 9.10. PROBLEM

A natural-gas-fired car-bottom furnace is to be built for heating 175 000 pounds of steel ingots from 50 F to 2150 F in 16 hr. Using formulas and data from this book and *References 51 and 52* as well as from refractory suppliers' data, select hearth, sidewall, and roof construction. Then calculate heat loss, heat storage, and coldface temperatures for the selected hearth, wall, and roof.

| *Given:* | |
|---|---|
| Maximum outside wall surface temperature | 210 F |
| Inside furnace dimensions | 14'w × 22'l × 9'h |
| Assumed hotface temperature | 2350 F |

## Solution

**TABLE 9.6.**

| Refractory | k[a] | Density lb/ft$^2$ | Interface temps | Stored Btu/ft$^2$ | Thickness | Heat loss hr[b] Btu/ft$^2$ hr[b] |
|---|---|---|---|---|---|---|
| HEARTH | | | 2350 F[c] | | | |
| 3000 F castable | 11.60 | 145 | | 50 900 | 9" | |
| | | | 2046 F[d] | | | |
| 2200F super duty fireclay | 10.66 | 147 | | 22 960 | 4.5" | 220r, 175c 395t |
| | | | 1880 F[d] | | | |
| 2000 F insulating firebrick | 1.08 | 31 | | 2 550 | 4.5" | |
| | | | 250 F[e] | | | |
| SIDEWALLS | | | 2350 F[c] | | | |
| 3000 F plastic | 6.83 | 142 | | 50 880 | 9.5" | |
| | | | 1905 F[d] | | | |
| 2300 F insulating firebrick | 2.09 | 35 | | 4 300 | 4.5" | 195r, 125c 320t |
| | | | 1214 F[d] | | | |
| 1900 F block insulation | 0.64 | 18 | | 424 | 2" | |
| | | | 250 F[e] | | | |
| ROOF | | | 2350 F[c] | | | |
| 3000 F castable 57% $Al_2O_3$ | 9.08 | 142 | | 48 050 | 9" | |
| | | | 1890 F[d] | | | |
| 2200 F light wt insul. castable | 2.71 | 65 | | 3 940 | 2" | 250r, 205c 455t |
| | | | 1555 F[d] | | | |
| 1950 F insulating castable | 0.70 | 27 | | 810 | 2" | |
| | | | 250 F[e] | | | |

[a] Conductivity, Btu/ft$^2$hr°F/ft.
[b] r = by radiation, c = by convection, t = total.
[c] hot face.
[d] interface.
[e] cold face.

## 9.10. PROJECT

Arrange the following concerns in order of importance—in your opinion, for your furnaces: Cleanliness ___ Customer relations ___ Employee relations ___ Energy conservation ___ Fire Prevention ___ Fuel cost ___ Furnace productivity ___ Personnel productivity ___ Pollution minimization ___ Product quality ___ Public Relations ___ Safety ___ Training ___ Other ___

Discuss the order with associates, supervisors, and management; then agree on a consensus for your organization and put it into practice.

# GLOSSARY

**ablative heat transfer** (as applied to melting metals) = the heating, melting, and flowing away of surfaces of ingots, sows, pigs, and scrap metal, exposing more solid metal for further melting—as on a dry hearth melter or with charge piled above the liquid bath surface in a reverberatory melter.

**absorptivity** = ability of a surface to absorb radiant energy, expressed as a decimal compared to the absorptive ability of a black body, absorptivity of which is 1.0. *See* **emissivity** and **emittance** for comparison.

**accordion effect** = a *domino effect* or *control wave effect*, usually referring to load temperature patterns through the length of a continuous furnace. If the temperature were shown by a series of vertical lines down the length of the furnace, with those lines closer together where the load temperature is high and spaced widely apart where the temperature is lower, it would look like pleats in the side of the windbox of a piano accordion. The same effect is noticeable when viewing the traffic on a busy highway from the air after a delay has been cleared.

**acf,** or **actual cubic feet**, or **acfh** = actual cubic feet per hour = volume or volume flow rate of a gas, at a specified temperature and pressure situation.

**adiabatic flame temperature** = "hot mix temperature" = the theoretical or calculated temperature of a flame resulting from complete combustion with a stoichiometric air–fuel mixture in a perfectly insulated (adiabatic) chamber so that all the combustion energy is absorbed by the combustion gases.

**adjustable thermal profile** or **ATP** = a burner with changeable flame length and character for better temperature uniformity across wide furnaces. (*See* sec. 2.6.)

**afterburner** = a burner installed in a furnace exhaust system to incinerate combustibles in the flue gas. A form of incinerator.

**air break** = See **barometric damper.**

**air-fuel firing** = conventional combustion using atmospheric air, as opposed to oxy-fuel firing.

**air/fuel ratio** = the reciprocal of fuel/air ratio. Usually expressed as a quotient of volumes (e.g., 10 $ft^3$ air/1 $ft^3$ gas = 10, or 10:1, or 10 to 1). Air/fuel ratio should be controlled with air flow as the primary variable (i.e., with fuel following air flow to avoid producing a rich furnace atmosphere).

**anchor** = an alloy or ceramic holding device for castable, rammed, or gunned refractory walls and roofs.

**annealing** = heat treating to remove stresses, soften, refine grain structure, and/or produce a specific microstructure.

**annular orifice** = a primary flow-measuring device consisting of a targetlike plate in the center of a round pipe or duct with a fluid flowing through the annulus around the periphery. Advantages over the traditional concentric orifice are (1) shorter upstream straight run required, and (2) avoiding pileup of liquid or solids in the bottom of the pipe. The principle is the same as for a concentric or a segmental orifice, but the flow coefficients are different.

**anomaly** = a deviation from the common rule, type, arrangement, or form.

**arch** = the top closure of a furnace or flue, built in the form a curve or arc of a circle to put the refractories that form it in compression (because refractory strength in tension and in bending is lower). Sometimes termed a vault, crown, or roof. Loosely used for a flat (suspended) furnace roof. A 'jack arch' is a flat arch with brick shapes that put themselves in compression, as in a curved arch.

**atm** = **atmosphere** = (1) pressure exerted by a **standard atmosphere** on the surface of the earth at sea level at lat. 45°N latitude, which is 29.92 in. Hg or 760 mm Hg or 14.696 psia, or (2) the chemical make-up of the gases within a furnace, as an oxidizing atmosphere or a reducing atmosphere.

**ATP burners** = adjustable thermal profile burners, manually or automatically adjustable to change the heat release pattern of the combustion reaction. (*See* section 2.6.)

**available heat** = the heat that is left available for heating the load and balancing wall, conveyor, and opening losses after the stack loss is subtracted from the gross heat input. It represents the best possible efficiency for a furnace. It can be calculated from estimates of flue gas exit temperature and %excess air.

**avg** = average.

**baffle** = a solid deflector in a furnace or duct to divert flow or partially block flow of a fluid or of radiant heat.

**bake** (refractories) = to remove moisture and to stabilize chemical reaction by subjecting a substance to heat (usually low temperature).

**banana, banana-ing** = (steel mill and forging slang describing) the curving of a piece of load because of uneven heating. Usually overheating the top, causing the top side of the piece to slowly hump upward due to greater thermal expansion and plasticity of higher temperature areas.

**bar, billet, bloom** = pieces of metal, square or rectangular in cross section, 1.5 to 12 in. across (0.04 to 0.3 m across) and 1 to 60 ft long (0.3 to 18 m long). These three terms may be used interchangeably, except that a bloom is generally 8 in. (0.20 m) or larger and a billet generally smaller than 5 in. (0.13 m). In contrast, see **slab.**

**barber poling** = an unwanted uneven spiral heat distribution on round mill products, often occurring in the process of making seamless pipe and tube using a rotary hearth furnace wherein the rounds rest on the hearth, creating a cold line of contact along the length of the round. In the piercing operation as the round is twisted, its cold line is twisted, resulting in a spiral line that looks like a barber pole.

**barometric damper** = a vertical stack with a side inlet furnace flue just above an open-ended bottom. Hot gases entering the stack create a natural convection updraft, pulling in cold air through the open bottom, thus "killing" the stack's draft.

**batch** = the load charged into a glass melter or frit smelter. *See* **load.**

**batch furnace** = **in-and-out furnace** = **a periodic kiln** = a heating chamber into which a load is charged, heated to process temperature, cooled, and then unloaded. The load stays stationary, and the temperature cycles with time. Examples: periodic kiln, shuttle kiln, cover annealer (bell) furnace, box furnace, slot furnace, car bottom furnace, elevator furnace. An intermittent or non-steady-state process. In contrast, see **continuous furnace**. A plot of temperatures versus time for a batch-type furnace will be similar to a plot of temperatures versus distance through a continuous furnace for the same load and process.

**bath** = liquid or molten material in a melting furnace. Or the chamber of a melting furnace that holds molten metal. In an open hearth furnace, the section where the furnace charge is melted and the heat is worked and alloyed.

**bell furnace** = a liftable furnace whose floor remains fixed, especially in ceramic kilns and cover annealing furnaces (opposite of an elevator furnace).

**Bernoulli equation** = a form of the 'general energy equation' = law of conservation of energy, applied to thermal and fluid flow situations. Particularly, illustrating the interconversion of kinetic (velocity) energy and pressure energy. *Also see* **Venturi.**

**betw** = between.

**billet** = See bar, billet, bloom.

**black body** = an emitter or receiver of radiation (usually solid) with maximum capability to emit or receive heat or light radiation (i.e., an absorptivity of 1.0 and an emissivity of 1.0). This is a theoretical concept used as a basis by which to measure or compare radiation emitting and absorbing capabilities of various materials and surface conditions. Usually applied to solids, but also used for liquids, vapors, gases, clouds of particles, and flames.

**blast** = air, or pressurized air supply.

**blast furnace** = a shaft furnace (refractory-lined, vertical cylindrical furnace) for melting charged material (scrap steel, limestone, and other) and for reduction of iron ore to iron by burning coke or charcoal with blast air injected through tuyeres at various levels. The objective is to produce cast iron pigs or molten feed to an open hearth or electric arc furnace for making steel. (*See* fig. 4.17.)

**blast furnace gas** = offtake gas from a blast furnace, comprised of CO, $H_2$, $CO_2$, and $N_2$, with a heating value ranging from 70 to 110 Btu/scf.

**blast furnace stove** = a very tall, steel cylindrical structure encapsulating checkers and a combustion chamber for heating air (blast) to 2500 F (1370 C) for combustion to improve blast furnace productivity. The fuel for the stoves (generally in groups of three or four) is usually blast furnace gas enriched with other fuels as necessary to achieve desired temperature at the checker inlets and subsequently in the air blast.

**bloom** = See **bar, billet, bloom, slab.**

**bloom down** = an intermediate product between rollings when two or more steps are required to achieve high surface quality (e.g., 32 in. square ingot to 13 in. × 17 in. bloom down to 8 in. square billet.

**blower** = a high-pressure fan used to push air through burners, lances, or nozzles. May be integral with the burner or piped with distribution manifolds to banks of burners, lances, or nozzles. In industrial furnaces, the blowers are usually centrifugal fans that develop air pressures from 0.5 to 3 psi (3.5 to 20.7 kPa).

**blown refractory** = **gunned refractory** = furnace lining material that is installed by being sprayed on the interior of furnace walls and roof.

**blue water gas** = a manufactured gaseous fuel made by passing steam over incandescent coke. Its gross heating value ranges from 260 to 300 Btu/ft. The 'water gas reaction,' $C + H_2O \rightarrow H_2 + CO$ is hazardous because of its high carbon monoxide content. **Carbureted water gas** has some oil vapor added to raise the calorific value to about 530 Btu/ft$^3$.

**boiler** (a steam generator) = a furnace, or combustion chamber, combined with a heat exchanger for the purpose of converting feedwater to steam.

**bot** = bottom.

**bottom-fired furnace** = heating chamber in which burners are positioned to fire beneath the load, as in top- and bottom-fired steel reheat furnaces, or vertically up through the hearth, as in the case of some refinery and chemical process industry heaters.

**box furnace** = **in-and-out furnace** = a kind of batch furnace, generally with a charge door on one or both ends.

**breeching** = large flue gas duct, often connecting flue to stack.

**bridge wall** = a refractory dam, as to prevent slag from entering a flue, or a radiation shield to separate zones of different temperatures to reduce flue gas temperature, thus improve available heat. Historically, to prevent a coal bed from spilling into the product material.

**brnr** = *See* **burner.**

**buckstays** = vertical I-beams or channels along the sides of a furnace to support the roof and strengthen the furnace shell.

**bullnose** = a curved refractory construction (often cast or rammed) designed to cause a change in flow direction of furnace gases, as the edge of a baffle, curtain wall, or bridge wall, to reduce cases of refractories heated on multiple sides, which reduces refractory life.

**bung** = furnace roof sections, sometimes designed for easy removal to allow for repairs, slag removal, or in continuous furnaces, for threading strip or strands.

**burner** = **brnr** = an assembly of air, oxygen, and fuel orifices that delivers those fluids to the burner quarl or combustion chamber with velocities and directions that position the flame in the desired location and so that continuous self-sustained ignition is accomplished. The aerodynamic design of the burner passages determines the flame character (size, shape, velocity, luminosity, completeness of combustion, and noise and pollution minimization). Most burner assemblies include atomizing, mixing, proportioning, piloting, and flame-monitoring devices. Many are designed specifically to enhance either radiation or convection heat transfer within a furnace.

**burner tile** = **burner quarl** = the refractory-lined hole through a combustion chamber wall, through which air and fuel are injected, and/or a burner flame is fired. The quarl is usually designed to enhance flame stability by adding the minimum ignition energy required to begin and sustain chemical reaction. The burner tile also may influence the flame character. The inside passage of a quarl may be cylindrical or conical, diverging or converging.

**burner tunnel** = refractory construction under top- and bottom-fired furnaces to permit burners to fire under the charged load (which may be on piers or skid rails). The term 'burner tunnel' is sometimes thought to mean 'burner tile' or 'quarl,' which is part of the burner.

**burning of metal** = steel surface that has been above 2500 F (1370 C) long enough for oxidation of intercrystalline boundaries. When steel has been burned and rolled, the surface will be full of cracks, often necessitating its scrapping.

**c** = **specific heat**, (*see also*).

**C** = **Celsius** (formerly centigrade) temperature scale (This book uses C for an actual temperature level, such as water boils at 100 C. Use °C only to indicate a temperature change or temperature difference. See **degree mark** and **T**.

**C-to-C** = c-to-c = center to center, or centerline to centerline.

**calcine** (refractories) = the process of applying a relatively high temperature heat to a mineral-based substance to oxidize it and remove moisture.

**car, car-bottom, car-hearth, lorry-hearth** = refractory-covered bottom of an industrial furnace or shuttle kiln or tunnel kiln, generally mounted on wheels, usually on rails for quick, easy loading and unloading.

**Carbon dioxide, $CO_2$** = a product of complete combustion of carbon, usually from a hydrocarbon fuel.

**Carbon monoxide, CO** = a product of incomplete combustion of carbon, usually from a hydrocarbon fuel.

**cast refractory** = castable refractory material that can be poured into forms or molds to form furnace wall, roof, and hearth linings or burner quarls, or piers.

**castable** = a kind of refractory that can be poured in place in a manner similar to concrete, often into a form or mold.

**catenary** = *See* sec. 4.3.

**CC** = **cc** = center to center. Also "(on) centers" = center to center.

**cf** = cubic foot or cubic feet. **Cfm** = cubic feet/minute. **cfh** = cubic feet per hour. **acfh** = actual cu ft /hr, as opposed to **scfh** = standard cfh. (*See* **stp, standard air.**)

**$CH_4$** = methane, the principal component of natural gas.

**channeling** = a fluid flow phenomenon in which some parts of a stream, or a near-stagnant mass of a fluid moves faster than the surrounding fluid. *See* sec. 5.11.3.2.

**charge** = load, batch, material, metal, pieces, product, stock, or ware that may be placed in a furnace, oven, or kiln primarily for heat processing. Not to be confused with materials to be heated as an intermediate objective such as walls, hearth, roof, muffles, radiant tubes, immersion tubes, furnace gases, air, water, or other heat-transfer media.

**checker** = **checkerwork** = a latticework of refractory shapes that serves as a heat-storage reservoir in a regenerative air preheater such as used on open-hearth furnaces, large glass melting furnace, soaking pits, coke oven batteries, blast furnace stoves, and reheating furnaces.

**chimney** = a refractory or metallic stack for conveying furnace waste gases (after heat recovery equipment) to the atmosphere.

**chimney effect** = **draft,** which see = natural convection effect on furnace pressure.

**chipping** = removal of product surface defects by cutting tools, manual or power-driven; similar to **scarfing,** which see.

**C.I.** = **ci** = cast iron.

**city gas** = a manufactured gaseous fuel made from coal. Its gross heating value is about 540 Btu/ft$^3$. Similar to 'towne gas' and blue water gas.

**CO** = **carbon monoxide** (poisonous), a product of incomplete combustion (pic) of carbon or a hydrocarbon fuel.

**$CO_2$** = **carbon dioxide**, a product of complete combustion (poc).

**cobble** = a section of product that did not enter a set of rolls (for any of many reasons), most often due to low-temperature or nonuniform heating. The cobble becomes scrap.

**col** = column.

**combustion chamber** = space where combustion takes place. Sometimes a 'dog house' or 'Dutch oven' appendage to the main furnace, but commonly within the furnace itself. Modern flame stability and flame-characterizing science have minimized the need for separate combustion chambers.

**combustion efficiency** = *See* sec. 5.1.

**computer modeling** (with reference to rolling mill production) = a method for developing automatic control systems for furnace zone temperatures to minimize fuel input. Problems may result because of changing production rates. For example, a mill's production level was raised from 70 to 90% after the furnace zone inputs had been stable for 30 min. The products in the furnace were being heated at the 70% rate, so the furnace zone inputs had to be raised to about 100% to attain the 90% rate quickly. The entering pieces were then heated at the 100% rate (above the mill rate), resulting in higher than desirable product temperatures. Changing mill production rates can cause instability swings because of the "flywheel effect" of a large furnace and load. More smaller zones in a furnace can minimize instability, but decisions should not be made without advice from experienced persons familiar with operating problems.

**concentric orifice** = a primary flow-measuring device consisting of a plate with a center hole through which the fluid is accelerated. The resultant pressure drop infers the energy required to accelerate the flow; thus, measuring the pressure drop provides a means for calculating the volume flow rate. The principle is the same as for an annular or a segmental orifice, but the flow coefficients are different.

**conduction** = a means of heat transfer by direct internal molecular contact. (Most often applied to solids, but conduction is actually a part of convection in gases and liquids.)

**conductivity** = (in this book) thermal conductivity, k = the ability of a material to conduct heat, measured in Btu/hr, joules/hr, or kW flowing through a square foot or square meter of cross-sectional area, and through a foot, inch, or meter thickness with one degree (F, C, K) of temperature difference across that thickness. In the United States, the refractory and insulation industries use Btu in./$ft^2$hr°F. Most others use Btu/($ft^2$) (hr) (°F/ft) or Btu ft/$ft^2$hr°F, which is sometimes abbreviated as Btu/ft hr°F.

**continuous furnace** = a tunnel kiln = tunnel furnace = a heating chamber wherein loads are moved through temperature zones continuously or intermittently. Examples: conveyor furnaces—pusher, walking beam, roller, chain belt; rotary hearth furnaces; tunnel kilns, enameling tunnels; rotary drum dryers, calciners, incinerators; Herreshoff multilevel furnaces; fluidized bed furnaces wherein the bed material is the load. In contrast, see **batch furnaces**. A plot of temperature versus distance/time through a continuous furnace will be similar to a plot of temperature versus time for a batch-type furnace for the same heating load.

**control wave effect** = *See* **accordion effect.**

**convection** = transfer of heat by moving masses of fluid (gas or liquid). Convection is conduction followed by stream movement, and its rate of heat transfer is dependent on: (1) thermal conductivity of the stationary fluid covering the solid surface, (2) Reynolds number (ratio of momentum forces to viscous forces); thus, velocity is the major variable, often to the 0.6 to 0.8 power, (3) temperature difference between the bulk stream and the solid, and (4) the area of solid surface contacted by the moving fluid. Convection currents occur in a fluid because of mechanical agitation (forced convection) or differences in density at different temperatures (natural convection).

**conveyor furnace** = a continuous furnace with material-moving apparatus such as rollers, chain belt, pusher, walking beam, or suspended hooks from a moving chain.

**couple** = thermocouple, a type of temperature sensor.

$c_p$ = specific heat at constant pressure, Btu/lb°F or calories/g°C.

**cpi** = chemical process industries.

**cracking** = the process of breaking or polymerizing hydrocarbon molecules so that they recombine into both lighter and heavier molecules. Thermal cracking involves the use of high temperatures in the absence of air. Catalytic cracking uses lower temperatures and pressures in the presence of a catalyst.

**crown** = the refractory roof of a furnace or kiln, especially if arched and/or over a glass bath. *See* **arch, roof.**

**C.S.** = **CS** = **cs** = carbon steel, steel alloyed with a small amount of carbon.

**cullet** = scrap pieces of glass recycled to a furnace for melting.

**cure** (refractories) = to stabilize the chemical reaction in concrete and cement-based masonry and refractory materials by subjecting them to heat below 200 F (93 C).

**curtain wall** = a baffle or wall to separate firing zones.

**cutback period** = the time elapsed in a furnace from when the temperature control begins to cut back the input (the "cutback point") to where the load is sufficiently "soaked out" (i.e., uniformly heated) for the process.

**C/W** = **(c to c)/w** = **spacing ratio,** which see for more details.

**cycle time** = the time from the beginning of load charging to completion of its discharge. This does not include normal equipment maintenance between cycles, such as hearth scale removal after load discharge.

**damper** = a type of valve used to control flow in large ducts, usually for air or flue gas. May be metal or refractory, and of a variety of configurations such as butterfly, clapper, coolie hat, guillotine, and louver. Often automatically power-actuated and counterbalanced, with mechanical advantage mechanisms.

**degree mark** (°) or **degrees** = a unit of measure for change or difference in angular position or change or difference in temperature. The convention used in this book is to omit the degree mark (°) with a temperature level (e.g., water boils at 212 F or 100 C), and to use the degree mark only with a temperature difference, change, or gradient [e.g., the difference, $\Delta T$, across an insulated oven wall was 100°F, or the temperature changed (rose or fell) 15°F in an hour. *See also* ***T***.
Conversion units *for temperature change or difference* are:

°F = (9/5) (°C), °C = (5/9) (°F), °R = 95 (°K), °K = 5/9°R.

Conversion units *for temperature level* are:

$$\begin{array}{lll} F = (9/5)\ C + 32. & F = (9/5)\ (K - 255.4), & \text{or } F = (9/5)\ (K - 459.7). \\ C = (5/9)\ (F - 32). & K = (5/9)\ (F + 459.7), & \text{or } K = (5/9)\ (F + 255.4). \\ R = (9/5)\ C + 491.7. & R = F + 459.7, & \text{or } F = R - 459.7. \\ C = (5/9)\ R - 273.2. & K = C + 273.15, & \text{or } C = K - 273.2. \end{array}$$

**delay** = an unscheduled mill stoppage.

**delayed mixing** = intentional slow mixing of air and fuel, usually to produce a long or luminous flame.

**delta *P*, delta *T*** = $\Delta P$, $\Delta T$ = a difference in pressure, or difference in temperature.

**design security factor** = *See* **security factor, safety factor.**

**destructor** = incinerator.

**detached flame** = a less stable form of flame, having a feed speed greater than the flame speed of the air-fuel mixture, resulting in the flame not appearing to begin until some distance downstream from the burner tile or burner nozzle, where the feed speed has fallen to flame speed or less.

**dfg** = **dry flue gas** = flue gas less its water vapor. This does not actually happen (unless there is no hydrogen in the fuel). It is simply a easy way to figure heat balances and flue gas analyses.

**diffusivity** = (in this book) **thermal diffusivity** = the ability of heat to diffuse through a material = k/c ρ = thermal conductivity divided by specific heat and density (or thermal conductivity divided by volume specific heat).

**direct-fired** = describes a combustion heating process in which the products of combustion contact the surfaces of the load being heated.

**diverter** = a baffle or device in a nozzle-mix burner for the purpose of causing the combustion air to change direction relative to the fuel stream to improve the burner stability or to reduce emissions.

**domino effect** = a reaction caused by a similar preceding reaction. Controls wave effect. *See* **accordian effect.**

**donut (doughnut) rotary hearth furnace** = *See* **rotary hearth furnace.**

**downcomers** = offtakes from a manifold or furnace (often broadened to include offtakes in any direction).

**downdrafting** = a furnace configurtion with burners at the top and flues at the bottom. This prevents runaway hot gas columns between stacked loads.

**downfiring** = the direction of burners or flames, but most importantly, the initial direction of flow of the combustion gases (often with high-velocity burners and top flues for full circulation).

**draft** = chimney effect = a breeze = the pressure difference that causes an air movement.

**draw, drawing** = (1) withdrawing from a furnace, (2) a tempering (heat-treating) process, or (3) a shaping process in which metal is pulled through a die.

**drier** = **dryer** = a low-temperature oven for removing water or other volatiles from a load. May be box, continuous, rotary drum.

**dropout** = (1) a system used for removing billets, blooms, or slabs from a reheat furnace, prior to modern extractors, (2) the whole apparatus by which the pieces are moved by pusher onto water-cooled skids and through the furnace to slide down by gravity through a door, then to the roll table, or (3) the door or opening through which loads are discharged from a furnace.

**dross** = **oxide**, such as is formed in a nonferrous metal melting furnace. Generally, it floats on top of a liquid metal.

**dry, drying** = to remove moisture from a substance. Also a form of masonry construction that does not use mortar, cement, or other binding materials.

**dryer** = *See* **drier.**

**dryout time** = a long, slow heating time required to eliminate moisture from a just-cast ceramic or refractory product, or from a newly installed refractory wall, hearth, or roof. Usually longer than **warm-up time**. (*See also* Sec. 9.5.)

**ductility** = a measure of the ability of a metal to undergo permanent changes of shape without breaking its surface.

**efficiency** = useful output divided by input, expressed as %. See sec. 5.1. Some people use thermal efficiency, fuel efficiency, and furnace efficiency interchangeably. Combustion efficiency is a measure of how well a fuel is burned, and therefore a measure of atomizing or mixing effectiveness.

**%elongation** = 100% × (new length − original length)/original length.

**elevator** (elevated) **furnace** = a furnace that is fixed in an elevated framework, the loaded hearth of which is mechanically, electrically, or hydraulically raised up into the furnace. Opposite of a **bell furnace** (*see also*).

**ell** = elbow in a pipe or duct.

**elongation** = the extension that a material sample undergoes before it fractures.

**emissivity** = ***e*** = a measure or ability of a material to radiate energy = the ratio of the radiating ability of a given material to that of a black body. (A black body emits radiation at the maximum possible rate at any given temperature, and therefore has an emissivity of 1.0.) Emissivity denotes a property of the material whereas 'emittance' refers to an actual geometry or surface condition. The emissivity and **absorptivity** of most materials are nearly the same, and are often used interchangeably. In industrial heating engineering, it is usually the absorptivity that is of most concern.

**emittance** = the ability of a surface to radiate energy, compared to the rate for a "black body" (emittance of 1.0). In contrast, emissivity is a property of the bulk material, independent of geometry, but emittance refers to an actual shape and surface condition.

**end-fired** = firing burners parallel to the long axis of a furnace; normally countercurrent to the product movement.

**enh htg** = **enhanced heating** = use of high-velocity burners to add convection heat transfer and gaseous radiation by replacing stagnant cool gases from spaces between or below load pieces to increase heat transfer by convection and by gaseous radiation, and by "solids radiation" from better-heated hearth and piers—all for better temperature uniformity and productivity. (*See* sec. 7.5.)

**entry pressure loss** = the pressure drop required to accelerate a fluid stream through an opening or into a pipe or duct. The actual total loss is greater than just the pressure drop required to accelerate the fluid to the required velocity (*a*) because the flow stream lines take a Venturilike path inside the opening with a smaller cross section than the opening, requiring a greater velocity, and (*b*) due to the energy expended in unproductive eddy movement.

**eqn** = equation or formula.

**equivalence ratio** = Greek letter phi = a means of expressing fuel/air ratio = the actual amount of fuel expressed as a decimal ratio of the stoichiometrically correct amount of fuel.

**excess air** = 'xs air' = air supplied to a combustion reaction beyond that required for chemically complete (stoichiometric) combustion. Usually expressed as percentage of the stoichiometric air volume at standard temperature. Excess air usually results in an oxidizing atmosphere in a furnace. (Excess fuel makes a reducing atmosphere.)

**extractor** = a mechanism that reaches into a furnace under a load to be discharged, lifts or pulls it out, and places it on a device for delivery to the next process.

F = **Fahrenheit** temperature scale. (This book uses F for an actual temperature level, such as water boils at 212 F. Use °F only to indicate a temperature change or temperature difference. *See* **degree mark** and ***T***.

**fantail arch** = refractory span that connects the chamber above the checkers of a regenerative furnace to the bottom of the uptake.

**FB** = **F.B.** = **fb** = firebrick, made from fire clays, hydrated aluminum silicates with minor amounts of other minerals.

**fce** = **furnace,** (*see also*).

**fg** = **flue gas**, (*see also*). (**dfg** = dry fg = flue gas without its water vapor)

**fget** = flue gas exit temperature, or furnace gas exit temperature.

**fgr** = **flue gas recirculation** = flow of poc in a furnace. *Internal recirculation* increases the mass flow rate of poc, causing (*a*) a lowering temperature gradient along the flow path, thereby improving temperature uniformity of the furnace loads and (*b*) increased convection heat transfer to the loads because of increased velocity. *External recirculation* is more effective in reducing NOx emissions because the external gas is cooler. However, its fuel usage is greater. Both internal and external recirculation enhance convection heat transfer and lower NOx.

**film coefficient** = $h_c$ = heat transfer coefficient for convection. *See h* and heat transfer coeffcient.

**firebrick** = kaolin (natural or man-made). Colloquially, often refers to any simple form of refractory material.

**firebrick equivalent** = a means for comparing the insulating capabilities of various refractories and of composite walls by telling how many inches of an all-firebrick wall would be required to accomplish the same insulating capability.

**flame character** = the nature of a flame—size, shape, color, luminosity, velocity. See flame types in fig. 6.2.

**flame instability** = lack of **flame stability**, (see also). Flame instability is evidenced by sputtering, "motorboating," flameout, or lighting difficulty.

**flameless combustion** = a furnace condition wherein the combustion reaction has been diluted by internal flue gas recirculation of poc and inerts to the point where the reaction temperature is so low that the flame is invisible. The combustion reaction is at such a low temperature that it fails to supply energy for luminosity.

**flame safety system** = electronic monitoring and fuel shutoff control for stopping the flow of fuel to a furnace in the event of flame failure, utilizing ultraviolet flame sensors (or infrared sensors in low-temperature chambers such as ovens or boilers).

**flame stability** = reliability, ease of lighting. In a furnace combustion chamber, a stable flame is one that keeps burning despite significant excess or deficiency of air or low-combustion chamber temperature or pressure. Opposite of **flame instability,** (see also).

**flame types** = *See* sec. 6.2.2.

**flow nozzle** = a gradually converging metering orifice that causes less total pressure drop than a thin-plate orifice, but more pressure loss than a Venturi meter.

**flue** = the opening in a furnace through which the poc exit the furnace and enter the stack or, sometimes, enter a breaching connecting to a stack. (See sec. 2.6.4.)

**flue gas** = **fg** = poc + xs air (including tramp air) or poc + excess fuel and/or partially burned fuel, sometimes called "waste gas" or "stack gas."

**flue gas exit temperature** = **furnace gas exit temperature** = **fget.**

**flue gas recirculation** = **fgr** (*see also*).

**flue loss** = heat lost up the flue as the heat content of the flue gases. Often calculated separately as dry flue gas loss + wet (or latent) fg loss.

**flux** (in this book) = heat flux = rate of heat flow per unit area = $q = Q/A$. Typical units are Btu/ft$^2$hr, joules/m$^2$hr, or kW/m$^2$.

**forced draft** = a method of conveying air supply, wherein combustion air is pushed through the burners and furnace by a 'forced draft fan' or blower that develops a positive pressure in a combustion air system, by converting velocity pressure into static pressure.

**forehearth** = refractory-lined feeder, channel, and final conditioning zone that delivers molten glass to the forming equipment. It is usually kept hot by burners in the roof or high in the sidewalls.

**forging** = hammering or pressing a piece of hot metal into a desired shape.

**fourth power effect** = of absolute temperature on radiant heat transfer. *See* Stephan-Bolzmann Law.

**fpm** = feet per minute.

**fps** = feet per second.

**front-fired** = firing a furnace with burners in the front wall, or load-discharge end, counterflow to product movement of a continuous furnace. (*See* **front-fired continuous furnace.**)

**front-fired continuous furnace** = a steady-state heating chamber in which the burners are located near the load-discharging end and aimed toward the load-charging end. (Similarly, the "front" of a burner is the end from which the flame exits; thus, the "back end" of a furnace or of a burner is the cooler end.)

**fuel/air ratio** = **F/A** = the reciprocal of air/fuel ratio, A/F, usually expressed in volume units. F/A sometimes means a ratio control system in which fuel flow is adjusted to follow changes in air flow (fuel primary control). It is usually safer to have airflow lead with fuel flow following (air primary control).

**fuel efficiency** = *See* **efficiency,** and sec. 5.1.

**fuel-fired** = heated by combustion of a fuel, as opposed to electrically heated.

**furnace** = **fce** = a combustion chamber, often including heat-exchange surfaces, or load-handling equipment. In this book, usually a refractory-lined enclosure wherein products are heated for industrial processing. May also mean an oven, kiln, heater, incinerator.

**furnace efficiency** = *See* **efficiency,** and sec. 5.1.

**furnace heat release** = See sec. 3.2.

**furnace pressure** = the gauge pressure in a combustion chamber = pressure greater or less than ambient (atmospheric) pressure outside the furnace. The furnace pressure rises with elevation within the furnace, and undulates considerably where flow cross section changes, especially near burners and flues. The furnace pressure may be positive or negative, and is zero at the 'neutral pressure plane'. *See* **neutral pressure plane.**

**furnace shell** = the steel plates, or other containment materials, that encompass the insulation and refractory lining of a furnace or oven heating chamber.

**fuse** (refractories) = to combine disparate substances by heating them to their melting points, as in welding.

**gas beam** = **gas blanket** = **gas cloud** = poc thickness = a measure of gas radiation capability.

**gas gravity** = the density of a gas relative to that of standard temperature (stp) air. For example, the density of stp air is 0.073 lb/ft$^3$, but its "gas gravity" is 1.0.

**gas radiation** = radiation from triatomic molecules, such as $CO_2$ and $H_2O$—as opposed to "solids radiation" for radiation from solids. (*See* chap. 2.)

**ghv** = gross heating value. *See* **heating value.**

**gray body** = a material or surface that emits and receives thermal radiation evenly over a wide spectrum of wavelengths and frequencies (i.e., has broadband emissivity and absorptivity), as opposed to spikes at specific wave lengths on a spectrograph as with gas radiation.

**grinding** = removal of product surface defects by motor-driven, abrasive wheels.

**gross heating value** = *See* **heating value.**

**gunned refractory** = blown refractory = furnace lining material that is installed by being sprayed on the interior of furnace walls and roof.

**h** (See **heat transfer coefficient**) $h_c$ = convection coefficient or 'film coefficient'. $h_r$ = radiation coefficient. $h_i$ = inside. $h_o$ = outside.

**$H_2$** = **hydrogen**, a flammable gas that burns to water vapor, $H_2O$. Hydrogen flames are usually invisible, highly reactive, forming acids, but usually considered non-polluting.

**head** = driving force, difference in potential, as an electromotive force or voltage or difference in pressure (head of water above an opening at the bottom of a dam that determines the flow through the opening). *See also* **thermal head.**

**hearth** = the floor of a furnace or kiln on which the product or its supporting piers or kiln furniture rests. The weight of the furnace load is supported by the hearth, which may be laterally movable as in a car-bottom furnace or vertically movable as in a bell furnace.

**heat content** = enthalpy. See sec. 5.9.2 for poc, and tables A1, A2, A4, A7, A8, and A11 of *reference 51.*

**heater** = furnace, in the chemical processing industries (cpi), including refineries.

**heat exchanger effectiveness** = *See* sec. 5.11.3.1 and 5.11.3.2.

**heat flow** = **heat flow rate** = thermal energy transmitted per unit of time (e.g., Btu/hr, watts, joules/second).

**heat flux** = rate of heat flow per unit area. $q = Q/A$. Typical units are Btu/ft$^2$hr, joules/m$^2$hr, or kW/m$^2$. *See reference 52*, pp. 317–327.

**heating capacity** (of a furnace) = weight of load that can be heated in unit time through a specified temperature range without overheating. *See also* **specific heating capacity**, which may be heating capacity per unit of hearth area or per unit of furnace volume. In contrast, *see* **heating rate**.

**heating minutes per inch** = heating time in minutes divided by product thickness in inches = rules of thumb heating times required for various heating processes—used before calculation of heating curves became very effective.

**heating rate** (of a furnace) = weight of load actually heated per unit of time. *See also* **specific heating rate,** which may be heating rate per unit of hearth area or per unit of furnace volume. In contrast, *see* **heating capacity**.

**heating value,** = ***hv*** = the heat obtained from combustion of a specified amount of fuel and its stoichiometrically correct amount of air, when both start at 60 F (16 C) and end being cooled to 60 F (16 C). **Gross or higher heating value** = ***hhv*** = the total heat release. **Net or lower *lhv*** = hhv minus the latent heat of vaporization of the water vapor formed by the combustion of hydrogen in the fuel. In the United States, hv is assumed to be hhv unless otherwise specified. In European practice, nhv or lhv is normally used.

**heat needs** = a term used in this book to summarize *all the 'available' heat input required* by a furnace, except the flue gas loss (the heat content of the flue gases).

**heat recovery** = getting back the heat energy that might otherwise be lost up the stack of a furnace, boiler, heater, incinerator, kiln, or oven. Heat recovery can be accomplished by addition of an unfired load preheat section, waste heat boiler, or air preheater (recuperator or regenerator). Some engineers consider oxygen enrichment and oxy-fuel firing as forms of heat recovery.

**heat recovery effectiveness** = heat exchanger effectiveness. (*See* sec. 5.11.3.1 and 5.11.3.2.)

**heat transfer** = delivery or transmission of thermal energy.

**heat transfer coefficient** = $U$ or $h$ = heat flux per degree of $\Delta T$ = heat transfer rate per degree of $\Delta T$ and per unit of area. 1 Btu/ft$^2$hr°F = 5.67 W/°Km$^2$. See **overall coefficient of heat transfer, U.** $h_c$ = convection coefficient or 'film coefficient'. $h_r$ = radiation coefficient. $h_i$ = inside. $h_o$ = outside.

**heat transfer rate** = flow rate of thermal energy, $Q = qA$, in units such as Btu/hr, kW, J/s. See *reference 52*, pp. 317–327.

**heat treating** = (broadly) a heating process that makes products more useful. Specifically for metals, heating to change crystalline structure to improve hardness, ductility, strength, and/or to relieve internal stresses from casting, working, or welding.

**heat-up time**—May mean curing time for newly placed refractories (Sec. 9.5) or the time to bring a load to working temperature.

**heat zone** = generally the temperature control zone above or below the load(s), and before the soak or equalization zone of a steel reheat furnace. May be end-, top-, or side-fired.

**Herreshoff multilevel furnace** = a vertical cylindrical furnace with many circular hearths attached to a central vertical drive shaft, and with plows to move granular load material across each hearth to expose all particles to furnace gases and to cause them to eventually drop to the next hearth level. Burners fire horizontally below and between the hearths. Used for drying sewage sludge, and for drying and pyrolizing ores.

**Hg** = **mercury** = A reading of 1" Hg on a mercury manometer = 3.386 kPa = 345.4 mm $H_2O$ = 7.859 ounces per square inch (osi).

***hhv*** = higher or gross heating value. *See* discussion under **heating value**.

**higher heating value**–*See* discussion under **heating value**.

**high-fire period** = The period in a batch process when maximum input is desired to achieve the furnace temperature setpoint.

**high-speed heating** = (usually implies) use of high thermal head or impingement.

**high temperature** = **hi temp** (as related to industrial heat processing) above 1400 F (760 C). See **T** = **temperature.**

**hi temp** = **high temperature**–*See* interpretation for this book under **temperature**.

**hotface** = the inner surface (or hotter face) of a furnace wall, roof, or hearth.

**ht** = heat.

**htg** = heating

**hydrogen** = $H_2$ = A highly flammable gas that burns to water vapor, $H_2O$. Hydrogen flames are usually invisible, highly reactive, and acid-forming, but usually considered nonpolluting. Its extremely low gas density allows it to permeate porous materials.

**hysteresis** = a phenomenon exhibited by a system in which the reaction of the system to changes is dependent upon its past reactions to change.

**ID** = id = inside diameter or inside dimension (e.g., of a pipe, tube, or duct). Also induced draft, as in **ID fan.**

**IDs** = inside dimensions.

**impingement heating** = high-velocity convection heat transfer by flame or hot poc gases actually contacting the load surface.

**in-and-out furnace** = a batch-type furnace that is charged and discharged through the same doors. *See* **Batch furnace.**

**incinerator** = a furnace for burning (oxidizing) waste materials—solid, liquid, or gaseous—a destructor, an afterburner.

**indirect-fired** = describes a furnace configuration in which the poc do not contact the load (e.g., with a muffle or with radiant tubes).

**Inconel** = a trade name for International Nickel Company alloys resistant to temperature and corrosion.

**indexing** = aligning billets, blooms, or slabs as they enter a furnace so that they are centered on the furnace centerline, or (in a longitudinal furnace) so that all of either the left or right ends are lined up equidistant from, one sidewall, or (in a rotary furnace) so that the outer ends of all the billets are close to, but equidistant from, the inside surface of the outer wall.

**induced draft** = a method for conveying flue gas, wherein combustion air is pulled through the burners and poc through the furnace by an **induced draft fan**, which develops more negative pressure (more suction) in the combustion system than can be created by natural draft alone.

**inerts** = gases and materials that are not capable of combustion reactions, including those already oxidized (e.g., $N_2$, $CO_2$).

**ingot** = a large metal casting to be rolled or forged to another size and shape. An ingot may be square, rectangular, or round in cross section and may weigh from 500 to 500 000 pounds (227 to 227 000 kg).

**instability** = opposite of flame stability, which see.

**in.wc = "wc = "wg = "$H_2O$ = inches of water column** on a water manometer, or 'water gauge,' a measure of pressure. 1.73 in. wc = 1 ounce per square inch (osi) *See reference 52*, pp. 318, 322.

**k** = thermal conductivity. *See* **conductivity.**

**K = Kelvin** = absolute Celsius temperature scale = C + 273.15 = (5/9 F) + 255.37. This book uses K for an actual temperature level, such as water boils at 273.15 K. Use °K only to indicate a temperature change or temperature difference. *See* **degree mark** and **T**.

**kiln** = a furnace for processing ceramic or other nonmetallic substances.

**kk** = a thousand thousand = 1 million. (Do not use m, M, or MM for million because those are official SI abbreviations for other specific units of measurement.)

**kPa** = **kiloPascal** = unit of pressure = 1000 $Nm^2$ = 0.01 bar = 0.145 psi = 4.02 in. wc.

**kW** = **kilowatt** = a unit of power, or measure of heat flow rate. 1 kW = 1000 J/s = 3412 Btu/hr = 1.341 hp = 859.8 kcal/h.

**kWh** = kilowatt hour = a unit of energy.

**lag time** = time-lag = the elapsed time required for the temperature of the center or bottom of a piece of furnace load (product) to reach the same temperature as the heated outer surface of the product. Understanding and predicting the many aspects of the heat-soaking (diffusion) phenomena has led to the modern use of furnace heating curves (*See* chap. 8).

**lance** = a **tuyere** (which see) with a tubular extension into a combustion chamber for feeding air, oxygen, or fuel into the combustion reaction.

**latent heat** = thermal energy absorbed or given off by a substance without changing its temperature, as when melting, solidifying, evaporating, condensing, or changing crystalline structure. "Latent flue loss" refers to the heat lost up the flue in the form of evaporated water formed by the combustion of hydrogen (from fuel).

**lean** = fuel-lean = air-rich = oxidant-rich = oxidizing (opposite of rich, reducing).

**lean fuels** = fuels with low calorific value, or fuels that contain low percentages of carbon and hydrogen, or major percentages of inerts (usually from upstream combustion reactions with less than stoichiometric combustion air).

**lehr** = a heat treating furnace (oven) for relieving stresses in glassware.

***lhv*** = **lower heating value** = net heating value. Whereas gross or higher heating value (*hhv*) is the total heat release, net or lower *hv* is *hhv* minus the latent heat of vaporization of the water vapor formed by the combustion of hydrogen in the fuel. In the United States, *hv* is assumed to be *hhv* unless otherwise specified. In European practice, *nhv* or *lhv* is normally used.

**lintel** = a horizontal beam support for refractory wall or roof; may be water cooled.

**LMTD** = log mean temperature difference, which see. *See reference 51*, p. 128.

**LNI** = low NOx injection.

**load** = furnace load = batch, charge, metal, pieces, product, stock, ware, work, or or any material placed in a furnace, kiln, melter, or oven—primarily for heat processing. Not to be confused with materials to be heated as an intermediate objective such as tubes, immersion tubes, furnace gases, air, water, or other heat-transfer media, or product supports (piers, stools, kiln furniture).

**log mean temperature difference** = **LMTD** = a term used in evaluating heat exchanger performance = (greatest $\Delta T-$ least $\Delta T$)/ln (greatest $\Delta T$/least $\Delta T$). [$\Delta T$ = delta $T$ = temperature difference.] *See* pp. 126–128 of *reference 51*.

**loopers** = rollers, the control of which helps maintain tension in a rolling mill and controls stress between mill stands.

**lorry furnace** = **car-furnace** = **car-bottom furnace**. *See* **car.**

**low temperature** = (as related to industrial heat processing) below about 1400 F (760 C). *See* ***T*** = **temperature.**

**M** = **mega** = millions. (Do not use old-fashioned Roman numerals for thousands, which are k in modern SI units.)

**manifold** = a pipe arrangement for delivering a fluid from one source to several use-points, similar to a plenum, but the latter implies a more generously sized distribution box; a header pipe; a bustle pipe.

**manifold door** = a furnace opening that is bricked up loosely to permit easy entry for repairs or slag removal.

**manipulator** = a machine for handling a piece of product in and out of a furnace, including charging, positioning in the furnace, removing from the furnace, and positioning for forging. This equipment may be suspended from an overhead crane runway, ride on tires, or rails in the floor.

**manometer** = a device for measuring pressure, most commonly U-tube, but also inclined, and well-type (single tube).

**melt** = to heat a substance from a solid state to a liquid state. Also, in the ametals industry, the amount of a substance melted in a single load.

**meltdown situation** = **runaway** = **snowballing** = an out-of control situation that could lead to major overheating.

**metal line** = the surface of molten load—metal or glass—or the elevation thereof.

**midrange temperature** = *See* ***T*** = **temperature.**

**mirror effect** = (as from molten scale). *See* sec. 8.3.1.

**modulus of rupture** = **MOR** = the maximum stress per unit area that a specific specimen can withstand without breaking.

**mol** = **mole** = molecule. In stoichiometric calculations, a pound mol of a gas has a volume of 379 cf at stp, and weighs its molecular weight in pounds; therefore, the stp density of oxygen is 32/379 = 0.0844 $lb/ft^3$.

**MOR** = *See* **modulus of rupture.**

**Morrison tube** = the first pass, usually a large corrugated alloy steel pipe, of a fire-tube boiler. It contains the flame and poc and is surrounded by feedwater that is to be boiled.

**mtph** = metric tons per hour. [1 metric ton = 1 tonne = 1000 kg = 2205 pounds]

**muffle** = a gas-tight enclosure that protects the pieces of a furnace load from contact with poc; often full of an inert gas. A muffle reduces fuel efficiency because it constitutes added resistance to heat flow. Most modern furnaces enclose the flames in radiant tubes, and fill the furnace chamber outside the tubes with inert gas. A 'semi-muffle' is not gas tight, and only for the purpose of preventing uneven heat transfer.

N or $\mathbf{N_2}$ = **nitrogen** = an inert gas, comprising about 80% of air and a large part of poc, unless using oxygen enrichment.

**net heating value** = ***nhv*** = lower heating value, *lhv*. See ***lhv*** .

**neutral pressure plane** = zero pressure 'plane' = balanced pressure 'line' (invisible), or level at which the pressure inside a furnace is exactly equal to the pressure outside the furnace at the same elevation. Usually not really a 'plane,' but an invisible 'surface' rumpled by burner jet and draft effects. See sec. 6.6.1.

$\mathbf{nm^3/h}$ = normal cubic meters per hour, a unit of volumetric flow rate, equal to 37.9 scfh. $nm^3$ is standardized at 0 C, 760 mm Hg, dry air or gas. A standard $ft^3$ is defined at 60 F, 30"Hg, saturated air or gas.

**normal air** = European near-equivalent of U.S. "standard air", (*see also*).

**NOx** = $NO_x$ = nitrogen oxides, specifically defined by the U.S. EPA as $NO + NO_2$. NOx is formed in some combustion reactions, particularly with flame temperatures above 2800 F. To minimize NOx formation, the mixing aerodynamics and thermodynamics of flames must be designed (*a*) to have the chemical burning take place in stages so that some burning occurs below combustion chamber temperature and the balance only slightly above furnace gas temperature and (*b*) to minimize the concentrations of free N and O ions. NOx formation is largely dependent on combustion reaction temperature.

**Nusselt equation** = a method for predicting forced convection heat transfer coefficients (film coefficients). (*See* sec. 2.5.2.)

**Nusselt number** = $h_c L/k$. A dimensionless ratio of convection to conduction capabilities. (*See* sec. 2.5.2.)

**O = $O_2$ = oxygen** = a highly reactive gas, responsible for combustion (burning), oxidation, slagging, and drossing of materials if exposed to high temperatures.

**OD = od** = outside diameter or outside dimension (e.g., of a pipe, tube, or duct).

**offtakes** = downcomers = distribution pipes from a manifold.

**Orsat** = a flue gas analysis instrument, originally by use of absorption chemical liquids. Primarily for $CO_2$, but also $O_2$ and CO. Now a generalized term for any type of flue gas analysis.

**osi** = ounces per square inch, a measure of pressure. 16 osi = 1 psi. (*See reference 52*, pp. 318, 322.)

**oven** = (especially in the United States) a low-temperature furnace or kiln, usually less than 1400 F or 790 C. Exception is a coke oven which operates above 2200 F (1200 C). In some countries, an oven is any furnace.

**overall coefficient of heat transfer** = $U = Q/A\,\Delta T$, in Btu/ft$^2$hr°F or kW/°Cm$^2$. See also **heat transfer coefficient.** Whereas $h$ is usually specifically for one mode of heat transfer, $U$ includes the combined effects of several resistances in series and in parallel, for example, $1/U = 1/[(1/h_i) + (x/k) + (1/h_o)]$ which covers three resistances in series, and in which $h_i$ and $h_o$ can include $h_c + h_r$, two resistances in parallel. $h_c$ = convection coefficient or 'film coefficient'. $h_r$ = radiation coefficient. $h_i$ = inside. $h_o$ = outside.

**overfill** = rolled material that more than fills the passes, creating bulges on the product at the pass line.

**oxidizing atmosphere** = a condition in a furnace or kiln wherein the furnace gases contain more free oxygen than reducing gases, so that the load in the furnace or kiln would tend to be oxidized or corroded. Also termed an air-rich or fuel-lean atmosphere. Opposite of a reducing atmosphere.

**oxy-fuel firing** = a system for operating a burner with 100% oxygen instead of air (which has only 20.9% oxygen).

**oxygen** = $O_2$ = a highly reactive gas, responsible for combustion (burning), oxidation, slagging, and drossing of materials if exposed to high temperatures.

**oxygen enrichment** = burning fuel with a mixture of air and commercially 'pure' oxygen (anywhere from 20.9 to 100% oxygen) to improve efficiency, to produce a higher flame temperature, or to reduce flue gas volume. Oxy-fuel firing (100% oxygen, no nitrogen) still has the NOx-making high temperature, but supposedly lacks the nitrogen to form NOx. Unfortunately, small amounts of nitrogen may be included in supposedly pure oxygen and in fuels, and they enter a combustion reaction as tramp air.

**Pa** = **Pascal**, a unit of pressure = 0.00202 oz/in.$^2$ (or osi). *See* pp. 318 and 322 of *reference 52.*

**peel bar** = a mechanism for pushing a billet or bloom endwise out of a side discharge reheat furnace. It consists of a long ram driven by a motor or a cylinder. Similar to an **extractor** (which see), but pushes instead of lifting or pulling.

**periodic kiln** = a batch or in-and-out furnace, a heating chamber in which loads remain without any conveyor movement for a period of heating time (i.e., a furnace which is periodically loaded and unloaded; perhaps, periodically fired and cooled).

**PIC** = pressure indicating controller.

**pic** = products of incomplete combustion, such as CO, OH, or aldehydes. The pic are often mixed with some **poc**.

**pickling** = immersion of metal parts in a (sometimes hot) chemical solution to remove surface scale, thereby exposing defects.

**pier** = a support for a load in a furnace, oven, or kiln for the purpose of enhancing convection and radiation heat transfer to the bottom and sides of the load(s), and to reduce heat loss from the loads to the hearth. Also used for these purposes are pillars, posts, stanchions, skid rails, walking beams, kiln furniture, "stools," "chairs," and conveyors.

**pileup** = an accident in a furnace, resulting in an accumulation of unfinished product, often damaged, similar to a 'wreck' in a ceramic tunnel kiln.

**pilot** = a small flame used to light a larger burner. An *interrupted pilot*, sometimes called an *ignition pilot*, is automatically spark ignited each time that the main burner is to be lighted. It burns during the flame-establishing period and/or trial for ignition period and is automatically cut off (interrupted) at the end of the main burner flame-establishing period while the main burner remains on. Interrupted pilots are usually preferred/required for industrial heating operations.

**pit** = (1) surface indentation (imperfection) caused by scale being rolled into the surface of the metal or (2) Short talk for a soaking pit furnace.

**plastic** = **plastic refractory** = a kind of refractory material having plasticity (which see), such as rammable refractories.

**plasticity** = the ability of a solid to be strained beyond its elastic limit, and thus to suffer permanent deformation, without fracture.

**plenum** = a windbox, or a generously sized distribution manifold.

**poc** = products of combustion (usually assumed stoichiometric or lean combustion—$CO_2$, $H_2O$, $N_2$, and $O_2$—unless specified as **pic** = products of *in*complete combustion. Should be specified as dry or wet (containing water vapor). May also contain excess air, tramp air, excess unburned fuel, or a variety of pic.

**polymerization** (as applied to fuels) = See **cracking.**

**pop scale** = metal oxide scale that explodes off the surface of cold billets or slabs as they enter a hot furnace.

**power stack** = a furnace exhaust system that uses mechanical energy, in addition to natural draft, to remove poc from the furnace and flue system. The gases may be pulled through a hot fan (induced draft fan) or inspirated by the Venturi effect of an air jet.

**ppb** = parts per billion. **ppm** = parts per million. Both must be specified as by volume (most common) or by weight. One ppm = 0.0001%.

**pr** = pressure, pres, or press. For units of pr, *see* pp. 318, 322 of *reference 52.*

**Prandtl Number** = $c\mu/k$ = a dimensionless ratio of fluid properties that affect heat flow. *See* sec. 2.5.2.

**preheat zone** = temperature control zone(s) above or below the product of a steel reheat furnace, before the main heat zone. May be top-fired, bottom-fired, side-fired, end-fired, or a combination of these.

**pressure drop, pressure change** = $\Delta P$ or $\Delta p$ (with respect to place or time).

**producer gas** = a manufactured gaseous fuel made by burning coal under reducing conditions. Gross heating value ranges from 117 to 499 Btu/ft$^3$ (4.36 to 18.6 MJ/m$^3$), but average around 150 Btu/ft$^3$ (5.60 MJ/m$^3$) with hot fuel gas.

**product** = the load being manufactured by heat processing. *See* **load, charge, ware, stock, batch.**

**productivity** = "the output of goods and services relative to the inputs of resources, human and nonhuman, used in the production process" per "Understanding Productivity" by John Kindrick, John Hopkins University Press, 1977 [*reference 79*]. Examples of uses in this book: pallets of bricks/MJ of gross fuel input, or dollars worth of finished pipe/man-hour, or yearly tons waste incinerated/million dollars of incineration plant capital investment.

**products of combustion** = poc = flue gases (Stack, exhaust, or exit gases may be cooler and diluted, or mixed with poc of other furnaces). The poc are usually assumed to be poc, on their way to or through a flue, heat recovery device, pollution reduction equipment, or stack. They consist of $CO_2$, $H_2$, and $N_2$, but also may include $O_2$, CO, $H_2$, aldehydes, and other complex hydrocarbons, and sometimes particulates, sulfur compounds, and nitrogen compounds. See also **pic**.

**psf** = **pounds per square foot** (pressure, or hearth coverage).

**psi** = pounds per square inch (pressure, stress, or strain). 1 psi = 144 psf. *See reference 52,* pp. 318, 322. 1 psi = 6.895 kPa = 51.72 mm Hg = 27.71"wc.

**pulse combustion** = a ramjetlike burner system used in some mass-produced domestic furnaces, utilizing a pressure wave to compress and mix the fuel and the air.

**pulsed firing** = pulse firing = pulsed-controlled combustion = controlling heat input rate by turning some burners to off or very low instead of modulating the input rate to all burners in a zone. The ratio of time-on to time-off is modulated to lower the fuel use rate of a furnace or kiln—often combined with step-firing. (Not 'pulse combustion,' the ramjetlike burner system used in some mass-produced domestic furnaces.)

**pusher furnace** = a continuous furnace, in which the conveying mechanism pushes billets, blooms, or slabs of rectangular cross section on a smooth hearth or on skid rails.

**QED** = quod erat demonstrandum = Latin for "What was to be shown." The closing statement at the end of a geometric proof, or the solution to a problem.

**quarl** = a burner 'tile,' the refractory-lined hole or 'port' through a combustion chamber wall, through which air and fuel are injected and/or a burner flame is fired. The quarl is usually designed to enhance flame stability by adding the minimum ignition energy required to begin and sustain chemical reaction. The burner tile may influence the flame stability and character. The inside passage of a quarl may be cylindrical or conical, diverging or converging.

**quenching** = very quick reduction of temperature of a metal to increase its hardness and tensile strength. This cooling can be done with air, water, brine, or oil. Normally, quenching of carbon steels is followed by tempering to prevent cracking and to improve toughness. To quench, to martensite, the cooling to 400 F (204 C) should be accomplished in less than 30 sec.

**R** = **Rankine** temperature scale. This book uses R for an actual temperature level, such as water boils at 492 R. Use °R only to indicate a temperature change or temperature difference. *See* **degree mark**.

**RA** = a trade name or specification for products of Rolled Alloys, Inc.

**rabbit ears** = a pair of ducts external to a steel reheat furnace, conveying gases from the bottom to top or top to bottom depending on flue exit locations. To provide sufficient flow cross section, these ducts usually extend out from both sides of the furnace, hence looking like rabbit ears.

**radiant tube** = a tubular muffle through which a burner is fired for indirect heating of furnace loads. The metal alloy or ceramic tube wall transfers heat to the load without poc contact by a combination of radiation and convection from its outer surface. This provides process heating with reduced risk of scale formation or damaging reactions on the product surface. A prepared atmosphere (friendly to the material being heated) may be piped into the furnace space outside the radiant tubes.

**radiation** = a mode of heat transfer in which the heat travels in straight lines at the speed of light without heating the intervening space (except it will heat triatomic gas molecules such as $CO_2$ and $H_2O$). Heat can be radiated through a vacuum, through many gases, and through a few liquids and solids. *See* **gas radiation, solids radiation,** chap. 2.

**radn** = radiation.

**rammed refractory** = refractory material that is installed using an air or hand hammer. Some such refractories also are sprayed (gunned) into place.

**rate of heat absorption** = **RHA** = heat flux rate received by a furnace load, usually in Btu/ft$^2$hr.

**recirculating oven** = a low-temperature furnace using an internal or external recirculating fan to enhance convection heat transfer, uniformity, and fuel economy by directing the warm air and poc over the loads, often at considerable velocity.

**recup** = recuperator, or recuperative.

**recuperator** = a piece of equipment that makes use of the energy in hot flue gases to preheat combustion air. The poc gases and airflow are in separate adjacent passageways so that heat is transferred from the hot exhaust gases (source), through a separating, conducting wall, to the cold air (receiver).

**recuperator effect** (also regenerator effect) = the bonus gain from preheating air, by virtue of the more intense heat transfer from a hotter flame in addition to the savings from having the combustion air preheated so that less fuel is used in getting the air and fuel up to flame temperature.

**reducing atmosphere** = rich atmosphere, nonoxidizing, purposely used for protection of some metals and ceramic materials. It may be created by utilizing reducing combustion (e.g., incomplete combustion, insufficient air). Opposite of an oxidizing atmosphere.

**refr** = **refractory** = heat-resistant material used to line combustion chambers and furnaces. May be in prefired shape form (bricks), cast, rammed, or gunned.

**regen** = regenerative or regenerator.

**regenerative furnace** = a furnace and associated regenerator, especially a furnace with a pair of refractory checkerworks for storage and recovery of waste heat from poc.

**regenerator** = a cyclic heat interchanger that alternately receives heat from gaseous combustion products and transfers that heat to air for combustion.

**regenerator effect** (also recuperator effect) = the bonus gain from preheating air, by virtue of the more intense heat transfer from a hotter flame in addition to the savings from having the combustion air preheated so that less fuel is used in getting the air and fuel up to flame temperature.

**reheat furnace** = (primarily) a continuous steel heating furnace used to reheat cooled billets, blooms, or slabs for primary or secondary rolling.

**reverberatory furnace** = any large heating chamber wherein radiation reverberates from walls and roof to the load, especially open hearth and other melting furnaces.

**Reynolds Number** = a dimensionless ratio of kinetic (momentum) forces to viscous forces = $\rho VD/\mu$. *See* sec. 2.5.2.

**RHA** = **rate of heat absorption** = heat flux rate received by a furnace load, usually in Btu/ft$^2$hr.

**rich** = **reducing** = **fuel rich** = **air lean** or air starved, containing pic.

**rider flue** = an arched flue-way that supports a checkerwork, serving as a windbox for cold air being pushed up through the checkers, or a collection plenum for hot poc being pulled down through the checkerwork.

**rolling efficiency** = the percentage of the scheduled time actually operated.

**roof** = the top refractory cover of a furnace. May be flat, arched, or crowned, and removable or fixed. *See* **arch, ceiling, crown.**

**roof burners** = type E ("flat flame") burners that spread their flame radially. Care must be observed to prevent any condition that would let these flames fire forward (downward), melting the scale or metal of the load(s).

**rotary drum furnace** = a furnace shaped like a large hollow tube, usually slightly inclined to cause granular matter to tumble as it is rotated from the high end to the lower (discharge) end. Mostly used for drying or calcining. See sec. 4.2.

**rotary hearth furnace** = a furnace shaped like a merry-go-round or carrousel. Mostly used for steel reheating for heat treating, rolling, or forging. *See* sec. 4.6.1.2, 4.6.3, and 6.4.1. Small rotary hearth furnaces are usually single zone furnaces consisting of a disclike hearth all across the diameter. *Donut rotary hearth furnaces* have a hole in the middle with an inner wall as well as an outer wall. Some equipment may be placed in the center "hole," but access and working conditions are poor in the hot "hole in the donut."

**runaway** = a control condition that accelerates (snowballs) out of control.

**safety factor** = should refer only to matters of human body safety, but this term is often used by designers to refer to a design multiplier or design margin that they put on their calculations to cover unknowns, estimates, and changes with time. In this book, those are termed **security factors,** (*see also*).

**saggers** = refractory boxes or holders for small parts being heated in an oven or kiln. Usually perforated or open sided and with "feet" to serve as spacers to allow hot gas flow through the small load pieces.

**Sankey diagram** = a visual aid to understanding the disposal of heat released in a furnace, oven, boiler, or kiln—by use of arrows of widths proportional to the magnitude of the heat flow.

**scale** = an oxide that forms on metals, often clinging to the surface of the metal from which it formed. With steel, it is a mixture of FeO, $Fe_2O_3$, and $Fe_3O_4$.

**scarfing** = removal of steel surface problems with oxy-fuel torches. *See also* **chipping** and **grinding**.

**scf** = standard cubic feet, a measure of gas volume at 60 F (16 C) and 1 atmosphere of pressure. 1 scf = 1728 standard cubic inches. *See* p. 324 of *reference 52*.

**screen burners** = a row of burners located at the dropout or other points of air inleakage on a steel reheat furnace to counter the air velocity pressure and thereby practically eliminate ambient air inleakage (tramp air).

**SD** = **sd** = **super-duty** = the best quality of fireclay brick.

**secondary air** = the second stream of air to be mixed with fuel in, at, or near a burner. *See also* **tertiary air**. In an air-atomizing oil burner, the atomizing air might be considered to be primary air and the combustion (or main) air to be secondary air. In an open burner (some air induced by draft), all air through the burner (atomizing and combustion air) may be considered to be primary air, and all through the register to be secondary air.

**sect** = section.

**security factor** = a *multiplier* used in design to allow for the user overloading the equipment and to allow for questionable information or unknowns used in the design. Specifically in furnace design, a "fudge factor" to allow for overstating heat availability due to understating flue gas temperature, and to allow for future problems that may increase heat losses, and for future growth and demand. It has been reduced over the years for cost reasons, but should not normally be below 1.25. This is sometimes called a "safety factor," but the design (security) factor usually does not apply to matters of human body safety, for which "safety factor" should be reserved. In contrast, a safety *margin* or security *margin* is an *additive* amount—not a multiplier.

**segmental orifice plate** = a primary flow metering device, the flow restriction being like a dam across a segment of the duct's cross-sectional area. The principle is similar to that of a concentric or annular orifice, but the flow coefficients of all three are different. Downstream tap locations also are different. The dam or solid-segment portion of a segmental orifice plate should be at the top of the pipe to minimize the effect of liquids or solids accumulation on the upstream side. In contrast, *see* **concentric orifice** and **annular orifice**.

**semimuffle** = a refractory partial enclosure around load pieces to assure more uniform temperature (not for protection from reactive contact with poc, as with a full muffle or radiant tube). Semimuffles are used less since the advent of a variety of flame shapes that can assure more uniform heat distribution.

**sensible heat** = thermal energy, the addition or removal of which results in a change of temperature (able to be sensed) as opposed to latent heat, which can be added or withdrawn without changing the material's temperature (as in freezing, melting, condensing, or vaporizing).

**setpoint** = the value chosen to be maintained by an automatic controller (e.g., set point temperature or selected air/fuel ratio, or selected pressure to be controlled).

**sfc** = specific fuel consumption, such as Btu/ton.

**sfr** = **specific fuel rate** = amount of fuel consumed per hour, or per hour and per unit of hearth area, or per hour and per unit of furnace volume, OR **specific fuel requirement** (or required) per ton of product, in Btu/ton, or Btu/mton, or kcal/mton.

**SI** = **Systeme International d'Unites** = the world-wide system of units (except in the United States), an outgrowth of the metric system. For conversion factors between US and SI units, see pp. 245–252 of *reference 51* and pp. 317–127 of *reference 52*.

**side-fired furnace** = a heating chamber with burners fired through its sidewalls. In a continuous furnace, firing across the direction of product movement.

**skelp** = narrow hot-rolled steel strip, mainly for making butt-welded pipe in $\frac{1}{2}$ in. to 4 in. pipe size, for which wall thicknesses run 0.12 to 0.327 in. and widths 8.25 to 17.5 in.

**skid block** = a very wear-resistant refractory hearth material alongside skid rails, or skid rails themselves, generally made from fused refractories for maximum wear resistance.

**skid rail** = metal support, often water cooled, on which rectangular billets, blooms, or slabs are pushed or walked through a furnace.

**slab** = a semifinished, oblong metal block continuously cast or forged or rolled from an ingot, usually for further rolling into plate, sheet, or strip. Typically 2 in. to 10 in. thick (50 to 250 mm) by 24 in. to 60 in. wide (600 to 1500 mm) and up to 40 ft long. In contrast, *see* **bloom, billet, bar**.

**slag** = a metal oxide, or by-product of a blast furnace (from molten limestone).

**slag pocket** = bottom of the uptake or downtake of an open hearth, soaking pit, or reheat furnace, having a large manifold door on the casting side, for slag storage during operation.

**slag seal** = a refractory seal or dam used to prevent flow of molten slag into a flue, which could block gas flow to the flue and thus require shutdown of the furnace.

**snowballing** = **runaway** = **meltdown situation** = a loss of control (such as accelerating temperature) as with making a snowball in which each step enables the next to add more.

**soaking pit** = soak pit = a refractory-lined furnace with a combustion system used to heat large, heavy pieces such as ingots, slabs, or bloom downs.

**soak time** = added time in a furnace for temperature equalization throughout a load.

**soak zone** = a final area of a continuous reheat furnace in which time is allowed for the stock temperature to equalize by conduction.

**solids radiation** = radiation from solid bodies such as refractories, other loads in a furnace, and soot particles—as opposed to "gas radiation" from gases. *See* chapter 2.

***sp*** = static pressure, as opposed to velocity pressure or total pressure ($sp + vp$).

**spacing ratio** = ($c$ to $c$)/$w$ = center-to-center distance divided by width. If there is no space between pieces, this spacing ratio is 1.0. If there is a 3 in. space between 6 in. wide pieces, their spacing ratio is (6 + 3)/6 = 1.5.

**specific fuel rate** = **sfr** = amount of fuel consumed per hour, or per hour and per unit of hearth area, or per hour and per unit of furnace volume, or **specific fuel requirement** (or required) per ton of product, in Btu/ton, or Btu/mton, or kcal/mton.

**specific heat** = ***c*** = heat absorbed by a unit weight of a material when its temperature is raised one degree. 1 Btu/lb°F = 1 cal/gram°C. For gases, differentiate between $c_p$ at constant pressure and $c_v$ at constant volume. The $c_p$ is used in furnace work.

**specific heating capacity** (of a furnace) = weight of load that a furnace can heat uniformly per hour (over an extended period) and per unit of hearth area or per unit of furnace volume (e.g., pounds/ft$^2$hr, or pounds/ft$^3$hr).

**SS** = **ss** = **stainless steel**, (*see also*).

**stack** = a pipe, duct, or chimney, often refractory-lined, to convey furnace exhaust gases away from personnel, usually through the roof of the building. *See* sec. 2.6.4.

**stack effect** = the result of hot air rising in a furnace—creating a negative pressure at the bottom of a furnace or a stack.

**stack gas** = **flue gas** = furnace waste gases that have passed through the flue and heat recovery equipment, and entered the stack or chimney.

**staged air** = air added to a combustion reaction in stages. For example, a dual-fuel (or combination) burner may have atomizing air as the primary air stage, 1st stage combustion air as the 2nd stage combustion air, and 2nd stage combustion air as tertiary air. Pilot air is not usually considered part of any of the above. Staging is sometimes accomplished with peripheral air or fuel jets around a burner proper to reduce NOx formation by lengthening a flame (delayed mixing), which results in a lower average-reaction temperature.

**stainless** = **stainless steel** = a class of steel alloys capable of resisting oxidation or loss of desirable properties with high temperature or in corrosive atmospheres.

**standard air** (in the United States) = air at standard temperature and pressure, which is 60 F and 14.696 psia and saturated (100% relative humidity). In Europe, "normal air" is at 0 C, 760 mm Hg, and dry (0% relative humidity).

**static pressure** = the pressure pushing outward on the inside of a tank wall. (Very different from velocity pressure, which see.) Total pressure is static pressure + velocity pressure.

**Stefan-Boltzmann Law** = the 4th power effect of absolute temperature on radiation heat transfer rate.

**stepped firing** = A timing system for a series of boilers, furnaces, or burners originally for extending their life by rotating the unit(s) in use so that no one unit would be worn out faster than the others. It applies only when not all units are needed at one time. Burner step firing also is used to improve temperature uniformity within a kiln or furnace during less than 100% input periods. North American Mfg. Co. has patented a ("StepFire") control system for furnaces and kilns combining pulsed firing and stepped firing. *See also* sec. 2.6.4.

**stock** = furnace load = batch, charge, metal, pieces, product, ware, work, or any material placed in a furnace, kiln, melter, or oven—primarily for heat processing. Not to be confused with materials to be heated as an intermediate objective such as tubes, immersion tubes, furnace gases, air, water, or other heat-transfer media.

**stoichiometric** = (when referring to combustion, flame, or air/fuel ratio) = chemically correct, perfect, ideal (i.e., no excess fuel or oxidant).

**stove** = *See* **blast furnace stove.**

**stp** = standard temperature (60 F, 15.56 C) and pressure (14.696 psi, 760 mm Hg). *See also* discussion of **standard air**.

**stp velocity** = (stp volume)/(area of the flow path).

**stp volume** = actual volume × (stp absolute temperature/actual absolute temperature or = actual volume × (actual density/stp density).

**surging** = **pulsation** = fan or blower instability, alternately delivering large and small flow rates, sometimes causing noise, physical damage, and unreliable burner flames. Caused by operating at a volume output rate below that of maximum pressure. When the fan's discharge pressure drops below the downstream duct pressure (zero volume flow), followed by a reverse flow until the duct pressure drop below the fan's output pressure. This causes a sudden second reversal—to forward flow again—and thus begins cycling, or surging, which may be amplified if resonant conditions exist. Fans with large discharge volumes at high pressure produce greater surging noise and damage. Air reversal through burners has caused explosions in large air ducts supplying burners.

**suspended roof** = a furnace roof that is supported from above to put no strain on the furnace sidewalls (as brick arches and crowns do). The refractory roof is suspended from a steel superstructure with steel clips holding refractory anchors embedded in the roof refractory.

**szt** = soak zone temperature.

**T** = **tee** = pipe tee, duct tee = a junction in a pipe or duct fluid conveying system that is shaped like the letter T. It may be used for one incoming stream splitting into two streams, or two incoming streams joining into one outgoing stream—similar to a Y or wye fitting, but the T would have more pressure drop.

***T*** = **temp** = **temperature level** (must be specified as C, F, K, or R) = a measure of molecular velocity. A measure of the accumulation of heat (thermal energy). [The practice within this book is to use the degree mark, °, only when describing a temperature change, or specifying a temperature difference ($\Delta T$), the driving force (potential) in heat flow. Examples: water freezes at 32 F or 0 C. The temperature difference ($\Delta T$) between the refractory and the load was 900°F or 500°C. The temperature dropped 45°F (or 25°C) overnight. In this book, "very high temperature" usually means >2300 F (>1260 C), "high temperature" = 1900–2300 F (1038–1260 C), "midrange temperature" = 1100–1900 F (593–1038 C), and "low temperature" = <1100 F (<593 C). See *reference 52*, p. 322 for temperature level conversion formulas. *See* **degree** for temperature change or difference formulas. **Warning:** Do not confuse $T$ with $t$, which is thickness or time, not temperature.

$t$ = thickness, or time.

**tank** = a refractory-lined holder for molten glass or zinc, which constitutes the lower portion of a glass melter, galvanizing "kettle," or liquid salt bath.

**td** = **t/d** = **turndown** = **turndown ratio,** (see also).

**temperature** = ***T*** (*see also*).

**temperature control** = *See* chap. 6. *See also* **accordion effect**.

**temperature sensor** = **T-sensor** = such as a thermocouple (T/c, or tc)—for observation, input control, or high-limit protection.

**tempering** = a heat-treating process used after quenching steel to martensite, which is very hard and brittle. In tempering, the steel is normally heated to 1000 F to 1260 F to reduce stresses, improve ductility, and increase toughness.

**tertiary air** = a third supply of air to a burner, introduced downstream from the secondary air. Example: a dual-fuel low NOx burner with staged air might have atomizing air as the primary air, combustion (or main) air as the secondary air, and the staged air as the tertiary air. *See also* **secondary air**.

**thermal conductivity** = $k$ = a measure of a material's ability to conduct heat, measured in Btu/hr or joules/hr flowing through a unit of cross sectional area (square foot or square meter) and through a unit thickness (ft, in., m) with 1° (F, C) of temperature difference across that thickness. In the United States, refractory and insulation industries use Btu in./ft$^2$ hr°F. Most others use Btu ft/ft$^2$hr°F.

**thermal efficiency** = *See* sec. 5.1. Care must be used in differentiating between thermal efficiency and combustion efficiency, furnace efficiency, fuel efficiency, and heating (or heat transfer) efficiency. They may not be synonymous. *See* the discussion under **efficiency**.

**thermal head** = the difference in temperature between the source of heat (furnace refractory or poc) and the receiver of heat (the furnace load). Increasing this difference in potential increases the rate of heat transfer.

**thermal turndown** = achieving a lower effective input to a furnace by adding excess air through burners—in effect, turning down the thermal efficiency when a lower minimum input is required than achievable by valve-throttling turndown. One way to accomplish temperature control by thermal turndown is to hold the air flow constant while reducing fuel input.

**thumb guide** = "rule of thumb" downgraded from a 'rule' to a 'guide.' (Coauthor Reed does not have a lot of respect for 'rules of thumb' because one must remember all the limiting conditions on which they are based.) They should be used only when no other option exists.

**tile** = (usually a burner tile or quarl) = the refractory-lined hole through a combustion chamber wall through which air and fuel are injected, and/or a burner flame is fired. The quarl is usually designed to enhance flame stability by adding the minimum ignition energy required to begin and sustain chemical reaction. The burner tile also may influence the flame character. The inside passage of a quarl may be cylindrical or conical, diverging or converging. Not to be confused with **burner tunnel**.

**time-lag** = *See* **lag time.**

**time/temperature (T/t) curve** = load heating curve—such as derived by the Shannon Method.

**top-fired furnace** = a heating chamber with burners firing above the load. These may be horizontally fired burners high in the sidewalls, or longitudinally fired from the end walls, or in a sawtooth roof, or vertically fired "roof burners" such as type E flat-flame burners. (See fig. 6.2.)

**tpc** = tons per cycle.

**tpd** = tons per day; tonnes per day.

**tph** = tons per hour, assumed US ton = 2,000 pounds, unless specified as British (2,240 lb, long tons) or mtph, metric (2,205 pounds).

**track time** = the elapsed time between end of pouring of ingots and the end of charging the ingots into a soaking pit or furnace.

**tramp air** = air that leaks into a furnace, perhaps not helping the combustion reaction or the heating process, and generally increasing temperature nonuniformity.

**triatomic molecules** = molecules having three atoms, such as $CO_2$ and $H_2O$, which are capable of radiating heat when in the gaseous state. $SO_2$ also is triatomic, but is bad for pollution and corrosion reasons.

**T/s** = **T-sensor** = such as a thermocouple (T/c, or tc)—for observation, input control, or high-limit protection.

**tufa** = a porous limestone from calcium carbonate, or solidified bubbled lava—similar to insulating fire brick.

**turndown** = turndown ratio = high-fire rate/low-fire rate. *See also* thermal turndown.

**tuyere** (pronounced 'tweer') = the opening through which a blast of air, oxygen, or fuel is fed into a blast furnace or cupola. If it has a tubular extension into the furnace, it would be termed a "lance."

**UBC** or **ubc** = used beverage containers—a major source for some aluminum melting operations.

**uptake** = any outlet connection of a processing vessel that conveys gas or products of combustion upward. In the case of regenerator checkers, the large refractory duct that connects the fantail duct with the furnace above. If the flue gases flow downward out of a furnace into a slag pocket, it is called a "downtake."

**U-tube** = a tube bent into the shape of the letter *U*. Often used as radiant tubes or in a heat exchanger. Also a type of manometer.

**variable frequency drive** = **VFD** = an energy-saving way to control speed or input by controlling electric motor speed (rpm)—applied to fans, blowers, exhausters, compressors for air and fuel, and to load pumps and conveyors.

**velocity pressure, *vp*** = the pressure drop necessary to accelerate a fluid (gas or liquid) to a certain velocity. When a fluid is in motion at some velocity, the velocity pressure is the pressure rise that was required to raise it to that velocity. (Compare with static and total pressure.)

**Venturi** = a converging and then diverging flow nozzle, used for metering and for creating a suction such as in eductors and ejectors.

**Venturi effect** = suction created by conversion of pressure energy to kinetic (velocity) energy.

**vertical furnace** = a heating chamber in which long loads are suspended vertically to prevent bending from their own weight during heating.

**very high temperature** = *See* ***T*** (**temperature**).

**VFD** = **variable frequency drive** (*see also*).

**vitiated air** (pronounced vish'-ee-ate-ed) = air containing less than 20.9% oxygen.

**vitrify** = the application of high heat to a substance to cause chemical change and physical change (including temporary liquifaction) resulting in a glasslike or ceramic material.

*vs.* = versus, against, opposite—as in a temperature-*vs.*-time (*T vs. t*) curve or graph.

**W** = **watt** or watts (*see also*).

*w* = width or weight.

**walking beam** = a conveying mechanism that advances pieces through a furnace at a selected intermittent, but regular, rate by lifting every piece, advancing it, and lowering it onto stationary holder.

**walking beam furnace** = a heating chamber with loads placed on insulated and water-cooled longitudinal "beams," moved by a "walking beam" mechanism with top and bottom firing. Usually a steel reheat furnace.

**walking hearth furnace** = a heating chamber with loads placed on large refractory slabs for product advancement, with top firing only. The refractory surface of a walking hearth is generally similar in construction to the main hearth.

**ware** = load or product in a ceramic kiln.

**warm-up time** or heat-up time = the necessary slow heating of a furnace that has been allowed to cool below its normal operating temperature. Much shorter than **dryout time,** which see. *See* sec. 9.5.

**washed, washing** = melting scale on the surface of a steel workpiece. One might think that letting the scale be washed away is a good way to remove scale, but the temperature required to "wash" scale is so high that more scale will form almost instantly. Washed steel is caused by temperatures exceeding 2490 F (1365 C) and/or flame impingement.

**washings** = melting scale on the surface of a workpiece, especially on ingots—generally caused by flame impinging on the loads.

**waste gas** = (as related to furnaces) flue gas or stack gas.

**waste heat boiler** = a steam generator heated by waste flue gases passing through it, from some adjacent process heater. A heat recovery device, usually with no burners or fuel consumption.

**water column** = **wc** = a measure of pressure, referring to the height of a column of water in a water manometer. *See also* in. **wc**.

**watt** = a unit of power, or rate of energy flow, generation, or consumption = 1 J/s = 1 N·m/s. See p. 320 of *reference 52*. Named after James Watt, inventor of the steam engine, 1882.

**wave effect**–See **accordion effect**

**"wc** = inches of water column = in wc = "water gauge = "wg. = "$H_2O$ = inches of water column height in a water manometer, or 'water gauge,' measure of pressure. 1.73"wc = 1 osi (ounce per square inch). *See* pp. 318 and 322 of *reference 52* or pp. 246, 248, and 249 of *reference 51*. *See also* in. wc.

**work** = See **batch, charge, load, product, stock, ware, workpiece.**

**work hardening** = changing the mechanical properties of a metal by physically "working" it (e.g., bending, rolling, stretching). Ductility is reduced; strength is increased.

**workpiece** = See **batch, charge, load, product, stock, ware.**

**Wye** = Y-shaped pipe or duct fitting.

$x$ = Distance through which heat is conducted. *See* **overall coefficient of heat transfer.** *X* also means a pipe or duct fitting (cross) shaped like the letter $x$. × is also used to mean multiplied by as in a 9 ft × 12 ft × 7 ft ID furnace.

**xs** = excess. As in **xs air, xs fuel.**

**Y** = wye pipe or duct fitting—takes less pressure drop than a T.

**yield** = fraction or percent of the total charged weight that becomes shipped product.

**yield point elongation** = the point in tensile testing where the elasticity of the test piece is deformed and does not return to its original shape and dimension when the strain is removed (i.e., beyond its elastic limit).

## Greek Letters

- $\Delta$ delta = change, difference, gradient (e.g., $\Delta$p for pressure difference or $\Delta$t for temperature differential.
- $\lambda$ lambda = wavelength.
- $\rho$ rho = density, lb/ft$^3$ or kg/m$^3$. Not to be confused with “specific gravity” which is a ratio of densities, usually relative to water for liquids and relative to air for gases. For example, the density of stp water is 62.4 pounds/ft$^3$, but its specific gravity is 1.0. In contrast, *see* **gas gravity**.
- $\sigma$ sigma = the Stefman-Boltzmann constant for radiation. See Sec. 2.3.3 and 2.3.4.

## Mathematical and Other Symbols

| Symbol | Meaning |
|---|---|
| + or & | plus, and, added to |
| $-$ | minus, less than, subtracted |
| $\pm$ | plus or minus |
| $\times$ | times, multiplied by. Note: ( ) ( ) is the same as ( ) $\times$ ( ) or ( ) $\cdot$ ( ). Also means “by” as in 9 ft $\times$ 12 ft rug. |
| / | divided by, per, for each, over, ___ |
| = | equals, or equal to |
| $\neq$ | “not equal to” or “unequal to” |
| $\sim$ | similar to, about, proportional to |
| $<$ | less than |
| $>$ | greater than |
| $\therefore$ | therefore |
| @ | at |
| # | pound weight (0.4536 kg); or number (sometimes abbreviated ‘no.’) or quantity |
| % | percent, of 100. |
| ' | feet (0.3048 m); or quotation begin (') or end ('); or apostrophe. |
| " | inch (25.4 mm); or quotation begin or end; or ditto (meaning same as above). |
| ° | *See* **degree mark**. |

# REFERENCES AND SUGGESTED READING

(In alphabetical order by *author*.)


1 American Society for Testing and Materials ASTM S110: "Standard Practice for Use of the International System of Units (SI)." IEEE/ASTM SI 10–1997.

2 American Society of Mechanical Engineers, 1956 ASME Transactions, pp. 177–192. Sherman, R.A.: "Radiation from Luminous and Nonluminous Natural Gas Flames."

3 Association of Iron and Steel Engineers: "Making, Shaping, and Treating of Steels" (originally by USSteel Corp.); AISE, Pittsburgh, PA, 1998.

4 Bartok and Sarofim: "Fossil Fuel Combustion—A Source Book"; John Wiley & Sons, New York, NY, 1991.

5 Baukal, C.E.: "Oxygen-Enhanced Combustion," CRC Press, Boca Raton, FL, 1998.

6 Bennett, C.O. and Myers, J.E.: "Momentum, Heat, and Mass Transfer," 3rd ed., McGraw-Hill, New York, NY, 1982.

7 Bhowmik, A.K.: "Maintenance Spells Extended Life for Chimneys and Stacks," Plant Engineering 9–3–92.

8 Bloom, F.S.: "Rate of heat Absorption of Steel," Iron and Steel Engineer, 1955.

9 Borman, G.L. and Ragland, K.W.: "Combustion Engineering," McGraw-Hill, New York, NY. 1998.

10 Bosworth, R.C.L.: "Heat Transfer Phenomena," John Wiley & Sons, New York, NY, 1952.

11 Brooks, G.: "Materials Processing II," McMaster University, Montreal, Quebec, Canada, 2000.

12 Brunner, Calvin R.: "Handbook of Incineration Systems," McGraw-Hill, New York, NY, 1991.

13 Caspersen, L.: "Next Generation Insulating Products Cut Energy Consumption," Industrial Heating Journal, Feb., 2001.

14 Ceramic Industry(journal), pp. 53–55, Feb. 1994.

15 Clark, F.H.: "Metals at High Temperatures," Reinhold Publ. Co., New York, NY, 1950.

16 CRC Press: "Handbook of Chemistry and Physics," Boca Raton, FL, 1993.

17 Drew Chemical Corp.: "Principals of Industrial Water Treatment," 1987.

18 Essenhigh, R.H.: "An Introduction to Stirred Reactor Theory Applied to Design of Combustion Chambers," in Palmer and Beer: "Combustion Technology" pp. 389–391, Academic Press, New York, NY 1974.

19 Faraday, Michael: "The Chemical History of a Candle." Cherokee Publishing Co., Marietta, GA. ISBN 0-87797-209-5, 1861.

20 Ganapathy, V.: "Applied Heat Transfer," John Wiley & Sons, New York, NY, 1982.

21 Gilchrist, J. D.: "Fuels, Furnaces and Refractories," Pergamon Press, New York, NY, 1977.

22 Glinkov, M.S and Glinkov, G. M: "A General Theory of Furnaces," Mir Publishers, Moscow, 1980.

23 Gubareff, G.G., Jansson, J.E., Torberg, R.H.: "Thermal Radiation Properties Survey," in Orisik, M.N.: p. 103, "Radiative Transfer," Wiley-Interscience, 1973.

24 Guenther, Rudolph: "Glass Melting Tank Furnaces," Society of Glass Technology, Sheffield, England, 1958.

25 Guyer, E.C. (Ed.): "Handbook of Applied Thermal Design," Part 10 (by R. J. Reed), Taylor and Francis, Philadelphia, PA, 1999.

26 Harbison-Walker Refractories Company: "Modern Refractory Practice," 5th ed., Pittsburgh, PA, 1992.

27 Hottel, H.C. and Egbert, R. B.: "The Radiation of Furnace Gases," ASME Transactions, May 1941.

28 Hougen, G.A., Watson, K.M., Ragatz, R.A.: "Chemical Process Principles," John Wiley & Sons, New York, NY, 1959.

29 Howden Buffalo, Inc.: "Fan Engineering," 9th ed., 1999.

30 Hoyle, C.J.: "Combustion Characteristics of Fuels for Glass Melting," Glass Journal, Feb. 1989.

31 Iron & Steel Institute: "Reheating for Hot Working," I&SI, London, 1968.

32 Industrial Heating Equipment Assn.: "Combustion Technology Manual," 5th ed. Figure 3, p. 326. IHEA, Arlington, VA, 1994.

33 Industrial Heating Journal: Thermal Processing Databook, Dec., 2000, pp. 37–113.

34 Iron and Steel Institute: "Reheating for hot working," 1968.

35 Jones, J.C.: "Combustion Science," Millennium Books, Newtown, NSW, Australia, 1993.

36 Karlekar, B.V. and Desmond, R. M.: "Heat Transfer," West Publishing Co., St. Paul, MN, 1982.

37 Khan, Y.U., Lawson, D.A., Tucker, R. J.: "Analysis of Radiative Heat Transfer in Ceramic-Lined and Ceramic-Coated Furnaces," pp. 21, 26. Institute of Energy journal, March 1998.

38 Kindrick, J.: "Understanding Productivity," Johns Hopkins University Press, Baltimore, MD, 1977.

39 Krivandin, V. and Markov, B.: "Metallurgical Furnaces," Mir Publishers, Moscow, 1977/1980.

40 Kutz, Myer, (Ed.): "Mechanical Engineers' Handbook," chapters 57–60, 69 (by R. J. Reed), John Wiley & Sons, New York, NY, 1996.

41 Lukasiewicz, M.A.: "Industrial Combustion Technologies," GRI, (now Gas Technology Institute), Des Plaines, IL, 1986.

42 Malloy, J.F.: "Thermal Insulation," Reinhold Book Corp., New York, NY, 1969.

43 Marino, P.: "Numerical Modelling of Steel Tube Reheating in Walking Beam Furnaces." Proceedings of 5th European Conference on Industrial Furnaces and Boilers, Volume II, INFUB, Portugal, 2000.

44 McAdams, W.H.: "Heat Transmission," 3rd ed., McGraw-Hill, New York, NY, 1954.

45 McGraw-Hill: " Dictionary of Scientific and Technical Terms," McGraw-Hill, New York, NY, 1994.

46 McGraw-Hill: "Perry's Chemical Engineers Handbook," 5th ed., McGraw-Hill, New York, NY, 1973.

47 National Fire Protection Assn., Quincy, MA: "Standard for Ovens and Furnaces, NFPA 86," 2001.

48 National Fire Protection Assn., Quincy, MA: "Flammable and Combustible Liquids Code Handbook," 1993.

49 New York State Energy Research and Development Authority: "Energy Efficiency in the Galvanizing Industry." Summarized in North American Mfg. Co.'s Application Report R-Gal-1, 7–88.*

50 Niessen, W.R.: "Combustion and Incineration Processes," Marcel Dekker, New York, NY, 1978.

51 North American Mfg. Co.: "Combustion Handbook, Volume I," 3rd ed., 2001.*

52 North American Mfg. Co.: "Combustion Handbook, Volume II," 3rd ed., 1995.*

53 North American Mfg. Co.: "Incineration of Hazardous, Toxic, Mixed Wastes," 1993.*

54 North American Mfg. Co.: "Practical Pointers," 1989.*

55 North American Mfg. Co.: "Handbook Supplement 146a: Applying Automatic Controls to Furnace Dampers," 1998.*

56 North American Mfg. Co.: " Handbook Supplement 146b: Throttled Air Jet Dampers—Sizing, Installation," 1998.*

57 North American Mfg. Co.: "Handbook Supplement 230: Industrial Flame Types", 1997.*

58 North American Mfg. Co.: "Handbook Supplement 247: Stack Gas Dew Points," 1990.*

59 North American Mfg. Co.: "Handbook Supplement 260: Combustion Equipment Problem Workshop B-3," 1990.*

60 North American Mfg. Co.: "Handbook Supplement 280: Manifold Size Checking," 1985.*

61 Osekoski, A.J.: "Selecting Refractories for PM and MIM Sintering Furnaces," Parts 1 and 2 in Industrial Heating Journal, Apr.–May, 2001.

62 Ozisik, M.N.: "Radiative Transfer," John Wiley & Sons, New York, NY, 1973.

63 Palmer, H.B. and Beer, J.M.: "Combustion Technology," pp. 389–391, Academic Press, New York, NY, 1974.

64 Peray, K. and Waddell, J.: "The Rotary Cement Kiln," Chemical Publishing, New York, NY, 1972.

65 Peyton, K.B.: "Fuel Field Manual," Nalco/Exxon, 1998.

66 Pfaender, H.G.: "Schott Guide to Glass," Chapman and Hall, New York, NY, 1994.

67 Pincus, A.G.: "Melting Furnace Operation in the Glass Industry," Magazines for Industry, New York, NY, 1980.

68 Pritchard, R.: "Handbook of Industrial Gas Utilization," Van Nostrand Reinhold, New York, NY, 1977.

69 Process Heating: "Improved Moisture Control Saves . . . ," Energy, page 24, June, 2000.

*North American Mfg. Co., 4455 East 71st Street, Cleveland, OH 44105. Tel. 216-271-6000.

70 Raznjeric, K.: "Handbook of Thermodynamic Tables and Charts," McGraw-Hill, New York, NY, 1973.

71 Reed, R.J.: "Fitting Flames to Furnaces and Load," North American Handbook Supplement 176*; 1980.

72 Remmey, G. Bickley, Jr.: "Firing Ceramics," World Scientific Publishers, River Edge, NJ, 1994.

73 Reynoldson, R.W.: "Heat Treatment in Fluidized Bed Furnaces," ASM International, Metals Park, OH 44073, 1993.

74 Rosenow, W.M.: "Handbook of Heat Transfer," McGraw Hill, New York, NY, 1998.

75 Ruark, R.: "Making the Connection—The Role of Kiln Management," Ceramic Industry Journal, July 2000.

76 Ruark, R.: "What to Avoid when Buying a Kiln," Ceramic Industry Journal, Jan. 2000.

77 Scholes, S.R./Greene, C.H.: "Modern Glass Practice," 7th ed., ISBN 0-436-0612-6, CBI, Boston, MA, 1975.

78 Segeler, C.G.: "Gas Engineers Handbook," Industrial Press, New York, NY, 1965.

79 Selvendy, G.: "Handbook of Industrial Engineering," Wiley-Interscience, New York, NY, 1982.

80 Sherman, R.A.: "Radiation from Luminous and Nonluminous Natural Gas Flames," ASME Transactions, 1956, pp. 177–192.

81 Siegel, R. and Howell, J.R.: "Thermal Radiation Heat Transfer," Hemisphere, New York, NY, 1992.

82 Singer, F. and Singer, S.: "Industrial Ceramics," Chemical Publishing Co. New York, NY, 1963.

83 Taplin, H.R.: "Combustion Efficiency Tables," Fairmount Press, Lilburn, GA, 1991.

84 Traub, D.: "Drying Files, Control, Part 2," Process Heating, May 2000.

85 Trinks, W. and Mawhinney, M.H.: "Industrial Furnaces, Volume I," 5th ed., John Wiley & Sons, New York, NY, 1961.

86 Watson, J.: "Why Heat Recovery is a 'Natural' for Radiant Tube Furnaces." Heat Treating, Feb. 1983. Also North American Handbook Supplement 204.*

87 Whitaker, S.: "Forced Convection Heat Transfer Correlations for Flow in Pipes, Past Flat Plates, Bundles;" AIChE, 18, No. 2, p. 361, 1972.

88 Yuan, W.W. and Tien, C.L.: "A Simple Calculation Scheme for Luminous Flame Emissivity," The Combustion Institute's 16th Symposium, Pittsburgh, PA, pp. 1471–1487, 1977.

*North American Mfg. Co., 4455 East 71st Street, Cleveland, OH 44105. Tel. 216-271-6000.

# INDEX

## C

**D**

## G

Q



R



S

www.ingramcontent.com/pod-product-compliance
Lightning Source LLC
LaVergne TN
LVHW021226120826
844863LV00010B/105